# Engineering Drawing

## SIXTH EDITION

## A.W. BOUNDY

AssDipMechEng, MPhil

McGraw Hill

Boston   Burr Ridge, IL   Dubuque, IA   Madison, WI   New York
San Francisco   St. Louis   Bangkok   Bogotá   Caracas   Kuala Lumpur
Lisbon   London   Madrid   Mexico City   Milan   Montreal   New Delhi
Santiago   Seoul   Singapore   Sydney   Taipei   Toronto

# McGraw·Hill Australia

A Division of The McGraw·Hill Companies

---

**National Library of Australia Cataloguing-in-Publication data:**

Boundy, A. W. (Albert William).
Engineering drawing 6e.

Includes index.
ISBN 0 07 471043 5 (set).

1. Mechanical drawing.  2. Mechanical drawing—Problems,
exercises, etc.  I. Boundy, A. W. (Albert William)
Sketchbook to accompany Engineering drawing.  II. Title.

604.2

---

Published in Australia
**McGraw-Hill Australia Pty Ltd**
**Level 2, 82 Waterloo Road, North Ryde NSW 2113 Australia**
Sponsoring Editor: Michael Tully
Production Manager: Jo Munnelly
Editor: Catherine Dunk
Marketing Manager: Sharon-Lee Lukas
Designer: Ramsay Macfarlane
Typesetter: Post Pre-Press Group
Illustrator: Shelly Communications
Proofreader: Trish Fox
Indexer: Puddingburn Publishing Services
Printed by: Kyodo Printing Co. (Singapore) Pte Ltd

# Contents

# Preface

The sixth edition of Engineering Drawing has two features which make it easier for students to understand concepts and obtain the necessary drawing practice required in engineering drawing courses. First, lengthy explanatory detail has been minimised and a step-by-step method of instruction used wherever possible. Second, the problem format used is that normally found in examination questions, giving students essential practice in this approach.

A large number and variety of problems follow each drawing topic, reinforcing the principles explained in the text. The range of difficulty in the problem sets allows students to plan a complete instructional and practical program to a depth consistent with prescribed course objectives.

Reference tables commonly used by drafters have been included so that students may become familiar with their use when solving problems. These tables, along with other information, make this book a valuable reference for practising drafters and engineers.

Recognising the importance of Computer Aided Drawing (CAD), a section called 'CAD corner' has been included at the end of each chapter. It comprises a series of questions relating to the chapter material for students to attempt in their CAD laboratory. Rick Bradshaw of Douglas Mawson Institute of TAFE (Panorama) has provided this valuable addition to the book.

This edition has been revised to conform to current Australian Standards and incorporates suggestions from a number of reviewers who are using this text in TAFE and University engineering courses. It also addresses much of the content of the TAFE national curriculum modules: EA061 Engineering graphics, EA701 Engineering drawing—detail, EA702 Engineering drawing—development and pipework, EA703 Engineering drawing—structural and mechanical, NBB12 Engineering drawing interpretation 1, NM44 Engineering drawing interpretation 2, and EA780 Design for manufacture. A number of the diagrams and tables in this book are based on Australian Standard (AS) 1100 (Technical Drawing)

Parts 101 (General Principles) and 201 (Mechanical Drawing) and the publishers acknowledge with thanks the consent of Standards Australia for their reproduction.

The author wishes to acknowledge contributions to the sixth edition of Bantex Pty Ltd, for allowing the reproduction of brochure photographs of Linex drawing equipment, Jasco Pty Ltd for the photo of the Rotring pen set, and reviewers from the TAFE and University sectors who made valuable recommendations for changes and/or improvements to the fifth edition. In particular:

| | |
|---|---|
| Michael Smit | Logan Institute of TAFE (Meadowbrook) |
| Trevor McCall | Moreton Institute of TAFE (Mt Gravatt) |
| Geoffrey Chesterton | Brisbane Institute of TAFE (Gateway) |
| Ross Wiley | Regency Institute of TAFE |
| William Wolstoncroft | Douglas Mawson Institute of TAFE (Panorama) |
| Leigh Clark | TAFE Tasmania (Launceston) |
| Morris Deans | Royal Melbourne Institute of TAFE |
| Soullis Tavrou | Swinburne Institute of TAFE (Hawthorn) |
| David White | Box Hill Institute of TAFE (Elgar) |
| Vic Ilic | University of Western Sydney (Nepean) |
| John Eastwell | University of Southern Queensland |
| Harold Greer | University of Southern Queensland |
| Zlatko Gradinscak | Royal Melbourne Institute of TAFE |
| Peter Doe | University of Tasmania |
| Glen Krause | Logan Institute of TAFE (Logan) |
| David Coulter | Yeronga Institute of TAFE |
| Len Farron | Yeronga Institute of TAFE |

# Introductory
# *&* standards information

After studying this chapter you should be able to do the following:

- select a range of instruments suitable for drafting

- complete an engineering drawing using standard paper sizes, layouts and methods

- use standard symbols related to dimensioning, sectioning, welding, surface texture, common engineering features and linear and geometry tolerances

- understand the principles of linear and geometry tolerancing, sectioning of orthogonal views and dimensioning engineering drawings

# 1.1 Standard abbreviations

The abbreviations in Table 1.1 have been selected from the more comprehensive list found in AS 1100 Parts 101, 201 and 501, and are those which are commonly used on mechanical and structural engineering drawings.

**TABLE 1.1** ► Standard abbreviations

| TERM | ABBREVIATION | TERM | ABBREVIATION | TERM | ABBREVIATION |
|------|--------------|------|--------------|------|--------------|
| **A** | | circular hollow section | CHS | external | EXT |
| abbreviation | ABBR | circumference | CIRC | **F** | |
| absolute | ABS | coating | CTG | figure | FIG |
| across flats | AF | coefficient | COEF | fillister head | FILL HD |
| acrylic | ACRY | cold-rolled steel | CRS | flange | FLG |
| addendum | ADD | column | COL | flat | FL |
| amendment | AMDT | computer-aided design | | floor | FLR |
| approximate | APPROX | and drafting | CAD | | |
| arrangement | ARRGT | computer-aided | | **G** | |
| assembly | ASSY | engineering | CAE | galvanise | GALV |
| assumed datum | ASSD | computer-aided | | galvanised iron | GI |
| audio frequency | AF | manufacture | CAM | galvanised-iron pipe | GIP |
| automatic | AUTO | concentric | CONC | general arrangement | GA |
| auxiliary | AUX | concrete | CONC | general-purpose outlet | GPO |
| average | AVG | contour | CTR | geometric reference | |
| | | corner | CNR | frame | GRF |
| **B** | | counterbore | CBORE | grade | GR |
| bearing | BRG | countersink | CSK | grid | GD |
| benchmark | BM | countersunk head | CSK HD | ground | GND |
| block | BLK | cross-recess head | C REC HD | | |
| bottom | BOT | cup head | CUP HD | **H** | |
| bracket | BRKT | cylinder | CYL | head | HD |
| brick | BK | | | height | HT |
| brickwork | BKW | **D** | | hexagon | HEX |
| Brinell hardness number | HB | dedendum | DED | hexagon head | HEX HD |
| brass | BRS | detail | DET | hexagon-socket head | HEX SOC HD |
| bronze | BRZ | diagonal | DIAG | high strength | HS |
| building | BLDG | diagram | DIAG | high-tensile steel | HTS |
| | | diameter | DIA | hollow section | |
| **C** | | diametral pitch | DP | circular | CHS |
| cadmium plated | Cd PL | diamond pyramid hardness | | rectangular | RHS |
| capacity | CAP | number (Vickers) | HV | square | SHS |
| casing | CSG | dimension | DIM | horizontal | HORIZ |
| cast iron | CI | direct current | DC | hot rolled steel | HRS |
| cast-iron pipe | CIP | distance | DIST | | |
| cast steel | CS | drawing | DRG | **I** | |
| cement | CEM | | | inside diameter | ID |
| centre line | CL | **E** | | integrated circuit | IC |
| centre of gravity | CG | each | EA | internal | INT |
| chamfer | CHAM | earth | E | intersection point | IP |
| channel | CHNL | elevation | ELEV | | |
| cheese head | CH HD | electric, electrical | ELEC | **J** | |
| chrome plated | CP | electromotive force | EMF | joint | JT |
| circle | CIRC | engine, engineering | ENG | junction | JUNC |
| circuit | CCT | equivalent | EQUIV | | |
| | | | | **L** | |
| | | | | least material condition | LMC |

| TERM | ABBREVIATION | TERM | ABBREVIATION | TERM | ABBREVIATION |
|---|---|---|---|---|---|
| left hand | LH | pattern | PATT | spigot | SPT |
| length | LG | pipe | P | spotface | SF |
| level | LEV | pitch-circle diameter | PCD | spring steel | SPR STL |
| longitudinal | LONG | phosphor bronze | PH BRZ | square | SQ |
| lubricate | LUB | plate | PL | square head | SQ HD |
| **M** | | position | POSN | square hollow section | SHS |
| machine | M/C | positive | POS | stainless steel (corrosion- | |
| malleable iron | MI | prefabricated | PREFAB | resistant steel) | CRES |
| material | MATL | pressure | PRESS | standard | STD |
| maximum | MAX | pressure angle | PA | Standards Association of | |
| maximum material | | **Q** | | Australia | SAA |
| condition | MMC | quantity | QTY | steel | ST |
| mechanical | MECH | **R** | | surface level | SL |
| mild steel | MS | radius | RAD | switch | SW |
| minimum | MIN | raised countersunk head | RSD CSK HD | symmetry | SYM |
| miscellaneous | MISC | rectangular | RECT | **T** | |
| modification | MOD | rectangular hollow | | tangent point | TP |
| modulus of elasticity | E | section | RHS | taper flange beam | TFB |
| modulus of section | Z | relief valve | RV | temperature | TEMP |
| moment of inertia | I | reference | REF | tensile strength | TS |
| mounting | MTG | regardless of feature size | RFS | thread | THD |
| mushroom head | MUSH HD | required | REQD | tolerance | TOL |
| **N** | | right hand | RH | true position | TP |
| negative | NEG | Rockwell hardness A | HRA | true profile | TP |
| nickel plated | NP | Rockwell hardness B | HRB | typical | TYP |
| nominal | NOM | Rockwell hardness C | HRC | **U** | |
| nominal diameter | DN | rolled-hollow section | RHS | undercut | UCUT |
| nominal size | NS | rolled-steel angle | RSA | universal beam | UB |
| north | N | rolled-steel channel | RSC | universal column | UC |
| not to scale | NTS | roughness value | $R_a$ | **V** | |
| number | NO | round | RD | vertical | VERT |
| **O** | | round head | RD HD | volume | VOL |
| octagon | OCT | **S** | | **W** | |
| opposite | OPP | schedule | SCHED | wood | WD |
| outside diameter | OD | screw | SCR | wrought iron | WI |
| overall | OA | section | SECT | **Y** | |
| **P** | | sheet | SH | yield point | YP |
| parallel | PAR | sketch | SK | **Z** | |
| part | PT | spherical | SPHER | zinc plated | Zn PLT |
| | | spherical radius | SR | | |

# 1.2 Drawing instruments

In order to achieve a high standard or, indeed, even a reasonable standard of drafting and sketching ability, a student must invest many hours of practice as well as have a knowledge of the particular graphical technique being used.

To produce satisfactory graphical communications, drafters need a range of quality pencils, pens and instruments. The complete range of equipment available is very broad. At the start students should select a range of basic instruments which will produce good results and keep costs to an affordable level.

## BASIC INSTRUMENTS

The following are essential, and should be purchased by all students of graphical communication before commencement of study and drawing practice:

- pencils
- scale rule
- eraser and erasing shield
- two triangular set squares (45° and 30°–60°) and an adjustable set square
- set of drawing instruments—for lead only
- set of french curves (3) or a flexible curve
- protractor, 180° or 360° (preferred)
- circle and ellipse templates
- adhesive tape or drawing board clips

For those students who are progressing on to ink work, the following additional and/or alternative instruments are required:

- technical pen set (sizes 0.25 mm, 0.35 mm, 0.5 mm, 0.7 mm)
- lettering guides (sizes 0.25/2.5 mm, 0.35/3.5 mm, 0.5/5 mm, 0.7/7 mm)
- two triangular set squares (45° and 30°–60°) and an adjustable set square with raised or recessed edges
- circle template with raised or recessed edges
- an ellipse template with raised or recessed edges

All students may be required to attempt drawing exercises as homework and will need access to a drawing board and tee-square. The size of drawing board will depend on the paper size used, which will probably be A2 or A3, but students should consult their teacher before purchasing these items. A brief description of each of the above items follows.

## PENCILS

There are three types of pencils to choose from: wood-cased, clutch and fine-point. Wood-cased pencils are not recommended as it is difficult to keep the lead sharp enough to produce a quality drawing. Clutch pencils are often preferred by professional drafters but the removable lead must be properly sharpened, otherwise the quality of the drawing will be poor. The third type, the fine-point, can be purchased according to the size of lead required, for example 0.3 mm, 0.5 mm, 0.7 mm and 0.9 mm. The lead never needs sharpening and for this reason is the best type for student use.

Leads are available in a range of hardness values such as the H and B ranges. The H range is the hard range, while B is the soft range. Perhaps the best for outlines is an H or 2H; for light lines a 4H is more suitable. Many students develop enough skill to produce all types of line with the one grade of pencil (H or 2H) simply by varying the pressure.

## SCALE RULE

A suitable scale rule is one graduated in metric units, flat or triangular in section and tapering towards the ruling edges. Four scales are available on the flat rule (Fig. 1.1) and six on the triangular rule (Fig. 1.2; one on either side of each of the ruling edges). Many scales are available—1:1, 1:2, 1:5, 1:10, 1:50, 1:100 and so on.

**FIGURE 1.1 ▶**
A flat section scale rule

**FIGURE 1.2 ▶**
A triangular section scale rule

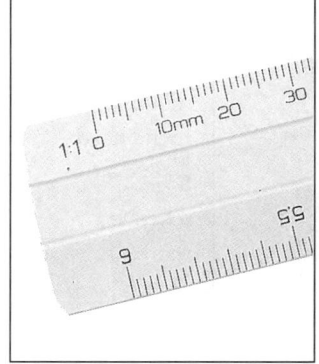

## ERASER AND ERASING SHIELD

Plastic erasers, which do not damage the surface of the drawing paper, are used. Their action is such that a multitude of small rolls are formed on the paper which pick up the graphite from the pencil line. Similar erasers are available for ink work.

The thin steel erasing shield (Fig. 1.3) is essential when selectively removing a line from a drawing area which is crowded with lines that are not to be removed. The erasing shield is suitable for both pencil and ink work.

**FIGURE 1.3** ▶ A thin steel erasing shield

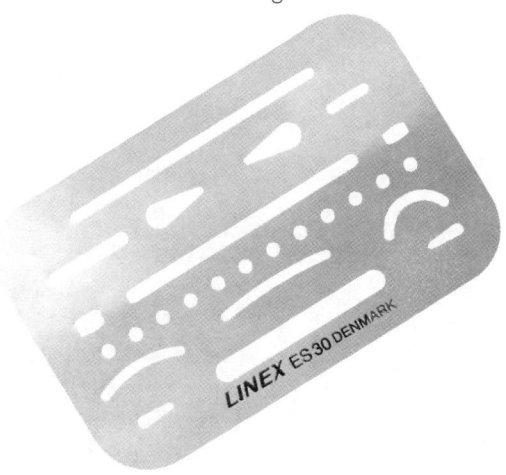

## SET SQUARES

Two triangular set squares are required; one with 45° angles and the other with angles of 30° and 60° (Fig. 1.4). They are made of clear plastic which is highly susceptible to buckling if left in the sunlight. For ink work, set squares can be purchased with a raised or recessed edge to prevent ink from running underneath as would happen with a flat surface. One edge of the set square is normally graduated in metric length units.

**FIGURE 1.4** ▶ Triangular set squares

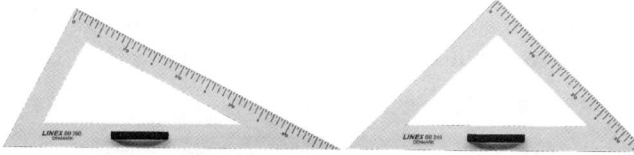

The adjustable set square (Fig. 1.5) allows a variety of angles to be drawn, especially when used in conjunction with the two triangular set squares.

**FIGURE 1.5** ▶ Adjustable set square

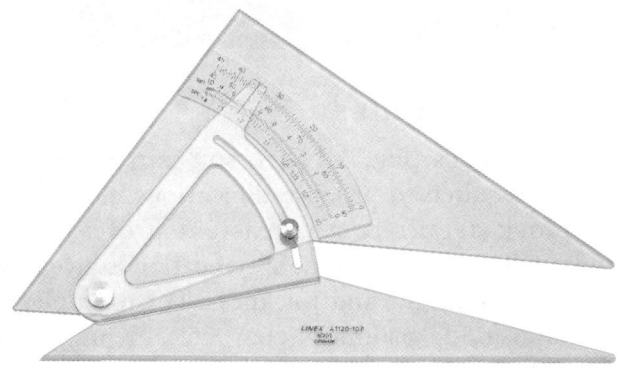

## DRAWING SET

Drawing sets are available in a large variety of configurations and are costly to purchase (Fig. 1.6). The advantage of buying a set is that instruments are kept in form-fit hollows in a case and are thus protected from damage. An ideal set contains a large masterbow or quick release bow compass, a small bow compass for small circles and an extension arm for the masterbow for drawing large circles. An attachment for holding technical pens is also often included.

**FIGURE 1.6** ▶ Drawing set

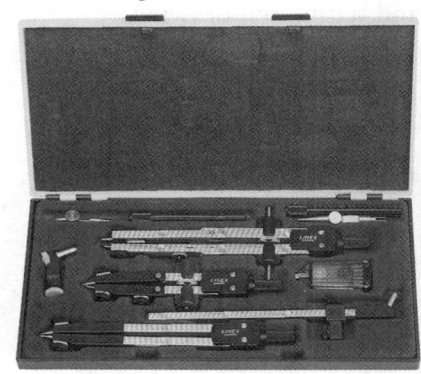

## FRENCH CURVES AND FLEXIBLE CURVE

A set of three French curves enables a drafter to draw a smooth curve through a series of plotted points. Using them requires practice, but good results can be achieved in both pencil and ink work. Curves with recessed edges are required for ink work. The flexible curve can be bent to form a smooth curve through a succession of plotted points. However, it must be used with care as undue force when bending may break the lead core and render it useless.

**FIGURE 1.7** ▶ Set of French curves (a) and flexible curve (b)

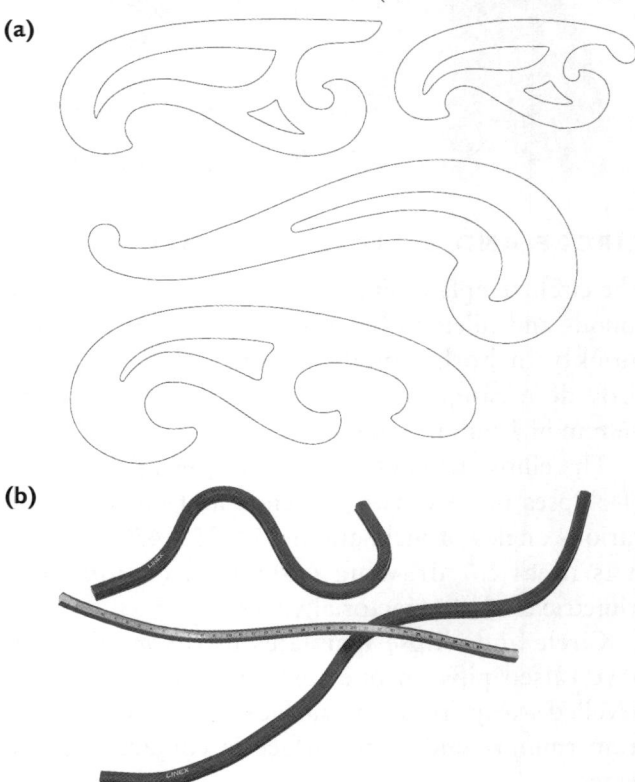

(a)

(b)

## PROTRACTOR

Protractors (Fig. 1.8) are made from clear plastic and are graduated in divisions of half a degree. There are two types: half circle (180°) and full circle (360°). The full circle is more useful as angles between 180° and 360° can be measured, which is necessary for surveying and setting-out drawings.

FIGURE 1.8 ► Examples of protractors

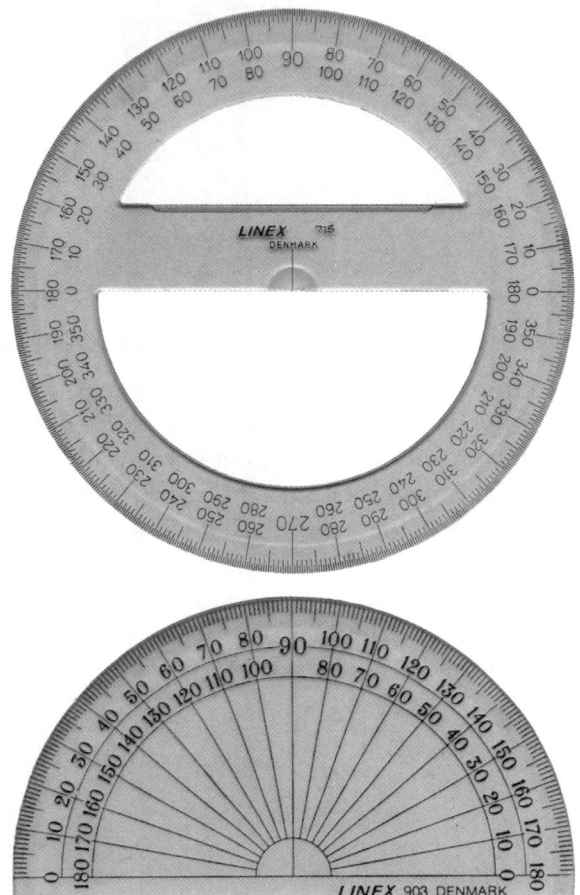

## CIRCLE AND ELLIPSE TEMPLATES

The circle template (Fig. 1.9(a)) enables small circles, rounds and fillets to be drawn easily, accurately and quickly in both pencil and ink. They normally provide a range of diameters from 1 mm up to 36 mm in 1 mm increments.

The ellipse template (Fig. 1.9(b)) enables a range of ellipses to be drawn in pencil and ink based on various angles of inclination: 25°, 35°, 45° and 60°. It is ideal for drawing isometric, oblique and trimetric circles on pictorial views.

Circle and ellipse templates used for ink work have raised pips on one surface or have the holes bevelled away from one surface. This prevents ink from running under the surface in contact with the paper.

FIGURE 1.9 ► Circle and ellipse templates

(a)

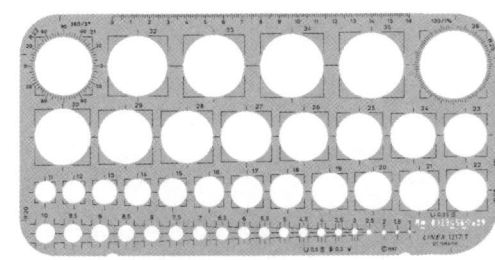

(b)

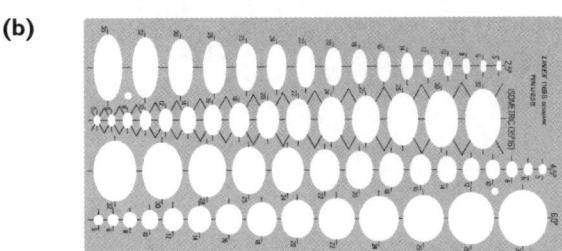

## TECHNICAL PEN SET

Drafters who produce drawings in ink must purchase a variety of pens which accurately draw lines of a constant and prescribed width. A reasonably priced pen set (Fig. 1.10) can be purchased which includes pens of various thicknesses; for example, 0.25 mm (thin line), 0.35 mm (medium line), 0.5 mm (outline) and 0.7 mm (borderline). Ink is contained in either refillable reservoirs or disposable cartridges. Lower priced pens are available with a fixed supply of ink. The pens are thrown away when the ink supply is exhausted.

FIGURE 1.10 ► Technical pen set

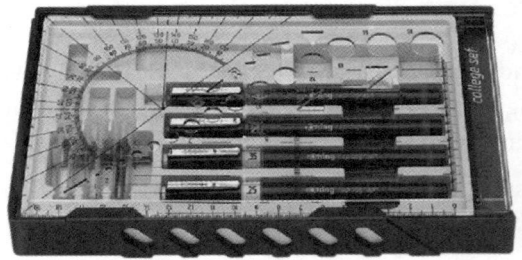

## LETTERING GUIDES

Lettering guides (Fig. 1.11) to suit particular size pens are required to produce dimensions, notes and title block details to a consistent professional standard. The lettering guide specification is given by the character height and thickness. For example, a 5/0.5 mm guide produces characters 5 mm high using a 0.5 mm drafting pen.

FIGURE 1.11 ► Lettering guide

ENGINEERING DRAWING

**FIGURE 1.12** ▶ Lettering and numerals suggested by Standards Australia

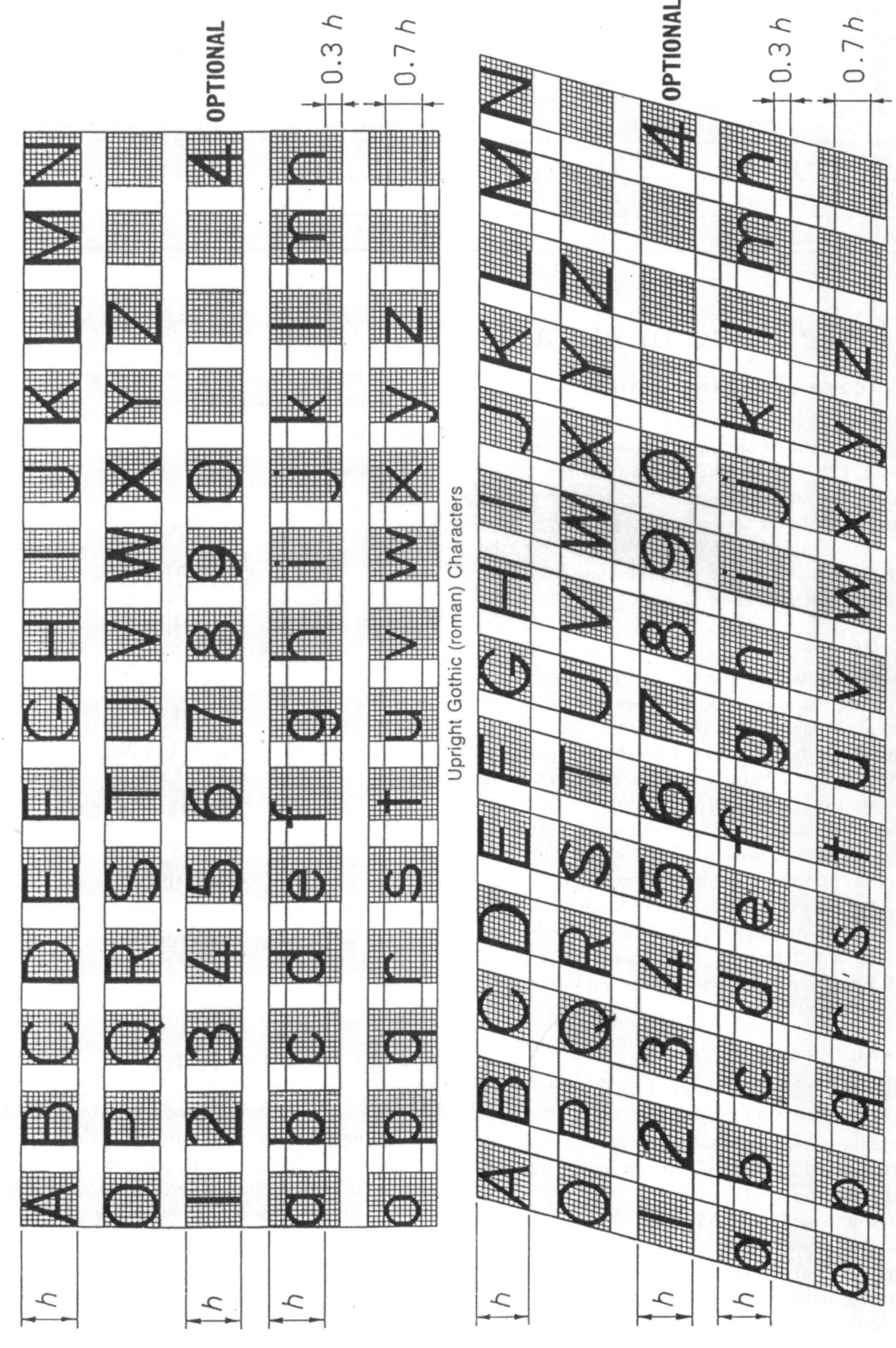

Upright Gothic (roman) Characters

Sloping Gothic (italic) Characters

Source: SAA HB7, Standards Australia, 1993. Reproduced with permission.

**TABLE 1.2** ► Minimum character height for sheet sizes

| USAGE (UPPER CASE ONLY) | CHARACTER HEIGHT (h), mm | |
| | SHEET SIZE | |
| | A0, B1 | A1, A2, A3, A4, B2, B3, B4 |
| --- | --- | --- |
| Titles and drawing numbers | 7 | 5 |
| Subtitles, headings, view and section designations | 5 | 3.5 |
| General notes, material lists, dimensions | 3.5 | 2.5 |

# 1.3 Letters and numerals

## CHARACTER FORM AND HEIGHT

For freehand characters the basic form of letters and numerals should be similar to those illustrated in Figure 1.12. The height ($h$) in millimetres of characters for various drawing sheet sizes should be as stated in Table 1.2 above.

For upper and lower case combinations, the minimum character height should be one size larger than that specified in Table 1.2.

## SPACING

Spacing is the widest gap which can be measured between any two consecutive letters and should be consistent throughout the word. Letter characters in a word should be spaced at twice the line thickness of the letters or 1 mm, whichever is greater. Some letter combinations which produce an inclined spacing, such as PA and AC in the word SPACING, have to be placed closer so that their spacing '$s$' is the same as between other letters (Fig. 1.13).

Spacing between words should be not less than 0.6 $h$ and not more than 2 $h$. For example, when using 2.5 mm characters, word spacing should be from 1.5 mm to 5 mm. Spacing between lines should be not less than 0.6 $h$.

## CHARACTER USAGE

The following general rules should apply when inserting characters on a drawing:

- Upper case letters should be used except for conventional applications, for example, mm, kg, kPa and so on.
- Only one style of character should be used throughout a drawing.
- Vertical characters should be used for titles, drawing and reference numbers.
- Underlining of lettering should be avoided.

- All characters on a drawing should be kept clear of lines except where a line precludes this requirement; the line may then be interrupted to accommodate characters.

**FIGURE 1.13** ► Spacing of characters

CORRECT

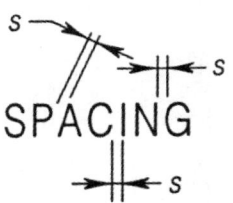

INCORRECT

SPACING EXCESSIVE

CHARACTERS TOO CLOSE

# 1.4 Types of line

The types of line which are commonly used in engineering drawings are illustrated in Table 1.3.

Figure 1.14 (page 10) includes examples of the use of nine types of lines, lettered to correspond with the types above (with the exception of type F).

1. The *visible outline* of the bracket, type A, is dark enough to make it stand out clearly on the drawing sheet. This line should be of even thickness (about 0.5 mm) and darkness.

**TABLE 1.3** ▶ Types of line

| TYPE | DESCRIPTION | DRAWING EXAMPLE | USAGE |
|------|-------------|-----------------|-------|
| A | continuous thick line | ———————— | to indicate visible outlines |
| B | continuous thin line | ———————— | for fictitious outlines, dimensions, projection, hatching and leader lines; also for the imaginary intersection of surfaces, revolved sections, adjacent parts, fold and tangent bend lines, short centre lines, and for indicating repeated detail |
| C | continuous thin freehand line | (freehand wavy line) | on part sectional boundary lines or to terminate a part view, and for short break lines |
| D | continuous thin ruled line with intermittent zigzag | (ruled line with zigzag) | to show a break on an adjacent member to which a component is attached; also to indicate a break in a long continuous series of lines on architectural or structural drawings |
| E | thin dashed line | (dashed line) $s = 1$ mm minimum $q = 2s$ to $4s$ | to show outlines of hidden features: <br>• for complete hidden features, the line should begin and end with a dash <br>• dashes should meet at corners <br>• where a hidden line is a continuation of a visible outline, it should commence with a space |
| F | medium dashed line | — — — — — — (proportions as for E) | in electrotechnology drawing only; for assemblies, boxes and other containers |
| G | thin chain line | (chain line) $s = 1$ mm minimum $q = 2s$ to $4s$ $p = 3q$ to $10q$ | to indicate centre lines, pitch lines, path movement, developed views, material for removal and features in front of a cutting plane |
| H | chain line, thick at the ends and at change of direction but thin elsewhere | ▬ — ▬ — ▬ — ▬ (proportions as for G) | to indicate a cutting plane for sectional views |
| J | thick chain line | ▬▬ ▬ ▬▬ ▬ ▬▬ (proportions as for G) | to indicate surfaces that must comply with certain requirements such as heat treatment or surface finish |
| K | thin double-dashed chain line | —— — — —— — — —— (proportions as for G) | to indicate adjacent parts, alternative and extreme positions of moving parts, centroidal lines and tooling profiles |

2. The *dimension, projection, hatching and leader line*, type B, is illustrated. Leader lines are of two types, one which terminates with an arrowhead at an outline and the other which terminates in a dot (4) within the outline of the part to which it refers. Leaders touching lines should be between 45° and 90° to the line. Other uses of type B lines are to partly outline the adjacent part to which the bracket is bolted and to represent fictitious outlines such as minor diameters of male threads and major diameters of female threads.

3. The *short break line*, type C, is drawn freehand to terminate part views and sections as shown. It is also used to sketch the curved break section used on cylindrical members.

4. A *ruled zigzag line*, type D, is used for long break lines which extend a short distance beyond the outlines on which they terminate.

5. The *hidden outline line*, type E, represents internal features which cannot normally be seen. A hidden outline should commence with a dash (1) except where it is a continuation of a visible outline (2), where there is a space first. Corners and junctions (3) should be formed by dashes.

6. The *centre line*, type G, denotes the axis of symmetrical views as well as the axis and centre lines of holes. Centre lines project a short distance past the outline. When produced further for use as dimension lines, they may revert to thin continuous (type B) lines. Type G lines may also be used to show the outline of material which has to be removed (not shown).

FIGURE 1.14 ▶ Use of different types of lines

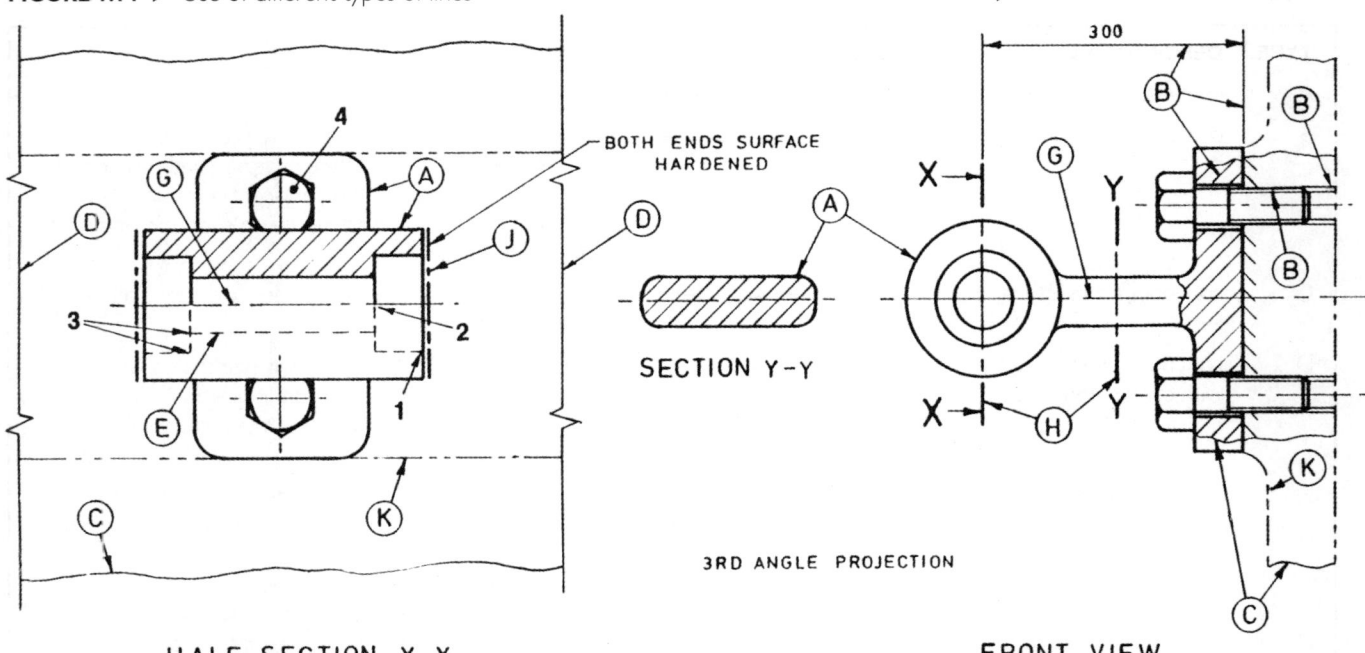

HALF SECTION X-X

SECTION Y-Y

3RD ANGLE PROJECTION

FRONT VIEW

BOTH ENDS SURFACE HARDENED

7. The *cutting plane of the section*, X-X, is represented by the type H line. Arrows are located at right angles to the thick ends of the line, and point to the direction in which the sectional view is being taken. In the case of the removed section, Y-Y, which merely shows the cross-sectional shape of the member, it is immaterial which direction the view is taken from, and the arrows may be left off the cutting plane.

8. *Surfaces requiring special treatment* such as heat treatment or surface finish may be indicated with a type J line drawn parallel to the profile of the surface in question.

9. When drawing a component where it is necessary to show its relationship to an adjacent part, the latter is outlined using a type K line. Other uses of this type of line are to indicate extreme positions of movable parts, and to outline tooling profiles in relation to work set up in machine tools.

## 1.5 Line thicknesses

Thicknesses for the various types of line are divided into specific groups according to the size of drawing sheet being used. Table 1.4 shows the metric sheet size, line type and thickness applicable in each case.

## 1.6 Scales

The scales recommended for use with the metric system are:

Full size      1:1
Enlargement    2:1, 5:1, 10:1, 20:1, 50:1
Reduction     1:2, 1:2.5, 1:5, 1:10, 1:20, 1:50, 1:100, 1:200, 1:500, 1:1000, 1:2000, 1:5000, 1:10 000

TABLE 1.4 ▶ Line thicknesses for various sheet sizes

| SHEET SIZE | LINE TYPE AND THICKNESS (mm) | | | | | | | | | |
|---|---|---|---|---|---|---|---|---|---|---|
| | A | B | C | D | E | F | G | H | J | K |
| **A0, B1** | 0.7 | 0.35 | 0.35 | 0.35 | 0.35 | 0.5 | 0.35 | 0.35 0.7 | 0.7 | 0.35 |
| **A1, A2, A3, A4, B2, B3, B4** | 0.5 | 0.25 | 0.25 | 0.25 | 0.25 | 0.35 | 0.25 | 0.25 0.5 | 0.5 | 0.25 |

## USE OF SCALES

Engineering drawings may be prepared full size, enlarged or reduced in size. Whatever size of scale is used, it is important that it be noted in or near the title block.

## INDICATION OF SCALES

When more than one scale is used, they should be shown close to the view(s) to which they refer and a note in the title block should read 'scales as shown'.

If a drawing has predominantly one scale, the main scale should be shown in the title block together with the notation 'or as shown' to indicate the use of other scales elsewhere on the drawing.

Sometimes it is necessary to use different scales on the one view, for example on a structural steel truss where the cross-sections of members are drawn to a larger scale than the overall dimensions of the truss. Such variations are indicated on the drawing, for example:

*Scales*

Member cross-sections    1:10
Truss dimensions            1:100

If a particular scale requirement needs to be used on a drawing, it may be shown by one of the following methods:

1.  a scale shown on the drawing, for example:

```
              0         10        20 cm
   10  8  6  4  2
```

2.  the word 'scale' followed by the appropriate ratio, for example SCALE 1:10
3.  the words SCALE: NONE in or near the title block, for example on pictorial drawings.

# 1.7 Sizes of drawing paper

## PREFERRED SHEET SERIES

Standards Australia has recommended that paper sizes be based on the 'A' series of the International Organization for Standardization (ISO), and these sizes are specified in AS 1100 Part 101. This series is particularly suitable for reduction onto 35 mm microfilm because the ratio is $1:\sqrt{2}$ for the sides of the paper (Fig. 1.15(a), page 12) and this ratio is also used for the microfilm frame.

Paper sizes are based on the A0 size, which has an area of 1 square metre. This allows paper weights to be expressed in grams per square metre.

The relationship between the various paper sizes is illustrated in Figures 1.15(a) and (b), where the application of the $1:\sqrt{2}$ side ratio can be seen. An A0 size sheet can be divided up evenly into the various other sizes simply by halving the sheet on the long side in each case. This is shown in Figure 1.15(c). The dimensions of metric sheets from size A0 to A4 are given in Table 1.5, together with appropriate border widths for each sheet size.

## NON-PREFERRED SHEET SERIES

The 'B' series of sheet sizes provides for a range of sheets designated by B1, B2, B3, B4, etc., which are intermediate between the A sizes. The relationship of the B and A sizes is shown in Figure 1.15(b); B sizes are in broken outline.

## ROLLS

The standard widths of rolls are 860 mm and 610 mm. Drawing sheets can be cut off the roll to suit individual drawings.

# 1.8 Layouts of drawing sheets

Standard layouts for drawing sheets of all sizes are given in AS 1100 Part 101. Figures 1.16 and 1.17 (page 13) show typical layouts of A1 and A2 sheets, illustrating the paper size, drawing frame with microfilm camera alignment marks and zoning by grid referencing details. Additionally Figure 1.17 includes a parts list and a revisions table. The layout of Figure 1.16 is suitable for detail drawings, while that of Figure 1.17 is suitable for multidetail and assembly drawings.

## SHEET FRAMES (BORDERLINES)

It is usual for each sheet to be provided with a drawing frame a short distance in from the edge of the paper. Drawing frames are standardised for the various sizes of paper, and Figure 1.15(d) and Table 1.5 detail this information.

## TITLE BLOCK

The title block represents the general information source for a drawing. It is normally placed in the bottom right-hand corner of the drawing frame.

**FIGURE 1.15** ▶ Paper sizes

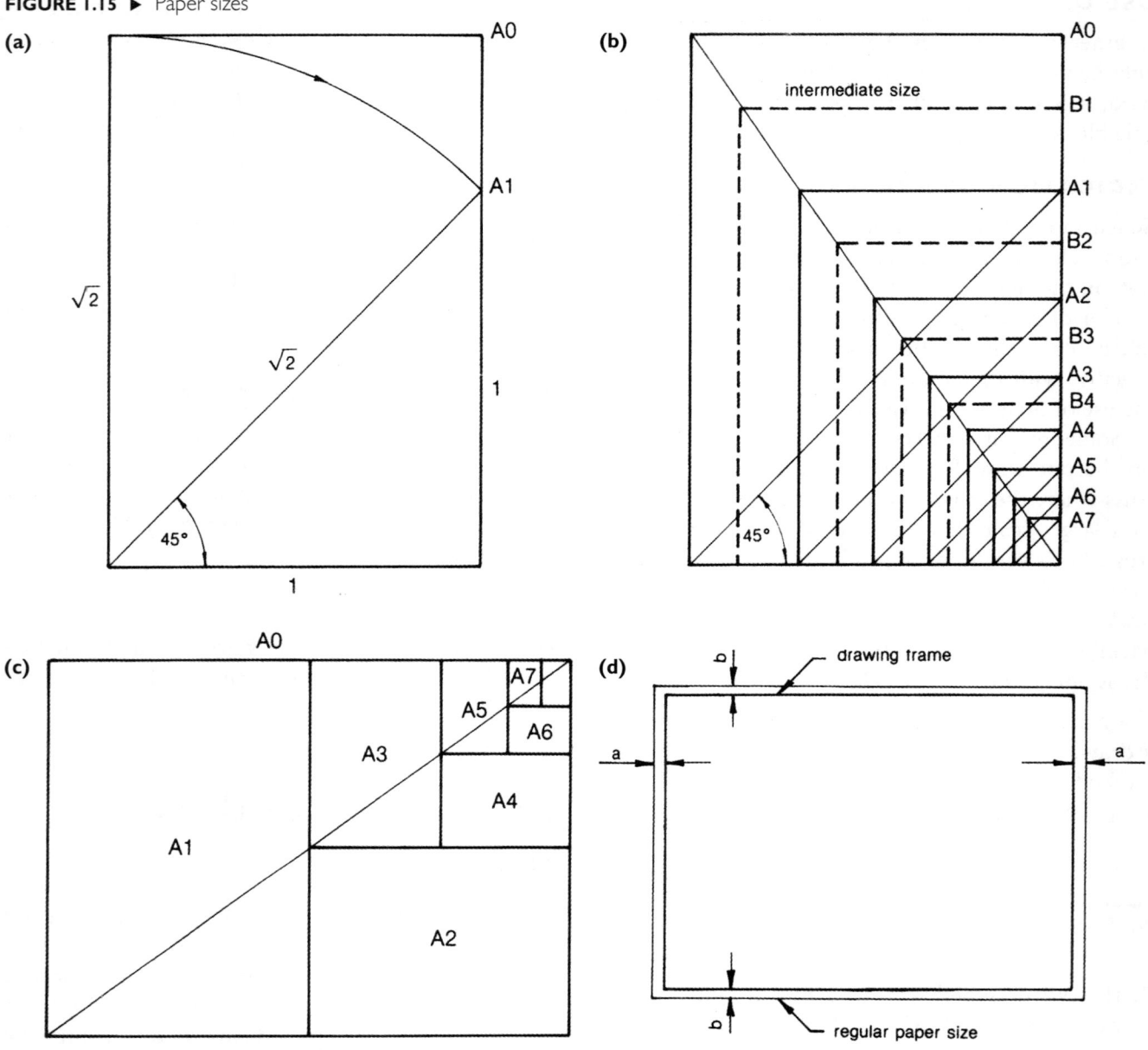

**TABLE 1.5** ▶ Drawing frames without a filing margin

| PAPER SIZE | BORDER WIDTH (mm) | | DIMENSIONS OF DRAWING SHEETS (mm) |
|---|---|---|---|
| | **BOTH SIDES** | **TOP AND BOTTOM** | |
| | **A** | **B** | |
| A0 | 20 | 20 | 1189 × 841 |
| A1 | 20 | 20 | 841 × 594 |
| A2 | 10 | 10 | 594 × 420 |
| A3 | 10 | 10 | 420 × 297 |
| A4 | 10 | 10 | 297 × 210 |

NOTES:

1. The sides of the metric drawing paper sheets are in the ratio of 1:√2. Area of the A0 size sheet = 1 m².

2. If a filing margin is required, usually on the left hand side of the sheet, the dimension 'a' on the left is increased by 10 mm.

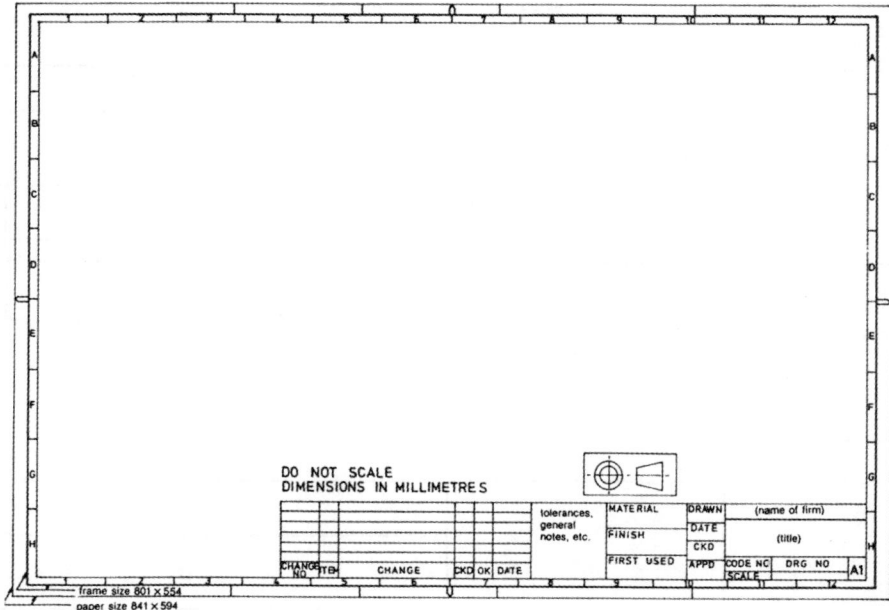

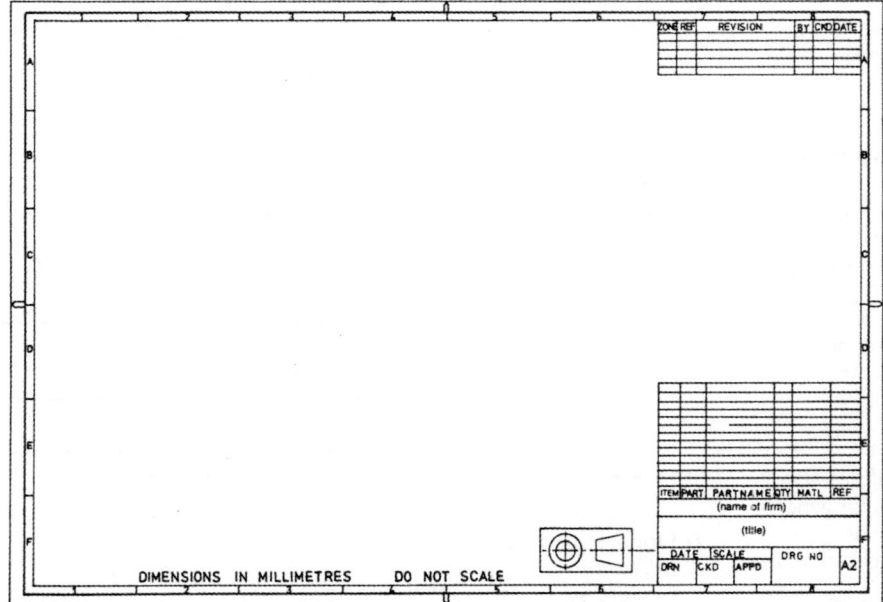

Figure 1.18 (page 14) illustrates title block dimensions for various sheet sizes, together with the type of information which should be contained in the title block and its location.

## MATERIAL OR PARTS LIST

When a drawing includes a number of parts on the one sheet, or when an assembly drawing of a number of parts is required, a tabulated list of the parts is attached to the top of the title block and against the right-hand drawing frame as shown in Figure 1.17.

The list should give the following information:

1. the item or part number
2. the part name
3. the quantity required
4. the material and its specification
5. the drawing number of each individual part
6. other information considered appropriate

A separate standard drawing sheet may be used to set out a parts list alone when it is desirable or when the list is very large. Such a list should be provided with a standard title block and a revisions table. For further details see AS 1100 Part 101.

FIGURE 1.18 ▶ Typical title blocks for various size sheets

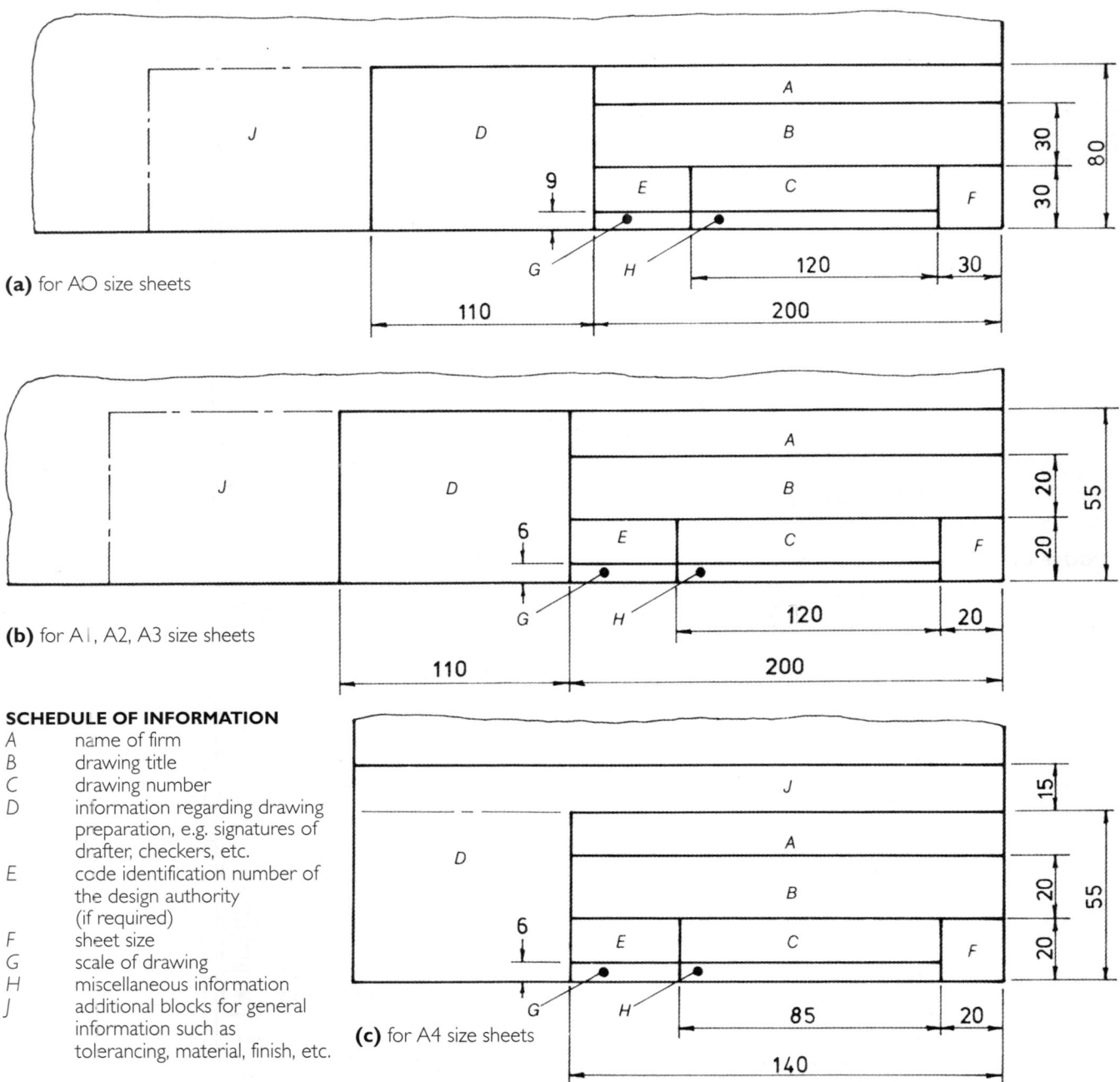

**(a)** for A0 size sheets

**(b)** for A1, A2, A3 size sheets

**(c)** for A4 size sheets

**SCHEDULE OF INFORMATION**

| | |
|---|---|
| A | name of firm |
| B | drawing title |
| C | drawing number |
| D | information regarding drawing preparation, e.g. signatures of drafter, checkers, etc. |
| E | code identification number of the design authority (if required) |
| F | sheet size |
| G | scale of drawing |
| H | miscellaneous information |
| J | additional blocks for general information such as tolerancing, material, finish, etc. |

## REVISIONS TABLE

A table of revisions is located normally in the upper right-hand corner of the drawing frame as shown in Figure 1.17.

The ability to effect revisions or modifications to existing drawings is an important requirement in all drawing and design offices. In many instances when only minor revisions are required, it is much easier to revise an existing drawing than to create a new one. However, such revisions must be tabulated to record existing details of the feature as well as the revision.

Each change should be identified by a symbol such as a letter or number placed close to the revision on the drawing. The letter or number may (but need not) be encircled on the drawing. Reference is made to the symbol in the tabulated details of the change as shown in Figures 6.1 (page 217) and 6.2 (page 219). Drawings so revised should be given a new issue number or letter situated in the title block adjacent to the drawing number.

For example in Figure 6.1 (page 217) the drawings of the jack body and spindle are numbered 17644A and 17645A respectively. Similarly the assembly of the machine screw jack in Figure 6.2 (page 219) is renumbered as 17646A.

If a particular revision affects the interchange-ability of a part, the revised part should be allocated a new drawing number.

## ZONING

Drawings may be divided into zones by a grid reference system based on numbers and letters as shown in Figures 1.16 and 1.17. Zoning is located inside the drawing frame.

The purpose of a grid reference system is to assist location of detail. It is particularly useful on large drawings.

Horizontal zones are designated by capital letters starting with A, reading from top to bottom. Vertical zones are designated by numbers reading from left to right.

The number of zones and the widths of zone margins to be used on various sheet sizes are detailed in Table 1.6 (page 16). Further use of zoning is shown in Figure 6.2, where a revision in the table designates a change of thread form (Whitworth to metric), and the reference C2 is a grid reference indicating the position on the sectional view of the thread in question, that is symbol Ⓐ.

# 1.9 Dimensioning

### DIMENSIONING SYMBOLS

Symbols are used to indicate geometrical features relating to a dimension. Symbol proportions are related to the height (h) of the characters used on a particular drawing. Table 1.7 shows symbols in common use.

### DIMENSION AND PROJECTION LINES

These lines are thin, continuous type B lines drawn outside the outline wherever possible.

*Projection lines* are used as follows:

1. to project from one view to another in order to transfer detail (refer to Figure 3.4)
2. to allow dimensions to be inserted—projection lines indicate the extremities of a dimension

*Dimension lines* are necessary to indicate the extent of a measurement.

Figure 1.19 shows the use of projection and dimension lines with appropriate measurements indicating dimension line spacing, projection line gap and extension.

**FIGURE I.19** ▶ Use of projection and dimensioning lines

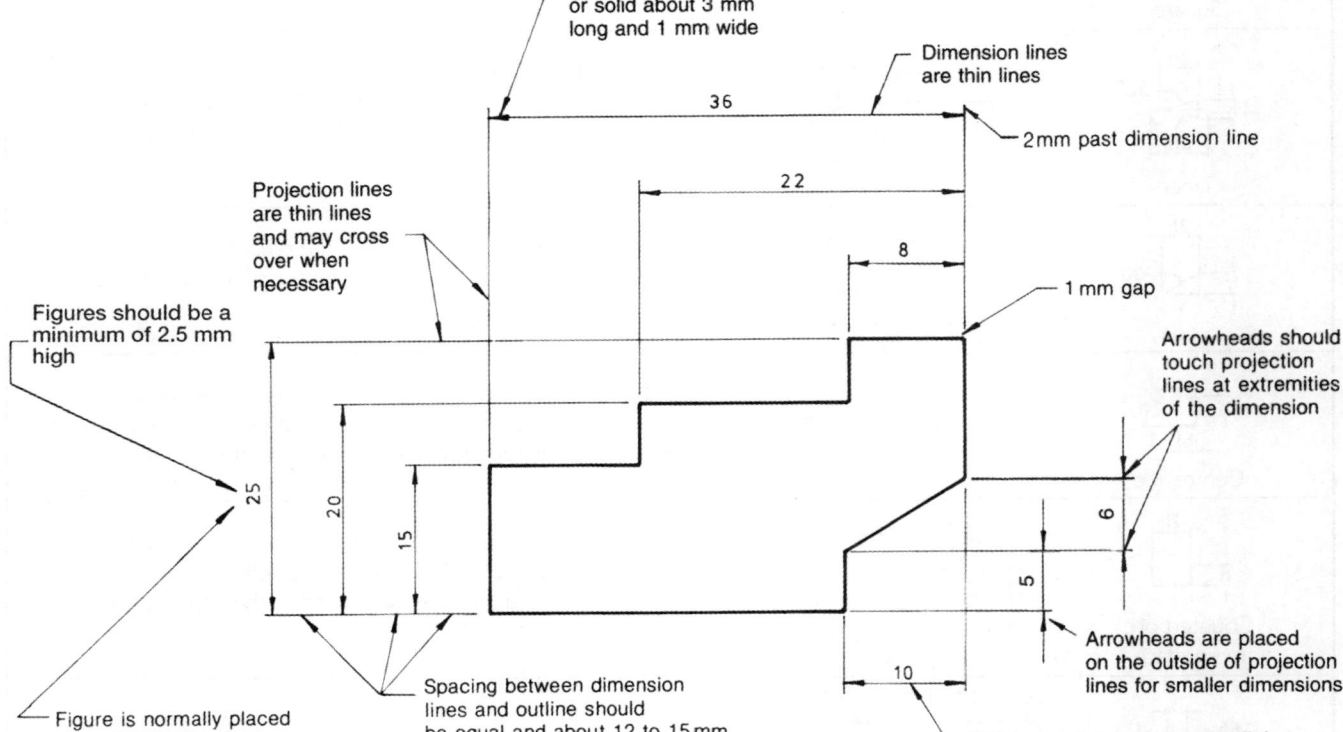

**TABLE 1.6** ▶ Details of grid references

| DETAIL | SIZE OF DRAWING | | | | |
|---|---|---|---|---|---|
| | **A0, B1** | **A1, B2** | **A2, B3** | **A3, B4** | **A4** |
| **Number of vertical zones designated (1, 2, etc.)** | 16 | 12 | 8 | 6 | 4 |
| **Number of horizontal zones designated (A, B, etc.)** | 12 | 8 | 6 | 4 | 4 |
| **Width of margins for grid reference (mm)** | 10 | 7 | 7 | 5 | 5 |

**TABLE 1.7** ▶ Dimensioning symbols

| SYMBOL | DRAWING CALLOUT | DESCRIPTION |
|---|---|---|
| Diameter | Ø12 | The symbol is placed in front of the dimension and indicates that it refers to the diameter of a circle, hole, cylinder or other circular feature (Fig 1.25) |
| Radius | R10 | The symbol is placed in front of the dimension and indicates that it refers to the radius of part of a circle (Fig. 1.26). |
| Square | □12 | The symbol is placed in front of the dimension and indicates that it refers to the width across the flats of a square section (Fig. 1.29) |
| Taper | 1:6 | The symbol is placed in front of the slope ratio of the taper and indicates a taper and its direction. The centre line of the symbol is parallel to the axis or plane of symmetry of the tapered feature. |
| Slope | 1:10 | The symbol is placed in front of the ratio of the slope and indicates the slope direction. The base of the symbol is parallel to the datum plane. |
| Centre line | | The symbol is placed adjacent to or on the centre line and indicates the centre line of a part, feature or group of features. It is not normally used when the centre line is obvious. |
| Counterbore or spotface — Spotface | Ø6 ⊔ Ø12 | The symbol is placed after the hole diameter and before the spotfaced diameter. If a depth symbol follows the second diameter a counterbore is indicated to the depth specified (Figs 1.34 and 1.35). |
| Depth — Counterbore — Hole | Ø4 ⊔ Ø8 ↧6    Ø6 ↧8 | The symbol is placed in front of the dimension. |

TABLE 1.7 ▶ Dimensioning symbols (continued)

| SYMBOL | DRAWING CALLOUT | DESCRIPTION |
|---|---|---|
| Countersink | ⌀6 ⌵ ⌀12 × 90° | The symbol is placed after the hole dimension and indicates that the hole has to be countersunk out to the diameter immediately following the symbol and at the angle stated (Fig. 1.33). |
| Curved surface or arc length | Circumferential (25) (22) Chordal 20 | The symbol is placed above the dimension and is used in conjunction with a curved dimension line drawn parallel to the curved surface. Dimension end lines indicate the extent of the dimension as well as the surface being dimensioned. Chordal dimensioning uses a straight line parallel to the chord length, in this case on the top surface joining the hole centres. |
| Spherical diameter S⌀ | S⌀ 20 | The symbol is placed in front of the dimension and indicates the diameter of a spherical surface (Fig. 1.28(a)). |
| Spherical radius SR | SR 20 | The symbol is placed in front of the dimension and indicates the radius of a spherical surface (Fig. 1.28(b)). |
| Surface texture | line representing the surface (a) (b) (c) | (a) is the basic texture symbol, and means that the surface may be produced by any production process<br>(b) is a modified symbol which indicates that the surface must be produced by a machining process<br>(c) is a modified symbol which indicates that the surface must be left in the state resulting from a preceding manufacturing process. |
| Third angle projection | 3RD ANGLE PROJECTION | Either the symbol or the drawing callout '3RD ANGLE PROJECTION' should be displayed prominently on each drawing sheet. |
| First angle projection | 1ST ANGLE PROJECTION | Either the symbol or the drawing callout '1ST ANGLE PROJECTION' should be displayed prominently on each drawing sheet. |

Figure 1.20 (page 18) illustrates correct and incorrect methods of employing centre lines and projection lines for dimensioning purposes.

## LINEAR DIMENSIONS

These should preferably be expressed in millimetres. It is not necessary to write the symbol 'mm' after every dimension. A general note such as 'all dimensions are in millimetres' in the title block is sufficient.

## ANGULAR DIMENSIONS

Angular dimensions should be stated in degrees; in degrees and minutes; or in degrees, minutes and seconds; for example 36.5°, 36°30', 36°29'30". A zero should be used to indicate an angle less than one degree, for example 0°30'0.5".

**FIGURE 1.20** ▶ Use of centre and projection lines in dimensioning

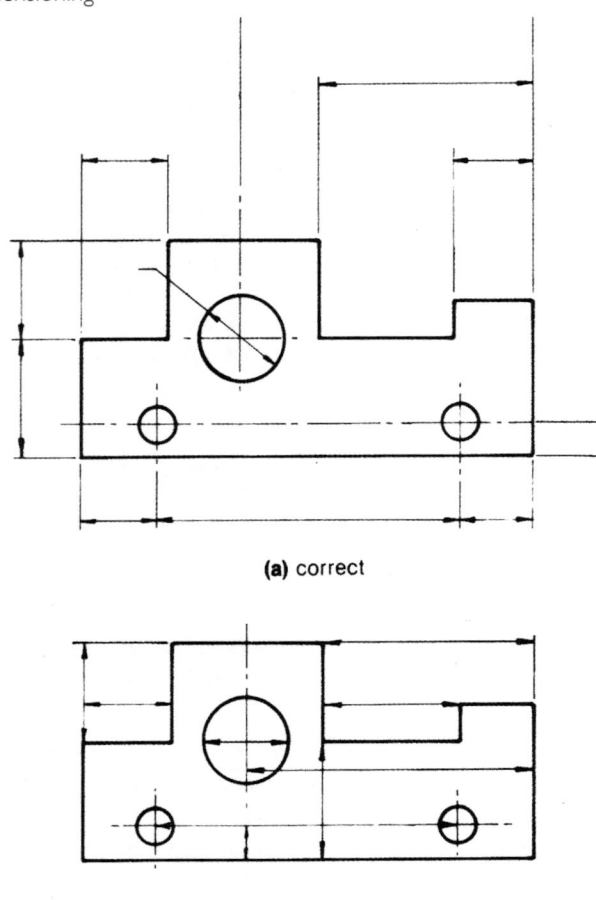

**(a)** correct

**(b)** incorrect

## METHODS OF DIMENSIONING

Two methods of indicating dimensions are in common use:

1. *unidirectional*, where the dimensions are drawn parallel to the bottom of the drawing, that is, horizontal
2. *aligned*, where the dimensions are drawn parallel to the related dimension line and are readable from the bottom or right-hand side of the drawing

Dimensions and notes indicated by leaders should use the unidirectional method as illustrated in Figure 1.21.

## STAGGERED DIMENSIONS

Where a number of parallel dimensions are close together they should be staggered to ensure clear reading, as shown in Figure 1.22.

## FUNCTIONAL DIMENSIONS

Some dimensions are essential for the proper operation or function of a component. These are called *functional dimensions* and are always inserted on the component detail drawing. Functional dimensions may also be toleranced if necessary to ensure a proper working relationship with mating parts on assembly.

**FIGURE 1.21** ▶ Methods of dimensioning

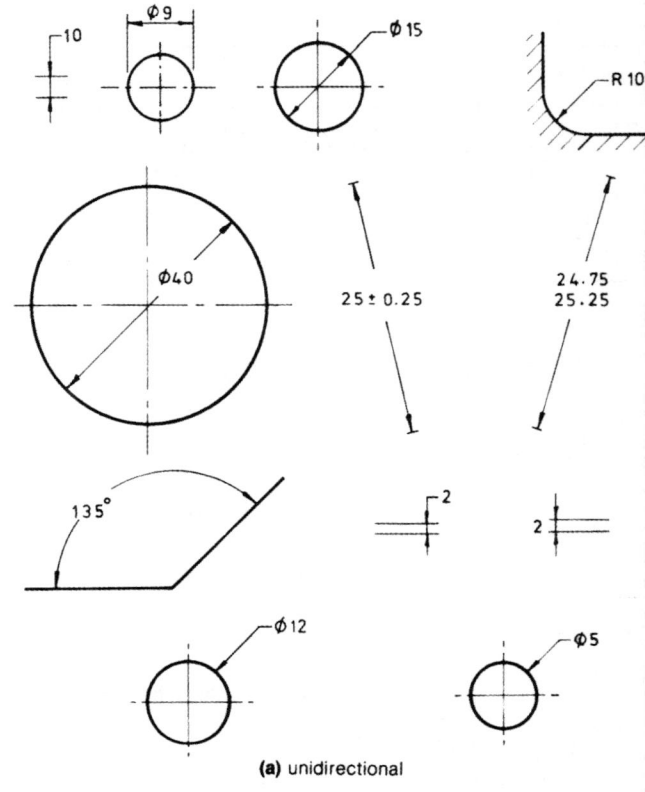

**(a)** unidirectional

**(b)** aligned

**FIGURE 1.22** ▶ Use of staggered dimensions

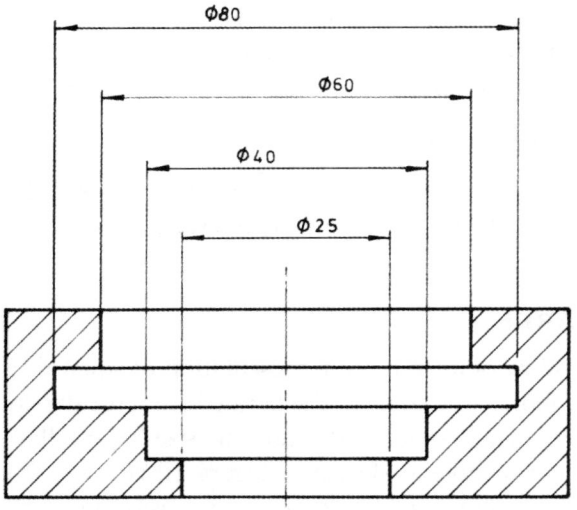

**FIGURE 1.23** ▶ Use of overall dimensions

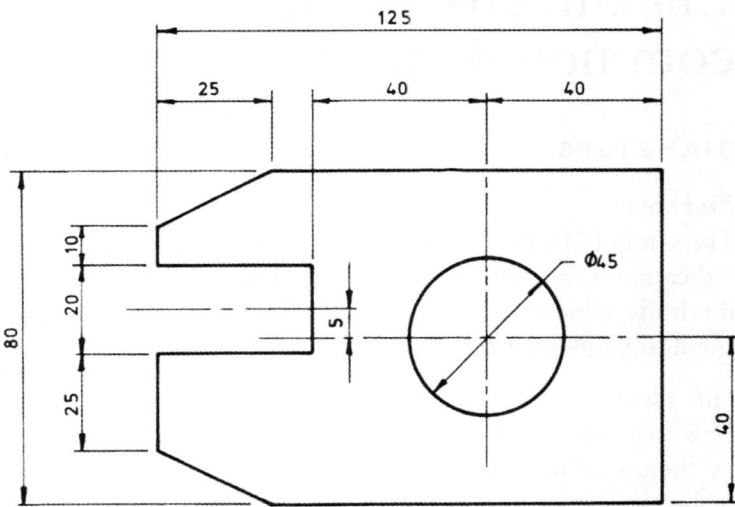

## OVERALL DIMENSIONS

When a length consists of a number of dimensions, an overall dimension may be shown outside the dimensions concerned (see Fig. 1.23). The end projection lines are extended to allow this. When an overall dimension is shown, however, one or more of the dimensions that make up the overall length is omitted. This is done to allow for variations in sizes that may occur during production. The omitted dimension is always a non-functional dimension, that is, one which does not affect the function of the product. Functional dimensions are those necessary for the operation of the product; these dimensions are essential.

## AUXILIARY DIMENSIONS

When all the dimensions that add up to give an overall length are functional and/or convenient for manufacture, the overall dimension may be added as an auxiliary dimension. This is indicated by enclosing the dimension in brackets.

Auxiliary dimensions are never toleranced and are in no way binding as far as machining operations are concerned. Figure 1.24 illustrates the use of an auxiliary dimension, namely (100).

If the overall length dimension is important, then one of the intermediate dimensions is redundant, for example the dimension 42 from the left-hand end (Fig. 1.24). This dimension may be inserted as an auxiliary.

## DIMENSIONS NOT TO SCALE

When it is desirable to indicate that a dimension is not drawn to scale, the dimension is underlined with a continuous, thick type A line, for example:

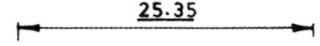

## DIMENSIONS NOT COMPLETE

Where a dimension is defining a feature that cannot be completely inserted on a drawing (for example, for a large distance or diameter) the free end is terminated in a double arrowhead pointing in the direction the dimension would take if it could be completed:

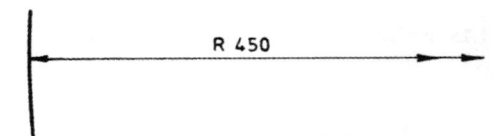

**FIGURE 1.24** ▶ Use of auxiliary dimensions

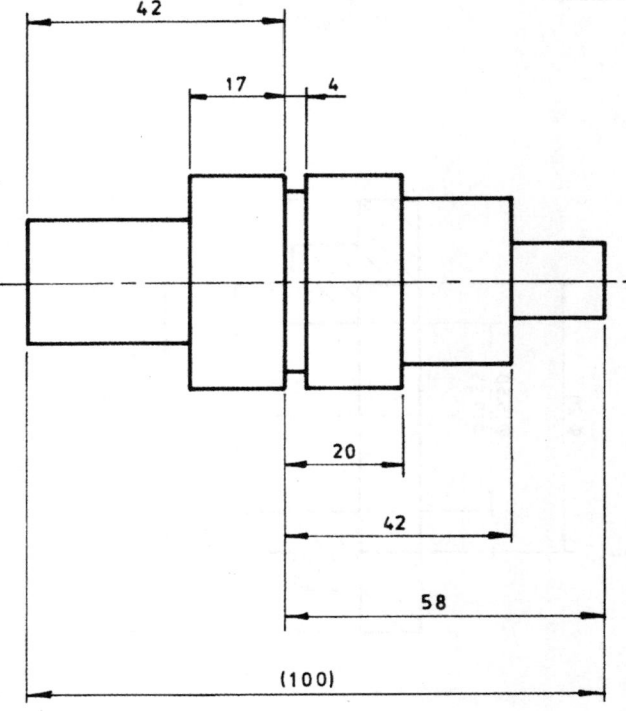

# 1.10 Dimensioning common features

## DIAMETERS

### End view
The symbol Ø shall be used to precede the dimension indicating a hole or cylinder. See Figure 1.25(a) for methods which are used on circles ranging from small to large diameters.

### Side view
This may be indicated, as shown in Figure 1.25(b), by the use of the symbol Ø preceding the dimension or by the use of leaders which are at right angles to the outline in conjunction with the symbol Ø.

## RADII

Figure 1.26 illustrates methods of dimensioning these features. A radius dimension is preceded by the letter R. Leaders should pass through or be in line with the centres of arcs to which they refer.

## SMALL SPACES

Figure 1.27 illustrates methods of dimensioning small spaces which are too small to be dimensioned by normal methods.

## SPHERICAL SURFACES

These are dimensioned as shown in Figure 1.28. Note the distinction made between spherical diameters and spherical radii.

## SQUARES

AS 1100 recommends the representation of squares by the symbol □ preceding and separated from the dimension by a single space, indicating the size 'across flats' of a shaft or hole as shown in Figure 1.29(a) and (b). The Ⓔ symbol used on the hole dimension indicates the Envelope Principle described on pages 83 and 85.

Some practitioners advise dimensioning of a square shaft by the 'across-corners' method together with an additional end view, Fig. 1.29(c), claiming there is less chance of error when reading this drawing representation to determine the diameter to be turned preparatory to machining the flats of the square shaft.

**FIGURE 1.25** ▶ Diameters dimensioned on end view (a) and side view (b)

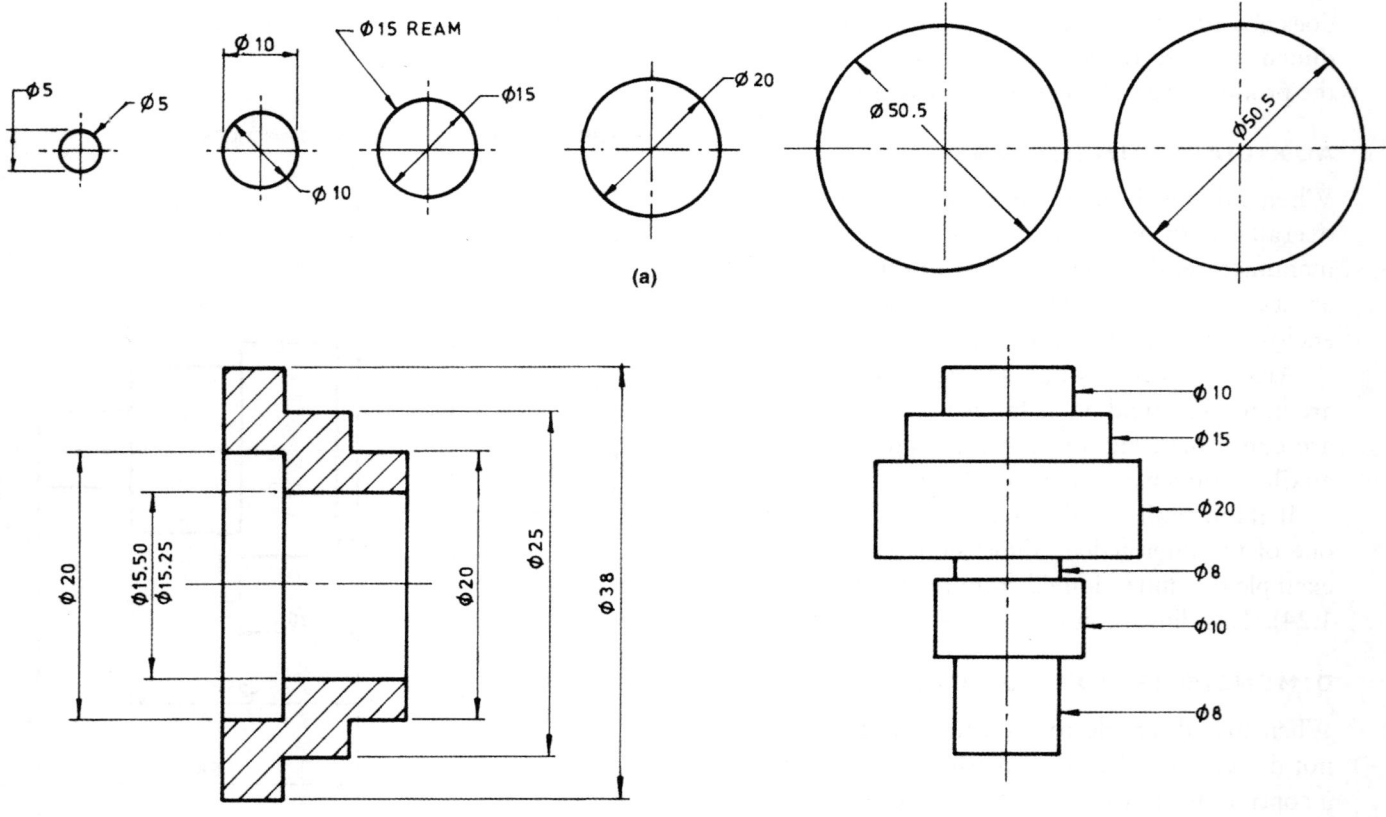

**FIGURE 1.26** ▶ Methods of dimensioning radii

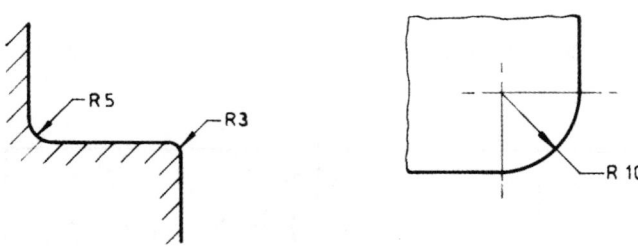

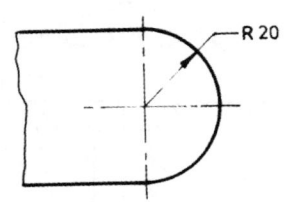

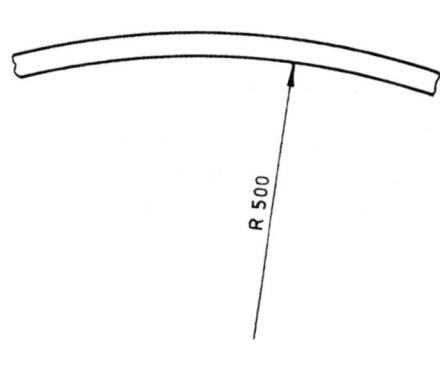

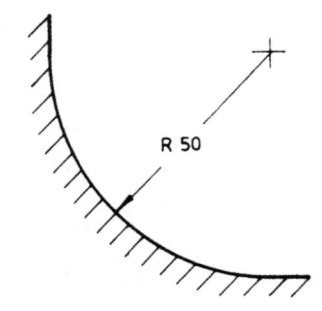

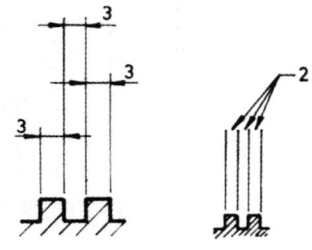

**FIGURE 1.27** ▶ Dimensioning small spaces

**FIGURE 1.28** ▶ Methods of dimensioning spherical surfaces

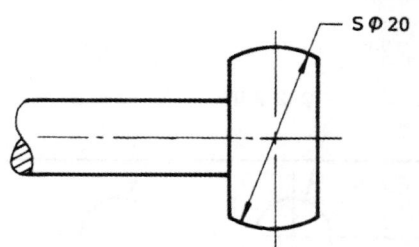

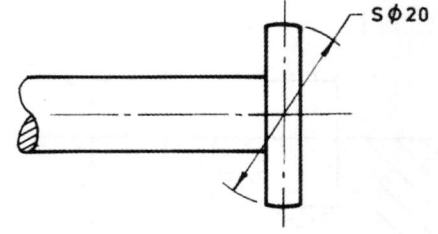

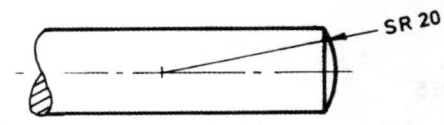

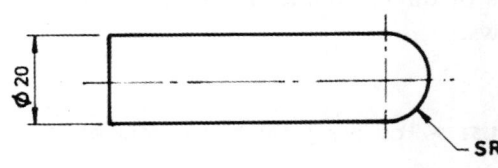

**FIGURE 1.29** ▶ Methods of dimensioning squares

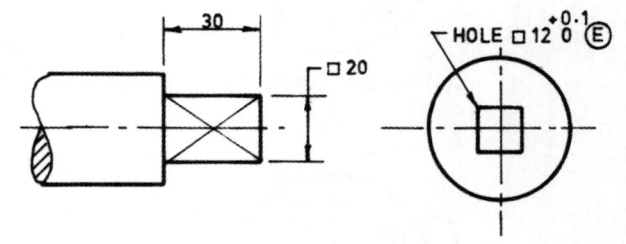

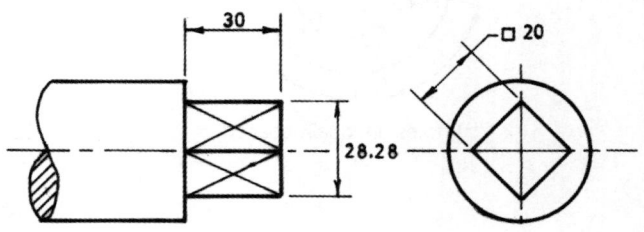

**FIGURE 1.30** ▶ Methods of dimensioning holes

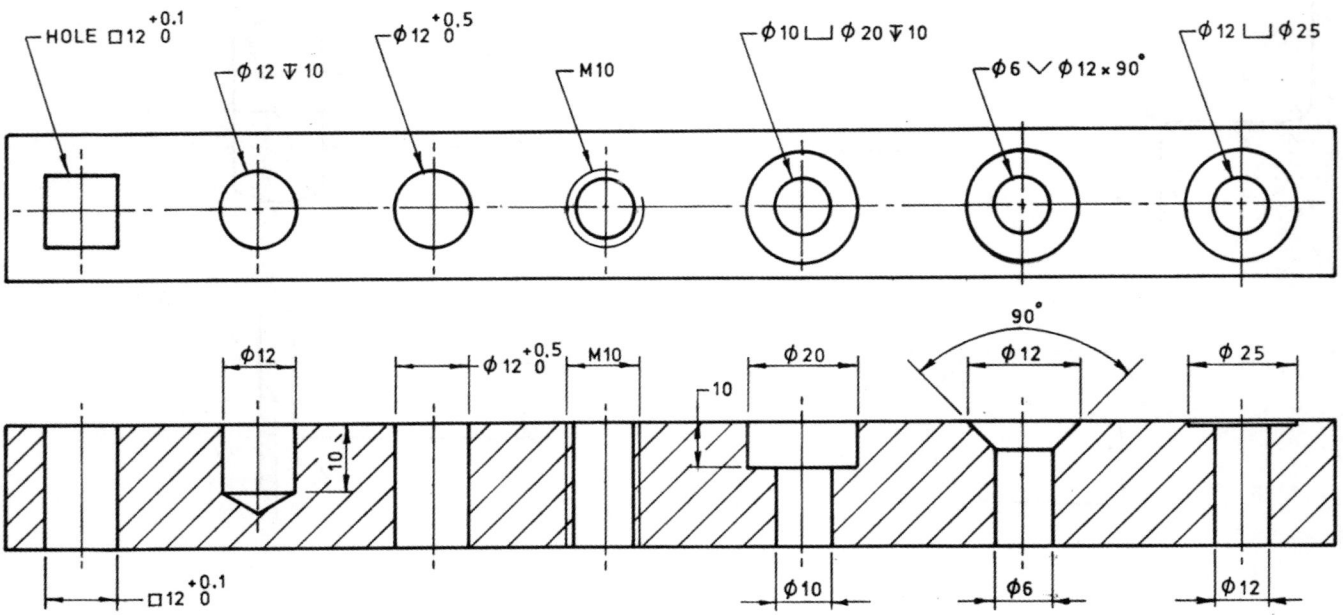

## HOLES

Holes either go right through a material or to a certain depth, and this must be specified as well as the diameter. If no indication is given, a hole is taken as going right through. Figure 1.30 above illustrates methods of dimensioning holes using both end and side views.

## POSITIONING HOLES

Holes may be positioned by specifying the diameter of pitch circles as shown in Figure 1.31 or by specifying rectangular coordinates of centre distances as shown in Figure 1.32.

**FIGURE 1.31** ▶ Positioning holes by angular dimensions

**FIGURE 1.32** ▶ Positioning holes by co-ordinate dimensions

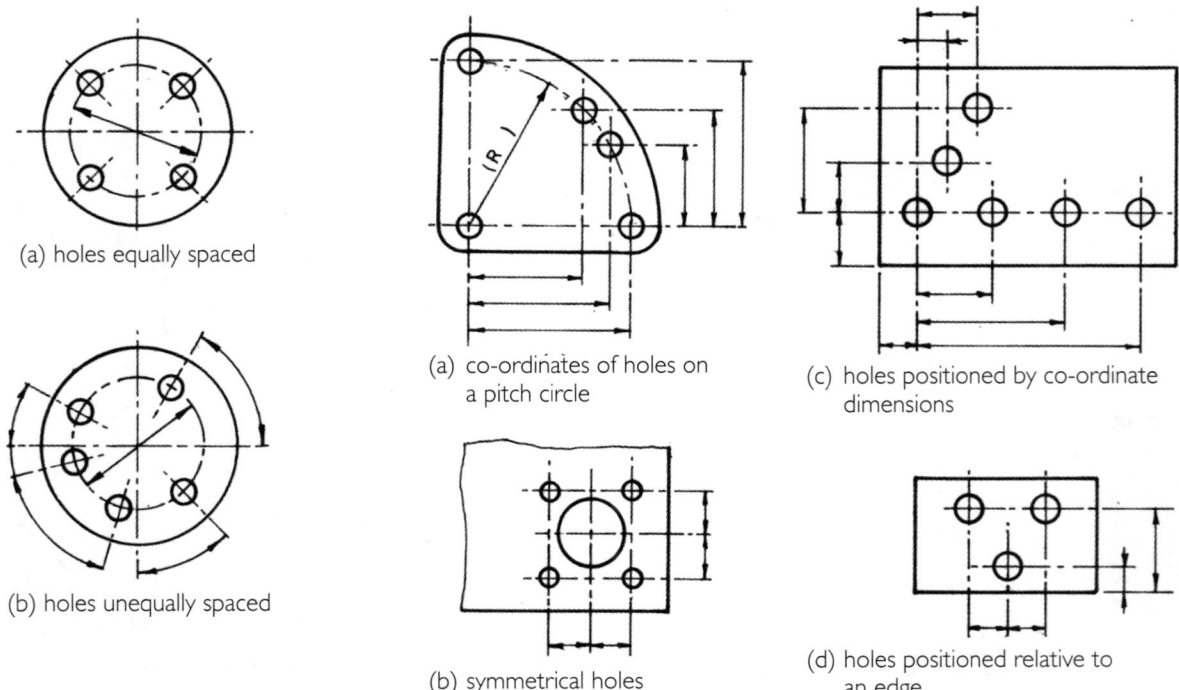

(a) holes equally spaced

(b) holes unequally spaced

(a) co-ordinates of holes on a pitch circle

(b) symmetrical holes

(c) holes positioned by co-ordinate dimensions

(d) holes positioned relative to an edge

## COUNTERSINKS

These may be dimensioned by one of the methods shown in Figure 1.33.

## COUNTERBORES

These may be dimensioned by one of the methods shown in Figure 1.34.

## SPOTFACES

These may be dimensioned by one of the methods shown in Figure 1.35.

## CHAMFERS

These may be dimensioned by one of the methods shown in Figure 1.36.

**FIGURE 1.33** ▶ Methods of dimensioning countersinks

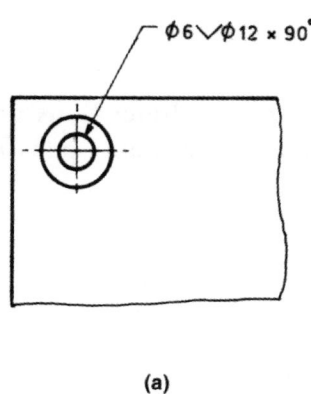

(a)

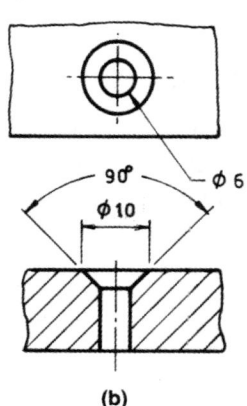

(b)

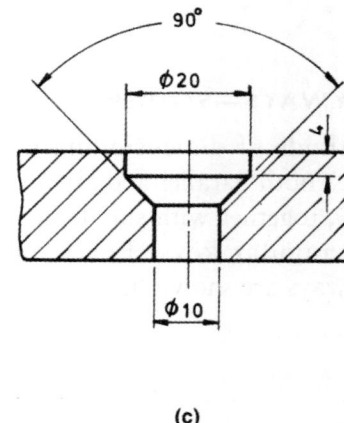

(c)

**FIGURE 1.34** ▶ Methods of dimensioning counterbores

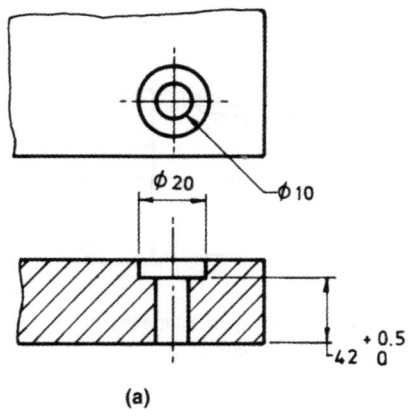

(a)

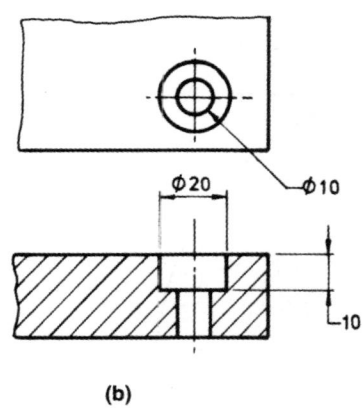

(b)

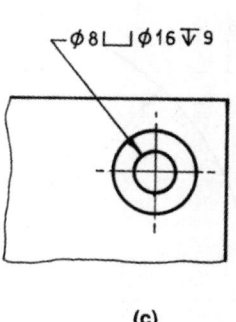

(c)

**FIGURE 1.35** ▶ Methods of dimensioning spotfaces

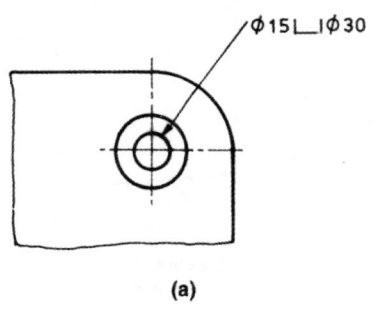

(a)

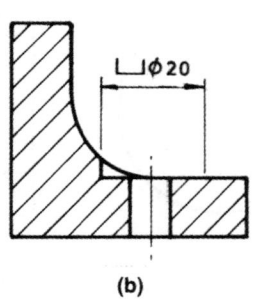

(b)

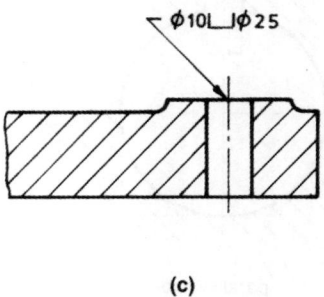

(c)

**FIGURE I.36** ▶ Methods of dimensioning chamfers

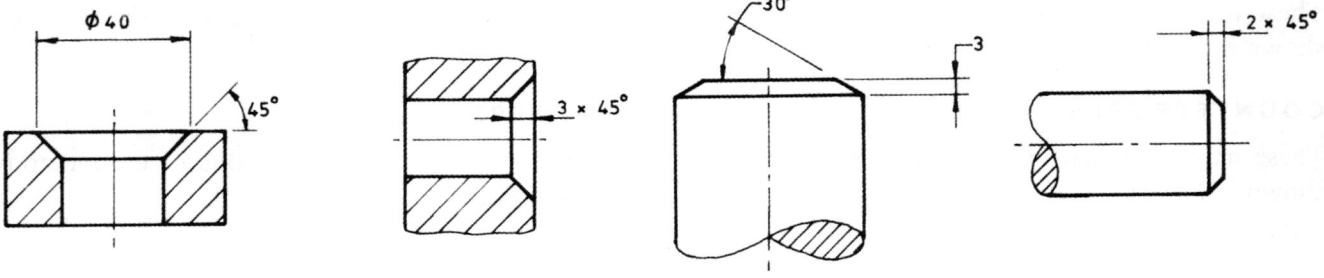

## KEYWAYS—SQUARE AND RECTANGULAR

Methods of dimensioning keyways in shafts and hubs, both parallel and tapered, are shown in Figure 1.37, together with suitable proportions for drawing rectangular keys. Enlarged details of key and keyways are shown in Figures 1.38 and 1.39.

Note: Tables 1.8 and 1.9 give dimensions and tolerances for square and rectangular parallel keyways.

**FIGURE I.37** ▶ Methods of dimensioning keys and keyways

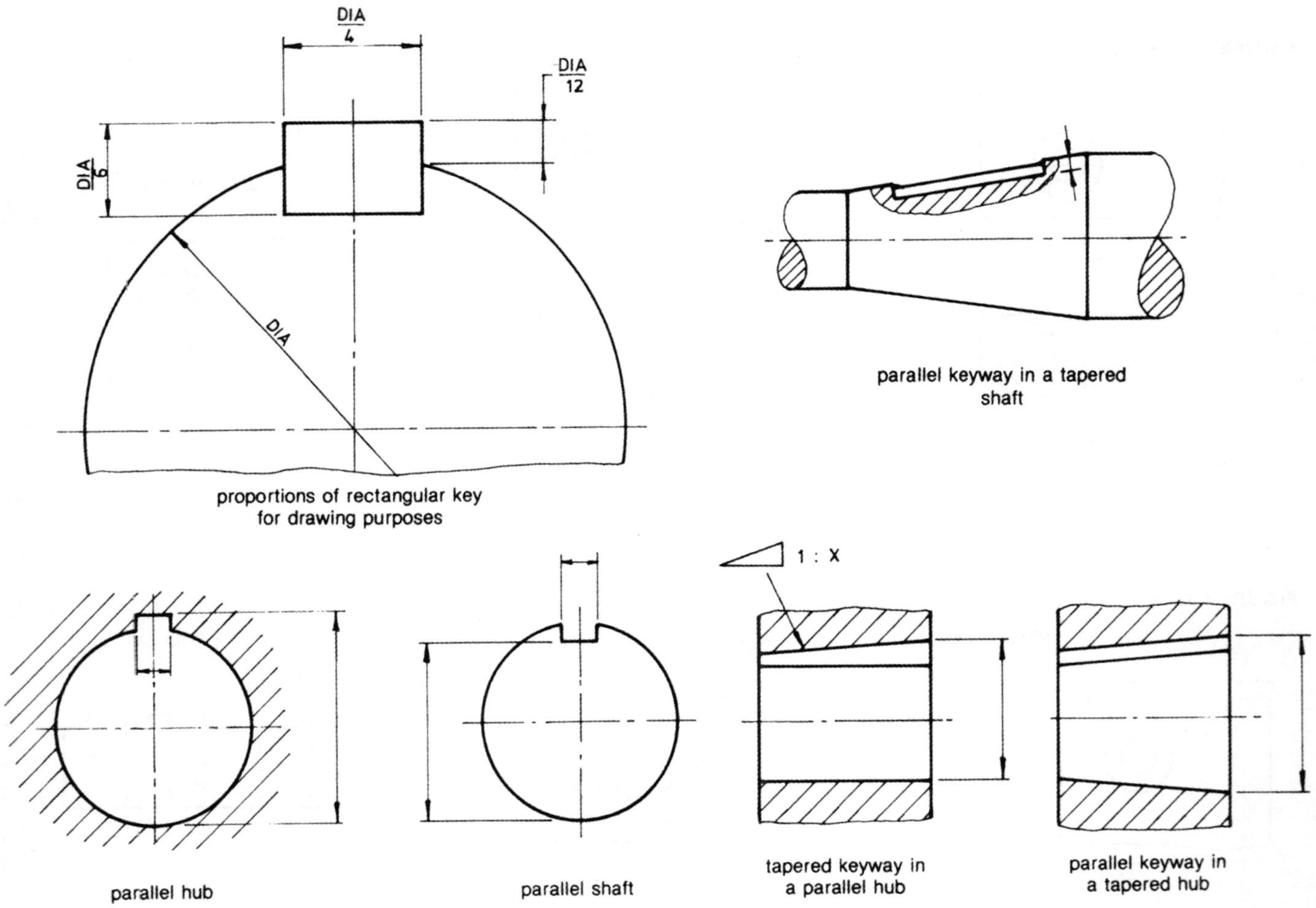

proportions of rectangular key
for drawing purposes

parallel keyway in a tapered
shaft

parallel hub

parallel shaft

tapered keyway in
a parallel hub

parallel keyway in
a tapered hub

**FIGURE 1.38** ▶ Enlarged detail of key and keyways

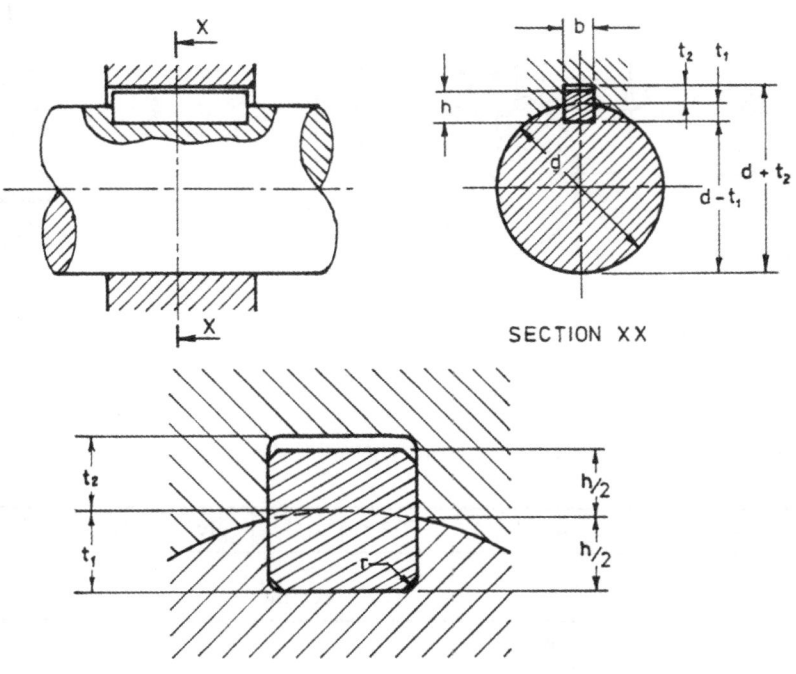

SECTION XX

**TABLE 1.8** ▶ Dimensions and tolerances for square parallel keyways

All dimensions in millimetres                                                                                                            As amended July 1974

| 1 | 2 | 3 | 4 | 5 | 6 | 7 | 8 | 9 | 10 | 11 | 12 | 13 | 14 | 15 |
|---|---|---|---|---|---|---|---|---|----|----|----|----|----|----|
| **SHAFT** | | **KEY** (see Note) | **KEYWAY** | | | | | | | | | | | |
| **NOMINAL DIAMETER d** (see Note) | | **SECTION b x h** WIDTH x THICKNESS | **WIDTH b** | | | | | | **DEPTH** | | | | **RADIUS r** | |
| | | | | **TOLERANCE FOR CLASS OF FIT** | | | | | **SHAFT $t_1$** | | **HUB $t_2$** | | | |
| | | | | **FREE** | | **NORMAL** | | **CLOSE AND INTERFERENCE** | | | | | | |
| **OVER** | **INCL.** | | **NOM.** | **SHAFT (H9)** | **HUB (D10)** | **SHAFT (N9)** | **HUB ($J_2$9)\*** | **SHAFT AND HUB (P9)** | **NOM.** | **TOL.** | **NOM.** | **TOL.** | **MAX.** | **MIN.** |
| 6 | 8 | 2 x 2 | 2 | + 0.025 | +0.060 | −0.004 | +0.012 | −0.006 | 1.2 | | 1 | | 0.16 | 0.08 |
| 8 | 10 | 3 x 3 | 3 | 0 | +0.020 | −0.029 | −0.012 | −0.031 | 1.8 | | 1.4 | | 0.16 | 0.08 |
| 10 | 12 | 4 x 4 | 4 | | | | | | 2.5 | +0.1 0 | 1.8 | +0.1 0 | 0.16 | 0.08 |
| 12 | 17 | 5 x 5 | 5 | +0.030 0 | +0.078 +0.030 | 0 −0.030 | +0.015 −0.015 | −0.012 −0.042 | 3 | | 2.3 | | 0.25 | 0.16 |
| 17 | 22 | 6 x 6 | 6 | | | | | | 3.5 | | 2.8 | | 0.25 | 0.16 |

\*The limits for tolerance $J_2$9 are quoted from BS 4500 (ISO limits and fits), to three significant figures.

NOTE: The relations between shaft diameter and key section given above are for general applications. The use of smaller key sections is permitted if suitable for the torque transmitted. In cases such as stepped shafts when large diameters are required, for example to resist bending, and when fans, gears and impellers are fitted with a smaller key than normal, an unequal disposition of key in shaft with relation to the hub results. Therefore, dimensions $d - t_1$ and $d + t_2$ should be recalculated to maintain the $h/2$ relationship.

The use of larger key sections which are special to any particular application is outside the scope of this standard.

**FIGURE 1.39** ▶ Enlarged detail of key and keyways

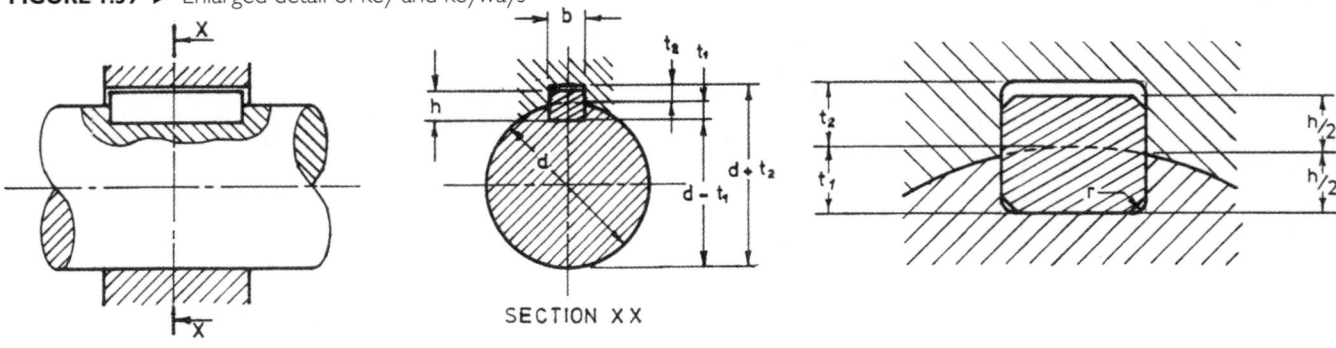

SECTION XX

**TABLE 1.9** ▶ Dimensions and tolerances for rectangular parallel keyways

All dimensions in millimetres

| 1 | 2 | 3 | 4 | 5 | 6 | 7 | 8 | 9 | 10 | 11 | 12 | 13 | 14 | 15 |
|---|---|---|---|---|---|---|---|---|---|---|---|---|---|---|
| SHAFT | | KEY (see Note) | KEYWAY | | | | | | | | | | | |
| NOMINAL DIAMETER *d* (see Note) | | SECTION *b x h* WIDTH x THICKNESS | WIDTH *b* | | | | | | DEPTH | | | | RADIUS *r* | |
| | | | NOM. | TOLERANCE FOR CLASS OF FIT | | | | | SHAFT $t_1$ | | HUB $t_2$ | | | |
| | | | | FREE | | NORMAL | | CLOSE AND INTERFERENCE | | | | | | |
| OVER | INCL. | | NOM. | SHAFT (H9) | HUB (D10) | SHAFT (N9) | HUB (J₂9)* | SHAFT AND HUB (P9) | NOM. | TOL. | NOM. | TOL. | MAX. | MIN. |
| 22 | 30 | 8 × 7 | 8 | +0.036 | +0.098 | 0 | +0.018 | −0.015 | 4 | | 3.3 | | 0.25 | 0.15 |
| 30 | 38 | 10 × 8 | 10 | 0 | +0.040 | −0.036 | −0.018 | −0.051 | 5 | | 3.3 | | 0.40 | 0.25 |
| 38 | 44 | 12 × 8 | 12 | +0.043 | +0.120 | 0 | +0.021 | −0.018 | 5 | | 3.3 | | 0.40 | 0.25 |
| 44 | 50 | 14 × 9 | 14 | | | | | | 5.5 | +0.2 | 3.8 | +0.2 | 0.40 | 0.25 |
| 50 | 58 | 16 × 10 | 16 | 0 | +0.050 | −0.043 | −0.021 | −0.061 | 6 | 0 | 4.3 | 0 | 0.40 | 0.25 |
| 58 | 65 | 18 × 11 | 18 | | | | | | 7 | | 4.4 | | 0.40 | 0.25 |
| 65 | 75 | 20 × 12 | 20 | +0.052 | +0.149 | 0 | +0.026 | −0.022 | 7.5 | | 4.9 | | 0.60 | 0.40 |
| 75 | 85 | 22 × 14 | 22 | | | | | | 9 | | 5.4 | | 0.60 | 0.40 |
| 85 | 95 | 25 × 14 | 25 | 0 | +0.065 | −0.052 | −0.026 | −0.074 | 9 | | 5.4 | | 0.60 | 0.40 |
| 95 | 110 | 28 × 16 | 28 | | | | | | 10 | | 6.4 | | 0.60 | 0.40 |
| 110 | 130 | 32 × 18 | 32 | | | | | | 11 | | 7.4 | | 0.60 | 0.40 |
| 130 | 150 | 36 × 20 | 36 | +0.062 | +0.180 | 0 | +0.031 | −0.026 | 12 | | 8.4 | | 1.00 | 0.70 |
| 150 | 170 | 40 × 22 | 40 | | | | | | 13 | | 9.4 | | 1.00 | 0.70 |
| 170 | 200 | 45 × 25 | 45 | 0 | +0.080 | −0.062 | −0.031 | −0.088 | 15 | | 10.4 | | 1.00 | 0.70 |
| 200 | 230 | 50 × 28 | 50 | | | | | | 17 | | 11.4 | | 1.00 | 0.70 |
| 230 | 260 | 56 × 32 | 56 | +0.074 | +0.220 | 0 | +0.037 | −0.032 | 20 | +0.3 | 12.4 | +0.3 | 1.60 | 1.20 |
| 260 | 290 | 63 × 32 | 63 | | | | | | 20 | 0 | 12.4 | 0 | 1.60 | 1.20 |
| 290 | 330 | 70 × 36 | 70 | 0 | +0.100 | −0.074 | −0.037 | −0.106 | 22 | | 14.4 | | 1.60 | 1.20 |
| 330 | 380 | 80 × 40 | 80 | | | | | | 25 | | 15.4 | | 2.50 | 2.00 |
| 380 | 440 | 90 × 45 | 90 | +0.087 | +0.260 | 0 | +0.043 | −0.037 | 28 | | 17.4 | | 2.50 | 2.00 |
| 440 | 500 | 100 × 50 | 100 | 0 | +0.120 | −0.087 | −0.043 | −0.124 | 31 | | 19.5 | | 2.50 | 2.00 |

*The limits for tolerance J₂9 are quoted from BS 4500 (ISO limits and fits), to three significant figures.

NOTE: The relations between shaft diameter and key section given above are for general applications. The use of smaller key sections is permitted if suitable for the torque transmitted. In cases such as stepped shafts when large diameters are required, for example to resist bending, and when fans, gears and impellers are fitted with a smaller key than normal, an unequal disposition of key in shaft with relation to the hub results. Therefore, dimensions $d - t_1$ and $d + t_2$ should be recalculated to maintain the *h/2* relationship.

The use of larger key sections which are special to any particular application is outside the scope of this standard.

## KEYWAYS—WOODRUFF

Methods of dimensioning Woodruff keyways in shafts and hubs, both parallel and tapered, are shown in Figure 1.40.

## TAPERS

Tapers are dimensioned by one of the four methods shown in Figure 1.41.

**FIGURE 1.40** ▶ Methods of dimensioning Woodruff keys

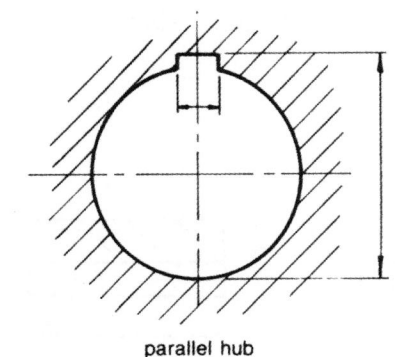

parallel hub

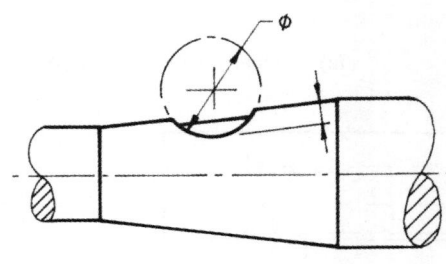

tapered hub

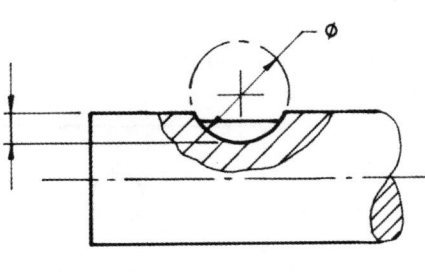

parallel shaft

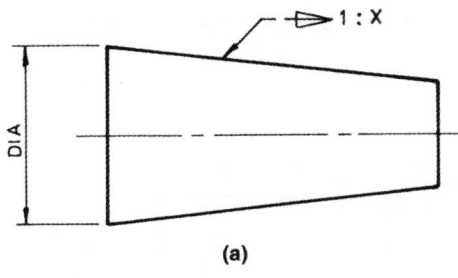

tapered shaft

**FIGURE 1.41** ▶ Methods of dimensioning tapers

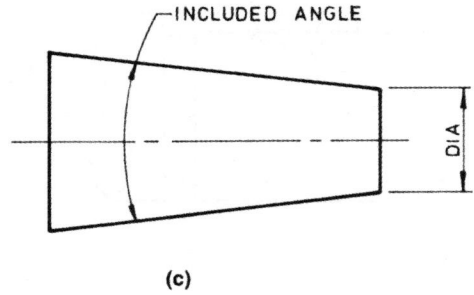

(a)

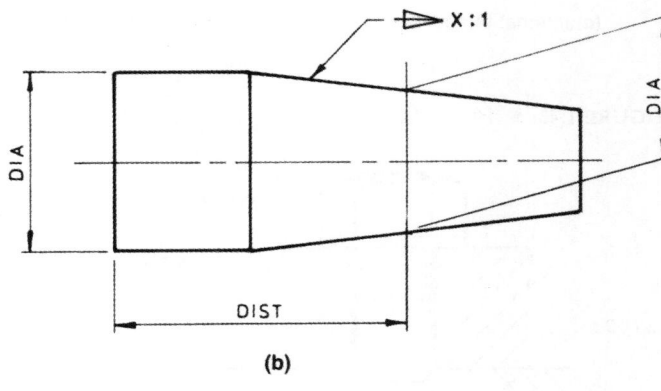

(b)

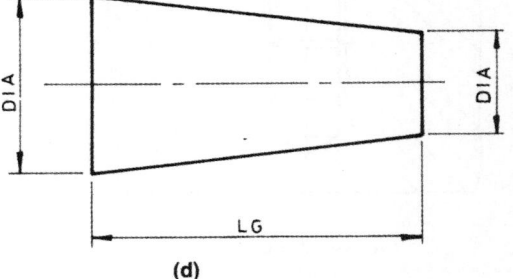

(c)

(d)

# 1.11 Screw threads

### GENERAL REPRESENTATION

The methods shown in Figure 1.42 are recommended for right-hand or left-hand representation of screw threads. The diameter (ØDIA) of a thread is the nominal size of the thread, for example for a 12 mm thread (M12, see p. 28), DIA = 12 mm.

### THREADS ON ASSEMBLY AND SPECIAL THREADS

Figure 1.43(a) illustrates the method of representing two threads in assembly. Figure 1.43(b) shows the assembly of two members by a stud mounted in one of them. Special threads are usually represented by a scrap sectional view illustrating the form of the thread, as shown in Figure 1.43(c).

**FIGURE 1.42** ▶ Methods of representing screw threads

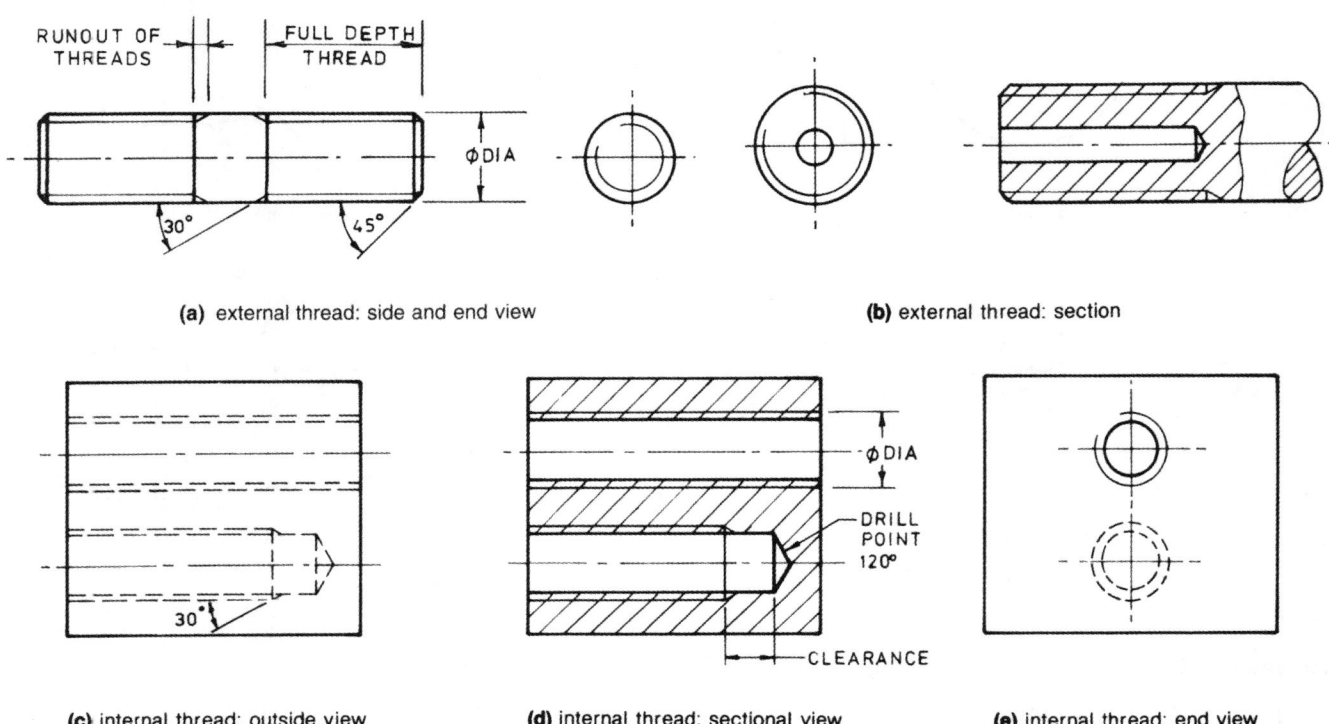

**(a)** external thread: side and end view

**(b)** external thread: section

**(c)** internal thread: outside view

**(d)** internal thread: sectional view

**(e)** internal thread: end view

**FIGURE 1.43** ▶ Methods of representing assembled and special threads

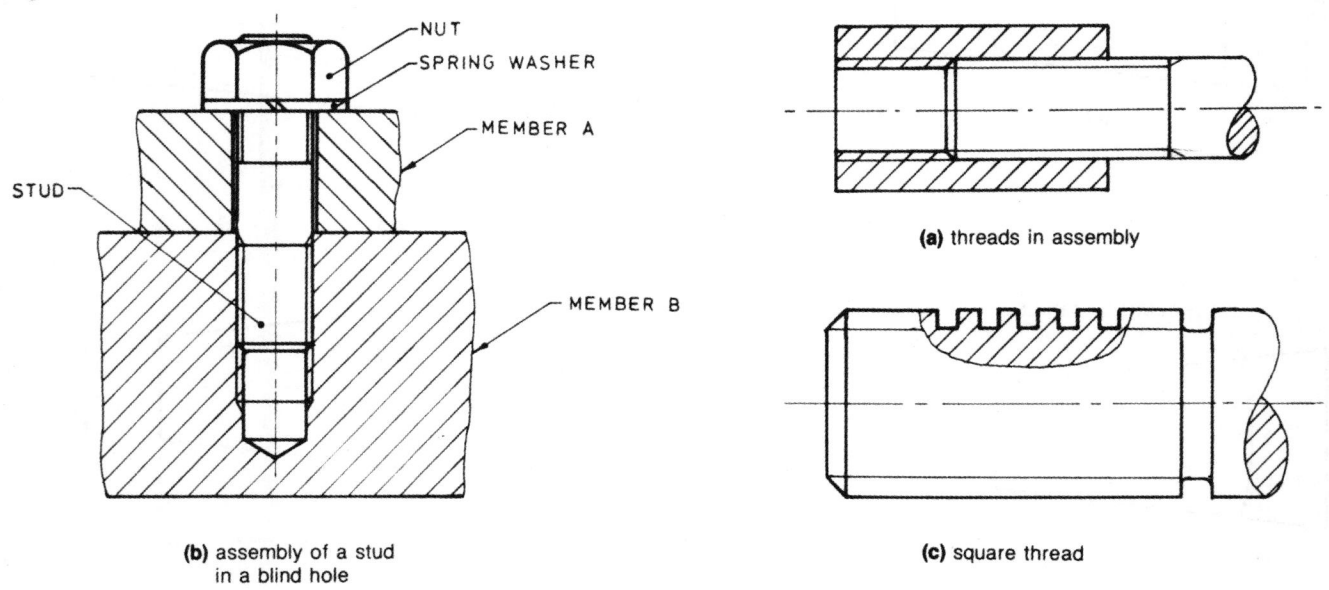

**(b)** assembly of a stud in a blind hole

**(a)** threads in assembly

**(c)** square thread

## DIMENSIONING FULL AND RUNOUT THREADS

When full and runout threads have to be distinguished, the methods of dimensioning shown in Figure 1.44 are recommended. Where there is no design requirement, the runout threads need not be dimensioned.

## DIMENSIONING METRIC THREADS IN HOLES

Figure 1.45 (below) shows various methods used to dimension threaded holes. The diameter of the thread is always preceded by the capital letter M, which indicates metric threads.

The coarse thread series is dimensioned simply by the letter M followed by a numeral, for example M12.

However, fine threads should show the pitch of the thread as well, for example M12 × 1.25. The term 6H contained in the thread dimensions of Figures 1.45(b), (c) and (d) refers to the grade of tolerance to be used in the manufacture of these threaded holes. This tolerance combined with a similar tolerance on the mating screw provides a certain 'fit' when the screw is assembled into the threaded hole. The system used for threaded 'fits' is the same as that used for plain shaft and hole 'fits' described on page 53.

If it is not important, the runout threads need not be dimensioned. However, in blind holes it is often important to have fully formed threads for a certain depth, and dimensioning must be provided to control this.

**FIGURE 1.44** ▶ Methods of dimensioning threaded members

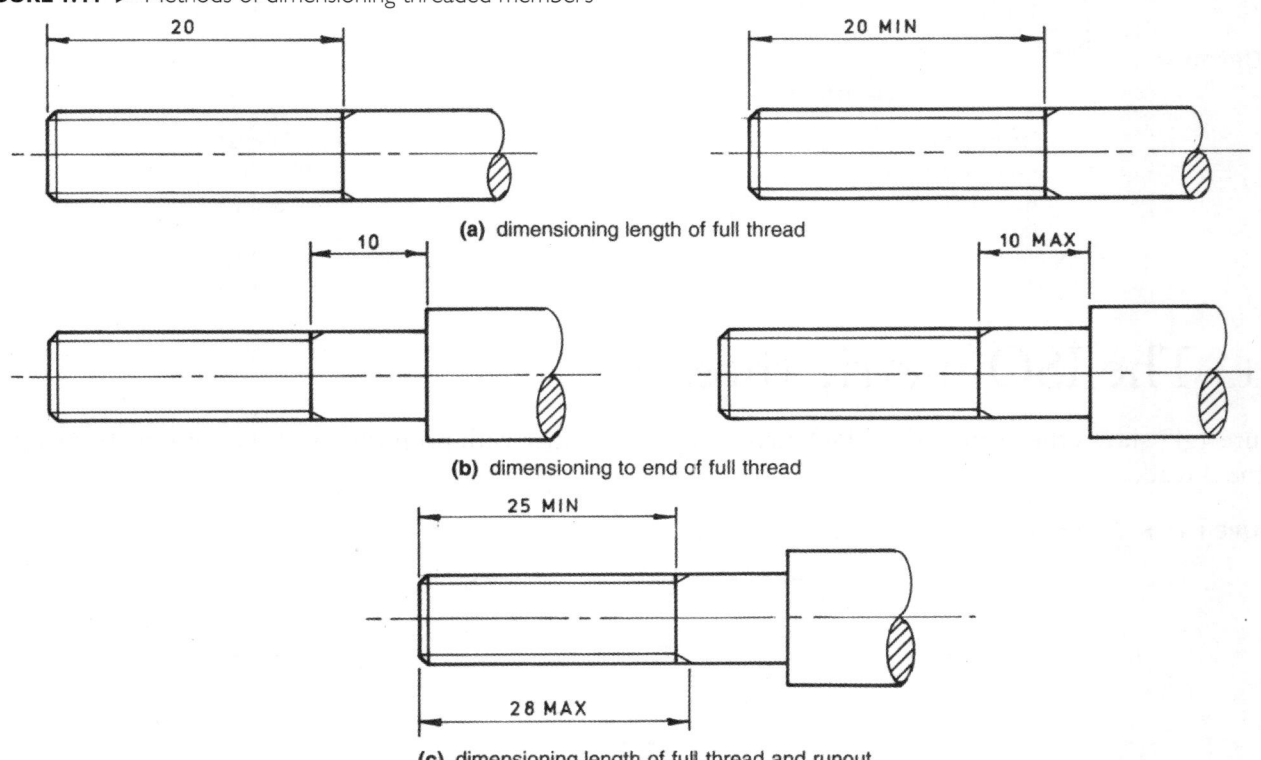

**(a)** dimensioning length of full thread

**(b)** dimensioning to end of full thread

**(c)** dimensioning length of full thread and runout

**FIGURE 1.45** ▶ Methods of dimensioning threads in holes

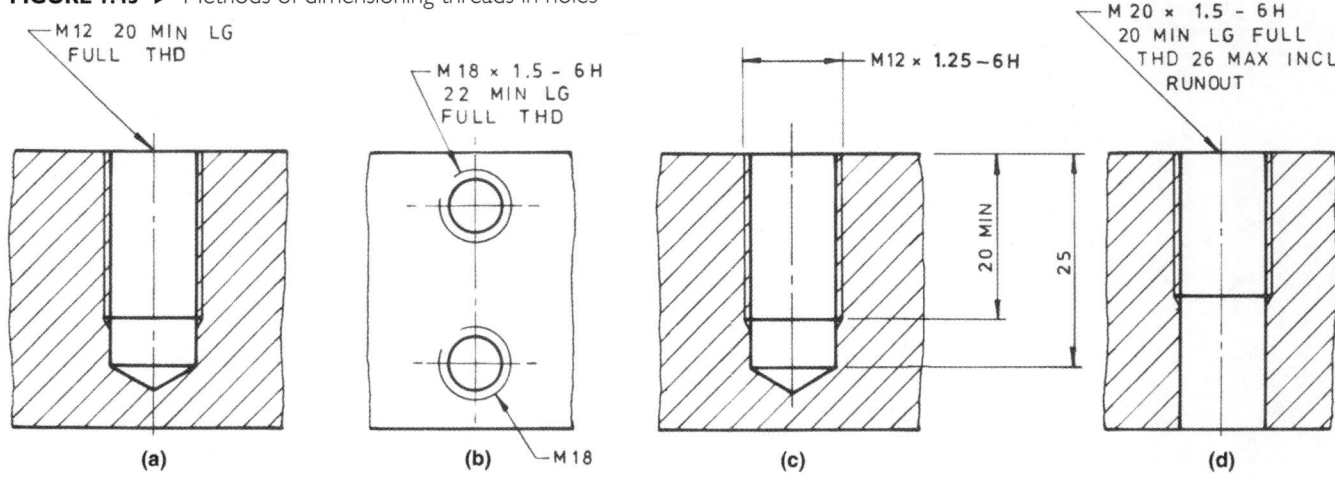

(a)      (b)      (c)      (d)

# 1.12 The Australian metric thread profile

Figure 1.46 shows the profile of the internal and external metric threads, which are suitable for single point screw cutting at maximum material condition according to AS 1721—1985. If the rounded projection is not used, a flat should be ground on the appropriate tool to the values of $W_n$ and $W_s$ and the corners will round off as the tool wears.

**FIGURE 1.46** ▶ Profile of the Australian metric thread

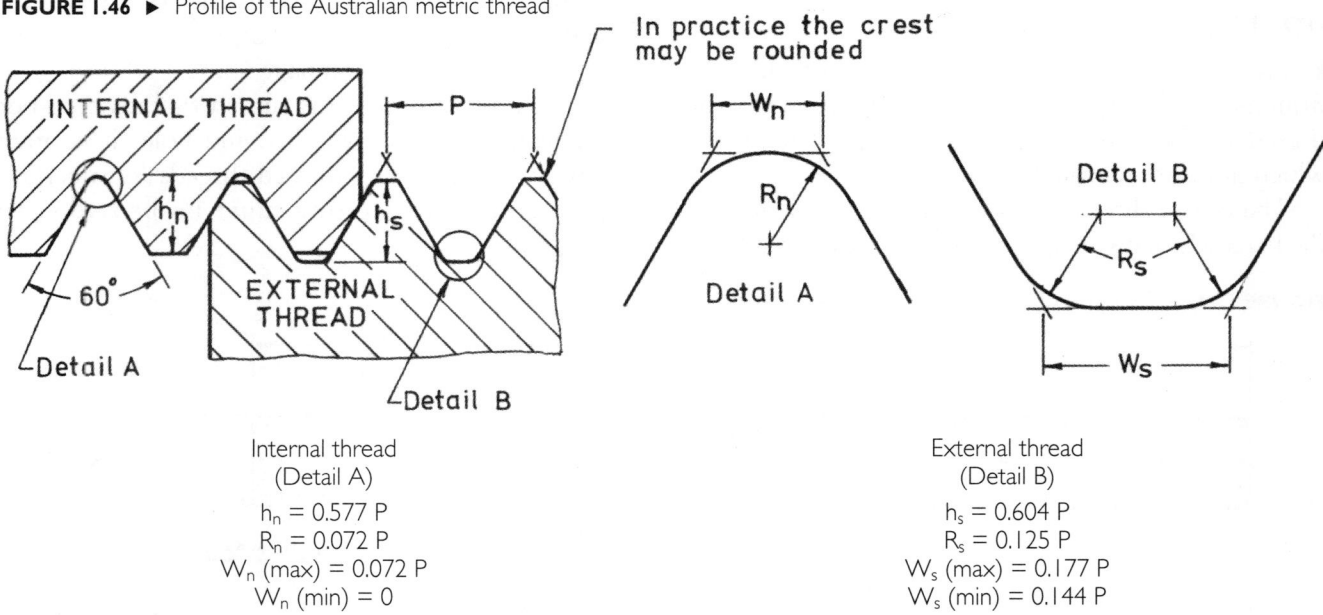

Internal thread
(Detail A)
$h_n = 0.577\ P$
$R_n = 0.072\ P$
$W_n\ (max) = 0.072\ P$
$W_n\ (min) = 0$

External thread
(Detail B)
$h_s = 0.604\ P$
$R_s = 0.125\ P$
$W_s\ (max) = 0.177\ P$
$W_s\ (min) = 0.144\ P$

# 1.13 The ISO metric thread

Figure 1.47 shows the profile of the ISO metric thread, together with proportions of the various defined parts of the thread.

**FIGURE 1.47** ▶ Basic profile and proportions of the ISO metric thread

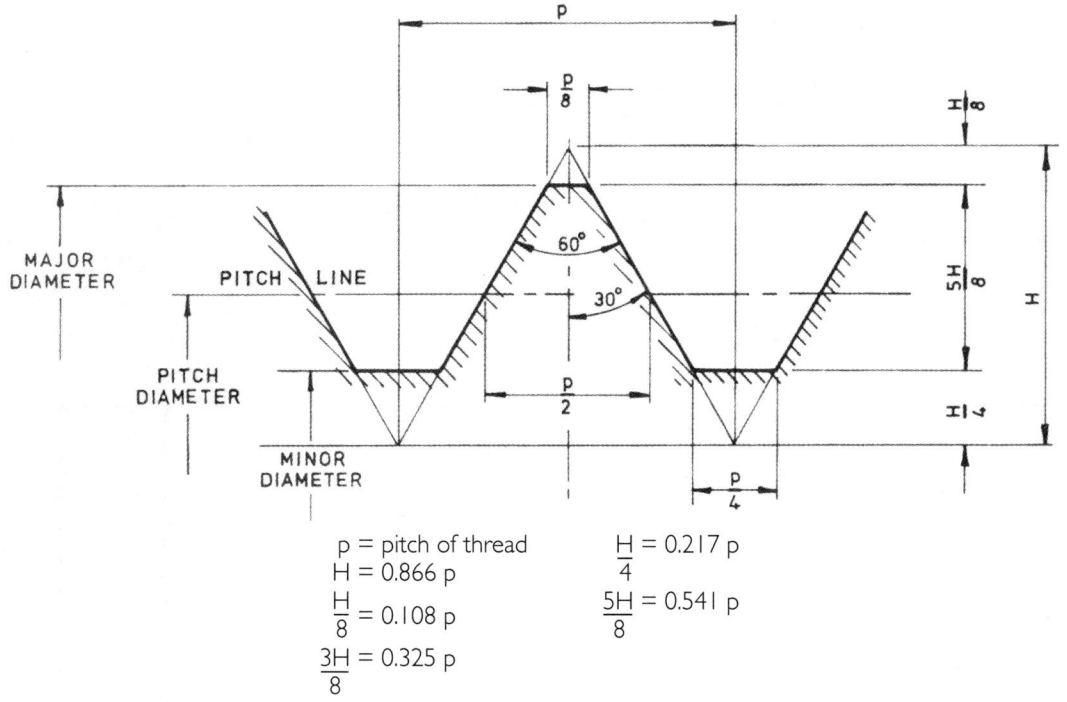

$p$ = pitch of thread
$H = 0.866\ p$
$\dfrac{H}{8} = 0.108\ p$
$\dfrac{3H}{8} = 0.325\ p$

$\dfrac{H}{4} = 0.217\ p$
$\dfrac{5H}{8} = 0.541\ p$

## GRAPHICAL COMPARISON OF METRIC THREAD SERIES

ISO metric threads are of two kinds: coarse and fine thread. A graphical comparison of these two series is shown in Figure 1.48.

## TAPPING SIZE AND CLEARANCE HOLES FOR ISO METRIC THREADS

Tapping sizes and clearance holes for metric threads are shown in Table 1.10 (page 32). In this table column 1 represents first and second choices of thread diameters. The sizes listed under second choice should be used only when it is not possible to use sizes in the first choice column.

The pitches listed in column 2 of Table 1.10 are compared on the graph in Figure 1.48. These pitches, together with the corresponding first and second choice diameters of column 1, are those combinations which have been recommended by the ISO as a selected 'coarse' and 'fine' series for screws, bolts, nuts and other threaded fasteners commonly used in most general engineering applications. Column 3 is the tapping size for the coarse and fine series. These values

represent approximately 83 per cent full depth of thread, and can be calculated simply by the formula:

$$\text{tapping drill size} = \text{outside diameter} - \text{pitch}$$
$$3.3 = 4 - 0.7$$

Sometimes the drill size has to be rounded off to the next largest stock drill size; this can be obtained from Table 1.11 (page 33).

Column 4 of Table 1.10 gives tapping sizes for coarse threads in mild steel only; these will give approximately 71 per cent of the full depth of thread. In most general engineering applications this depth of thread is sufficient and desirable for the following reasons:

1. Tapping 83 per cent depth of thread necessitates about three times more power than tapping 71 per cent depth of thread.

2. The possibility of tap breakage is greater as the depth of thread increases.

3. The 83 per cent depth of thread has approximately 5 per cent more strength than the 71 per cent depth of thread.

**FIGURE 1.48** ▶ Graphical comparison of metric threads

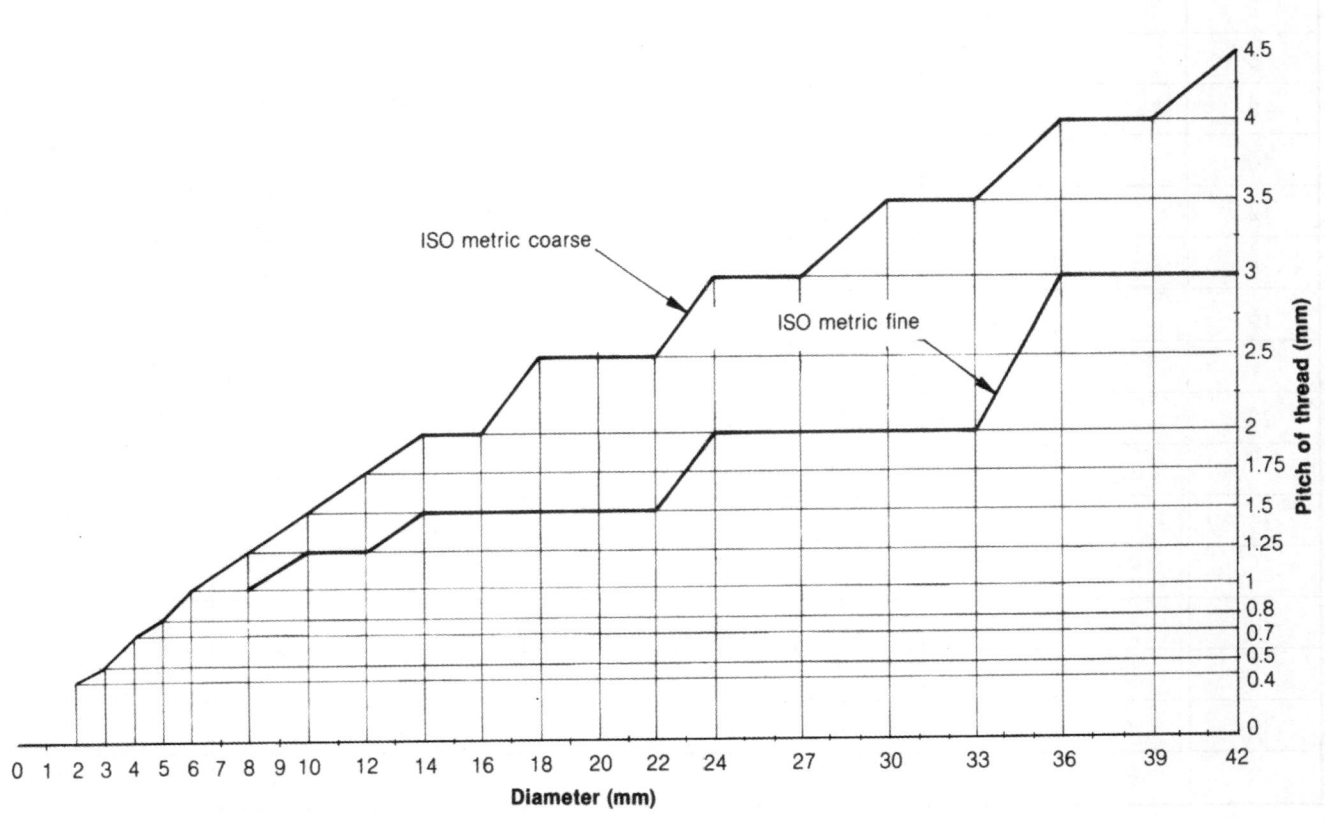

4. The amount of metal removed from a 71 per cent depth of thread is much less than that removed for 83 per cent depth of thread.

There are cases when a deeper thread is necessary, for example on machines and in situations where movement in the mating threads is to be kept to a minimum.

Column 5 of Table 1.10 gives three classes of clearance holes recommended for the various sizes of metric threads.

**TABLE 1.10** ▶ Tapping sizes and clearances (mm) for metric threads

| 1 | | 2 | | 3 | | 4 | 5 | | |
|---|---|---|---|---|---|---|---|---|---|
| NOMINAL DIAMETERS | | PITCH (mm) | | TAPPING SIZE DIAMETER (mm) | | TAPPING SIZE (71% FULL DEPTH COARSE THREADS IN MILD STEEL) | CLEARANCE HOLES | | |
| 1st choice | 2nd choice | coarse threads | fine threads | coarse threads | fine threads | | close | medium | coarse |
| 1.6 | | 0.35 | | 1.25 | | 1.3 | 1.7 | 1.8 | 2 |
| | 1.8 | 0.35 | | 1.45 | | 1.5 | 1.9 | 2 | 2.2 |
| 2 | | 0.4 | | 1.6 | | 1.65 | 2.2 | 2.4 | 2.6 |
| | 2.2 | 0.45 | | 1.75 | | 1.8 | 2.4 | 2.6 | 2.8 |
| 2.5 | | 0.45 | | 2.05 | | 2.1 | 2.7 | 2.9 | 3.1 |
| 3 | | 0.5 | | 2.5 | | 2.55 | 3.2 | 3.4 | 3.6 |
| | 3.5 | 0.6 | | 2.9 | | 2.95 | 3.7 | 3.9 | 4.1 |
| 4 | | 0.7 | | 3.3 | | 3.4 | 4.3 | 4.5 | 4.8 |
| | 4.5 | 0.75 | | 3.8 | | 3.8 | 4.8 | 5 | 5.3 |
| 5 | | 0.8 | | 4.2 | | 4.3 | 5.3 | 5.5 | 5.8 |
| 6 | | 1 | | 5 | | 5.1 | 6.4 | 6.6 | 7 |
| 8 | | 1.25 | 1 | 6.8 | 7 | 6.9 | 8.4 | 9 | 10 |
| 10 | | 1.5 | 1.25 | 8.5 | 8.7 | 8.6 | 10.5 | 11 | 12 |
| 12 | | 1.75 | 1.25 | 10.2 | 10.8 | 10.4 | 13 | 14 | 15 |
| | 14 | 2 | 1.5 | 12 | 12.5 | 12.2 | 15 | 16 | 17 |
| 16 | | 2 | 1.5 | 14 | 14.5 | 14.25 | 17 | 18 | 19 |
| | 18 | 2.5 | 1.5 | 15.5 | 16.5 | 15.75 | 19 | 20 | 21 |
| 20 | | 2.5 | 1.5 | 17.5 | 18.5 | 17.75 | 21 | 22 | 24 |
| | 22 | 2.5 | 1.5 | 19.5 | 20.5 | 19.75 | 23 | 24 | 26 |
| 24 | | 3 | 2 | 21 | 22 | 21.25 | 25 | 26 | 28 |
| | 27 | 3 | 2 | 24 | 25 | 24.25 | 28 | 30 | 32 |
| 30 | | 3.5 | 2 | 26.5 | 28 | 27 | 31 | 33 | 35 |
| | 33 | 3.5 | 2 | 29.5 | 31 | 30 | 34 | 36 | 38 |
| 36 | | 4 | 3 | 32 | 33 | 32.5 | 37 | 39 | 42 |
| | 39 | 4 | 3 | 35 | 36 | 35.5 | 40 | 42 | 45 |

**TABLE 1.11** ▶ Stock sizes of metric drills (mm)

| | | | | | | | | | | | | | | | |
|---|---|---|---|---|---|---|---|---|---|---|---|---|---|---|---|
| 0.32 | 0.68 | 1.1 | 1.8 | 2.5 | 3.4 | 4.8 | 6.2 | 7.6 | 9 | 10.4 | 11.8 | 13.2 | 15.5 | 19 | 22.5 |
| 0.35 | 0.7 | 1.15 | 1.85 | 2.55 | 3.5 | 4.9 | 6.3 | 7.7 | 9.1 | 10.5 | 11.9 | 13.3 | 15.75 | 19.25 | 22.75 |
| 0.38 | 0.72 | 1.2 | 1.9 | 2.6 | 3.6 | 5 | 6.4 | 7.8 | 9.2 | 10.6 | 12 | 13.4 | 16 | 19.5 | 23 |
| 0.4 | 0.75 | 1.25 | 1.95 | 2.65 | 3.7 | 5.1 | 6.5 | 7.9 | 9.3 | 10.7 | 12.1 | 13.5 | 16.25 | 19.75 | 23.25 |
| 0.42 | 0.78 | 1.3 | 2 | 2.7 | 3.8 | 5.2 | 6.6 | 8 | 9.4 | 10.8 | 12.2 | 13.6 | 16.5 | 20 | 23.5 |
| 0.45 | 0.8 | 1.35 | 2.05 | 2.75 | 3.9 | 5.3 | 6.7 | 8.1 | 9.5 | 10.9 | 12.3 | 13.7 | 16.75 | 20.25 | 23.75 |
| 0.48 | 0.82 | 1.4 | 2.1 | 2.8 | 4 | 5.4 | 6.8 | 8.2 | 9.6 | 11 | 12.4 | 13.8 | 17 | 20.5 | 24 |
| 0.5 | 0.85 | 1.45 | 2.15 | 2.85 | 4.1 | 5.5 | 6.9 | 8.3 | 9.7 | 11.1 | 12.5 | 13.9 | 17.25 | 20.75 | 24.25 |
| 0.52 | 0.88 | 1.5 | 2.2 | 2.9 | 4.2 | 5.6 | 7 | 8.4 | 9.8 | 11.2 | 12.6 | 14 | 17.5 | 21 | 24.5 |
| 0.55 | 0.9 | 1.55 | 2.25 | 2.95 | 4.3 | 5.7 | 7.1 | 8.5 | 9.9 | 11.3 | 12.7 | 14.25 | 17.75 | 21.25 | 24.75 |
| 0.58 | 0.92 | 1.6 | 2.3 | 3 | 4.4 | 5.8 | 7.2 | 8.6 | 10 | 11.4 | 12.8 | 14.5 | 18 | 21.5 | 25 |
| 0.6 | 0.95 | 1.65 | 2.35 | 3.1 | 4.5 | 5.9 | 7.3 | 8.7 | 10.1 | 11.5 | 12.9 | 14.75 | 18.25 | 21.75 | 25.25 |
| 0.62 | 1 | 1.7 | 2.4 | 3.2 | 4.6 | 6 | 7.4 | 8.8 | 10.2 | 11.6 | 13 | 15 | 18.5 | 22 | |
| 0.65 | 1.05 | 1.75 | 2.45 | 3.3 | 4.7 | 6.1 | 7.5 | 8.9 | 10.3 | 11.7 | 13.1 | 15.25 | 18.75 | 22.25 | |

# 1.14 Sectioning—symbols and methods

### GENERAL SYMBOL

A sectional view represents that part of an object which remains after a portion has been removed. It is used to reveal interior detail. Only solid material which has been cut is sectioned. The main types of sectional views used in mechanical drawing are illustrated on pages 34–38. As far as possible the general sectioning symbol (hatching) should be used (Fig. 1.49(a)).

A useful aid for drawing equally spaced hatching lines is shown in Figure 1.49(b).

**FIGURE 1.49** ▶ (a) General symbol for hatching; (b) aid for drawing hatching lines

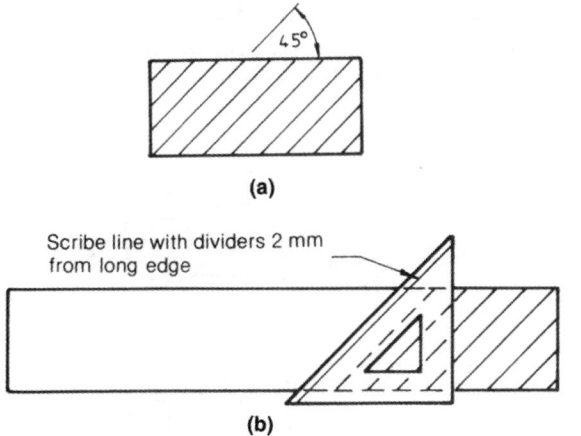

Scribe line with dividers 2 mm from long edge

**(a)**

**(b)**

### HATCHING LINES

These are thin lines (type B), and are normally drawn at 45° to the horizontal, right or left. If the shape of the section would bring the hatching lines parallel to one or more of the sides, another angle may be used (Fig. 1.50).

**FIGURE 1.50** ▶ General application of hatching lines

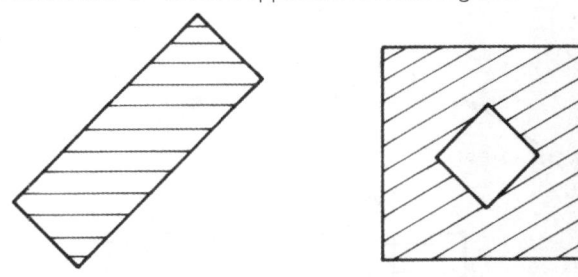

### ADJACENT PARTS

When sectioning adjacent parts the hatching on one part should be at right angles to the hatching on the other part (Fig. 1.51(a)). When more than two parts are adjacent, as in Figure 1.51(b), they may be distinguished by varying the spacing and/or the angle of the hatching lines.

### DIMENSIONS

Dimensions may be inserted in sectioned areas by interrupting the hatching lines, as shown in Figure 1.51(c).

**FIGURE 1.51** ▶ Special application of hatching lines

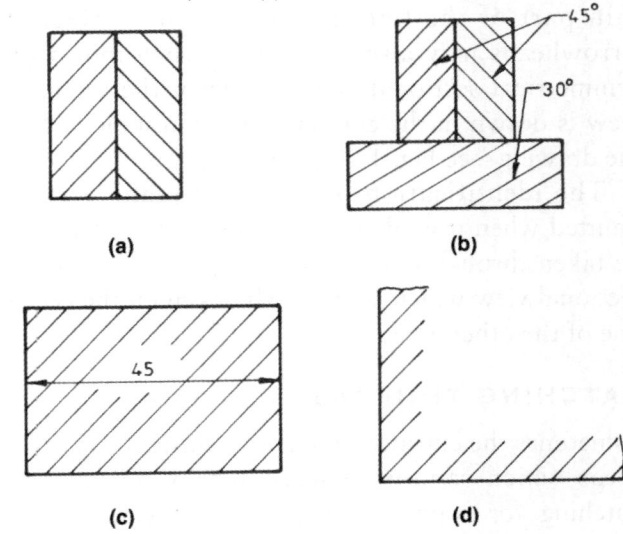

**(a)**

**(b)**

**(c)**

**(d)**

## LARGE AREAS

These can be shown sectioned by placing hatching lines around the edges of the area only, as in Figure 1.51(d).

## SECTIONAL VIEW AND CUTTING PLANE

Section cutting planes are denoted by a chain line (type H) drawn across the part as shown in the front view of Figure 1.52. Arrowheads indicate the direction of viewing.

**FIGURE 1.52** ▶ Representation of a sectional view

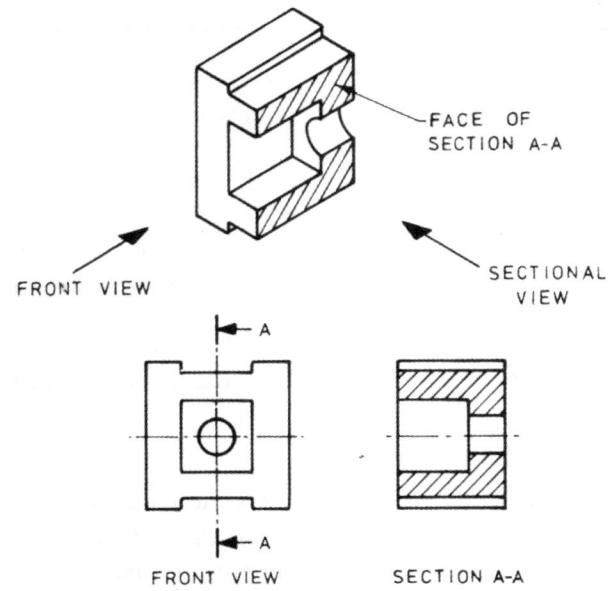

A specific cutting plane is identified by letters placed near the tail of the arrows, and the sectional view is identified by letters, separated by a hyphen, for example section A-A. Where only one cutting plane is used on a drawing, the letters may be omitted.

The chain line may be simplified by omitting the thin part of the line, if clarity is not affected. Arrowheads may also be omitted when indicating symmetrical sectional views or when the sectional view is drawn in the correct projection indicated on the drawing (see Fig. 1.14 (page 10)).

The identification of a cutting plane may be omitted when it is obvious that the section can only be taken through one location. Figure 1.53 shows a sectional view which is obviously taken on the centre line of the other view.

## HATCHING THIN AREAS

Sometimes the cutting plane passes through very thin areas which cannot be sectioned by normal 45° hatching, for example gaskets, plastic sheet, packing,

sheet metal and structural shapes. These areas should be filled in as shown in Figure 1.54(a).

If two or more thin areas are adjacent, a small space should be left between them (Fig. 1.54(b)).

**FIGURE 1.53** ▶ Sectional view with identification of cutting plane omitted

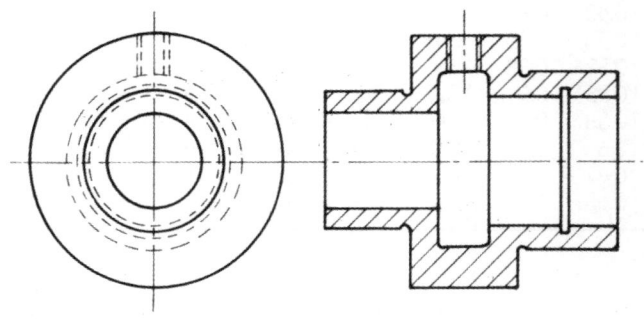

**FIGURE 1.54** ▶ Hatching of thin parts

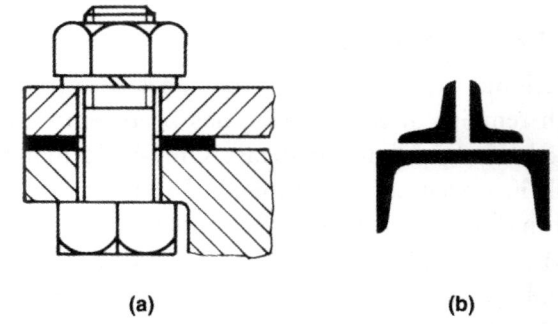

(a)                                        (b)

## EXCEPTIONS TO THE GENERAL RULE

As a general rule all material cut by a cutting plane is hatched in orthogonal views but there are exceptions. When the cutting plane passes through the centre of webs, shafts, bolts, rivets, keys, pins and similar parts, they are not shown sectioned but in outside view, as in Figure 1.55.

**FIGURE 1.55** ▶ Exceptions to the general rule of sectioning

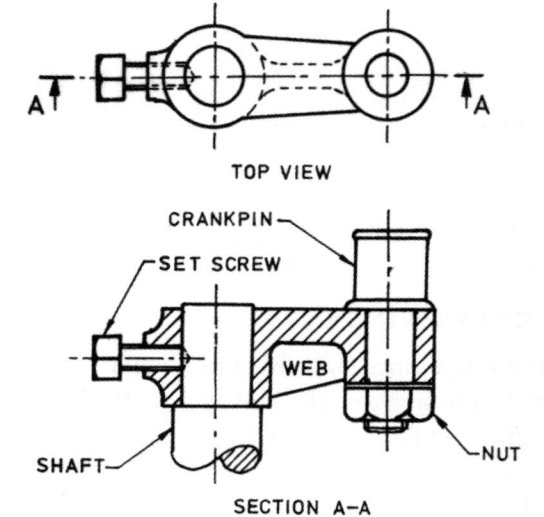

## INTERPOSED AND REVOLVED SECTIONS

The shape of the cross-section of a bar, arm, spoke or rib may be illustrated by a revolved or interposed section.

The interposed section has detail adjacent to it removed, and is drawn using a thick line (type A).

The revolved section has the cross-sectional shape revolved in position with adjacent detail drawn against the revolved view. It is drawn using a thin line (type B). Figure 1.56 illustrates these two sections.

**FIGURE 1.56** ▶ Revolved and interposed sections

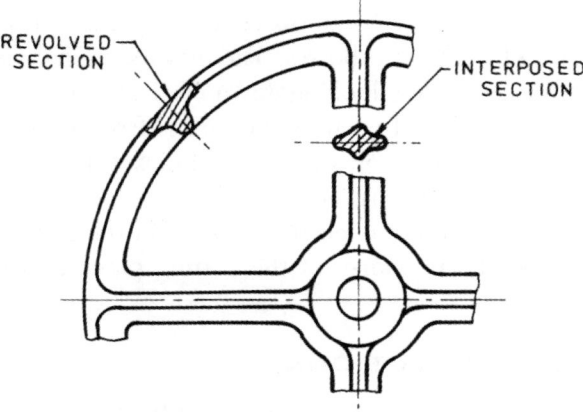

## REMOVED SECTIONS

These are similar to revolved sections except that the cross-section is removed clear of the main outline for the sake of clarity. The removed section may be located adjacent to the main view (Fig. 1.57) or away from it entirely. In the latter case it must be suitably referenced to the cutting plane to which it refers. The outline of a removed section is a heavy continuous line (type A).

**FIGURE 1.57** ▶ Removed sections

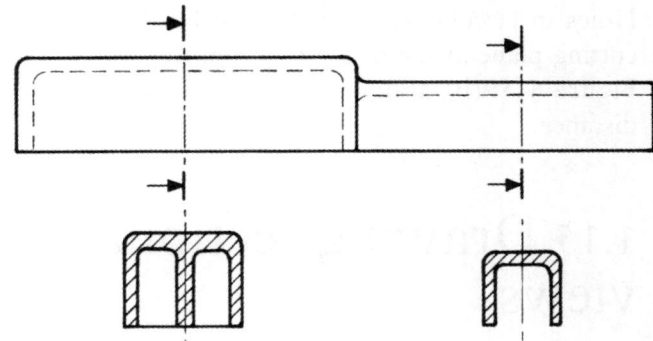

## PART OR LOCAL SECTIONS

Part or local sections may be taken at suitable places on a component to show hidden detail. The boundary of the sections is drawn freehand using a type C line, as in Figure 1.58.

**FIGURE 1.58** ▶ Part or local sections

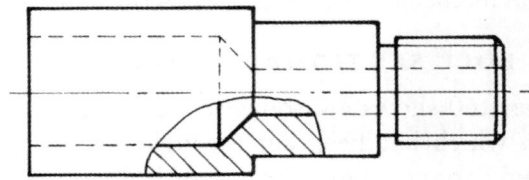

## ALIGNED SECTIONS

In order to include detail on a sectional view which is not located along one plane, the cutting plane may be bent to pass through such detail. The sectional view then shows the detail along the line of the bent cutting plane without any indication that the plane has been bent. The principle is illustrated in Figure 1.59(a). Note that when indicating the cutting plane on the front view, heavy lines are used where the plane changes direction.

**FIGURE 1.59** ▶ Aligned sections (note: in both examples the projection lines would not be shown on the finished drawing)

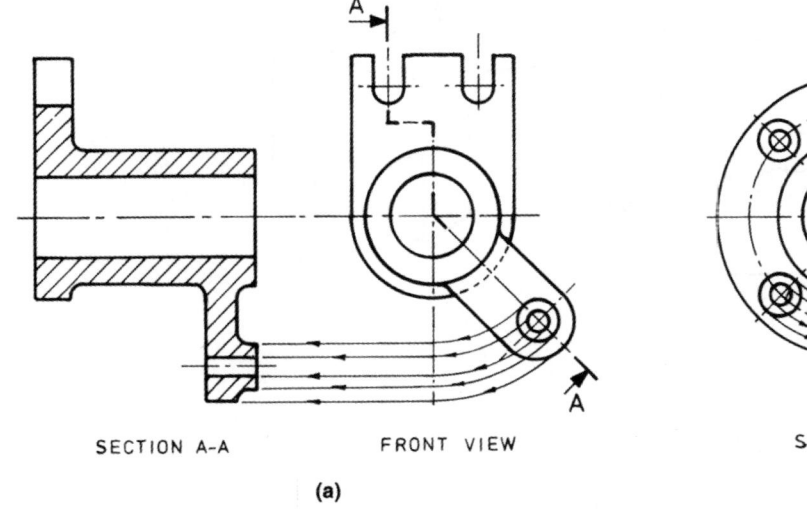

(a)

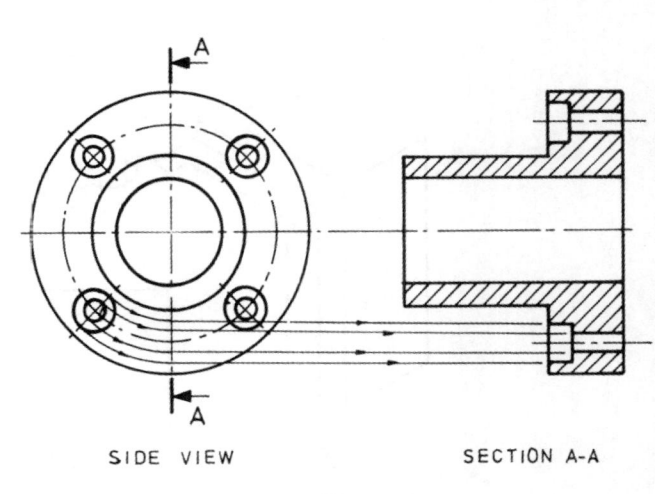

(b)

## DISPLACEMENT OF HOLES IN SECTION

Holes in circular elements should be shown in the cutting plane at the true pitch from the centre, as in Figure 1.59(b) above, rather than the projected distance.

# 1.15 Drawing sectional views

In most cases the normal outside views obtained from orthogonal projection are not sufficient to complete the shape description of an engineering component, both internally and externally. Hence other views of a different type must be drawn in conjunction with, or instead of, the normal external views. These special views are called sectional views and the main types used in mechanical drawings are described in this part.

## THE FULL SECTIONAL VIEW

Figure 1.60 shows an isometric view of a machined block which has been cut through the centre and moved apart. The shape and detail of the counterbored holes are revealed along the face of the cut. This is the purpose of the sectional view—to reveal interior detail. A normal view would be taken from position X.

Figure 1.61 shows the sectional view and a right side view taken from position Y in Figure 1.60. The course of the cutting plane is indicated by A-A on the side view. The direction of the arrows on the cutting plane A-A indicates the direction from which the section is viewed.

## THE OFFSET SECTIONAL VIEW

With a full sectional view, interior detail which lies along one plane only is revealed. Sometimes it is desirable to show detail which lies along two or more planes, and this is done by means of the offset sectional view.

Figure 1.62 is an isometric view of a shaft bracket which has been cut by an offset cutting plane to reveal the detail of the two bosses. The offset sectional view in this case is taken looking down on the bottom piece as shown. Figure 1.63 shows a normal front view and an offset sectional top view of the bracket; the course of the cutting plane is shown by A-A.

Note that there is no line shown on the sectional view where the course of the cutting plane changes direction.

## THE HALF SECTIONAL VIEW

This type of view is often used on objects which are symmetrical about a centre line. The cutting plane effectively removes a quarter of the object as shown in Figure 1.64 (page 38). The resulting view provides two views in one, as one half shows interior detail

**FIGURE 1.60** ▶ Pictorial view of section

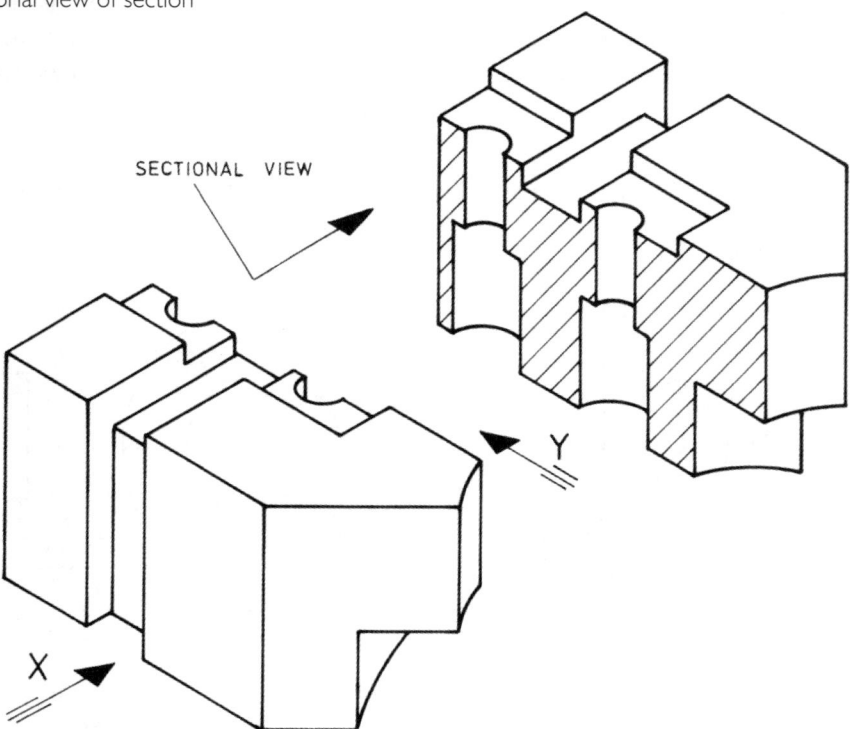

SECTIONAL VIEW

Y

X

**FIGURE 1.61** ▶ Orthogonal view of section

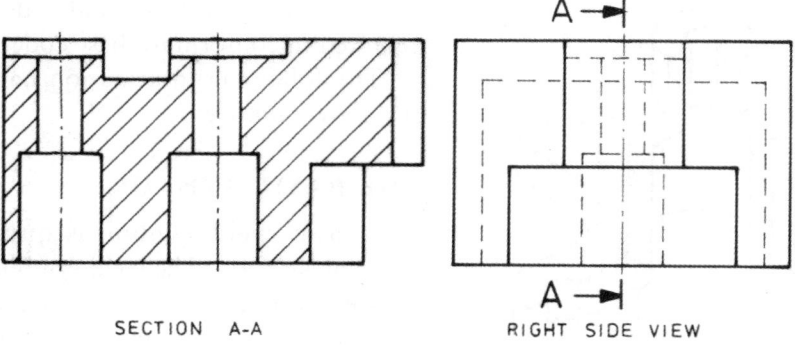

SECTION A-A

RIGHT SIDE VIEW

**FIGURE 1.62** ▶ Pictorial view of offset section

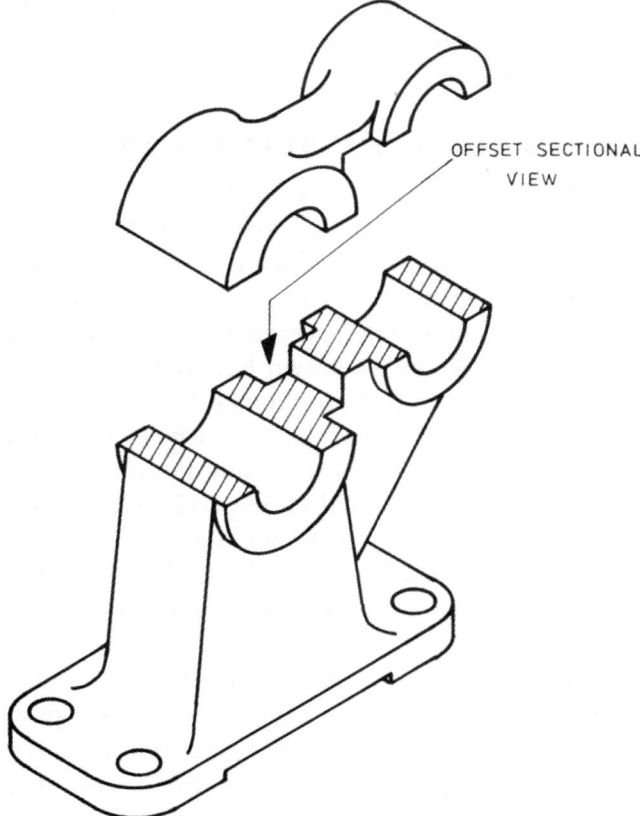

OFFSET SECTIONAL VIEW

**FIGURE 1.63** ▶ Orthogonal view of offset section

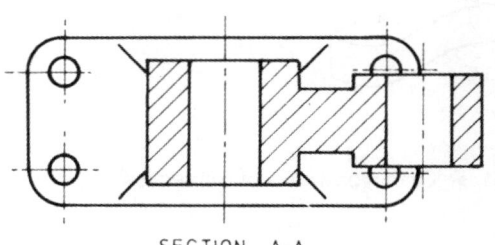

SECTION A-A

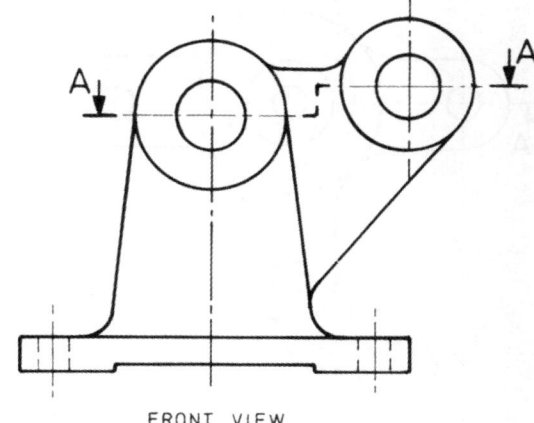

FRONT VIEW

and the other half shows external detail. This is illustrated in Figure 1.65 (page 38).

As with the offset sectional view, the division between the external half and the internal half of the view is not indicated by a full line, but by a centre line. Hidden outlines may be shown in the external half, particularly if it assists with the dimensioning of the view, for example diameters.

## RULES TO REMEMBER WHEN SECTIONING

1. A sectional view shows the part of the component in front of the cutting plane arrows. In third-angle projection the sectional view is placed on the side behind the sectioning viewing plane, while in first-angle projection it is placed on the side in front of the sectional viewing plane.

2. Material which has been cut by the cutting plane is hatched. Standard exceptions are given on page 34.

3. A sectional view must not have any full lines drawn over hatched areas. A full line represents a corner or edge which cannot exist on a face which has been cut by a plane.

4. As a general rule, dimensions are not inserted in hatched areas, but where it is unavoidable, it may be done as shown on page 33.

FIGURE 1.64 ▶ Pictorial view of half section

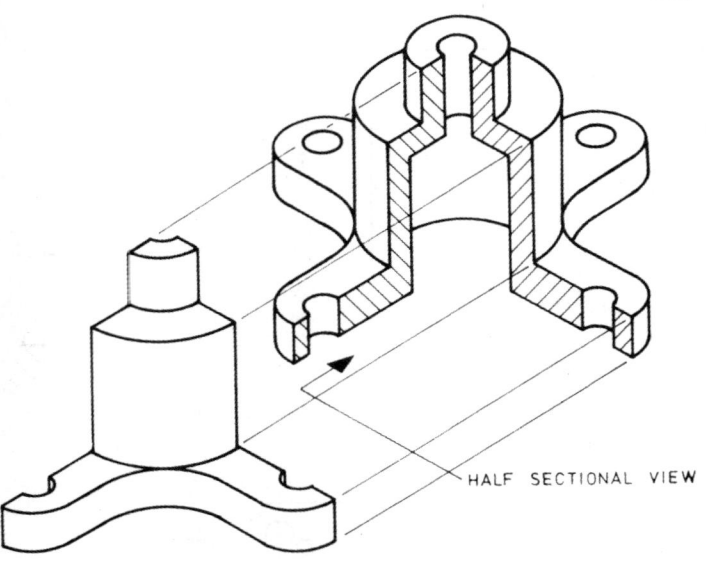

FIGURE 1.65 ▶ Orthogonal view of half section

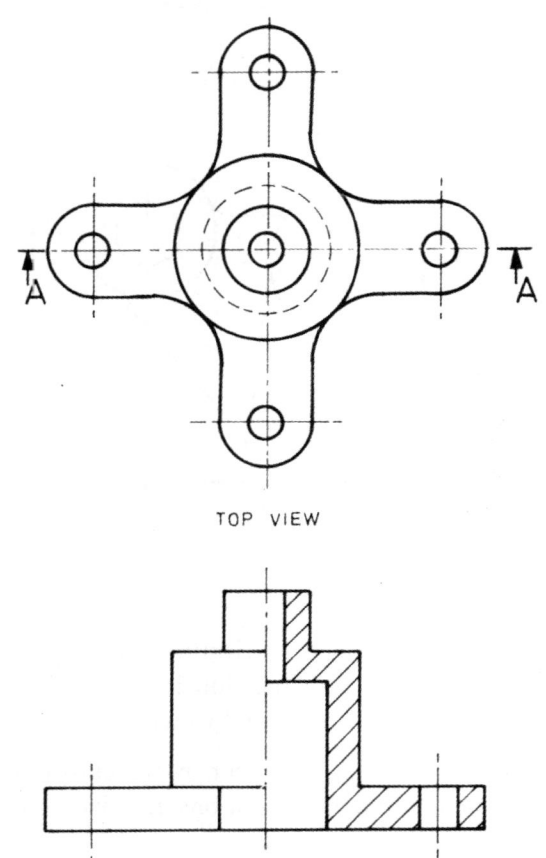

TOP VIEW

HALF SECTION A-A

# 1.16 Welding drafting

## WELDING STANDARDS

When representing welds on drawings, refer to AS 1101 (Graphical symbols for general engineering)

Part 3 (Welding and non-destructive examination) or to the various constructional codes where welding is required to conform to these codes.

The following information has been taken from the above standard.

### THE BASIC SYMBOL

The basic 'weld symbol' is quite distinct from a 'welding symbol'. The weld symbol indicates the type of weld only, whereas the welding symbol is a complete instructional symbol indicating the particular joint, the type of weld required and all supplementary instructions necessary to complete the welded joint. Table 1.12 illustrates basic 'weld symbols' and Figure 1.66 the makeup of the standard 'welding symbol'.

### THE STANDARD WELDING SYMBOL

The standard welding symbol, Figure 1.66, comprises an arrow which points to the welded joint, and a reference line which is 20 mm to 30 mm long and, when needed, a tail where specifications, procedures or other references may be indicated. The reference line is always drawn horizontally and the arrow is inclined to the reference line at least 30 degrees or more but less than 90 degrees and may point in any direction that best indicates a particular joint. A basic weld symbol (Table 1.12) occupies the centre position of the reference line, either above, below or on both sides depending on the requirements of the joint. (See Application of the standard welding symbol.)

Supplementary symbols, Table 1.13, may be attached to the welding symbol in certain positions to affect the basic weld outcome in many different ways. Each supplementary symbol is positioned around the reference line in its correct position so that such representations can be evaluated and acted on by any welding authority. Fig 1.66 shows the position of the supplementary elements and indicates the controlling feature of each application.

Reference to Tables 1.12, 1.13 and 1.16 and the letter designation table of welding processes, Table 1.14 (page 41), will assist in drawing the required welding symbol. Table 1.16 (page 43) details and illustrates the application of the basic welding symbols.

### WELDING TERMINOLOGY

Figure 1.67 (page 40) illustrates the standard terminology for various elements of fillet and butt welds.

## APPLICATION OF THE STANDARD WELDING SYMBOL

When applying the standard welding symbol, thought must be given as to whether the actual weld is situated on the same side of the joint as the arrow or on the other side.

Consider Figure 1.68 which illustrates two T joints 1 and 2 welded as shown at A and B respectively.

**TABLE 1.12** ▶ Basic symbols for welding

| Fillet | BUTT | | | | | | |
|---|---|---|---|---|---|---|---|
| | Square | V | Bevel | U | J | Flare-V | Flare-Bevel |
| | | | | | | | |
| Spot or Projection weld | Plug or slot weld | Seam weld | Backing run or weld | Surfacing | Stud weld | **FLANGE WELD** | |
| | | | | | | Edge | Corner |
| | | | | | | | |

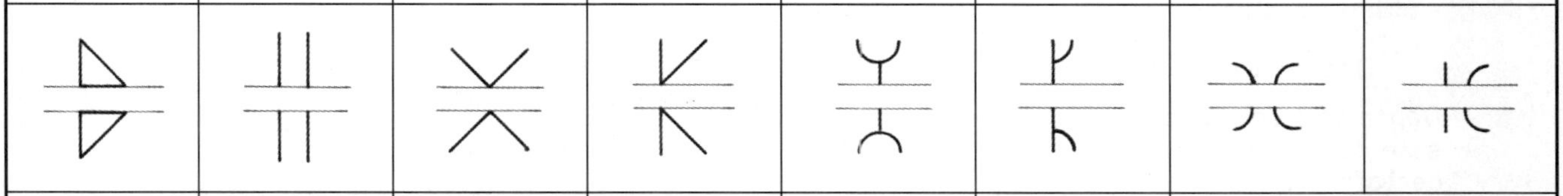

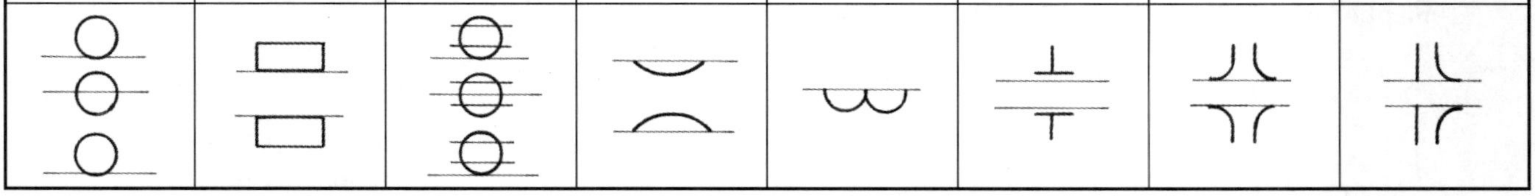

**TABLE 1.13** ▶ Supplementary symbols to be used with basic symbols

| Weld all around | Site weld | Complete penetration from one side | Blocking or spacer material | CONTOUR | | |
|---|---|---|---|---|---|---|
| | | | | Flush | Convex | Concave |
| | | | | | | |

EXAMPLES
OF USE
Weld finishes
C – Chipping
G – Grinding
M – Machining
R – Rolling
P – Peening

Fillet weld, all around, chipped finish

Square butt, arrow side complete penetration, flush contour both sides, non-mechanical means

flush

Convex both sides

concave

Flange corner weld, other side, finished concave from welding process

Seam weld arrow side, ground flush peened after welding

Side V butt weld, arrow side, ground flush by mechanical means with a backing run on the other side

Contour finishes without mechanical means

flush

convex

concave

Contour finishes by mechanical means

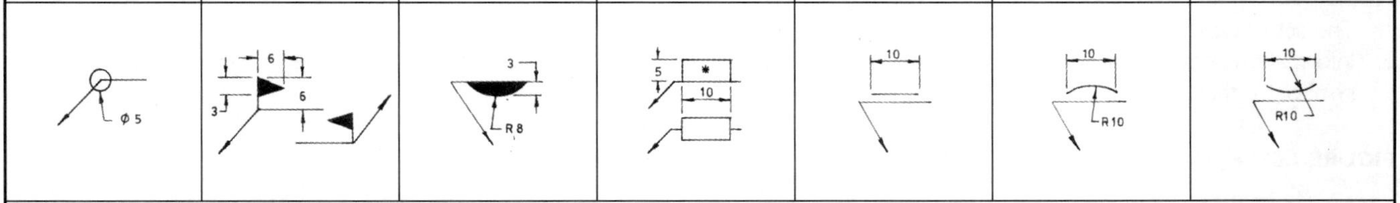

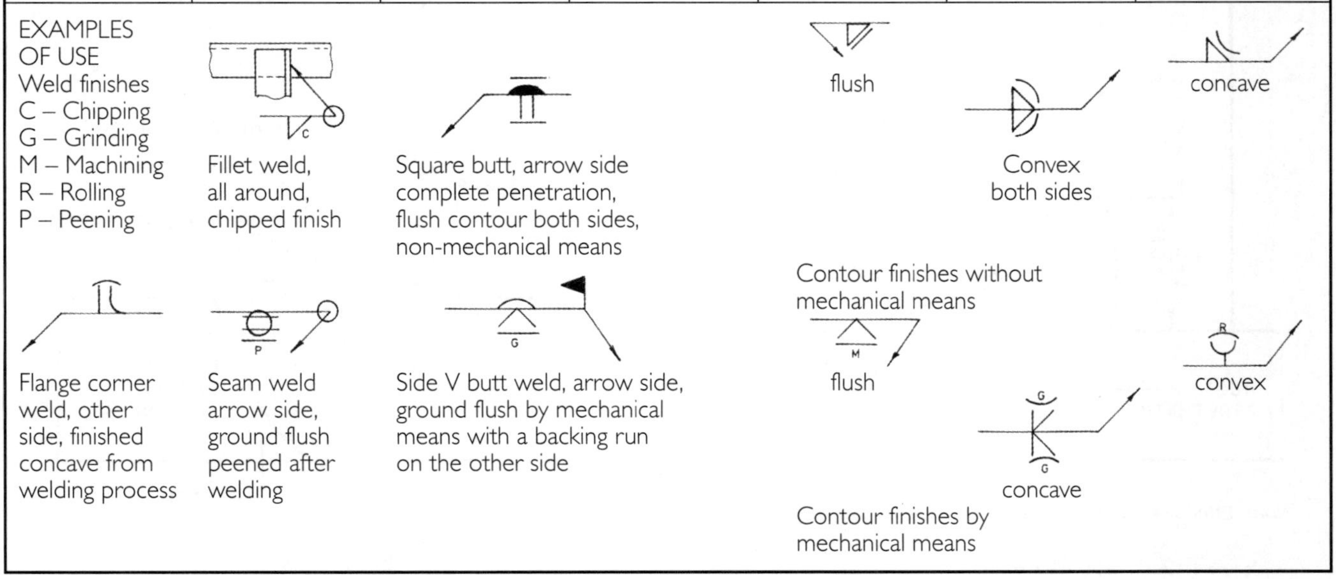

For weld A, the basic fillet symbol is placed underneath the reference line, indicating that the weld is on the arrow side of joint 1.

For weld B, however, the basic fillet symbol is placed above the reference line, indicating that the weld is on the other side of joint 2.

Wherever possible, the arrow should be positioned adjacent to the weld, as with joint 1, with the symbol underneath the reference line. Indicating a weld by the joint 2 method is justified when the weld side is inaccessible due to the other detail occupying the space. Table 1.15 (page 42) illustrates the basic weld position.

**FIGURE 1.66** ▶ Standard location of elements on a welding symbol

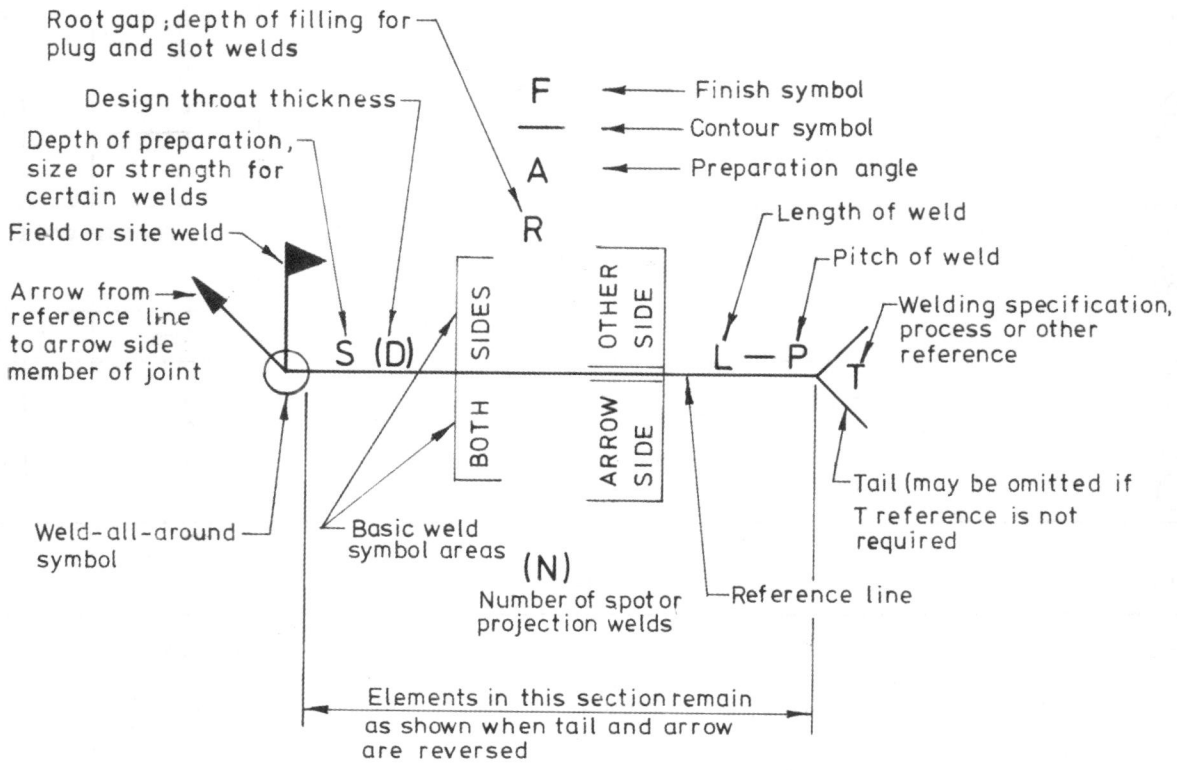

Notes:
1.  The actual weld length is to include allowance for starting and stopping the weld.
2.  Information relating to a particular weld, e.g. S, [D], F, A, R, L and P, must be placed on the same side of the reference line as the symbol for that weld.

**FIGURE 1.67** ▶ Welding terminology

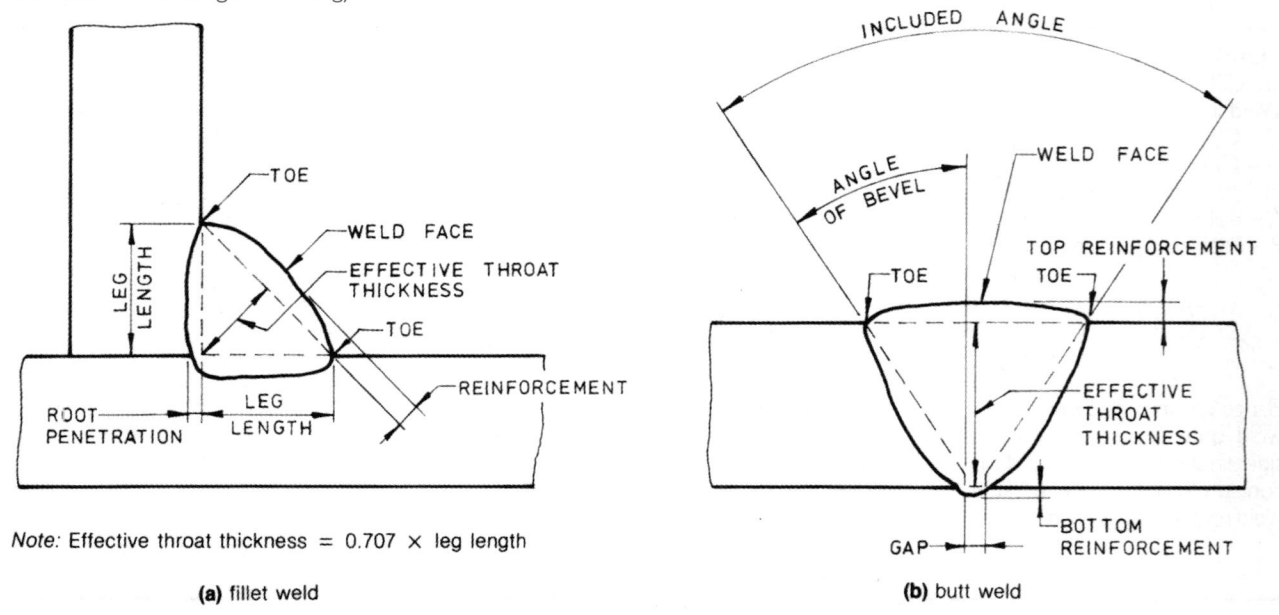

*Note:* Effective throat thickness = 0.707 × leg length

**(a)** fillet weld

**(b)** butt weld

**FIGURE 1.68** ▶ Representation of the welding symbol

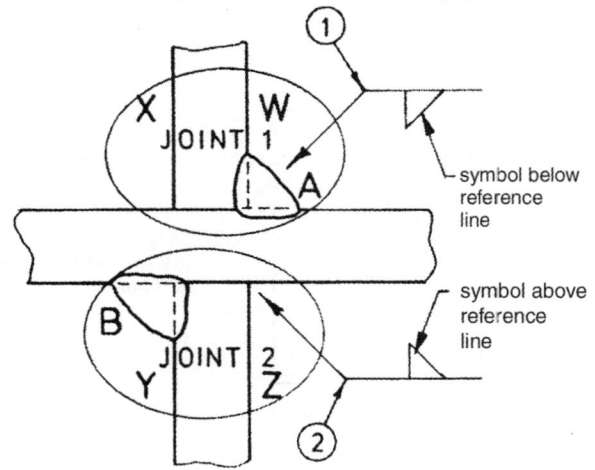

symbol below reference line

symbol above reference line

Arrow ①
W is called the *arrow* side of joint 1
X is called the *other* side of joint 1

Arrow ②
Z is called the arrow side of joint 2
Y is called the other side of joint 2

Note: Arrow ① bears no relation to arrow ②, as they refer to different joints.

**TABLE 1.14** ▶ Designation of welding and allied processes by letters

| WELDING AND ALLIED PROCESSES | DESIGNATION |
|---|---|
| **Arc welding** | |
| carbon-arc welding | CAW |
| flux cored arc welding | FCAW |
| gas metal arc welding | GMAW |
| gas tungsten arc welding | GTAW |
| manual metal arc welding | MMAW |
| plasma arc welding | PAW |
| stud welding | SW |
| submerged arc welding | SAW |
| **Resistance welding** | |
| high frequency resistance welding | HFRW |
| projection welding | RPW |
| resistance seam welding | RSEW |
| resistance spot welding | RSW |
| **Other welding processes** | |
| electron beam welding | EBW |
| induction welding | IW |
| laser-beam welding | LBW |
| oxy-acetylene welding | OAW |
| **Solid state welding** | |
| forge welding | FOW |
| friction welding | FRW |
| ultrasonic welding | USW |
| **Braze welding** | |
| flame brazing | FLB |
| induction brazing | IB |
| resistance brazing | RB |

## WELDING PROCEDURES

It is sometimes necessary to specify certain procedures or requirements about a weld. The standard symbol used in such cases should be provided with a tail as shown in Figure 1.66 (page 40), and the information inserted where shown, for example at T.

In order to control a welding process more fully, a procedure sheet may be added to the drawing. The sheet should contain the following general information:

1. type of material being welded
2. form of weld (to include plate preparation such as angle of bevel, root penetration, root radius, etc.)
3. set-up details such as welding position, alignment, gap required
4. number and order of runs
5. electrode size, type and make
6. electrical supply data such as polarity, current and voltage values
7. preheating requirements
8. pre- and post-weld cleaning procedures
9. preatment of joint after welding
10. preparation and/or procedures to apply in between runs.

## JOINT PREPARATION

The arrow may also be used to indicate when one plate only of a joint is to be prepared in welding single-bevel and single-J butt joints.

The arrow is cranked as shown in Figure 1.69, and points towards the plate which has to be prepared. The crank is omitted when the edge to be prepared is obvious, for example a T butt joint.

# 1.17 Surface texture

### INDICATION ON DRAWINGS

Symbols indicating the type of surface finish, production methods and/or required roughness of a surface are used on a drawing when this feature is necessary to ensure functionality, and then only on those surfaces which require it. Surface finish specification is not necessary when normal production process finish is satisfactory.

A symbol should be used only once for a given surface, and where possible on a view which shows the size and position of the surface in question.

### SURFACE TEXTURE TERMINOLOGY

Figure 1.70 (page 42) illustrates the standard terminology relating to surface texture. Surface roughness ($R_a$ value) is a measure of the arithmetical mean deviation of a short distance of the surface in question.

*(continued on page 45)*

**FIGURE 1.69 ▶** Use of cranked arrow

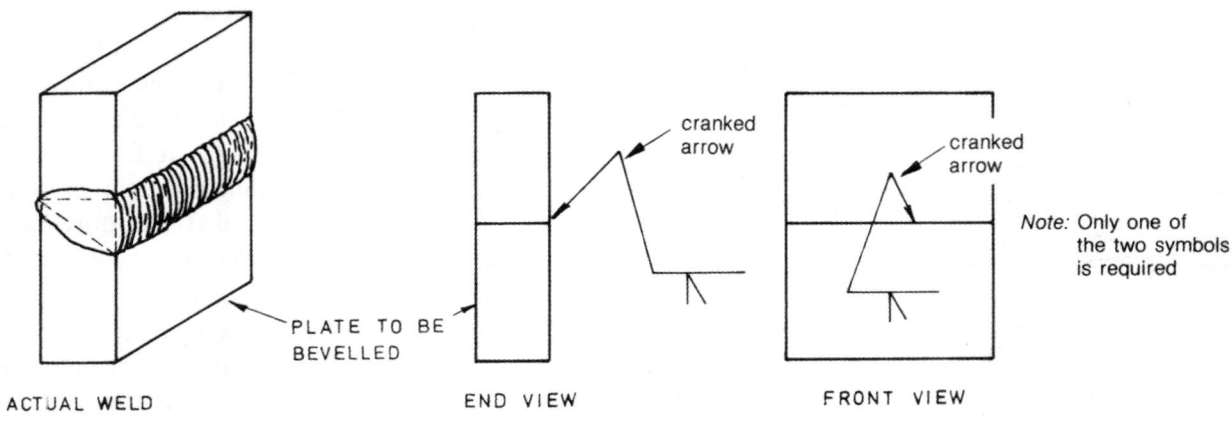

PLATE TO BE BEVELLED

ACTUAL WELD          END VIEW          FRONT VIEW

cranked arrow

cranked arrow

*Note:* Only one of the two symbols is required

**FIGURE 1.70 ▶** Surface texture terminology

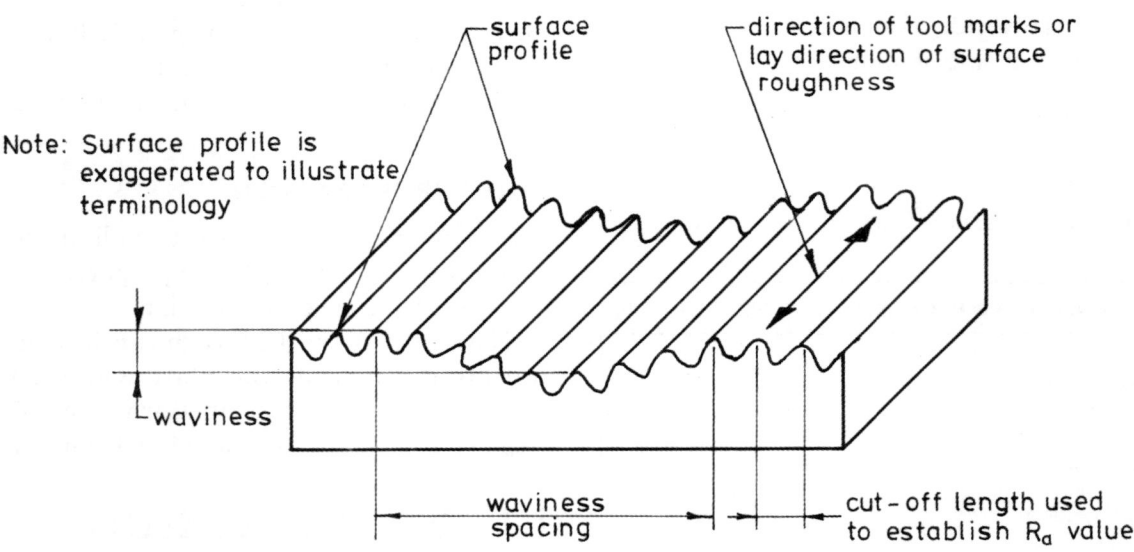

surface profile

direction of tool marks or lay direction of surface roughness

Note: Surface profile is exaggerated to illustrate terminology

waviness

waviness spacing

cut-off length used to establish $R_a$ value

**TABLE 1.15 ▶** Representation of basic weld position

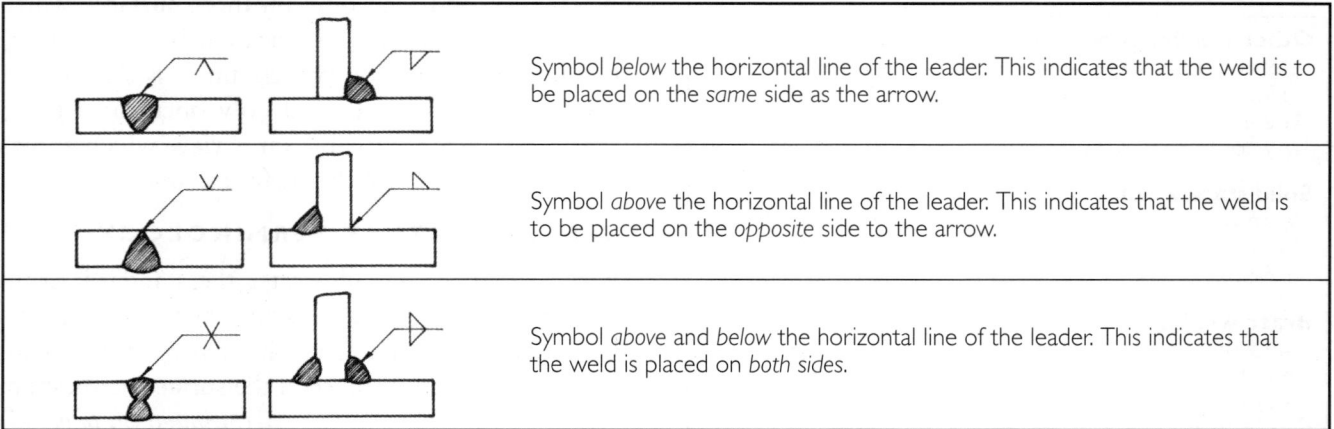

| | |
|---|---|
| | Symbol *below* the horizontal line of the leader. This indicates that the weld is to be placed on the *same* side as the arrow. |
| | Symbol *above* the horizontal line of the leader. This indicates that the weld is to be placed on the *opposite* side to the arrow. |
| | Symbol *above* and *below* the horizontal line of the leader. This indicates that the weld is placed on *both sides*. |

**TABLE 1.16** ▶ Application of welding symbols

| WELD TYPE | BASIC SYMBOL | WELD CROSS SECTION | DRAWING APPLICATION | DRAWING INTERPRETATION |
|---|---|---|---|---|
| **FILLET** | | | (a) (b) (c) | (a) fillet weld on the arrow side<br>(b) fillet weld on the other side<br>(c) fillet weld on both sides |
| **SQUARE-BUTT** | | | (a) (b) (c) | (a) butt weld on the arrow side<br>(b) butt weld on the other side<br>(c) butt weld on both sides (thick plate) |
| **V-BUTT** | | | (a) (b) (c) | (a) V-butt weld on the arrow side<br>(b) V-butt weld on the other side<br>(c) V-butt weld both sides, flush contour on the other side |
| **BEVEL-BUTT** | | | (a) (b) (c) | (a) bevel-butt weld on the arrow side<br>(b) bevel-butt weld on the other side<br>(c) bevel-butt weld on both sides |
| **U-BUTT** | | | (a) (b) (c) | (a) U-butt weld on the arrow side<br>(b) U-butt weld on the other side<br>(c) U-butt weld on both sides |
| **J-BUTT** | | | (a) (b) (c) | (a) J-butt weld on the arrow side<br>(b) J-butt weld on the other side<br>(c) J-butt weld on both sides |
| **FLARE-V-BUTT** | | | (a) (b) (c) | (a) flare-V-butt weld on the arrow side<br>(b) flare-V-butt weld on the other side<br>(c) flare-V-butt weld on both sides |
| **FLARE-BEVEL-BUTT** | | | (a) (b) (c) | (a) flare-bevel-butt weld on the arrow side<br>(b) flare-bevel-butt weld on the other side<br>(c) flare-bevel-butt weld on both sides |
| **PLUG OR SLOT** | | | (a) (b) (c) | (a) slot weld on the arrow side<br>(b) slot weld on the other side<br>(c) plug weld, arrow side filled to a depth of 10mm |

*(continued)*

TABLE 1.16 ▶ Application of welding symbols (continued)

| WELD TYPE | BASIC SYMBOL | WELD CROSS SECTION | DRAWING APPLICATION | DRAWING INTERPRETATION |
|---|---|---|---|---|
| SPOT OR PROJECTION | | Spot weld | Spot weld | The arrow points to the centre of the first spot weld of which there are four, 6 mm diameter (S) and spaced 25mm apart. RSW Resistance Spot Weld |
| | | Weld side | Projection weld | The arrow points to the centre line of the projection welds, numbering 4, which are welded on the other side and are evenly spaced along the 75mm section of the centre line. S = 6 |
| BACKING RUN OR WELD | | Main Weld side / Backin weld side | (a) (b) | (a) the backing weld symbol is on the other side while the V-butt symbol is on the arrow side (b) the backing weld symbol is on the arrow side while the V-butt symbol is on the other side |
| SEAM | | | | The arrow points to the centre line along which the weld runs. It only has arrow or other side significance as shown. The tail indicates the process to be used. |
| | | | | The resistance seam weld has no arrow or other side significance for this type of weld. The tail indicates the process to be used. |
| SURFACING | | Weld side | (a) (b) (c) | (a) height of surface along centre line to be 3 mm, no specified width (b) surface built up to 5 mm and finished flush (c) round surface built up 3 mm and finished smooth, e.g. turned |
| EDGE-FLANGE | | Weld side | (a) (b) | (a) edge-flange weld on the arrow side (b) edge-flange weld on the other side Note: a weld both sides symbol has no significance. |
| CORNER-FLANGE | | Weld side | (a) (b) | (a) corner-flange weld on the arrow side (b) corner-flange weld on the other side Note: a weld both sides symbol has no significance. |
| STUD | Reference line | | (a) (b) (c) | (a) six stud welds evenly spaced on Ø200 circle on the arrow side (b) stud weld on the other side (c) stud weld on both sides |

## SURFACE ROUGHNESS MEASUREMENT—$R_a$

The $R_a$ value may be defined as the average value of the departure (both above and below) of the surface from the centre line assessed over a selected cut-off length, also known as the 'sampling length'. Cut-off lengths commonly available on electronic roughness measuring instruments are 0.25 mm, 0.8 mm and 2.5 mm, selection depending on the process used to produce the surface.

Referring to Figure 1.71 the centre line is positioned so that, over the sampling or cut-off length chosen:

areas above the centre line = areas below the centre line

$$(A_1 + A_2 ... + A_N) = (B_1 + B_2 + ... + B_N)$$

Then to calculate surface roughness:

$$R_a = \frac{(A_1 + A_2 ... + A_N) + (B_1 + B_2 + ... + B_N)}{L}$$

$$= \frac{\text{sum of the areas above and below the centre line}}{\text{cut-off length}}$$

The preferred or standard values of $R_a$ in micrometres (μm) for drawing specification are as follows:

| 0.025 | 0.05 | 0.1 | 0.2 | 0.4 | 0.8 |
|-------|------|-----|-----|-----|-----|
| 1.6 | 3.2 | 6.3 | 12.5 | 25 | 50 |

**FIGURE 1.71** ▶ Surface roughness profile

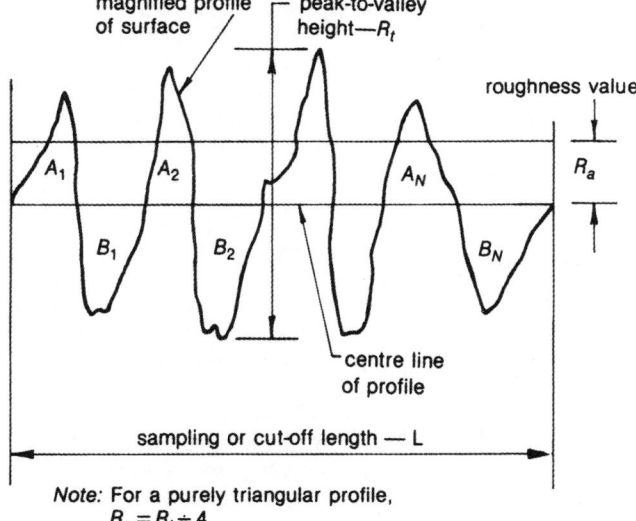

*Note:* For a purely triangular profile,
$R_a = R_t \div 4$

## THE STANDARD SYMBOL

Surface finish requirements are indicated on a drawing by means of a standard symbol consisting of a basic character which may have further information attached to it, depending on the finish requirements of the surface in question.

Figure 1.72(a) illustrates the basic symbol and its size in relation to other drawing characters.

Figure 1.72(b) illustrates the basic symbol with a crossbar indicating that a machined finish is necessary.

Figure 1.72(c) illustrates the basic symbol with a circle inset indicating that removal of material is not permitted.

Figure 1.72(d) indicates the type of information which may be required and where it is to be found on the basic symbol. All of this information is seldom required on the one symbol. Table 1.18 (page 46) illustrates typical applications of the symbol and its interpretation.

**FIGURE 1.72** ▶ Surface roughness symbol

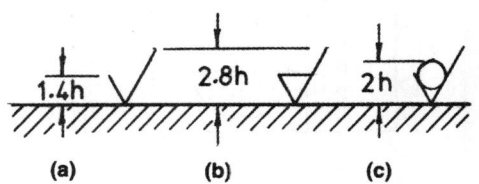

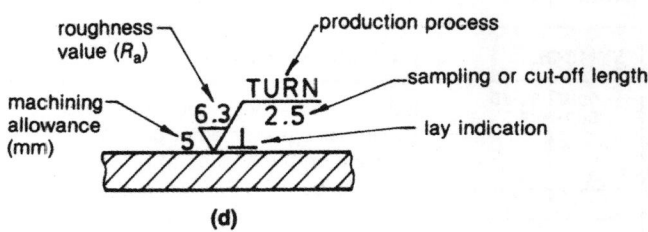

## SURFACE ROUGHNESS ($R_a$) APPLICATIONS

The ultimate finish of a component's surfaces is determined largely by their function or required appearance characteristics. The ability to produce various finishes is governed by the types of production processes available for component manufacture within a firm.

Designers and drafters must have a knowledge of the above factors before specifying roughness requirements. Table 1.19 (page 47) lists the standard roughness values, the processes which can produce them, their area of application and some indication of the relative cost associated with their production.

A more detailed specification of the roughness range applicable to various production processes is given in Table 1.20 (page 48).

## APPLICATION OF SURFACE TEXTURE SYMBOL TO DRAWINGS

The surface texture symbol should be located so it can be read from the bottom or right-hand side of the drawing. Symbols should be applied to the edge view of the surface in question, but extension and leader lines may also be used to apply the symbol.

Figure 1.73 illustrates correct methods of applying the symbol.

**FIGURE 1.73** ▶ Application of the surface texture symbol

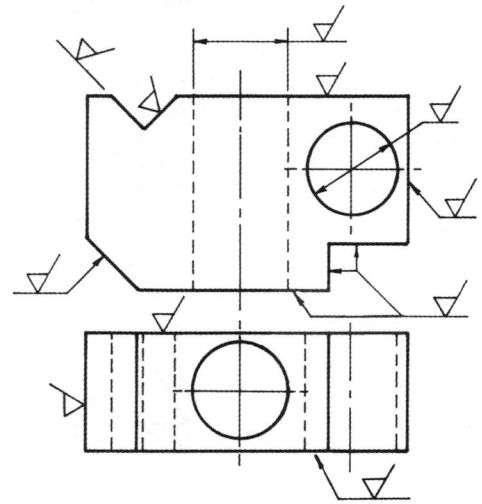

**TABLE 1.18** ▶ Applications of surface texture symbol

## ROUGHNESS GRADE NUMBERS

Where there is a possibility of misinterpretation due to using both metric and imperial units, surface roughness may be indicated by an equivalent surface roughness number shown in Table 1.17.

**TABLE 1.17** ▶ Roughness grade numbers

| ROUGHNESS VALUE ($R_a$) | | ROUGHNESS GRADE NUMBER |
|---|---|---|
| µm | µin | |
| 50 | 2000 | N12 |
| 25 | 1000 | N11 |
| 12.5 | 500 | N10 |
| 6.3 | 250 | N9 |
| 3.2 | 125 | N8 |
| 1.6 | 63 | N7 |
| 0.8 | 32 | N6 |
| 0.4 | 16 | N5 |
| 0.2 | 8 | N4 |
| 0.1 | 4 | N3 |
| 0.05 | 2 | N2 |
| 0.025 | 1 | N1 |

| SYMBOL | INTERPRETATION | SYMBOL | INTERPRETATION |
|---|---|---|---|
| CONTROLLED SURFACE | the basic symbol—consists of two unequal legs inclined at 60° and resting on the surface to be controlled | 0.4/ALL OVER | may be applied in the title block or as a note when a single value applies to all machined surfaces controlled |
| | used when machining is necessary to obtain the desired texture | 6.3/ ALL OVER EXCEPT WHERE OTHERWISE INDICATED | may be applied in the title block or as a note when a single value applies to the majority of machined surfaces; exceptions should be indicated on the individual surfaces concerned |
| | used when the surface texture is to remain as found from the last process and no material, e.g. a cast or forged part, is to be removed | TURN 2 1.6 | used to specify a turning allowance of 2 mm after which a surface texture of 1.6 µm is required |
| 6.3 3.2 | used to specify maximum and minimum limits of surface roughness obtained by any machining process | N4/ | used to specify the roughness value by a standard number equivalent to 0.2 µm or 8 µin |
| 25 6.3 | used to specify maximum and minimum limits of surface roughness obtained without machining | 3.2/ /M | used to specify a roughness value together with a multidirectional lay texture |
| MILL 3.2 | used to indicate a particular machining process and roughness value | 0.8 0.008–4 | used to specify a roughness value together with a waviness height of 0.008 mm and spacing of 4 mm |
| 1.6/2.5 | used to indicate a sampling length in millimetres and a machined surface texture | ✓ = 3.2 MILL | used when a surface texture specification is complicated and is required on a number of surfaces and space on the drawing is limited; the basic symbol is used on the surfaces in question and its meaning is clearly defined in note form on the drawing as shown |
| CADMIUM PLATE 1.6 0.8 | used to indicate roughness before and after surface treatment; note the use of type J line representing the surface after treatment | | |

**TABLE 1.19** ► Standard roughness values

| $R_a$ VALUE (μm) | | PROCESS AND APPLICATION |
|---|---|---|
| 0.025 or 0.025 and 0.05 or 0.05 | very fine quality surface finishes, costly to produce | This very smoothly finished surface is produced by fine honing, lapping, buffing or super-finishing machines. It is costly to produce and seldom required. It has a highly polished appearance, depending on the production process, and is normally used on precision instruments such as gauges, laboratory equipment and finely made tools. |
| 0.1 or 0.1 | | This is similar to the finer grades of finish and has much the same application. Very refined surfaces have this high degree of finish. It is produced by honing, lapping and buffing methods and is costly to produce. |
| 0.2 or 0.2 | | This fine surface is produced by honing, lapping and buffing methods. This texture could be specified on precision gauge and instrument work, and on high speed shafts and bearings where lubrication is not dependable. |
| 0.4 or 0.4 | | This fine quality surface can be produced by precision cylindrical grinding, coarse honing, buffing and lapping methods. It is used on high speed shafts, heavily loaded bearings and other applications where smoothness is desirable for the proper functioning of a part. |
| 0.8 or 0.8 | medium quality finishes, used where reasonable surfaces are required | This first-class machine finish can be easily produced on cylindrical, surface and centreless grinders but requires great care on lathes and milling machines. It is satisfactory for bearings and shafts carrying light loads and running at medium to slow speeds. It may be used on parts where stress concentration is present. It is the finest finish that it is economical to produce; below this costs rise rapidly. |
| 1.6 or 1.6 | | This good machine finish can be maintained on production lathes and milling machines using sharp tools, fine feeds and high cutting speeds. It is used when close fits are required but is unsuitable for fast rotating members. It may be used as a bearing surface when motion is slow and loads are light. This surface can be achieved on extrusions, rolled surfaces, die castings and permanent mould castings in controlled production. |
| 3.2 or 3.2 | | This medium commercial finish is easily produced on lathes, milling machines and shapers. A finish commonly used in general engineering machining operations, it is economical to produce and of reasonable appearance. It is the roughest finish recommended for parts subjected to slow speeds, light loads, vibration and high stress, but it should not be used for fast rotating shafts. This finish may also be found on die castings, extrusions, permanent mould castings and rolled surfaces. |
| 6.3 or 6.3 | rough finishes, used where quality surfaces are unimportant | This coarse production finish is obtained by taking coarse feeds on lathes, millers, shapers, boring and drilling machines. It is acceptable when tool marks have no bearing on performance or quality. This texture can also be found on the surfaces of metal moulded castings, forgings, extruded and rolled surfaces, and can be produced by rough hand filing or disc grinding. |
| 12.5 or 12.5 | | This surface is produced from heavy cuts and coarse feeds by milling, turning, shaping, boring, disc grinding and snagging. It can also be obtained by sand casting, saw cutting, chipping, rough forging and oxy cutting. This finish is rarely specified and is used only where it is not seen or its appearance is unimportant, e.g. on machinery, jigs and fixtures. |
| 25 | | This very rough finish is produced by sand casting, torch and saw cutting, chipping and rough forgings. Machining operations are not required as this finish is suitable as found, e.g. on large machinery. |

**TABLE 1.20** ► Roughness ranges for common production processes

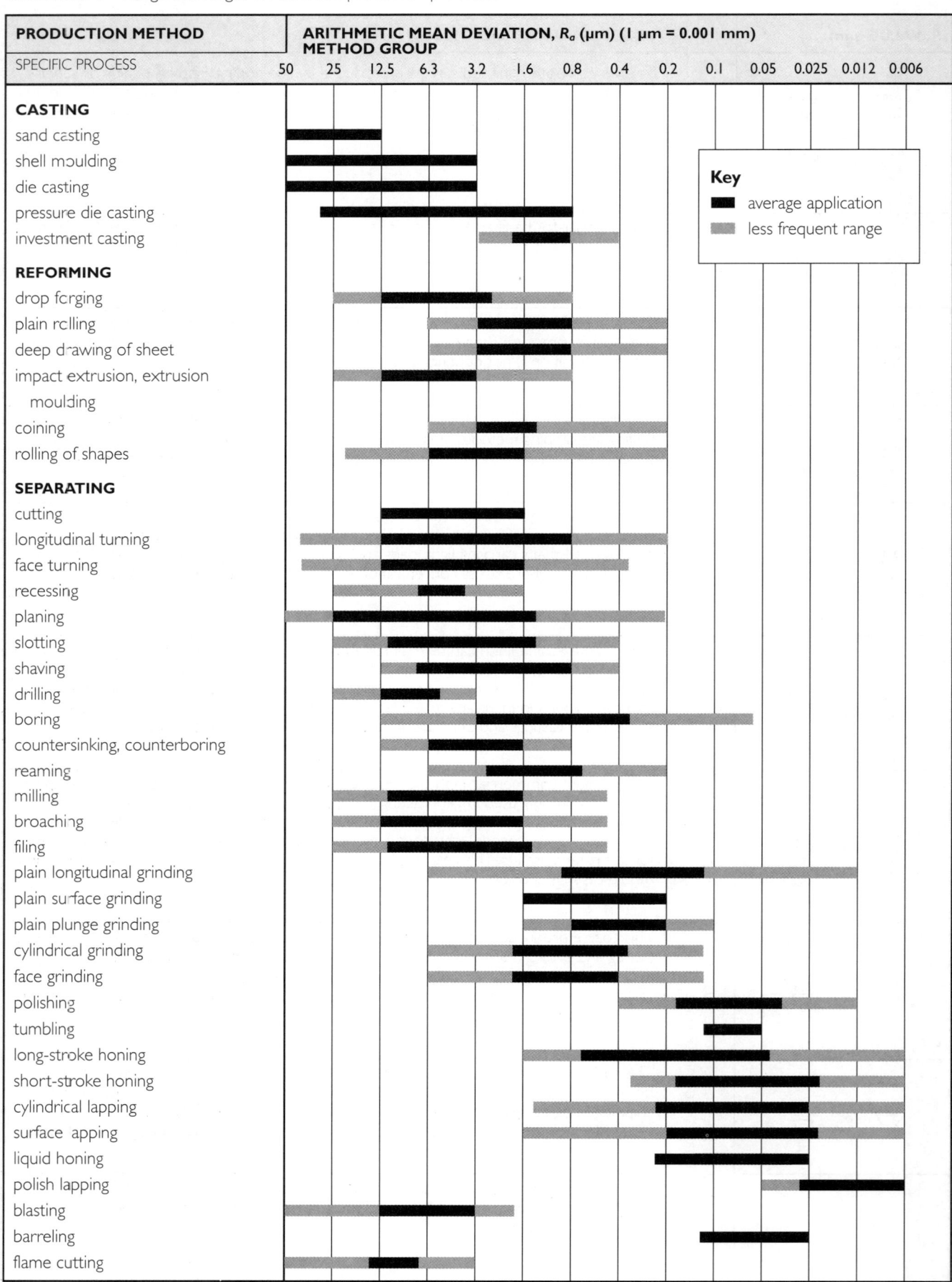

| PRODUCTION METHOD | ARITHMETIC MEAN DEVIATION, $R_a$ (μm) (1 μm = 0.001 mm) |
|---|---|
| SPECIFIC PROCESS | METHOD GROUP |

Scale values: 50, 25, 12.5, 6.3, 3.2, 1.6, 0.8, 0.4, 0.2, 0.1, 0.05, 0.025, 0.012, 0.006

**Key**
- ■ average application
- ▬ less frequent range

**CASTING**
- sand casting
- shell moulding
- die casting
- pressure die casting
- investment casting

**REFORMING**
- drop forging
- plain rolling
- deep drawing of sheet
- impact extrusion, extrusion moulding
- coining
- rolling of shapes

**SEPARATING**
- cutting
- longitudinal turning
- face turning
- recessing
- planing
- slotting
- shaving
- drilling
- boring
- countersinking, counterboring
- reaming
- milling
- broaching
- filing
- plain longitudinal grinding
- plain surface grinding
- plain plunge grinding
- cylindrical grinding
- face grinding
- polishing
- tumbling
- long-stroke honing
- short-stroke honing
- cylindrical lapping
- surface lapping
- liquid honing
- polish lapping
- blasting
- barreling
- flame cutting

A production process produces a regular pattern of tool marks on a surface; this feature is called the lay direction of the surface.

Table 1.21 illustrates the standard symbols used to represent various lay directions and their interpretations.

# 1.18 Representation of common features

Conventional representation of features which normally involve unnecessary drawing time and space is desirable on engineering drawings. Table 1.22 (pages 50–51) shows typical examples of features which normally are shown by convention. Table 1.23 (page 52) illustrates representation and proportions of bolts, nuts and screws. More detailed information on these features is given in AS 1100.201—1992.

**TABLE 1.21** ▶ Lay symbols

| LAY SYMBOL | DESCRIPTION | LAY SYMBOL | DESCRIPTION |
|---|---|---|---|
| = | direction of tool marks<br>Lay is *parallel* to the line representing the surface to which the symbol is applied. | C | direction of tool marks<br>Lay is generally *circular* relative to the centre of the surface to which the symbol is applied. |
| ⊥ | direction of tool marks<br>Lay is *perpendicular* to the line representing the surface to which the symbol is applied. | M | direction of tool marks<br>Lay is *multidirectional*, but generally has some kind of tool mark pattern. |
| X | direction of tool marks<br>Lay is *slanting* in both directions to the line representing the surface to which the symbol is applied. | R | direction of tool marks<br>Lay is approximately *radial* to the centre of the surface to which the symbol is applied. |

**TABLE 1.22** ▶ Drawing representation of common features

| FEATURE | DRAWING REPRESENTATION | FEATURE | DRAWING REPRESENTATION |
|---------|------------------------|---------|------------------------|
| knurling | straight · diamond (30°) | leaf springs | semi-elliptic · semi-elliptic with eyelets · semi-elliptic with centre band · semi-elliptic with centre band and eyelets |
| rolling element bearings | type A lines · type A lines · general method (thrust direction not required) · radial thrust transmission · axial thrust transmission · type A lines normal to angular force · angular thrust transmission | seals | general purpose representation · sealing direction shown · contour of seal shown |
| springs | cylindrical compression springs: normal view · section · schematic view · cylindrical tension springs: normal view · section · schematic view | breaks in long sections | solid round bar · tube · tube (sectioned) · solid square bar · square tube |

TABLE 1.22 ▶ Drawing representation of common features (continued)

| FEATURE | DRAWING REPRESENTATION | FEATURE | DRAWING REPRESENTATION |
|---|---|---|---|
| gears | spur gear    bevel gear<br><br>worm    worm wheel<br><br>spur gear assembly    section    worm gear assembly    section<br><br>bevel gear assembly<br><br>rack and pinion assembly | chain drive | |
| | | assembled bolts, screws, etc. | nut<br>spring washer<br>bolt head<br><br>Assembled bolts and nuts may be represented by across corner views to show clearances or to signify rotational capability. In the above case the nut is able to be rotated, hence it is shown across corners in both views. The bolt head cannot rotate and is drawn in true projection. The spring washer may also show 45° break lines on both views for clarity. |
| | | materials | concrete<br><br>timber (across grain)<br><br>timber (with grain)<br><br>water and other liquids<br><br>brickwork<br><br>earth |

**TABLE 1.23** ► Proportions of bolts, nuts and screws

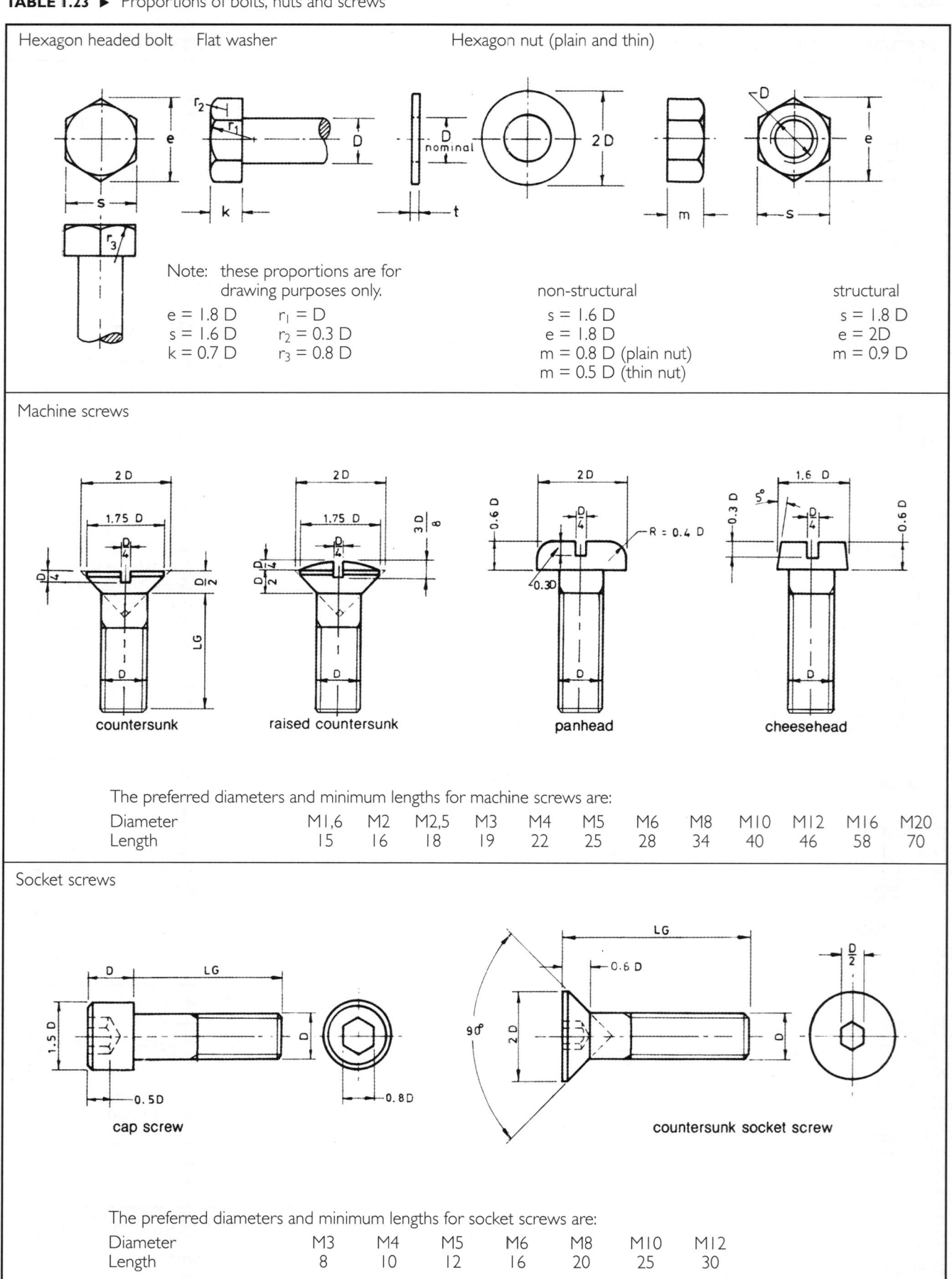

Hexagon headed bolt   Flat washer                    Hexagon nut (plain and thin)

Note: these proportions are for
drawing purposes only.

| | | non-structural | structural |
|---|---|---|---|
| e = 1.8 D | r₁ = D | s = 1.6 D | s = 1.8 D |
| s = 1.6 D | r₂ = 0.3 D | e = 1.8 D | e = 2D |
| k = 0.7 D | r₃ = 0.8 D | m = 0.8 D (plain nut) | m = 0.9 D |
| | | m = 0.5 D (thin nut) | |

$e = 1.8\,D$  $r_1 = D$
$s = 1.6\,D$  $r_2 = 0.3\,D$
$k = 0.7\,D$  $r_3 = 0.8\,D$

non-structural
$s = 1.6\,D$
$e = 1.8\,D$
$m = 0.8\,D$ (plain nut)
$m = 0.5\,D$ (thin nut)

structural
$s = 1.8\,D$
$e = 2D$
$m = 0.9\,D$

Machine screws

countersunk        raised countersunk        panhead        cheesehead

The preferred diameters and minimum lengths for machine screws are:

| Diameter | M1,6 | M2 | M2,5 | M3 | M4 | M5 | M6 | M8 | M10 | M12 | M16 | M20 |
|---|---|---|---|---|---|---|---|---|---|---|---|---|
| Length | 15 | 16 | 18 | 19 | 22 | 25 | 28 | 34 | 40 | 46 | 58 | 70 |

Socket screws

cap screw        countersunk socket screw

The preferred diameters and minimum lengths for socket screws are:

| Diameter | M3 | M4 | M5 | M6 | M8 | M10 | M12 |
|---|---|---|---|---|---|---|---|
| Length | 8 | 10 | 12 | 16 | 20 | 25 | 30 |

ENGINEERING DRAWING

# 1.19 Fits and tolerances

**INTRODUCTION**

This section, based on the International Organization for Standardization (ISO) system, introduces the engineering concept of sizing parts before fitting them together to achieve a desirable relative motion between them. Only more general applications will be considered; tolerance of form and position is dealt with on pages 69–85.

In manufacture it is impossible to produce components to an exact size, even though they may be classified as identical. Even in the most precise methods of production it would be extremely difficult and costly to reproduce a diameter time after time so that it is always within 0.01 mm of a given basic size. However, industry does demand that parts should be produced between a given maximum and minimum size. The difference between these two sizes is called the *tolerance*, which can be defined as the amount of variation in size which is tolerated. A broad, generous tolerance is cheaper to produce and maintain than a narrow, precise one. Hence one of the golden rules of engineering design is 'always specify as large a tolerance as is possible without sacrificing quality'. There are a number of general definitions and terms which are used, and these are described and illustrated below.

**SHAFT (FIG. 1.74)**

A *shaft* is defined as a member which fits into another member. It may be stationary or rotating. The popular concept is a rotating shaft in a bearing. However, when speaking of tolerances, the term shaft can also apply to a member which has to fit into a space between two restrictions, for example a pulley wheel which rotates between two side plates. In determining the clearance fit of the boss between the side plates, the length of the pulley boss is regarded as the shaft.

**HOLE (FIG. 1.74)**

A *hole* is defined as the member which houses or fits the shaft. It may be stationary or rotating, for example a bearing in which a shaft rotates is a hole. However, when speaking of tolerances, the term hole can also apply to the space between two restrictions into which a member has to fit, for example the space between two side plates in which a pulley rotates is regarded as a hole.

**NOMINAL SIZE**

This is the size by which an item is designated as a matter of convenience. Two examples of nominal sizes are: M20 screw thread and $75 \times 20$ flat bar stock. Nominal sizes in millimetres are expressed to the nearest whole number as in the examples given.

**BASIC SIZE (FIG. 1.74)**

This is the size from which the limits of size are derived by the application of the upper and lower deviations. It is the same size for both the shaft and hole for a given fit. It is usually equal to the nominal size.

**LIMITS OF SIZE (FIG. 1.74)**

These are the extremes of size which are allowed for a toleranced dimension. Two limits are possible: one the maximum allowable size called the 'upper limit of size' and the other the minimum allowable size called the 'lower limit of size'.

*Maximum Material Limit* (MML) is the designation applied to that of the two limits of size which corresponds to the maximum material size for the feature, that is:

- the maximum (upper) limit of size for an external feature (shaft)
- the minimum (lower) limit of size for an internal feature (hole).

*Least Material Limit* (LML) is the designation applied to that of the two limits of size which corresponds to the minimum material size for the feature, that is:

- the minimum (lower) limit of size for an external feature (shaft)
- the maximum (upper) limit of size for an internal feature (hole).

**DEVIATION (FIG. 1.74)**

This is the difference between the basic size and the actual size. The extremes of deviations are referred to as the *upper* and *lower deviations*.

The values given in Tables 1.24(a) (pages 56–57) and (b) (pages 58–59) are the upper and lower deviations for both shafts and holes.

**TOLERANCE (FIG. 1.74)**

*Tolerance* is defined as the difference between the maximum and minimum limits of size for a hole or shaft. It is also the difference between the upper and lower deviations.

**FIGURE 1.74** ▶ Designation of shaft and hole sizes and limits

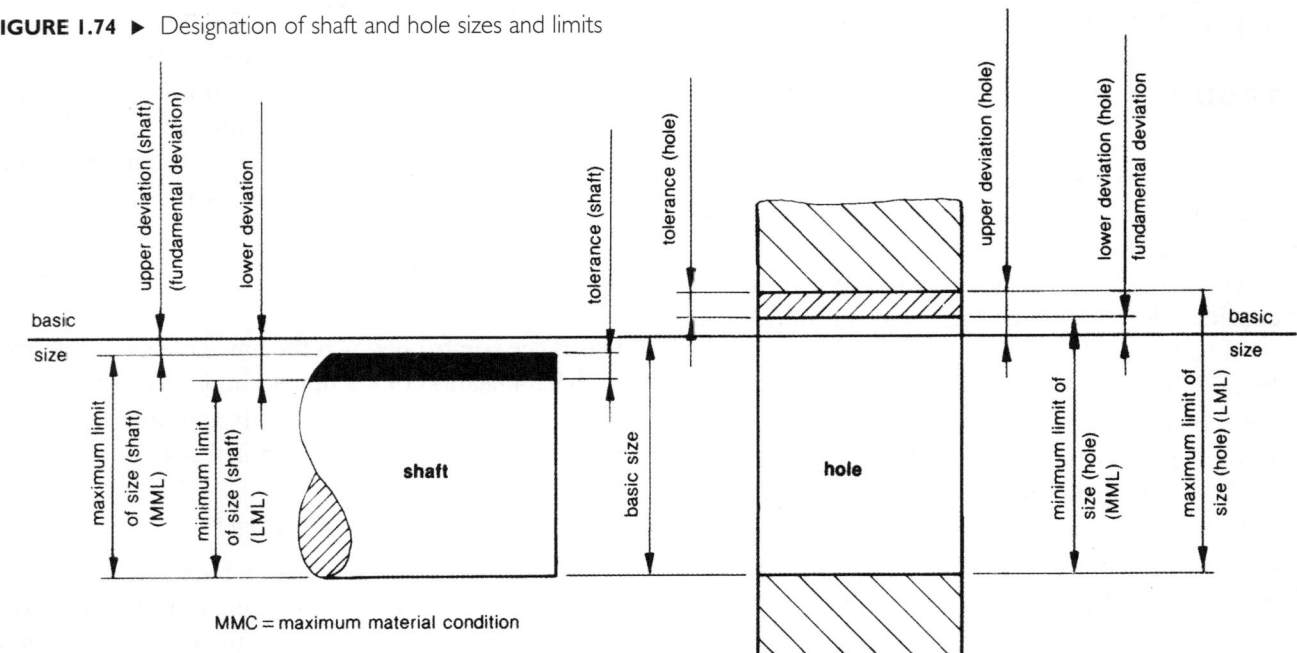

MMC = maximum material condition

**FIGURE 1.75** ▶ Comparison between the three classes of fit using the unilateral hole-basis system

**(a)** clearance fit    **(b)** transition fit    **(c)** interference fit

## FIT

A fit may be defined as the relative motion which can exist between a shaft and hole (as defined above) resulting from the final sizes which are achieved in their manufacture. There are three classes of fit in common use: *clearance, transition and interference.*

### Clearance fit (Fig. 1.75(a))

This fit results when the shaft size is always less than the hole size for all possible combinations within their tolerance ranges. Relative motion between shaft and hole is always possible.

The *minimum clearance* occurs at the maximum shaft size and the minimum hole size.

The *maximum clearance* occurs at the minimum shaft size and the maximum hole size.

Clearance fits range from coarse or very loose to close precision and locational. A few possible combinations are given in Tables 1.24(a) and (b).

### Transition fit (Fig. 1.75(b))

When the limits of sizes specified for a matching hole and shaft allow either a clearance fit or an interference fit, the class of fit is a transition fit.

54

In this case, the tolerance zones for the hole and shaft overlap (see Fig. 1.75).

Maximum clearance occurs at the maximum hole size and the minimum shaft size. Maximum interference occurs at the minimum hole size and the maximum shaft size. There are two common standard transition fits: one is called a 'light push fit' and the other a 'heavy push fit' (see Tables 1.24(a) and (b)).

### Interference fit (Fig. 1.75(c))

This is a fit which always results in the minimum shaft size being larger than the maximum hole size for all possible combinations within their tolerance ranges. Relative motion between the shaft and hole is impossible.

The minimum interference occurs at the minimum shaft size and the maximum hole size.

The maximum interference occurs at the maximum shaft size and minimum hole size. Two interference fits are given in each of Tables 1.24(a) and 1.24(b).

## ALLOWANCE (FIG. 1.75)

*Allowance* is the term given to the minimum clearance (called positive allowance) or maximum interference (called negative allowance) which exists between mating parts. It may also be described as the clearance or interference which gives the tightest possible fit between mating parts. It is also the fit that two mating features would have if they were both at their maximum material limits.

## GRADES OF TOLERANCE

To give a wide range of control over tolerance, provision has been made in the ISO system for 18 grades of tolerance, ranging from very fine for the lower numbers to extremely coarse for the larger numbers. Each grade is approximately 1.6 times as great as the grade below or finer than it. This ratio has been determined after extensive practical investigations, and is derived from the relationship $t = k f(d)$ where $t$ is the tolerance and is equal to a function of the diameter multiplied by the constant $k$. Different values of $k$ are used to provide a series of tolerance grades for various diameters.

The 18 grades are designated ITO1, ITO, IT1, IT2, up to IT16. The letters IT (which stand for 'International Tolerance') are omitted in tables and also when designating fits. The numerical values of these grades of fit for all diameters up to 3150 mm are given in AS 1654—1995 *ISO System of Limits and Fits*.

Figure 1.76 illustrates graphically a comparison between some of the grades (IT5 to IT13).

The grade actually represents the size of the tolerance zone and this in turn dictates the degree of accuracy of the machining process required to keep the size within the specified tolerance. Low grades require precision or toolroom machines with highly skilled labour. Coarse grades are much easier to maintain, and require cheaper machines and less skilled labour.

**FIGURE 1.76** ▶ Comparison of some grades of tolerance

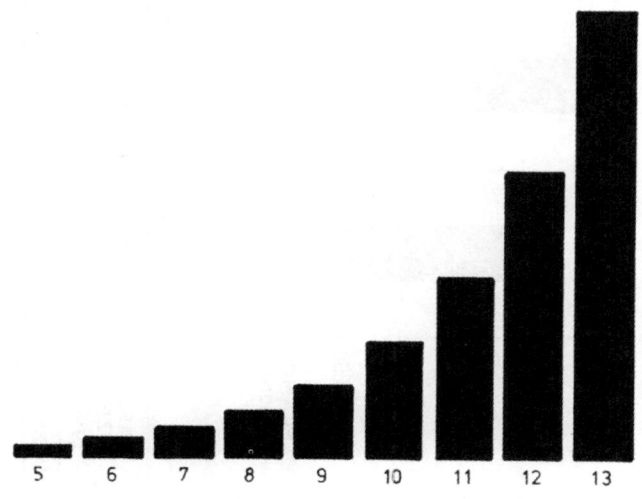

**TABLE 1.24(a)** ▶ A selection of fits—hole-basis system (deviations)

*This chart is to scale only for 20 mm basic size*

■ = shafts  ▨ = holes

Deviation cells are given as: **hole** (upper/lower) ; **shaft** (upper/lower). TOLERANCE unit = 0.001 mm.

| | | CLEARANCE FITS | | | | | | TRANSITION FITS | | | | INTERFERENCE FITS | |
|---|---|---|---|---|---|---|---|---|---|---|---|---|---|
| | | COARSE TOLERANCE | LOOSE | EASY | NORMAL RUNNING | PRECISION RUNNING, SLIDING | AVERAGE LOCATION | LIGHT | HEAVY | | | PRESS FIT (FERROUS) | HEAVY PRESS FIT (NON-FERROUS) |
| | | | RUNNING FIT | | | | | PUSH FIT | | | | | |
| **BASIC SIZES (mm) OVER** | **TO** | H11/c11 | H9/d10 | H9/e9 | H8/f7 | H7/g6 | H7/h6 | H7/k6 | H7/n6 | H7/p6 | | H7/s6 | |
| 0 | 3 | +60/0 ; −60/−120 | +25/0 ; −20/−60 | +25/0 ; −14/−39 | +14/0 ; −6/−16 | +10/0 ; −2/−8 | +10/0 ; 0/−6 | +10/0 ; +6/0 | +10/0 ; +10/4 | +10/0 ; +12/+6 | | +10/0 ; +20/+14 | |
| 3 | 6 | +75/0 ; −70/−145 | +30/0 ; −30/−78 | +30/0 ; −20/−50 | +18/0 ; −10/−28 | +12/0 ; −4/−12 | +12/0 ; 0/−8 | +12/0 ; +9/+1 | +12/0 ; +16/+8 | +12/0 ; +20/+12 | | +12/0 ; +27/+19 | |
| 6 | 10 | +90/0 ; −80/−170 | +36/0 ; −40/−98 | +36/0 ; −25/−61 | +22/0 ; −13/−28 | +15/0 ; −5/−14 | +15/0 ; 0/−9 | +15/0 ; +10/+1 | +15/0 ; +19/+10 | +15/0 ; +24/+15 | | +15/0 ; +32/+23 | |
| 10 | 18 | +110/0 ; −95/−205 | +43/0 ; −50/−120 | +43/0 ; −32/−75 | +27/0 ; −16/−34 | +18/0 ; −6/−17 | +18/0 ; 0/−11 | +18/0 ; +12/+1 | +18/0 ; +23/+12 | +18/0 ; +29/+18 | | +18/0 ; +39/+28 | |
| 18 | 30 | +130/0 ; −110/−240 | +52/0 ; −65/−149 | +52/0 ; −40/−92 | +33/0 ; −20/−41 | +21/0 ; −7/−20 | +21/0 ; 0/−13 | +21/0 ; +15/+2 | +21/0 ; +28/+15 | +21/0 ; +35/+22 | | +21/0 ; +48/+35 | |

Shaft/hole diagram labels (tolerance in 0.001 mm, relative to Basic size): scale markings +150, +100, +50, Basic size, −50, −100, −150, −200, −250.

Clearance fits: H11 / c11, H9 / d10, H9 / e9, H8 / f7, H7 / g6, H7 / h6
Transition fits: H7 / k6, n6 / H7
Interference fits: p6 / H7, s6 / H7

| Over | 30 | 40 | 50 | 65 | 80 | 100 | 120 | 140 | 160 | 180 | 200 | 225 | 250 | 280 | 315 | 355 | 400 | 450 |
|---|---|---|---|---|---|---|---|---|---|---|---|---|---|---|---|---|---|---|
| To | 40 | 50 | 65 | 80 | 100 | 120 | 140 | 160 | 180 | 200 | 225 | 250 | 280 | 315 | 355 | 400 | 450 | 500 |
| | +59 / +43 | | +72 / +53 | +78 / +59 | +93 / +71 | +101 / +79 | +117 / +92 | +125 / +100 | +133 / +108 | +151 / +122 | +159 / +130 | +169 / +140 | +190 / +158 | +202 / +170 | +226 / +190 | +244 / +208 | +272 / +232 | +292 / +252 |
| | +25 / 0 | | +30 / 0 | | +35 / 0 | | +40 / 0 | | | +46 / 0 | | | +52 / 0 | | +57 / 0 | | +63 / 0 | |
| | +42 / +26 | | +51 / +32 | | +59 / +37 | | +68 / +43 | | | +79 / +50 | | | +88 / +56 | | +98 / +62 | | +108 / +68 | |
| | +25 / 0 | | +30 / 0 | | +35 / 0 | | +40 / 0 | | | +46 / 0 | | | +52 / 0 | | +57 / 0 | | +63 / 0 | |
| | +33 / +17 | | +39 / +20 | | +45 / +23 | | +52 / +27 | | | +60 / +31 | | | +66 / +34 | | +73 / +37 | | +80 / +40 | |
| | +25 / 0 | | +30 / 0 | | +35 / 0 | | +40 / 0 | | | +46 / 0 | | | +52 / 0 | | +57 / 0 | | +63 / 0 | |
| | +18 / +2 | | +21 / +2 | | +25 / 3 | | +28 / +3 | | | +33 / +4 | | | +36 / +4 | | +40 / +4 | | +45 / +5 | |
| | +25 / 0 | | +30 / 0 | | +35 / 0 | | +40 / 0 | | | +46 / 0 | | | +52 / 0 | | +57 / 0 | | +63 / 0 | |
| | 0 / -16 | | 0 / -19 | | 0 / -22 | | 0 / -25 | | | 0 / -29 | | | 0 / -32 | | 0 / -36 | | 0 / -40 | |
| | +25 / 0 | | +30 / 0 | | +35 / 0 | | +40 / 0 | | | +46 / 0 | | | +52 / 0 | | +57 / 0 | | +63 / 0 | |
| | -9 / -25 | | -10 / -29 | | -12 / -34 | | -14 / -39 | | | -15 / -44 | | | -17 / -49 | | -18 / -54 | | -20 / -60 | |
| | +25 / 0 | | +30 / 0 | | +35 / 0 | | +40 / 0 | | | +46 / 0 | | | +52 / 0 | | +57 / 0 | | +63 / 0 | |
| | -25 / -50 | | -30 / -60 | | -36 / -71 | | -43 / -83 | | | -50 / -96 | | | -56 / -108 | | -62 / -119 | | -68 / -131 | |
| | +39 / 0 | | +46 / 0 | | +54 / 0 | | +63 / 0 | | | +72 / 0 | | | +81 / 0 | | +89 / 0 | | +97 / 0 | |
| | -50 / -112 | | -60 / -134 | | -72 / -159 | | -84 / -185 | | | -100 / -215 | | | -110 / -240 | | -125 / -265 | | -135 / -290 | |
| | +62 / 0 | | +74 / 0 | | +87 / 0 | | +100 / 0 | | | +115 / 0 | | | +130 / 0 | | +140 / 0 | | +155 / 0 | |
| | -80 / -180 | | -100 / -220 | | -120 / -260 | | -145 / -305 | | | -170 / -355 | | | -190 / -400 | | -210 / -440 | | -230 / -480 | |
| | +62 / 0 | | +74 / 0 | | +87 / 0 | | +100 / 0 | | | +115 / 0 | | | +130 / 0 | | +140 / 0 | | +155 / 0 | |
| | -120 / -280 | -130 / -290 | -140 / -330 | -150 / -340 | -170 / -390 | -180 / -400 | -200 / -450 | -210 / -460 | -230 / -480 | -240 / -530 | -260 / -550 | -280 / -570 | -300 / -620 | -330 / -650 | -360 / -720 | -400 / -760 | -440 / -840 | -480 / -880 |
| | +160 / 0 | | +190 / 0 | | +220 / 0 | | +250 / 0 | | | +290 / 0 | | | +320 / 0 | | +360 / 0 | | +400 / 0 | |

**TABLE I.24(b)** ▶ A selection of fits—shaft-basis system (deviations)

*This chart is to scale only for 20 mm basic size*

▨ = holes   ■ = shafts

Deviations (unit = 0.001 mm). Hole deviations (upper / lower):

| Fit group | Fit | Designation | 0–3 | 3–6 | 6–10 | 10–18 | 18–30 |
|---|---|---|---|---|---|---|---|
| CLEARANCE FITS | COARSE TOLERANCE | C11 | +120 / +60 | +145 / +70 | +170 / +80 | +205 / +95 | +240 / +110 |
| CLEARANCE FITS (RUNNING FIT) | LOOSE | D10 | +60 / +20 | +78 / +30 | +98 / +40 | +120 / +50 | +149 / +65 |
| CLEARANCE FITS (RUNNING FIT) | EASY | E9 | +39 / +14 | +50 / +20 | +61 / +25 | +75 / +32 | +92 / +40 |
| CLEARANCE FITS | NORMAL RUNNING | F8 | +20 / +6 | +28 / +10 | +35 / +13 | +43 / +16 | +53 / +20 |
| CLEARANCE FITS | PRECISION RUNNING, SLIDING | G7 | +12 / +2 | +16 / +4 | +20 / +5 | +24 / +6 | +28 / +7 |
| CLEARANCE FITS | AVERAGE LOCATION | H7 | +10 / 0 | +12 / 0 | +15 / 0 | +18 / 0 | +21 / 0 |
| TRANSITION FITS (PUSH FIT) | LIGHT | K7 | 0 / −10 | +3 / −9 | +5 / −10 | +6 / −12 | +6 / −15 |
| TRANSITION FITS (PUSH FIT) | HEAVY | N7 | −4 / −14 | −4 / −16 | −4 / −19 | −5 / −23 | −7 / −28 |
| INTERFERENCE FITS | PRESS FIT (FERROUS) | P7 | −6 / −16 | −8 / −20 | −9 / −24 | −11 / −29 | −14 / −35 |
| INTERFERENCE FITS | HEAVY PRESS FIT (NON-FERROUS) | S7 | −14 / −24 | −15 / −27 | −17 / −32 | −21 / −39 | −27 / −48 |

Shaft deviations (upper / lower):

| Shaft | 0–3 | 3–6 | 6–10 | 10–18 | 18–30 |
|---|---|---|---|---|---|
| h11 | 0 / −60 | 0 / −75 | 0 / −90 | 0 / −110 | 0 / −130 |
| h9 | 0 / −25 | 0 / −30 | 0 / −36 | 0 / −43 | 0 / −52 |
| h7 | 0 / −10 | 0 / −12 | 0 / −15 | 0 / −18 | 0 / −21 |
| h6 | 0 / −6 | 0 / −8 | 0 / −9 | 0 / −11 | 0 / −13 |

BASIC SIZES (mm): OVER / TO — 0/3, 3/6, 6/10, 10/18, 18/30

| 30–40 | 40–50 | 50–65 | 65–80 | 80–100 | 100–120 | 120–140 | 140–160 | 160–180 | 180–200 | 200–225 | 225–250 | 250–280 | 280–315 | 315–355 | 355–400 | 400–450 | 450–500 |
|---|---|---|---|---|---|---|---|---|---|---|---|---|---|---|---|---|---|
| 0/-16 | 0/-16 | 0/-19 | 0/-19 | 0/-22 | 0/-22 | 0/-25 | 0/-25 | 0/-25 | 0/-29 | 0/-29 | 0/-29 | 0/-32 | 0/-32 | 0/-36 | 0/-36 | 0/-40 | 0/-40 |
| -34/-59 | -34/-59 | -42/-72 | -48/-78 | -58/-93 | -66/-101 | -77/-117 | -85/-125 | -93/-133 | -105/-151 | -113/-159 | -123/-169 | -138/-190 | -150/-202 | -169/-226 | -187/-244 | -209/-272 | -229/-292 |
| 0/-16 | 0/-16 | 0/-19 | 0/-19 | 0/-22 | 0/-22 | 0/-25 | 0/-25 | 0/-25 | 0/-29 | 0/-29 | 0/-29 | 0/-32 | 0/-32 | 0/-36 | 0/-36 | 0/-40 | 0/-40 |
| -17/-42 | -17/-42 | -21/-51 | -21/-51 | -24/-59 | -24/-59 | -28/-68 | -28/-68 | -28/-68 | -33/-79 | -33/-79 | -33/-79 | -36/-88 | -36/-88 | -41/-98 | -41/-98 | -48/-108 | -48/-108 |
| 0/-16 | 0/-16 | 0/-19 | 0/-19 | 0/-22 | 0/-22 | 0/-25 | 0/-25 | 0/-25 | 0/-29 | 0/-29 | 0/-29 | 0/-32 | 0/-32 | 0/-36 | 0/-36 | 0/-40 | 0/-40 |
| -8/-33 | -8/-33 | -9/-39 | -9/-39 | -10/-45 | -10/-45 | -12/-52 | -12/-52 | -12/-52 | -14/-60 | -14/-60 | -14/-60 | -14/-66 | -14/-66 | -16/-73 | -16/-73 | -17/-80 | -17/-80 |
| 0/-16 | 0/-16 | 0/-19 | 0/-19 | 0/-22 | 0/-22 | 0/-25 | 0/-25 | 0/-25 | 0/-29 | 0/-29 | 0/-29 | 0/-32 | 0/-32 | 0/-36 | 0/-36 | 0/-40 | 0/-40 |
| +7/-18 | +7/-18 | +9/-21 | +9/-21 | +10/-25 | +10/-25 | +12/-28 | +12/-28 | +12/-28 | +13/-33 | +13/-33 | +13/-33 | +16/-36 | +16/-36 | +17/-40 | +17/-40 | +18/-45 | +18/-45 |
| 0/-16 | 0/-16 | 0/-19 | 0/-19 | 0/-22 | 0/-22 | 0/-25 | 0/-25 | 0/-25 | 0/-29 | 0/-29 | 0/-29 | 0/-32 | 0/-32 | 0/-36 | 0/-36 | 0/-40 | 0/-40 |
| +25/0 | +25/0 | +30/0 | +30/0 | +35/0 | +35/0 | +40/0 | +40/0 | +40/0 | +46/0 | +46/0 | +46/0 | +52/0 | +52/0 | +57/0 | +57/0 | +63/0 | +63/0 |
| 0/-16 | 0/-16 | 0/-19 | 0/-19 | 0/-22 | 0/-22 | 0/-25 | 0/-25 | 0/-25 | 0/-29 | 0/-29 | 0/-29 | 0/-32 | 0/-32 | 0/-36 | 0/-36 | 0/-40 | 0/-40 |
| +34/+9 | +34/+9 | +40/+10 | +40/+10 | +47/+12 | +47/+12 | +54/+14 | +54/+14 | +54/+14 | +61/+15 | +61/+15 | +61/+15 | +62/+17 | +62/+17 | +75/+18 | +75/+18 | +83/+20 | +83/+20 |
| 0/-25 | 0/-25 | 0/-30 | 0/-30 | 0/-35 | 0/-35 | 0/-40 | 0/-40 | 0/-40 | 0/-46 | 0/-46 | 0/-46 | 0/-52 | 0/-52 | 0/-57 | 0/-57 | 0/-63 | 0/-63 |
| +64/+25 | +64/+25 | +76/+30 | +76/+30 | +90/+36 | +90/+36 | +106/+43 | +106/+43 | +106/+43 | +122/+50 | +122/+50 | +122/+50 | +137/+56 | +137/+56 | +151/+62 | +151/+62 | +165/+68 | +165/+68 |
| 0/-62 | 0/-62 | 0/-74 | 0/-74 | 0/-87 | 0/-87 | 0/-100 | 0/-100 | 0/-100 | 0/-115 | 0/-115 | 0/-115 | 0/-130 | 0/-130 | 0/-140 | 0/-140 | 0/-155 | 0/-155 |
| +112/+50 | +112/+50 | +134/+60 | +134/+60 | +159/+72 | +159/+72 | +185/+85 | +185/+85 | +185/+85 | +215/+100 | +215/+100 | +215/+100 | +240/+110 | +240/+110 | +265/+125 | +265/+125 | +290/+135 | +290/+135 |
| 0/-62 | 0/-62 | 0/-74 | 0/-74 | 0/-87 | 0/-87 | 0/-100 | 0/-100 | 0/-100 | 0/-115 | 0/-115 | 0/-115 | 0/-130 | 0/-130 | 0/-140 | 0/-140 | 0/-155 | 0/-155 |
| +180/+80 | +180/+80 | +220/+100 | +220/+100 | +260/+120 | +260/+120 | +305/+145 | +305/+145 | +305/+145 | +355/+170 | +355/+170 | +355/+170 | +400/+190 | +400/+190 | +440/+210 | +440/+210 | +480/+230 | +480/+230 |
| 0/-160 | 0/-160 | 0/-190 | 0/-190 | 0/-220 | 0/-220 | 0/-250 | 0/-250 | 0/-250 | 0/-290 | 0/-290 | 0/-290 | 0/-320 | 0/-320 | 0/-360 | 0/-360 | 0/-400 | 0/-400 |
| +280/+120 | +290/+130 | +330/+140 | +340/+150 | +390/+170 | +400/+180 | +450/+200 | +460/+210 | +480/+230 | +530/+240 | +550/+260 | +570/+280 | +620/+300 | +650/+330 | +720/+360 | +760/+400 | +840/+440 | +880/+480 |

## BILATERAL LIMITS

This is the name given to the maximum and minimum limits when they are disposed above and below the basic size respectively. This results in the tolerance zone straddling the basic size, for example a J7 hole of 50 mm basic diameter has a maximum limit of 50.014 and a minimum limit of 49.989 (taken from the ISO table). For holes and shafts designated JS and js respectively, the tolerance is equally disposed above and below the basic size (see Fig. 1.78, p. 61).

## UNILATERAL LIMITS

This name is given to limits when one limit of size is the basic size and the other limit is either above or below the basic size, for example all the holes in Table 1.24(a) have positive unilateral tolerances.

## FUNDAMENTAL DEVIATION OF TOLERANCE

The fundamental deviation is the deviation which is closest to the basic size and is used to locate the tolerance zone with respect to the basic size. It may be the upper or lower deviation, depending on the type of fit and whether the member is a shaft or a hole. The fundamental deviation determines the maximum and minimum amounts of clearance or interference which are possible for a particular size of tolerance zone. For example, Figure 1.77 illustrates two clearance fits in which the tolerance zones are identical in size. However, the maximum and minimum clearances possible for fit 1 have been reduced for fit 2 simply by reducing the fundamental deviation (minimum clearance) for the shaft, which has the effect of moving the shaft tolerance zone closer to the basic size. This is achieved in this case by designating the shaft as d11 instead of c11. (Note: The fundamental deviation for the hole H11 is zero.)

In the ISO system there are 28 fundamental deviations provided for each of the 18 grades of tolerance on both shafts and holes. These are designated by upper case letters for holes and lower case letters for shafts as shown below.

Holes   A, B, C, CD, D, E, EF, F, FG, G, H, JS, J, K, M, N, P, R, S, T, U, V, X, Y, Z, ZA, ZB, ZC

Shafts   a, b, c, cd, d, e, ef, f, fg, g, h, js, j, k, m, n, p, r, s, t, u, v, x, y, z, za, zb, zc

These letters represent a wide range of fundamental deviation positions varying from above to below the basic size for both shafts and holes. Figure 1.78 (page 61) illustrates graphically these positions for a 10 mm shaft and hole respectively using a grade 7 tolerance throughout. The JS hole and js shaft tolerance zone

positions are unlike the rest in that they provide symmetrical bilateral tolerances and hence have no fundamental deviation. Stated simply, this means that the tolerance zone is equally disposed above and below the basic size for both shaft and hole.

It will also be noticed that the H hole, which is featured in Table 1.24(a), is the only one which has the basic size at the lower limit. Also the h shaft is the only one which has the basic size at the upper limit. These two fundamental deviations (zero for both h shaft and H hole) enable a selection of fits to be made on either a *hole basis* or a *shaft basis*.

**FIGURE 1.77** ▶ Use of fundamental deviation

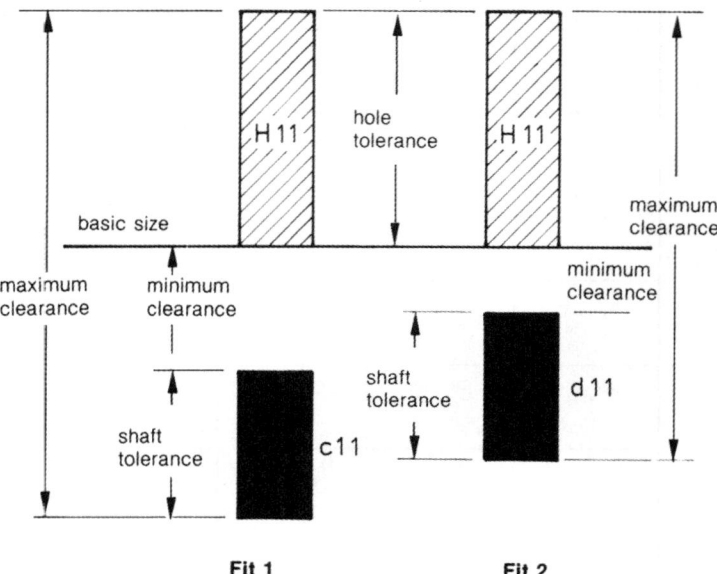

*Note:* fundamental deviation for hole H = 0
fundamental deviation for shafts c and d = minimum clearance

## THE HOLE-BASIS SYSTEM

Fits are obtained by regarding the hole as standard with a zero fundamental deviation and varying the fundamental deviations of the shafts to suit. The 18 grades of tolerance can still be applied to alter the size of the tolerance zones when required. Table 1.24(a) is based on this system which is also known as a *unilateral hole-basis system* because the disposition of the hole tolerance zones are all on the positive side of the basic size.

## THE SHAFT-BASIS SYSTEM

Table 1.24(b) is based on this system. In this case the fundamental deviation of the shaft, h, is zero, and the fits are obtained by varying the fundamental deviations of the holes as well as applying the 18 grades of tolerances. It is a *unilateral shaft-basis system* because the disposition of the shaft tolerance zones are all on the negative side of the basic size.

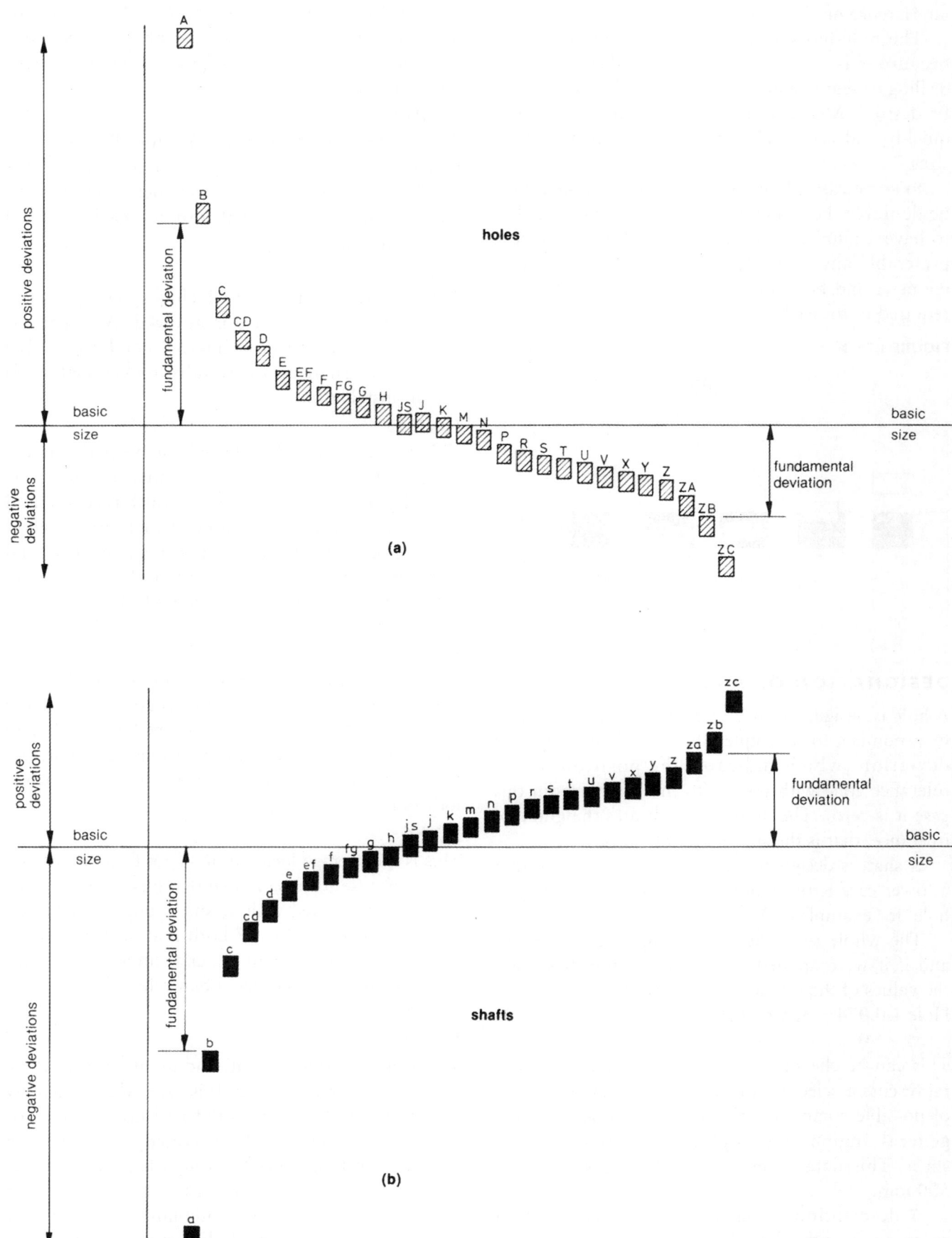

(a)

(b)

Figure 1.79 illustrates five classes of fit using this system, ranging from clearance on the left to interference on the right.

The hole-basis system is more commonly used because it is easier to produce standard holes by drilling or reaming and then turn the shaft to suit the fit desired. Measurements can also be made more quickly and accurately on shaft sizes than on hole sizes.

In some cases, however, a shaft-basis system may be desirable. For example, when a driving shaft has to have a number of different parts fitted to it, it is preferable and easier to keep the shaft a constant diameter and bore out the various parts to give the required fit for each.

**FIGURE 1.79 ▶** Five classes of fit using a shaft-basis system

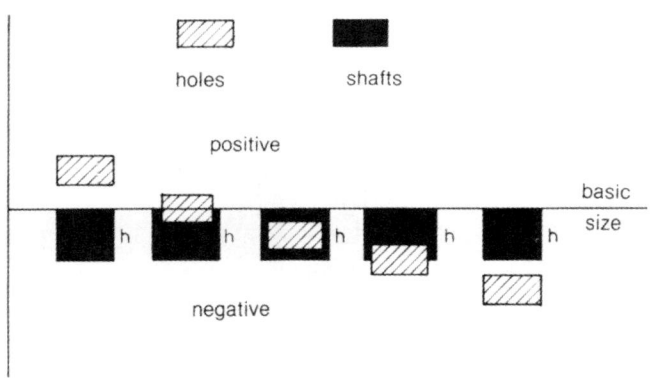

## DESIGNATION OF A FIT

A hole is designated by an upper case letter followed by a number, for example H9. H is the fundamental deviation, which indicates the position of the tolerance zone with respect to the basic size (in this case it is zero). The number 9 indicates the grade of tolerance, that is the size of the tolerance zone.

A shaft is designated in a similar way except that a lower case letter is used to distinguish it from the hole, for example d10.

The whole fit is therefore designated as H9/d10 and if it were applied to, say, an 80 mm basic size the values of the tolerance limits would be

Hole + 0.074    Shaft − 0.1
      0             − 0.22

This can be checked from Table 1.24(a). This table represents a selection of fits out of many thousands of possible combinations. These are suitable for the general engineering applications shown on the sheet. This data sheet covers all basic sizes up to 500 mm.

A description of each of the ten types of fit represented on the data sheet follows.

### H11/c11

This is a slack or coarse clearance fit which may be used where dirty conditions prevail and ease of assembly and disassembly are essential, for example, agricultural machinery, loose pulleys, very large shaft and bearing assemblies.

### H9/d10

This is a loose running fit suitable for idler gears and pulleys. It can also be used as a running fit for large bearing applications which are met in steel mills, large turbines, heavy metal forming machinery and similar installations.

### H9/e9

This is an easy running fit which is applicable where an appreciable tolerance is allowed. Applications include main bearings in internal combustion (IC) engines, camshaft bearings, valve rocker shafts and similar installations.

### H8/f7

This is the fit usually selected for normal running conditions. It is suitable for most applications requiring a reasonable quality fit which is economical and easy to produce. Rotating shaft bearings, gears running on shafts, fits of components in medium and light mechanisms, and general light to medium engineering applications are some of the uses of this class of fit.

### H7/g6

This is a precision running or sliding fit in which the clearance is small. It is only recommended for precision running assemblies where light loads and large variations in temperature are not encountered. It can also be used for spigot fits and other locational non-running fits.

### H7/h6

This is the average location or spigot fit used on non-running assemblies. It usually has a very small clearance associated with it, and is one of the closest possible clearance fits. If both mating features are made to their maximum material limit of size, they will both be the same size, often referred to as 'size-for-size'.

### H7/k6

This is a true transition fit, and on an average there will be no clearance found. It is used where assembly and disassembly are required and no vibration or relative movement can be tolerated, for example a gudgeon pin fitted into a piston, a handwheel keyed to a shaft, or similar applications. If both mating features are made to their minimum material limit of size, there will be a small clearance between them.

## H7/n6

This fit can give interference at one extreme and clearance at the other. However, on average it is a heavy push fit and is used in applications where a tight assembly is preferred, but a small clearance is also acceptable.

## H7/p6

This is a true interference fit used in pressing ferrous parts together. The amount of interference is small, and assemblies may be dismantled and reassembled with minimal damage to the surfaces.

## H7/s6

This is a heavy press fit used for permanent assembly of members. Pressing apart usually results in the scoring of the surfaces, especially if similar metals are used. Initial assembly may be achieved without damage to the surfaces by heating the hole and shrinking it on to the shaft. It is normally used on non-ferrous assemblies such as pressed in bushes, sleeves, liners, seats and the like.

## APPLICATION OF TOLERANCES TO DIMENSIONS

Tolerances should be specified in the case where a dimension is critical to the proper functioning or interchangeability of a component.

A tolerance can also be supplied to a dimension which can have an unusually large variation in size.

### General tolerances

These are generally quoted in note form and apply when the same tolerance is applicable all over the drawing or where different tolerances apply to various ranges of sizes or for a particular type of member. The following examples illustrate the use of general tolerances.

---

Tolerance except where
otherwise stated ± 0.125

---

Tolerances except where otherwise
stated on dimensions
| | |
|---|---|
| Up to 75 | ±0.075 |
| Over 75 up to 100 | ±0.125 |
| Over 100 up to 200 | ±0.25 |
| On angles | ±1° |

---

Tolerance on cast
thicknesses ± 15%

---

Tolerance unless otherwise stated
X ± 0.5
X.X ± 0.05
X.XX ± 0.02
X.XXX ± 0.002

---

General tolerances on non-critical linear dimensions may be selected from AS 1100 Part 201, which provides for a fine, medium, coarse and very coarse series of permissible deviations. Such deviations are shown in a general note for 'untoleranced' dimensions and may be selected from Table 1.25.

### Tolerancing angular dimensions

General tolerances on non-critical angular dimensions may be selected from Table 1.26. The dimensioning of angles using limits of size, bilateral limits and unilateral limits is shown in Fig 1.80 (a), (b) and (c) respectively.

---

**TABLE 1.25** ▶ General tolerances for linear dimensions

| NOMINAL DIMENSIONS (mm) | | 0.5 TO 3 | OVER 3 TO 6 | OVER 6 TO 30 | OVER 30 TO 120 | OVER 120 TO 315 | OVER 315 TO 1000 | OVER 1000 TO 2000 |
|---|---|---|---|---|---|---|---|---|
| **PERMISSIBLE DEVIATIONS IN MILLIMETRES** | **Fine series** | ±0.05 | ±0.05 | ±0.1 | ±0.15 | ±0.2 | ±0.3 | ±0.5 |
| | **Medium series** | ±0.1 | ±0.1 | ±0.2 | ±0.3 | ±0.5 | ±0.8 | ±1.2 |
| | **Coarse series** | | ±0.2 | ±0.5 | ±0.8 | ±1.2 | ±2 | ±3 |
| | **Very coarse** | | ±0.5 | ±1 | ±1.5 | ±2.5 | ±4 | ±8 |

TABLE 1.26 ▶ General tolerances for angular dimensions

| LENGTH OF THE SHORTER SIDE (mm) | UP TO 10 | OVER 10 TO 50 | OVER 50 TO 120 | OVER 120 TO 400 | OVER 400 |
|---|---|---|---|---|---|
| PERMISSIBLE DEVIATIONS IN DEGREES AND MINUTES | ±1° | ±0°30' | ±0°20' | ±0°10' | ±0°5' |

FIGURE 1.80 ▶ Tolerancing of angular dimensions

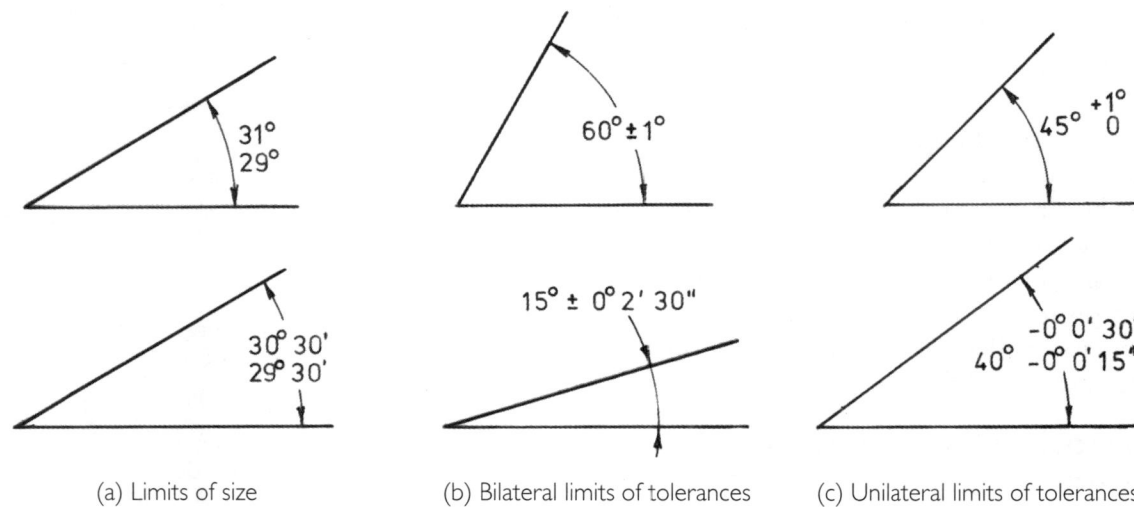

| (a) Limits of size | (b) Bilateral limits of tolerances | (c) Unilateral limits of tolerances |
|---|---|---|

## Individual tolerances

For tolerancing individual linear dimensions one of the following methods may be used. In some cases the fits are designated and values are taken from Table 1.24(a).

### Method 1—Limits of size (Fig. 1.81)

This is by specifying both limits of size. The maximum limit (called the upper limit of size) is placed above the dimension line, while the minimum limit (called the lower limit of size) is placed below the dimension line.

FIGURE 1.81 ▶ Method 1—specifying both limits of size

basic size 85 mm     fit H9/d10

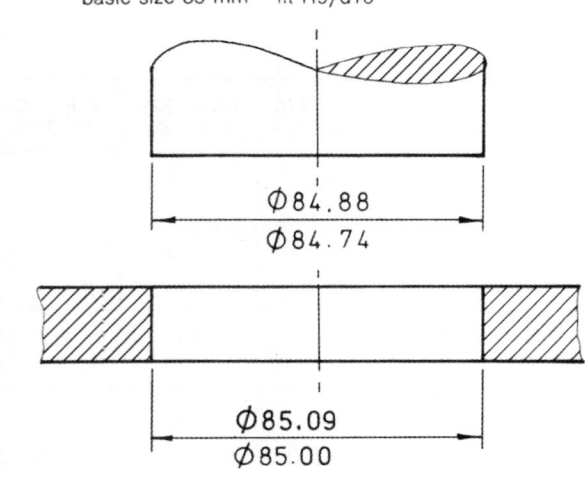

### Method 2—Bilateral tolerances (Fig. 1.82)

This method specifies the basic size (25) followed by the limits of tolerance above and below the basic size (± 0.25). This means that the upper and lower deviations have the same value, but opposite signs.

FIGURE 1.82 ▶ Specifying equally disposed limits (bilateral limits)

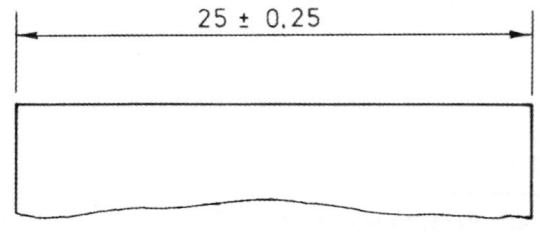

### Method 3—Unilateral tolerances (Fig. 1.83)

This method specifies the basic size followed by an allowable variation in one direction only. Hence the variation in size can occur above or below the basic size. In the hole-basis system the lower deviation of the hole is zero and in the shaft-basis system the upper deviation of the shaft is zero.

FIGURE 1.83 ▶ Specifying equally disposed limits (unilateral limits)

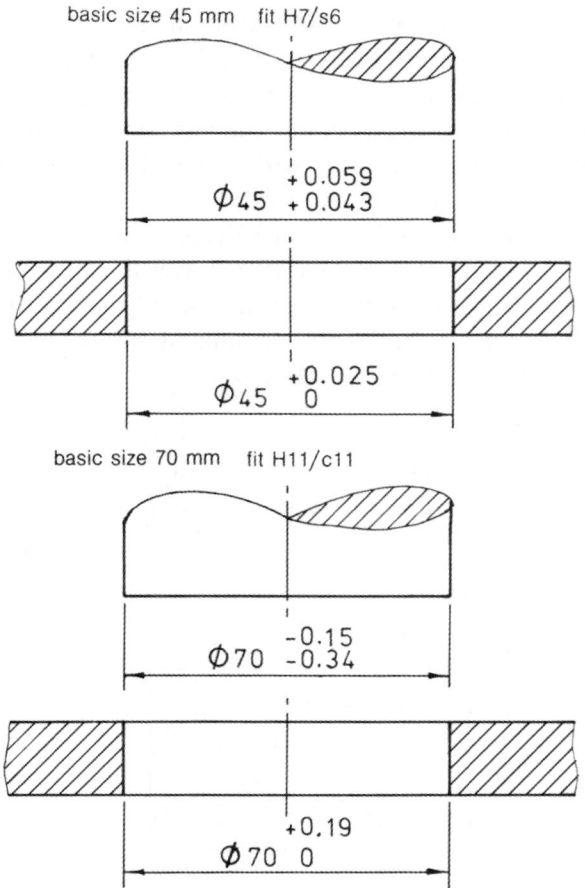

basic size 45 mm    fit H7/s6

$\phi$45 $^{+0.059}_{+0.043}$

$\phi$45 $^{+0.025}_{0}$

basic size 70 mm    fit H11/c11

$\phi$70 $^{-0.15}_{-0.34}$

$\phi$70 $^{+0.19}_{0}$

## METHODS OF DIMENSIONING TO AVOID ACCUMULATION OF TOLERANCES

*Chain dimensioning* can result in tolerances accumulating to such an extent as to make an overall tolerance impossible. This can be overcome by omitting one of the chain of dimensions as shown in Figure 1.84.

*Progressive dimensioning* from a fixed datum ensures that accumulation of tolerances will not occur. In Figure 1.85 this method is used in dimensioning all of the vertical surfaces from the left-hand end on the front view. Thus adjacent vertical surfaces, such as X and Y, have a space between them which is influenced by two toleranced dimensions. With chain dimensioning, this space would be controlled by one dimension.

On the top view the positions of the holes are dimensioned by the chain method using the bottom edge and the left-hand end as initial reference or datum surfaces.

Whichever method is used will depend on the relationship of functional dimensions and whether or not there are reference or datum surfaces from which it is desirable to refer these functional dimensions.

FIGURE 1.85 ▶ Progressive dimensioning

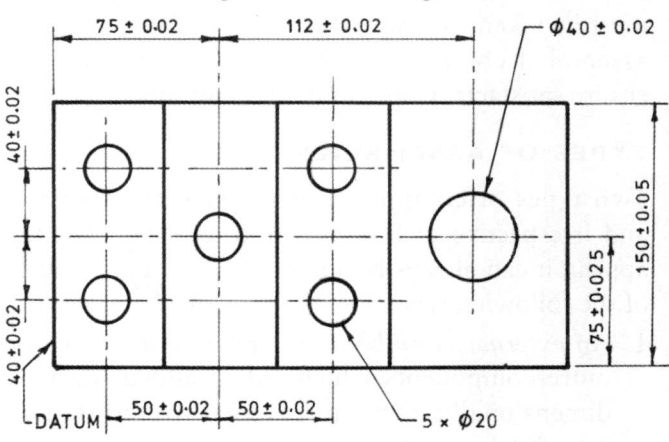

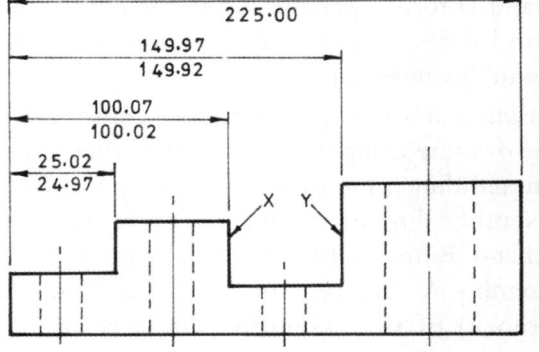

FIGURE 1.84 ▶ Chain dimensioning

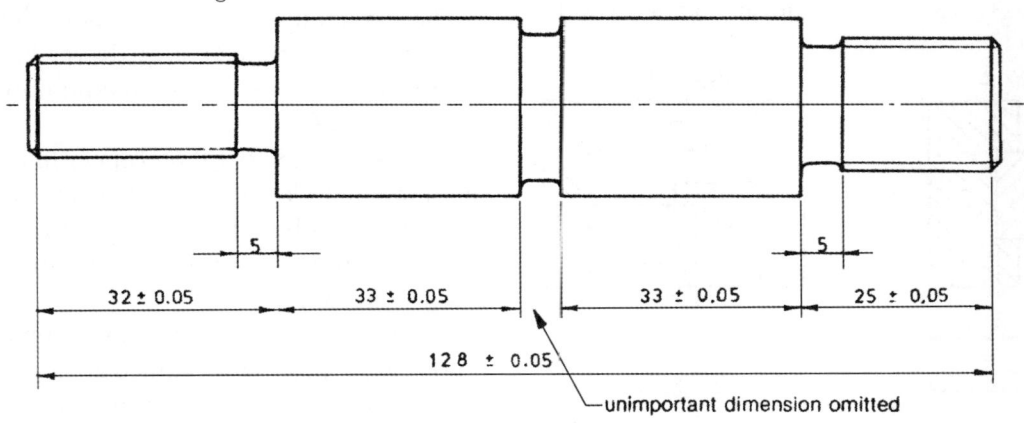

32 ± 0.05    33 ± 0.05    33 ± 0.05    25 ± 0,05

128 ± 0.05

—unimportant dimension omitted

# 1.20 Assembly of components

## INTRODUCTION

A mechanical assembly is a combination or 'fitting together' of components designed to perform a specific mechanical function. Each component has a finished size which lies within a specified tolerance. Because of the range of finished sizes allowable for each component, it follows that the overall dimension which encloses the assembly must be a function of the accumulation of tolerances of the individual components.

In the design of mechanical assemblies, great care must be taken to ensure that the cumulative effect of assembled component tolerances is controlled to ensure satisfactory operation of the product.

## TYPES OF ASSEMBLIES

Two types of component assemblies are possible, and irrespective of how involved an assembly may appear, it can always be analysed as one or the other of the following types:

1. An *external assembly* is a combination of two or more components which, when added together dimensionally, form an external overall dimension. For example, in Figure 1.86 components B, C and D form assembly A and the dimensions b, c and d respectively add together to give the assembly dimension a.

2. An *internal assembly* comprises a combination of one or more components added together to fit the internal dimension of the final component of the assembly. For example, in Figure 1.87 components B and C fit into component D to form assembly A. The type of fit (clearance or interference) of the assembly will determine the individual dimensions of b, c and d.

**FIGURE 1.86** ▶ External assembly

**FIGURE 1.87** ▶ Internal assembly

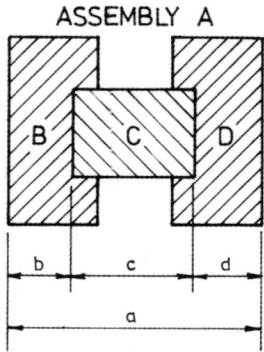

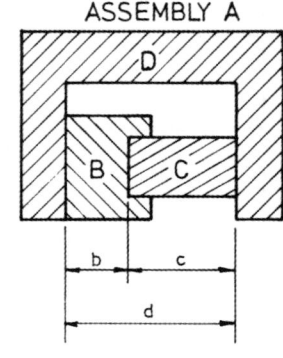

## COMPONENTS ASSEMBLED EXTERNALLY

Consider assembly A in Figure 1.88, which consists of three components B, C and D having dimensions b, c and d respectively with values of upper and lower limits of size as indicated. The upper and lower limits of the assembly dimension, a, are found by adding the upper and lower limits of the individual dimensions b, c and d:

10.05 + 20.1 + 10.05 = 40.2 (upper limit)

and 9.95 + 19.9 + 9.95 = 39.8 (lower limit)

It can also be seen that the tolerance of the assembly dimension a is equal to the sum of the individual dimension (b, c and d) tolerances:

| b | c | d | a |
|---|---|---|---|
| 10.05 | 20.1 | 10.05 | 40.2 |
| − 9.95 | −19.9 | − 9.95 | −39.8 |
| 0.1 + | 0.2 + | 0.1 = | 0.4 (assembly tolerance) |

**FIGURE 1.88** ▶ External assembly

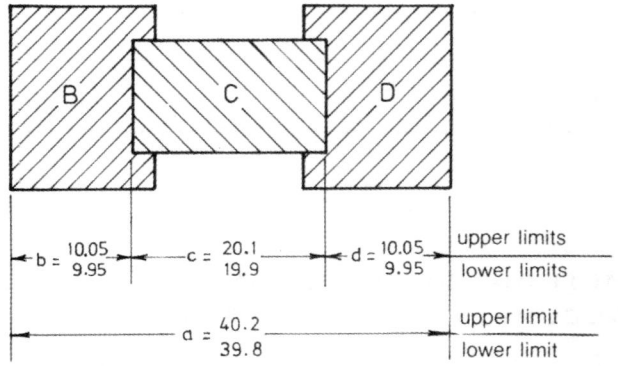

## COMPONENTS ASSEMBLED INTERNALLY

### Case 1

Consider assembly A in Figure 1.89, which consists of three components B, C and D having dimensions b, c and d respectively with values of upper and lower limits of size as indicated. It is necessary to determine the maximum (upper) and minimum (lower) limits of clearance between the three components.

The maximum clearance is found by subtracting the minimum combined sizes (lower limits) of components B and C from the maximum opening size (upper limit) of D:

35.5 − (14.95 + 19.9) = 0.65 (upper limit)

The minimum clearance is found by subtracting the maximum combined sizes (upper limits) of components B and C from the minimum opening size (lower limit) of D:

35.3 − (15.05 + 20.1) = 0.15 (lower limit)

In this case a positive clearance always results for all possible sizes of the three components.

FIGURE 1.89 ▶ Internal assembly—Case 1

ASSEMBLY A

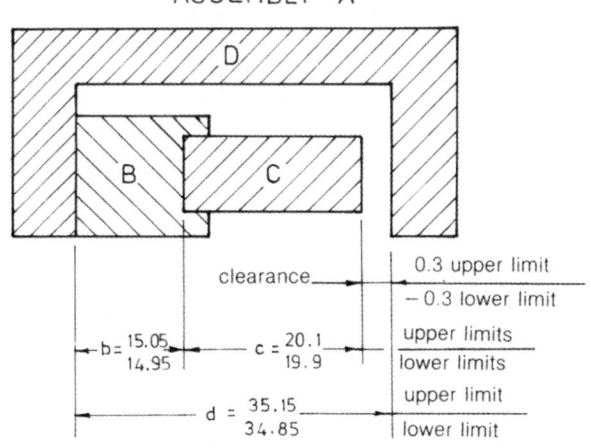

*Case 2*

This is similar to case 1, but dimension d has reduced limits. It is necessary to determine the maximum (upper) and minimum (lower) limits of clearance between the three components of assembly A shown in Figure 1.90.

The maximum clearance is found by subtracting the minimum combined sizes (lower limits) of components B and C from the maximum opening size (upper limit) of opening D:

$35.15 - (14.95 + 19.9) = 0.3$ (upper limit)

The minimum clearance is found by subtracting the maximum combined sizes (upper limits) of components B and C from the minimum opening size (lower limit) of opening D:

$34.85 - (15.05 + 20.1) = -0.3$ (lower limit)

The lower limit of the clearance is negative, so in fact the fit in this case ranges from 0.3 clearance at one extreme to 0.3 interference at the other extreme.

FIGURE 1.90 ▶ Internal assembly—Case 2

ASSEMBLY A

**FIGURE 1.91**

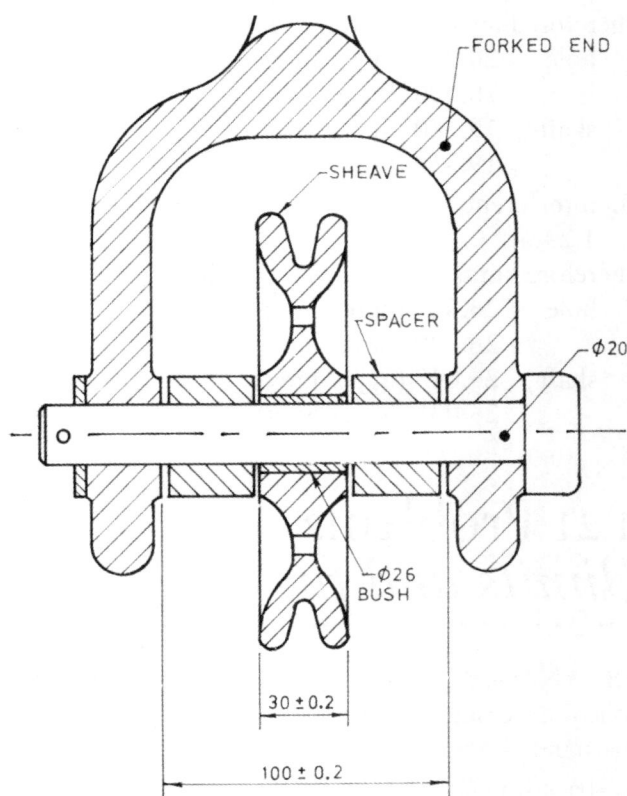

required to give a maximum and minimum total clearance of 1.50 and 0.50 mm respectively between the forked end, spacers and rope sheave.

Determine:

1. the upper and lower limit of size of each spacer
2. the limits of size of the fit of the bush and the spacers on the pin if a normal running fit is required
3. the fit of the non-ferrous bush in the sheave

Let X = upper limit of each spacer
    Y = lower limit of each spacer

1.  maximum = maximum opening –
    clearance      (minimum sheave +
                         2 × smallest spacer)
         1.50  = 100.20 – (29.80 + 2Y)
               = 100.20 – 29.80 – 2Y
         2Y   = 100.20 – 29.80 – 1.50
               = 68.9
          Y   = 34.45 (lower limit)

    minimum = minimum opening –
    clearance      (maximum sheave +
                         2 × largest spacer)
         0.50  = 99.80 – (30.20 + 2X)
               = 99.80 – 30.20 – 2X
         2X   = 99.80 – 30.20 – 0.50
               = 69.10
          X   = 34.55 (upper limit)

*Example*

A rope sheave block assembly is shown in Figure 1.91. Two spacers of equal widths and tolerance are

2. normal running fit = H8/f7 (Table 1.24(a))
therefore limits of size for 20 mm diameter are

    hole    20.033 (upper limit)
             20.000 (lower limit)
    shaft   19.980 (upper limit)
             19.959 (lower limit)

3. interference fit for non-ferrous = H7/s6 (Table 1.24(a))

therefore limits of size for 26 mm diameter are

    hole    26.021 (upper limit)
             26.000 (lower limit)
    shaft   26.048 (upper limit)
             26.035 (lower limit)

# 1.21 Problems (*limits and fits*)

**1.1**    Name the type of fit designated in each of the following cases, and write down the maximum and minimum clearance or interference as the case may be.

(a) basic size 65 mm, fit H7/g6, fit G7/h6
(b) basic size 284 mm, fit H7/p6, fit P7/h6
(c) basic size 25 mm, fit H7/k6, fit K7/h6

**1.2**    Write down values of the allowance for each of the six fits in question 1.1.

**1.3**    Give values of each fundamental deviation for both shafts and holes in the fits designated as follows:

(a) basic size 300 mm, fit H9/e9, fit E9/h9
(b) basic size 5 mm, fit H7/k6, fit K7/h6
(c) basic size 85 mm, fit H7/s6, fit S7/h6

**1.4**    A fit is specified as H9/e9 using the unilateral hole-basis system.

Specify the same fit using the unilateral shaft-basis system.

Using a basic size of 100 mm, write down both limits of size for the shaft and hole in each case.

**1.5**    A housing is to be bored out for a 50 mm outside diameter roller bearing. Name and designate the fit to be used, giving values for the upper and lower limits of size of the housing.

**1.6**    (a) Make a fully dimensioned detailed drawing or sketch of the bush shown in Figure 1.92. The method of tolerancing should be consistent throughout. (scale 2:1)
(b) Show separately the limits for the mating member in each case. What is the maximum and minimum clearance or interference in each case?

**FIGURE 1.92**

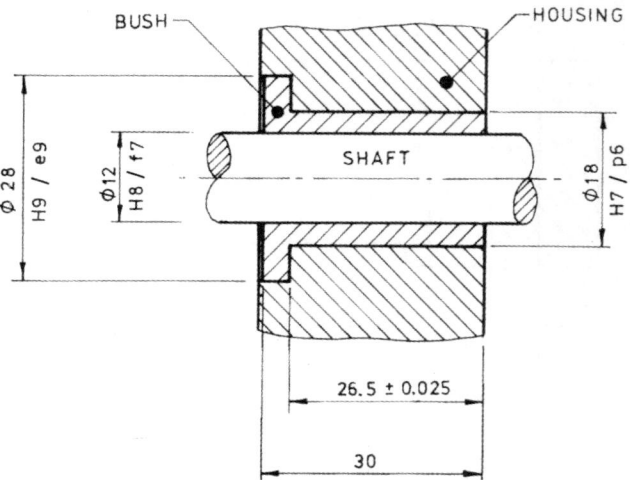

**1.7**    Figure 1.93 shows a knuckle joint consisting of a fork, a rod and a 10 mm diameter pin. The rod, which has a nominal width of 20 mm, is to have a loose clearance fit in the fork. The pin has a fit in the fork and rod designated by H/g6.

(a) What are the values of the maximum and minimum clearances for the fit of the rod into the fork?
(b) What are the limits of size for the pin and the pin holes in the rod and fork?
(c) What are the maximum and minimum amounts of relative lengthwise movement between the fork and rod resulting from the tolerances for the pin and its associated holes?

**FIGURE 1.93**

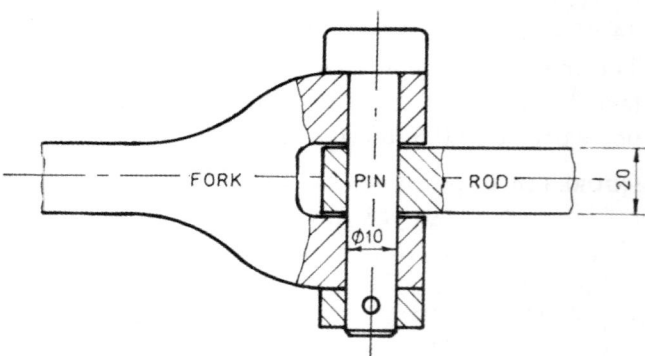

**1.8**    A 100 mm basic size shaft is to have the following five clearance fits located within its length. It is desirable to turn the shaft to one diameter for reasons of uniformity and ease of turning. What system can be used in order to accomplish this, and within what limits can the shaft be turned in order to achieve all of the fits?

D10/h9, E9/h9, F8/h7, G7/h6, H7/h6

**1.9**    The pulley assembly shown in Figure 1.94 has various fits designated. Scale off the correct basic sizes for these fits, determine both the hole and shaft limits in each case, and insert your answers in the table provided.

| FIT | HOLE LIMITS | | SHAFT LIMITS | |
|---|---|---|---|---|
| | UPPER | LOWER | UPPER | LOWER |
| H7/h6 | | | | |
| H7/p6 | | | | |
| H8/f7 | | | | |
| H9/d10 | | | | |
| H7/h6 | | | | |

**FIGURE 1.94**

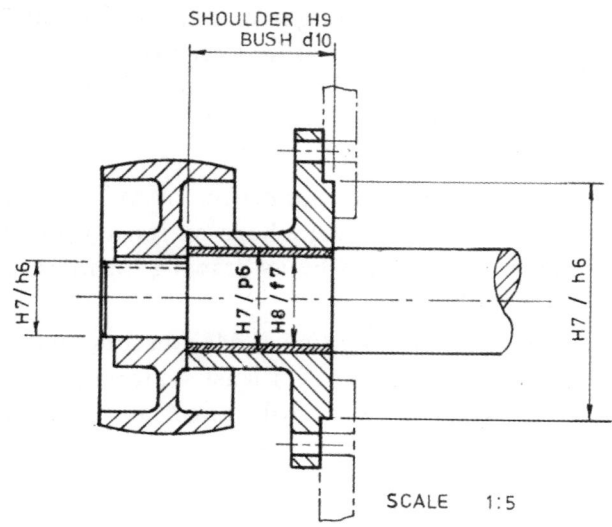

**1.10** Determine the maximum and minimum limits of size of the clearance X on the dog clutch shown in Figure 1.95.

**FIGURE 1.95**

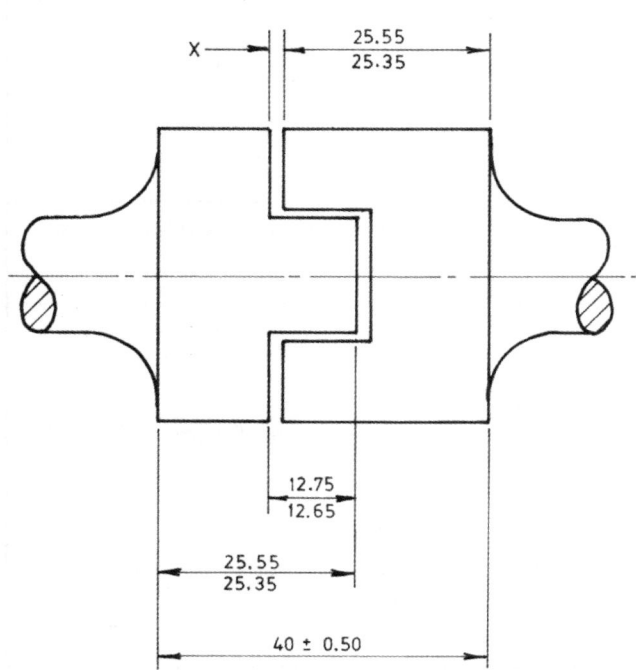

**1.11** The hole is assembled on the pin in Figure 1.96. Determine:

(a) the maximum and minimum distance X

(b) the maximum and minimum distance between surfaces A and B

**FIGURE 1.96**

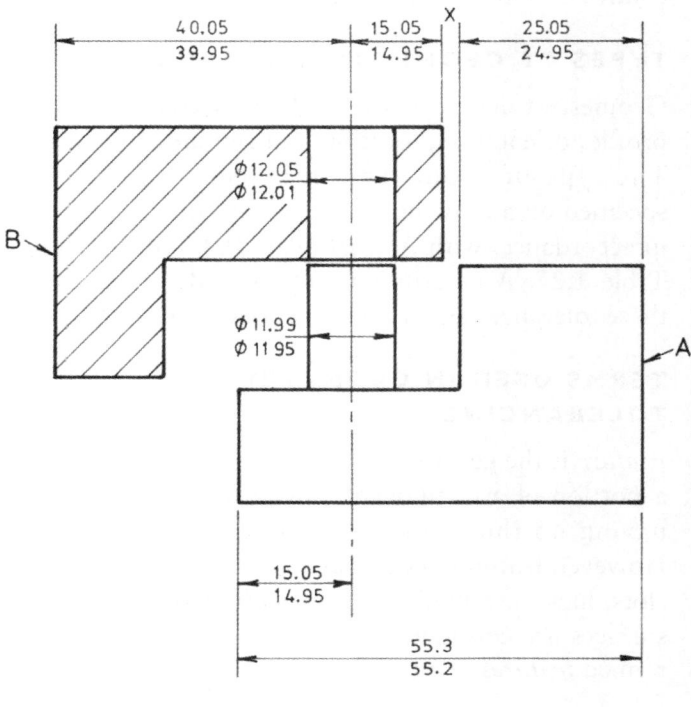

# 1.22 Geometry tolerancing

## INTRODUCTION

Size tolerancing is concerned with the sizing of dimensions. It facilitates producing elements of components (such as lengths, diameters, bores, recesses, keyways, etc.) as economically as possible while ensuring that when the component is produced and put to use it will be functional. However, size tolerancing takes no account of errors which may occur in the geometrical shape or form of the elements, and if such errors are present on a component to an excessive degree it can be rendered useless. For example, a shaft which may be within tolerance as far as the diameter dimension is concerned is quite useless if it is not acceptably straight within its length. The straightness of the shaft is a property imparted to it by the machining process (turning, grinding, etc.) which produced it.

In the aerospace, automobile and machinery manufacturing industries, where components are mass produced and interchangeability is essential, the control of both size dimensions and the geometrical shape of critical features is of prime importance to the design engineer and production controller. Just as important, size dimensions are allocated a size tolerance, so too are certain geometric features allocated a *geometry tolerance*. The application of both types of tolerances allows the designer complete control of both size and shape of components.

## TYPES OF GEOMETRY TOLERANCES

Geometry tolerances are used to specify the form, profile, orientation, location and runout of features. The type of geometry tolerance to be used is specified on a drawing by the use of symbols applied in accordance with AS 1100 Part 101 and shown in Table 1.27. A description and methods of applying these tolerances are given in the following pages.

## TERMS USED IN GEOMETRY TOLERANCING

*Feature* is the general term used to identify part of or a portion of a component. Single surfaces and lines having no thickness cannot have a feature size. However, features such as cylinders (shafts or holes), slots, lugs, rectangular parts (where two parallel flat surfaces are considered to form a single feature) are termed *features of size*.

In tolerancing features of size on mating components where ease of assembly is important, it should be realised that the least favourable condition for assembly occurs when the mating sizes are at the maximum material size allowable by the individual tolerance of each component. Greater variations in shape geometry can be accepted as the mating sizes approach their least material size.

Consider the pin and bush assembly with individual toleranced sizes shown in Figure 1.97(a). If both parts were everywhere at their maximum material size (largest shaft—smallest hole) of Ø 25.00 mm (Fig. 1.97(b)) each part would have to be perfectly round and straight in order to assemble. However, if the pin was at its least material size of Ø 24.98 mm, it could be bent up to 0.02 mm and still assemble with the smallest hole of Ø 25.00 mm (Fig. 1.97(c)).

*Maximum material condition* (MMC) occurs when a feature is everywhere at its maximum material size as allowed by its drawing tolerance, as in Figure 1.97(b).

*Least material condition* (LMC) occurs when a feature is everywhere at its least material size as allowed by its drawing tolerance, as in Figure 1.97(c).

**TABLE 1.27** ▶ Symbols for geometric characteristics

| GEOMETRIC CHARACTERISTIC | SYMBOL | TYPE OF TOLERANCE | APPLICATION |
|---|---|---|---|
| **Straightness** | —— | Form | For individual features |
| **Flatness** | ▱ | | |
| **Circularity** | ◯ | | |
| **Cylindricity** | ⌭ | | |
| **Profile of a line** | ⌒ | Profile | For individual or related features |
| **Profile of a surface** | ⌓ | | |
| **Angularity** | ∠ | Orientation | For related features |
| **Perpendicularity** | ⊥ | | |
| **Parallelism** | // | | |
| **Position including concentricity and symmetry** | ⊕ | Location | |
| **Circular runout** | ↗ | Runout | |
| **Total runout** | ↗↗ | | |

**FIGURE 1.97** ▶ Effect of combining linear and geometry tolerances

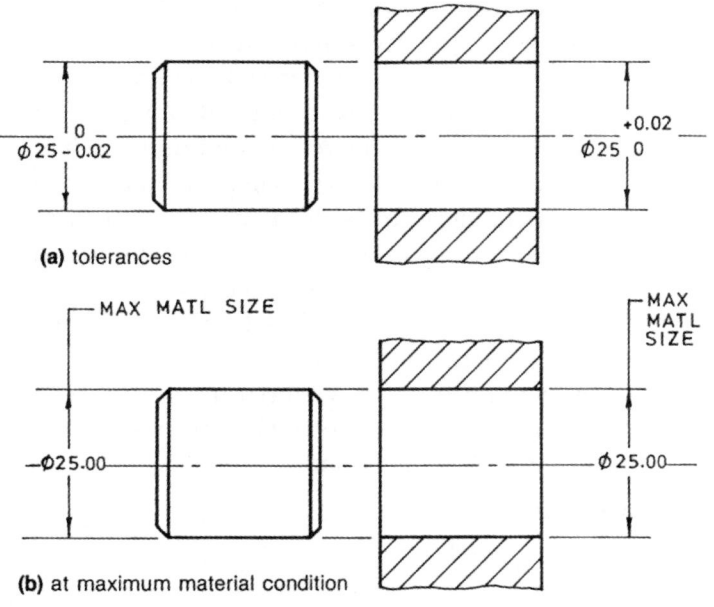

**(a)** tolerances

**(b)** at maximum material condition

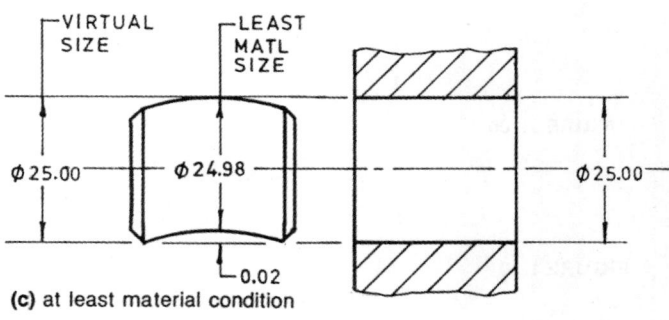

**(c)** at least material condition

*Virtual size* refers to the dimension of the overall envelope of perfect form which touches the highest points of a feature.

1. For a *shaft*, it is the maximum measured size plus the actual effect of form variations such as flatness, straightness, roundness, cylindricity and profile. For example, the bent pin of Figure 1.97(c) has a virtual size of 25.00 mm, which is the overall envelope size comprising the maximum measured size of 24.98 mm plus the straightness variation of 0.02 mm.

2. For a *hole*, it is the minimum measured size minus the actual effect of form variations such as flatness, straightness, roundness, cylindricity and profile. Figure 1.98 shows the virtual sizes of two individual holes.

*Datum* is a point, line, plane or other surface from which dimensions are measured or to which geometry tolerances are referenced. For measuring or manufacturing purposes a datum has an exact form and represents a fixed location.

*Datum feature* is a feature of a part such as an edge, surface or hole which forms the basis for a datum or is used to establish the location of a datum.

*Maximum material principle* recognises the fact that the allowable errors in geometry of two mating features may be allowed to increase as the size of the features decreases from the maximum material condition to the minimum material condition. Such allowance for geometry variation is governed by the symbol Ⓜ, which signifies control at the maximum material condition. The following are common uses of the symbol:

1. When the symbol Ⓜ is not used (Fig. 1.99), the allowable geometry tolerance applies regardless of feature size, and no relationship is intended to exist between the feature size and the geometry tolerance.

2. When the symbol Ⓜ is included with the geometry tolerance (Fig. 1.100), it means that when the feature is at its maximum material condition an extra geometry tolerance (0.05) may be allowed.

**FIGURE 1.98** ▶ virtual size of holes

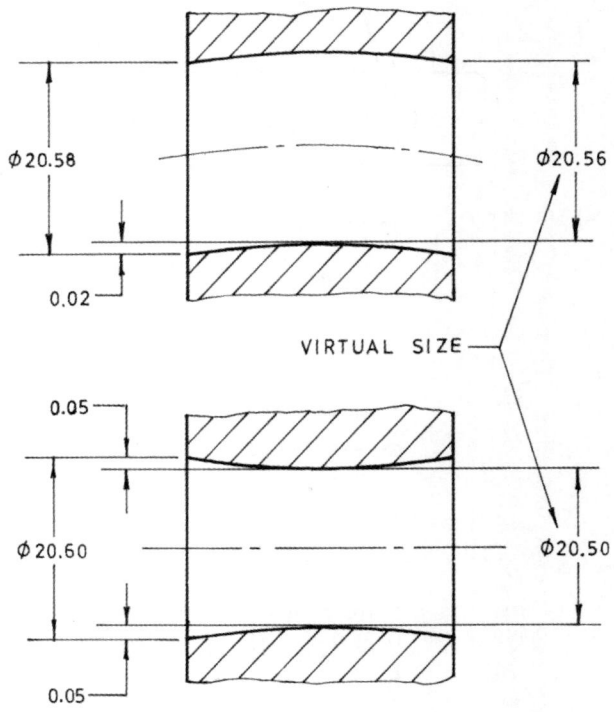

**FIGURE 1.99**

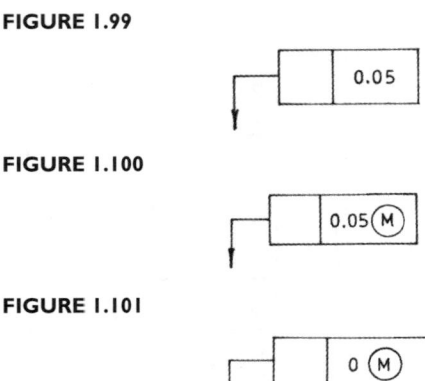

**FIGURE 1.100**

**FIGURE 1.101**

3. When the symbol Ⓜ is included with a geometry tolerance of zero (Fig. 1.101), it means that at the maximum material size the feature must be of perfect geometric form, but as the size decreases within tolerance the size difference may be allocated to the geometry error.

4. When the symbol Ⓜ is used with the hole positioning symbol (Fig. 1.102(a)), then at the correct position of the holes at their maximum material condition a position tolerance zone (0.05) is allowable for each hole axis (Fig. 1.102(b)). If, however, the holes are both at their least material condition, a tolerance zone equal to the sum of the hole tolerance (0.1) and the position tolerance (0.05) is allowable, giving a total (0.15) in which the axis of the hole must lie (Fig. 1.102(c)).

**FIGURE 1.102**

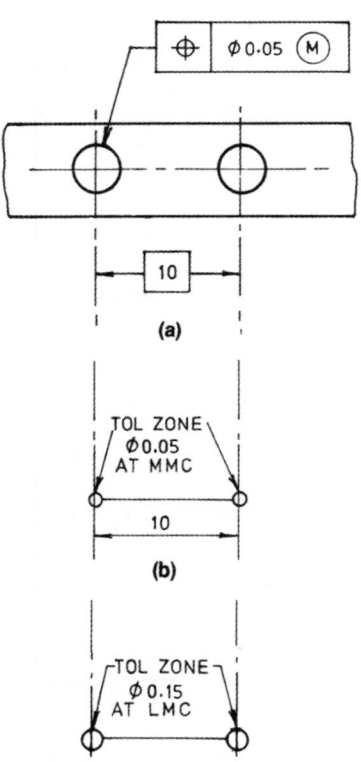

5. When specifying a geometry tolerance referred to a datum which is a feature of size, the checking of the feature is greatly simplified if the datum is specified on a maximum material condition basis because it permits the use of a fixed GO gauge to check the maximum material condition. Such is the case in Figure 1.103, where at maximum material condition of both diameters the axis of the larger diameter may vary in position displacement from the axis of the small diameter by an amount equal to 0.025 (0.05 ÷ 2) in any direction. At the minimum material condition of both diameters the variation in position displacement increases to 0.035 [(0.05 + 0.01 + 0.01) ÷ 2] in any direction.

**FIGURE 1.103**

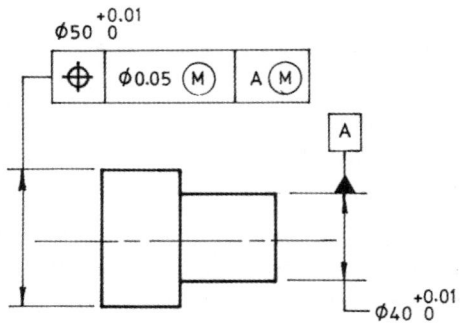

# 1.23 Methods of displaying geometry tolerances

Two methods are used to display geometry tolerances on a drawing:

1. the tolerance frame method
2. the tabular method

Method 1 is used when there are no more than three simple groups of geometry tolerances.

Method 2 is used when the group or groups of geometry tolerances are complex or are more than three in number. This method is not described in this book, but may be found in AS 1100 Part 101.

## TOLERANCE FRAME METHOD

The tolerance frame comprises a number of boxes. The frame should always be readable from the bottom of the drawing, and only those boxes necessary to complete the tolerance description are used. The symbol box and the tolerance value box

72

are always shown, and are the minimum necessary to state a geometry tolerance (Table 1.28 no. 1 (page 74)).

## DATUM FEATURE

A datum feature is a part of a component drawing (usually a surface or an axis) used as a reference from which to establish a required geometry tolerance. The datum is normally identified by an upper case letter in conjunction with the symbols shown (Table 1.28 no. 2).

# 1.24 Basic concepts of geometry tolerancing

Geometry tolerances are specified according to functional requirements. Methods of manufacture and inspection can also influence geometry tolerancing. The indication of geometry tolerances on a drawing should not imply the use of any particular method of production, measurement or gauging.

A *feature* is a specific part of a component, such as a point, line or surface. The feature can be an integral part of the component (for example, a flat surface) or a derived part (for example, the axis of a cylinder). A geometry tolerance applied to a feature defines the tolerance zone within which the feature shall be contained.

Depending on the feature to be toleranced and the way in which it is dimensioned, the tolerance zone will be one of the following:

- the space between two equidistant lines or two parallel straight lines
- the space within a circle
- the space between two concentric circles
- the space within a cylinder
- the space between two co-axial cylinders
- the space between two equidistant surfaces or two parallel planes
- the space within a sphere.

A toleranced feature may take any form or orientation within the tolerance zone, unless it is necessary to apply some restriction which is indicated by an explanatory note. A geometry tolerance applies to the whole extent of the indicated feature, unless otherwise specified.

Geometry tolerances assigned to features related to a datum do not limit the deviations of the datum feature itself. It may be necessary to specify tolerances of form for the datum feature as well.

# 1.25 Interpretation of geometry tolerancing

## FLATNESS

A flatness tolerance is used to control the flatness of a surface. The tolerance zone is a specified space separating two parallel planes between which the surface to be controlled must lie (Table 1.28 no. 3).

## STRAIGHTNESS

A straightness tolerance may be used to control:

1. straightness of a line
2. straightness of an axis in a single plane
3. straightness of the axes of solids of revolution

In case 1, the tolerance zone is a specified distance separating two parallel straight lines between which the line to be controlled must lie (Table 1.28 no. 4).

In case 2, the tolerance zone is a specified space separating two parallel straight lines between which the plane containing the surface must lie (Table 1.28 no. 5).

In case 3, the tolerance zone is a cylinder of specified diameter within which the axis of revolution of the cylinder must lie (Table 1.28 no. 6).

## PERPENDICULARITY

A perpendicularity tolerance is used to control the squareness of:

1. a line with respect to a datum line
2. an axis with respect to a datum plane
3. a surface with respect to a datum axis
4. a surface with respect to a datum plane

In case 1, the tolerance zone is a specified distance separating two parallel planes which are normal to the datum line. The line to be controlled must lie between the two parallel planes (Table 1.28 no. 7).

In case 2, the tolerance zone is a specified space separating two parallel planes which are normal to the datum plane. The axis to be controlled must lie between the two parallel planes or within a cylinder (Table 1.28 nos 8 and 9).

In case 3, the tolerance zone is a specified space separating two parallel planes which are normal to the datum axis. The surface to be controlled must lie between the two parallel planes (Table 1.28 no. 10).

In case 4, the tolerance zone is a specified space separating two parallel planes which are normal to the datum plane. The surface to be controlled must lie between the two parallel planes (Table 1.28 no. 11).

**TABLE 1.28** ▸ Symbols related to geometry tolerancing

| SYMBOL | DRAWING CALLOUT | INTERPRETATION |
|---|---|---|
| **frame**<br>2.5h<br>as needed | **1.**<br>feature identification (if required)<br>tolerance symbol<br>tolerance value<br>A<br>1 ⊕ Ø 0.1 B<br>group number (if required)<br>datum feature(s) (if required) | If the feature and group indentifications are not required, the left-hand and datum compartments may be omitted. |
| **feature**<br>2.5h<br>2.5h<br><br>**datum**<br>h  h | **2.**<br>A  or  A | The surface is identified as datum feature A. Either datum symbol may be used, but never both on the one drawing. |
| **flatness**<br>1.4h<br>1.4h / 60° | **3.**<br>▱ 0.02 | The surface indicated must lie between two parallel planes 0.02 mm apart.<br>0.02 |
| **straightness**<br>2h | **4.**<br>— 0.02 | Any actual line on the top surface, parallel to the plane of projection in which the tolerance frame is shown must lie between two horizontal parallel lines 0.02 mm apart.<br>0·02 |
| | **5.**<br>— 0.02 | Any position of the generator of the cylinder must lie between two parallel straight lines on the surface 0.02 mm apart.<br>0.02 |
| | **6.**<br>— Ø0.02<br>A | The axes of cylinder A only must lie within a cylindrical tolerance zone of 0.02 mm diameter.<br>Ø0.02 |

**TABLE I.28** ▶ Symbols related to geometry tolerancing (continued)

| SYMBOL | DRAWING CALLOUT | INTERPRETATION |
|---|---|---|
| **perpendicularity** | **7.** | The axis of the threaded hole must lie between two horizontal planes 0.05 mm apart, normal to the hole axis, datum A. |
| | **8.** | The axis of the upright member must lie between the intersection of four vertical planes mutually at right angles to datum plane A. |
| | **9.** | The axis of the cylindrical member must lie within a cylinder, 0.04 mm diameter, the axis of which is normal to the datum surface A. |
| | **10.** | The left-hand end of the part must lie between two planes 0.05 mm apart which are normal to the shaft axis, datum A. |
| | **11.** | The outside vertical face of the part must be between two planes 0.04 mm apart which are normal to the bottom surface, datum A. |

TABLE 1.28 ▶ Symbols related to geometry tolerancing (continued)

| SYMBOL | DRAWING CALLOUT | INTERPRETATION |
|---|---|---|
| **position** | **12.** | The axis of each hole must be contained within a cylinder 0.05 mm diameter through the thickness of the material, each cylinder having its axis in the true position specified by the dimensions. |
| | **13.** | The axis of each hole must be contained within a cylinder 0.05 mm diameter through the thickness of the material, each axis being in its true position in relation to the datum plane A and datum hole B. |
| | **14.** | The back vertical surface must lie between two parallel planes 0.05 mm apart and be symmetrically located about the true position of the surface in relation to the datum axis A and the datum plane B. |
| **position (concentricity)** | **15.** | The axis of the right-hand end of the shaft must lie within a cylinder 0.06 mm diameter which is co-axial with datum A, the axis of the left hand end of the shaft. |
| | **16.** | The axes of the left-hand and right-hand cylindrical ends of the shaft must lie within a cylinder 0.06 mm diameter. |
| | **17.** | The axis of the middle section of the shaft must lie within a cylinder 0.05 mm diameter which is co-axial with the common datum axis of the ends C and D. |

TABLE 1.28 ▶ Symbols related to geometry tolerancing (continued)

| SYMBOL | DRAWING CALLOUT | INTERPRETATION |
|---|---|---|
| **position (symmetry)** ⊕ | **18.** | The axis of the hole must lie between two parallel planes 0.05 mm apart which are equally spaced about the common median plane of the two end slots C and D. |
| | **19.** | The example is similar to the above except that the hole axis must comply individually with each of the two tolerances in two directions mutually at right angles. |
| | **20.** | The median plane of the tongue must lie between two parallel planes 0.05 mm apart which are equally spaced about the median plane of the datum width A. |
| | **21.** | The two median planes of the tongue on each end must lie between two parallel planes 0.04 mm apart. |
| **cylindricity** | **22.** | The cylindrical surface of the part must lie between two cylindrical surfaces co-axial with each other and a radial distance of 0.03 mm apart. |

**TABLE I.28** ▶ Symbols related to geometry tolerancing (continued)

| SYMBOL | DRAWING CALLOUT | INTERPRETATION |
|---|---|---|
| **profile (line)** <br> 1.4h | **23.** <br> R5  20  0.04 <br> BILATERAL TOLERANCES | The actual profile must lie between two profiles separated by a series of spheres 0.04 mm diameter whose centres lie on the theoretical profile. <br><br> CIRCLES ⌀0.04 |
| **(surface)** <br> 1.4h | **24.** <br> 0.03  BILATERAL TOLERANCES <br> spherical surface profile | The curved surface must lie between two surfaces separated by a series of spheres 0.03 mm diameter having their centres on a surface of correct profile shape. <br><br> SPHERES ⌀0.03 |
| | **25.** <br> R14  1.0 A B  A <br> 25 B <br> shape profile | The actual profile must lie within the 1 mm tolerance zone centred around the true profile. <br><br> R14  1.0 <br> True profile  25 <br> Datum B |
| | **26.** <br> A B 0.4  B  A <br> ⌀25.8/25.0  40  20° <br> conical feature profile | The actual conical profile must lie within an 0.4 mm diameter tolerance zone limited by the toleranced dimension. <br><br> 0.4  40  Datum B <br> ⌀25.8 MAX  ⌀25.0 MIN  20°  Datum Axis A |
| | **27.** <br> 0.5 A  ⌀13 <br> ⌀8.0/7.8 <br> 6 × R0 2MAX  M8 × 1.25 <br> 6 × 60°  A <br> polygon feature profile | The actual hexagon profile must lie within the 0.5 mm tolerance zone centred around the true profile. <br><br> 0 5 wide tolerance zone <br> One possible hexagon contour <br> Axis–Datum A <br> True profile <br> Enlarged View |

TABLE 1.28 ▶ Symbols related to geometry tolerancing (continued)

| SYMBOL | DRAWING CALLOUT | INTERPRETATION |
|---|---|---|
| **angularity** | 28. | The axis of the inclined hole must lie within a cylindrical tolerance zone 0.05 mm diameter which is inclined at 45° to the datum plane B in one direction and normal to it in the direction 90° to the first. |
| | 29. | The inclined surface of the part must lie between two parallel planes 0.04 mm apart which is inclined at 30° to the datum axis D in one direction and normal to it in the direction 90° to the first. |
| | 30. | The inclined surface of the part must lie between two parallel planes 0.06 mm apart which are inclined at 30° to the datum plane A in one direction and elements at right angles to this direction are parallel to datum plane A. |
| **parallelism** | 31. | The axis of the left-hand hole must lie between two sets of parallel planes spaced according to the dimensions given, mutually at right angles and placed as shown in relation to the hole axis, datum A. |
| | 32. | The axis of the left-hand hole must lie inside a cylindrical tolerance zone 0.03 mm diameter which has its axis parallel to the other hole axis, datum A. |
| | 33. | The top surface must lie between two parallel planes 0.3 mm apart and parallel to the hole axis, datum A. |

**TABLE 1.28** ▶ Symbols related to geometry tolerancing (continued)

| SYMBOL | DRAWING CALLOUT | INTERPRETATION |
|---|---|---|
| **parallelism** | **34.** | The two top surfaces must each lie between two parallel planes 0.3 mm apart, each set of parallel planes being parallel to the bottom surface, datum A. |
| **circularity** | **35.** | Any cross-section of the cylinder perpendicular to the axis must lie between two concentric circles of radial distance 0.02 mm apart. Note: The centre of the two concentric circles need not lie on the axis of the cylinder. |
| | **36.** | Any cross-section of the sphere passing through the centre must lie between two concentric circles in the same plane as the section and of radial distance 0.03 mm apart. Note: The centre of the two concentric circles need not coincide with the centre of the sphere. |
| **runout** | **37.** | The runout of the centre portion must not exceed 0.2 mm measured normal to the datum axis at any point along the surface. |
| **total runout** | **38.** | The total runout must not be greater than 0.3 mm in any traverse of the surface during a series of rotations about the common axis of ends C and D. |

## POSITION

A position tolerance is used to control the location of a feature by limiting its deviation from a specified true position.

Consider the commonly used co-ordinate method of tolerancing the hole centre from each side of the corner (Fig. 1.104(a)). The tolerance zone is a square of side 0.2 mm, which allows a maximum deviation of the centre along the diagonal of the square of $0.2 \times \sqrt{2} = 0.2828$ or $\pm 0.1414$ from the true centre position; this represents an increase of 0.0414 more than required.

The true position method stipulates that the maximum deviation of the centre from its true position is a circle of 0.2 mm diameter; thus the deviation is controlled in all directions around the true position (Fig. 1.104(b)).

The true position method may be used to control the true position of:

1. a hole axis (Table 1.28 nos 12 and 13)
2. a surface (Table 1.28 no. 14)

True position tolerancing should be applied when there are more than two features in a functional group, and the MMC (maximum material condition) principle should be specified where possible to make production of the features easier. MMC cannot be applied to a surface (case 2).

### POSITION (CONCENTRICITY)

A concentricity tolerance is used to control a condition in which two or more features such as circles, spheres, cylinders, cones or hexagons are required to share a common centre or axis.

A concentricity tolerance is a particular case of a positional tolerance. It controls the allowable variation in eccentricity of the axis of the feature being controlled, in relation to the axis of the datum feature, when the controlled feature and datum feature are meant to be concentric or coaxial.

The concentricity of various sections of a shaft that has two or more steps along its length is illustrated (Table 1.28 nos 15, 16 and 17).

### POSITION (SYMMETRY)

A symmetry tolerance is used to control a condition in which one or more features are symmetrically disposed either side of a centre line (axis) or centre plane (median) of another feature which is specified as the datum.

Symmetry tolerancing is a special case of position tolerancing. It has an advantage over the position symbol in that it indicates that the true position is symmetrical and thus eliminates the need for basic dimensions to interrelate the position of the features. It serves the same purpose on non-cylindrical features as concentricity serves on circular features.

The following cases are illustrated:

1. a hole axis is to be symmetrical, within tolerance, with a common surface represented by the median datum plane of two slot features (Table 1.28 no. 18), and similarly with two sets of median datum planes mutually at right angles (Table 1.28 no. 19)
2. a surface which represents the median plane of a feature is to be symmetrical, within tolerance, with the median plane of another feature as datum (Table 1.28 no. 20)
3. two surfaces representing a common median plane of two similar features are to be symmetrical, within tolerance (Table 1.28 no. 21)

### CYLINDRICITY

A cylindricity tolerance specifies a tolerance zone consisting of an annular space between two co-axial

**FIGURE 1.104** ▶ Methods of positional tolerancing hole centres

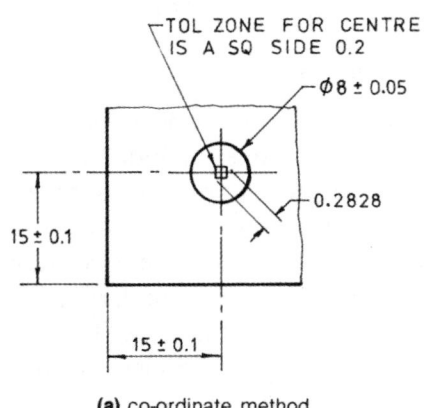

**(a)** co-ordinate method

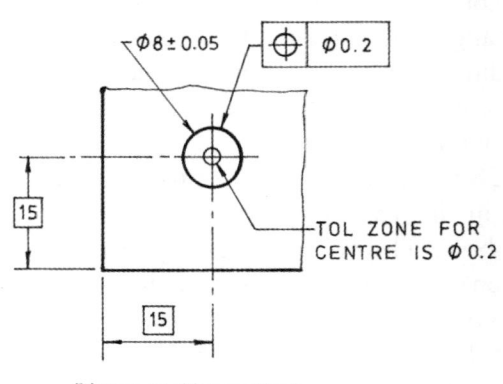

**(b)** true position method

cylinders having a difference in radii equal to the specified tolerance within which the entire cylindrical surface of the feature being controlled must lie.

Similarly to circularity, the axis of a cylindricity tolerance need not be co-axial with the axis of the cylindrical feature it controls (Table 1.28 no. 22).

## PROFILES

A profile tolerance may be applied to control the profile of a line (Table 1.28 no. 23) or the profile of a surface (Table 1.28 nos 24–27).

In the first case, the tolerance zone is a space between two lines which are drawn tangential to a series of spheres having their centres on the theoretically correct surface of the profile. This represents a bilateral (either side of centre) geometry tolerance. If a unilateral (all on one side of the theoretical profile) geometry tolerance is required it should be indicated on the drawing by a type J line (thick chain line) and dimensioned as in Figure 1.105.

**FIGURE 1.105** ▶ Unilateral profile geometry tolerance

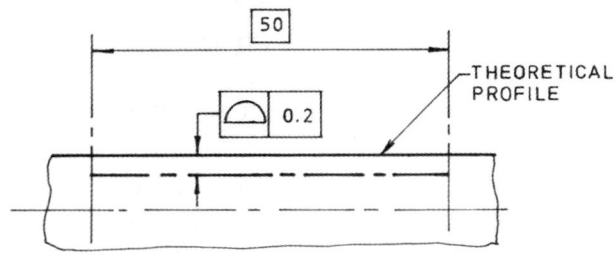

In the second case, Table 1.28 no. 24 represents the geometric control of a continuously changing surface contour, using bilateral tolerances (equally distributed about the correct profile).

Table 1.28 no. 25 shows how an extrusion with a semi-circular end is dimensionally controlled by specifying a uniform boundary of equal distribution within which the end surface must lie. The true profile is located relative to two reference datums.

Table 1.28 no. 26 shows a conical feature which has profile control of form and orientation. The cone has a basic angle and is related to datums referenced with basic dimensions. The tolerance zone applies all around the cone and is a uniform boundary equally disposed around the true profile.

Table 1.28 no. 27 illustrates the profile control of a polygon, in this case a hexagon representing the head of a bolt. The true hexagon profile has a tolerance zone referenced to the bolt diameter. The all around symbol extends the tolerance zone to all six sides of the hexagon. The tolerance zone limits the form, size, location and position of the hexagon.

## ANGULARITY

An angularity tolerance is used to control angular relationships of any angle, between straight lines (axes) or surfaces with straight line elements such as flat or cylindrical surfaces. The feature to be controlled may be a line (axis) or a surface, and the datum feature to which the controlled feature is referenced may also be a line (axis) or a surface.
The following cases are illustrated:

1. a hole axis is to be inclined to a datum surface by a specified angle in one direction and normal to it in another direction at right angles to the first (Table 1.28 no. 28)
2. a surface is to be inclined to a datum axis by a specified angle in one direction and normal to it in another direction at right angles to the first (Table 1.28 no. 29)
3. a surface is to be inclined to a datum surface by a specified angle in one direction and elements of both surfaces at right angles to this direction must be parallel (Table 1.28 no. 30)

## PARALLELISM

A parallelism tolerance is used to control the orientation of features related to one another by an angle of zero degrees. The feature to be controlled may be an axis or a plane surface, and the datum to which it is orientated may also be an axis or a plane surface. Generally the tolerance zone is the distance between two parallel planes which are parallel to the datum feature.
The following cases are illustrated:

1. an axis of a hole is to be parallel to the axis of another hole in two directions nominally at right angles (Table 1.28 no. 31) and in all directions (Table 1.28 no. 32)
2. a surface is to be parallel to the axis of a hole (Table 1.28 no. 33)
3. two separate surfaces are to be parallel to another surface (Table 1.28 no. 34)

## CIRCULARITY

A circularity tolerance specifies the width of an annular tolerance zone, bounded by two concentric circles in the same plane, within which the circumference of the feature must lie.

A circularity tolerance is not concerned with the position of the tolerance boundaries, for example its concentricity with a datum axis. This means that the centre of the concentric circles will not necessarily coincide with the axis (in the case of a cylinder) or

82

centre (in the case of a sphere) of the feature (Table 1.28 nos 35 and 36).

### RUNOUT

A runout tolerance represents the allowable deviation in position of a surface of revolution as a part is revolved about a datum axis.

There are two cases of runout: circular runout (usually referred to as 'runout') and total runout.

Runout concerns each circular element or cross-section and may be applied to cylinders (Table 1.28 no. 37), tapers and end surfaces where such a surface is at right angles to the axis of revolution.

Total runout is used to provide composite control of all the cross-sectional surface elements simultaneously. It also applies in the three cases stated above for runout (Table 1.28 no. 38).

# 1.26 Analysis of geometry tolerance on a drawing

The drawing of a cover plate (Fig 1.106, courtesy of Rolls Royce, page 84) illustrates the use of geometry tolerances together with linear tolerances to control the size and form of a cover plate designed to fit snugly into a housing and at the same time provide a fitting attachment base in the form of three equi-spaced threaded holes which are positioned with respect to datums A and B.

Datum B is the outside cylindrical surface indicated at maximum material condition (MMC), i.e. at $\varnothing69.991 + 0.009 = \varnothing70$. Datum A is the location surface of the attachment flange which is at right angles to datum B. The use of the $\textcircled{M}$ symbol means that at the maximum material condition of B at $\varnothing70$ the positions of the two sets of holes may vary 0.20 $(0.40 \div 2)$ in any direction. Note that each set of holes has a position tolerance of 0.40.

The drawing also illustrates the use of true position tolerancing with respect to the location of the sides of the angular flange, which has an unusual geometrical shape. In order to control the shape, the angles (18.5° and 40°) and sides locations (41.00 and 36.00) are dimensioned as true positions necessitating the use of geometry tolerances on the two sets of equi-spaced flats. The use of true position specification implies the absolute location of the flat surfaces of the flange and this must also be accompanied by geometry tolerancing of the flat features. True position tolerancing is also used to indicate the pitch circle diameters of the two sets of holes ($\varnothing50$ and $\varnothing86$).

Another measure of goemetrical control is that of the runout (Table 1.28 No. 37) of the surface which houses the three threaded holes, with respect to datums A and B to a tolerance of within 0.025. Such control is necessary to ensure that the true rotation of this surface and any fitting attached to it is within acceptable limits.

**Challenging Exercise** Draw a pictorial view of the cover plate either manually or using your CAD system.

# 1.27 Using geometry tolerances on a drawing

If a component drawing is to be complete in respect to tolerances generally, it should specify such size and geometry tolerances as are necessary to control the part completely in relation to its function. Geometry tolerances are used to control the form of component features such as straightness, perpendicularity, parallelism, circularity, position, etc.

### INTERPRETATION OF LIMITS OF SIZE FOR THE CONTROL OF FORM

If an individual feature is specified by means of limits of size without showing any geometry tolerances, then either of the following two interpretations apply:

*1. Limits of size with independency of size and form* In the normal situation when the limits of a feature are specified, the limits control only the size of the feature and provide no other control over its form.

This situation is known as the *principle of independency*, since the size and form are to be interpreted as being completely independent of one another.

Each specified size and geometry requirement on a drawing are met independently, unless a particular relationship is specified. Where no relationship is specified the geometry tolerance applies regardless of feature size and the two requirements are treated as unrelated. Consequently, if a particular relationship of size and form, or size and location, or size and orientation is required it must be specified on the drawing.

*2. Limits of size with dependency of size and form.* If an $\textcircled{E}$ symbol is attached to the limits of size of a feature, its size and form are both controlled. This situation is known as the *envelope principle* in which the size and form are to be interpreted as being dependent on one another.

FIGURE I.106

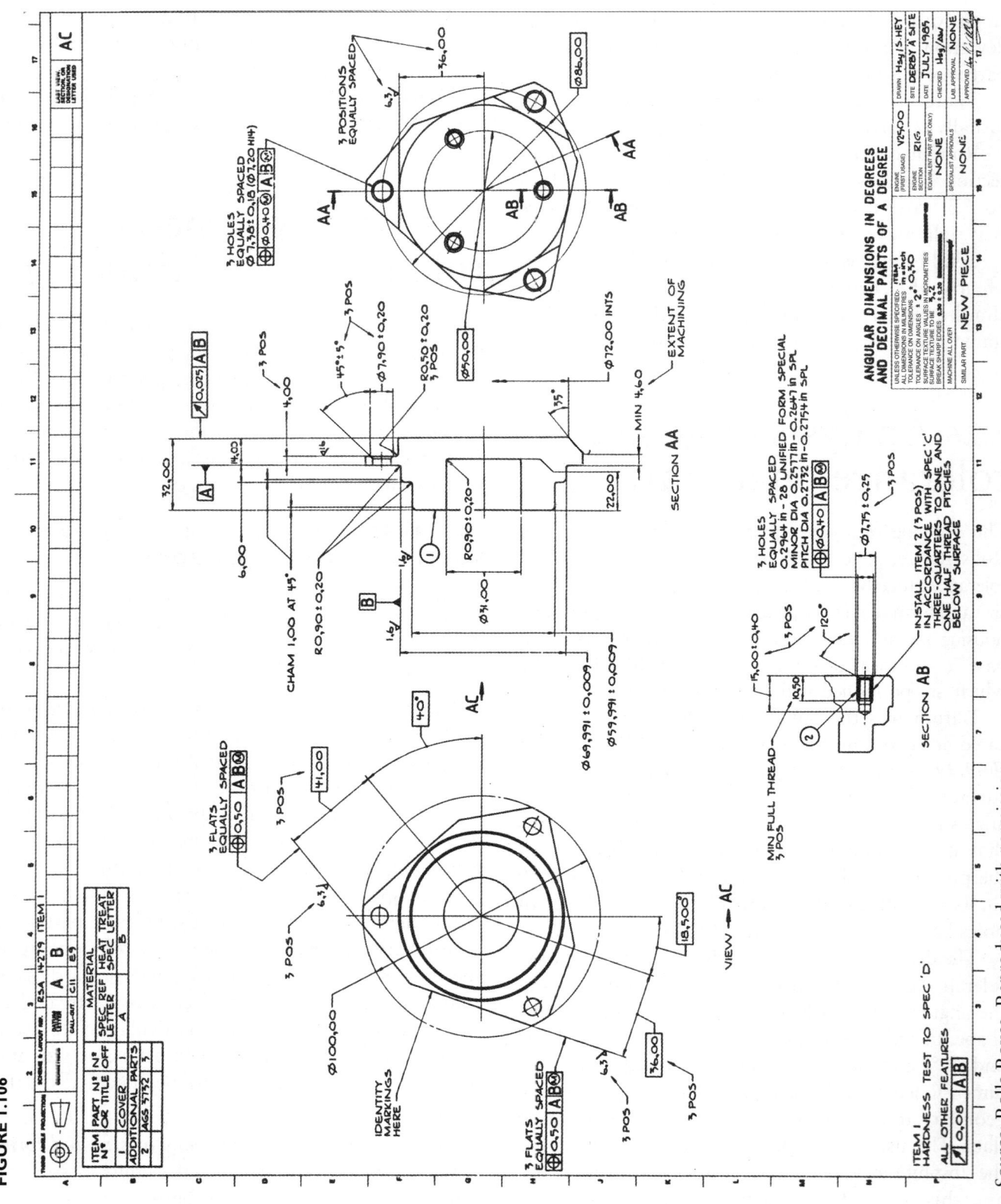

Source: Rolls-Royce. Reproduced with permission.

When the feature of size is defined by limits of size with an Ⓔ symbol attached, the maximum material limit of size (that is, the high limit of size for an external feature and the low limit of size for an internal feature) defines the boundary or envelope of perfect form for the relevant surface. If an individual feature of size is everywhere on its maximum material limit of size, it should be of perfect form.

With the Ⓔ specification, if the individual feature of size is not on its maximum material size, errors of form are permissible provided no part of the finished surfaces crosses the maximum material boundary or envelope of perfect form and the feature of size is everywhere in accordance with its specified limits of size.

For features of size having a geometry tolerance specification which includes an Ⓜ symbol in the geometry tolerance frame, there is also a dependency between the size and geometric form. This is discussed previously in the section on Maximum Material Principle, page 70–72.

# 1.28 Problems (*geometry tolerancing*)

**1.12** Apply geometry tolerances to the stepped shaft (Fig. 1.107) for the following cases:

(a) so that the common axis of the three cylinders is straight within 0.15 mm

(b) so that the axis of cylinder E only is straight within 0.15 mm

(c) so that the axes of cylinders F and G are straight within 0.15 mm

**FIGURE 1.107**

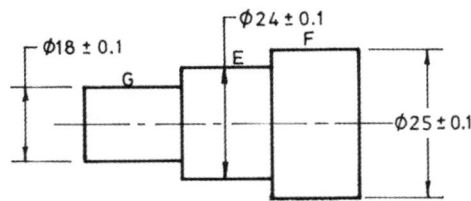

**1.13** With reference to the views of a rectangular bar (Fig. 1.108), sketch the tolerance zone which controls the axis.

**1.14** A shaft and hole assembly is designated as H8/f7 for a 25 mm shaft. What straightness tolerance may be applied equally to each feature at maximum material condition so that the clearance shall be not less than 0.01 mm?

**FIGURE 1.108**

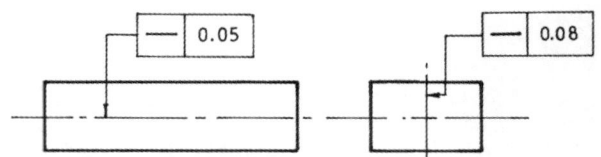

**1.15** Make a three-view orthogonal sketch of the part (Fig. 1.109) showing the following information:

(a) Surface A is a datum and must be straight within 0.2 mm over its length.

(b) Surface B is a datum and must be flat within 0.1 mm.

(c) Surfaces C and D are common datum features, and surface B is to be parallel to datum C-D within 0.1 mm.

**FIGURE 1.109**

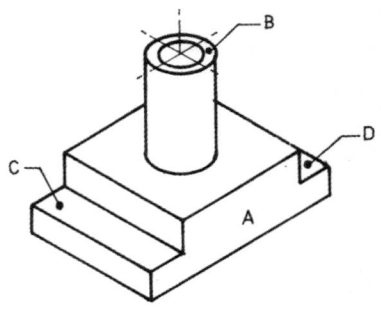

**1.16** Sketch two views of the part (Fig. 1.110, page 86) in good proportion, including the datums and geometry tolerances as indicated. Scale the drawing to obtain basic dimensions where required. Also indicate that the median plane of the slot is symmetrical with the median plane of the width, datum C, to within 0.05 mm.

**1.17** Sketch two views in good proportion of the part shown (Fig. 1.111) and include the following datums and geometry tolerances:

(a) The bottom is datum A.

(b) The back is datum B.

(c) The hole is perpendicular to the bottom within 0.05 mm.

(d) The back is perpendicular to the bottom within 0.08 mm.

(e) The top is parallel to the bottom within 0.1 mm.

(f) Surface C is to have an angularity tolerance of 0.15 mm with the bottom, and surface D is to be a secondary datum for this feature.

(g) The sides of the slot are to be parallel to each other within 0.1 mm.

**1.18** The hole shown (Fig. 1.112) is required to be positioned so that it never deviates from its true position by more than 0.14 mm in any direction at its maximum material condition.

FIGURE 1.110

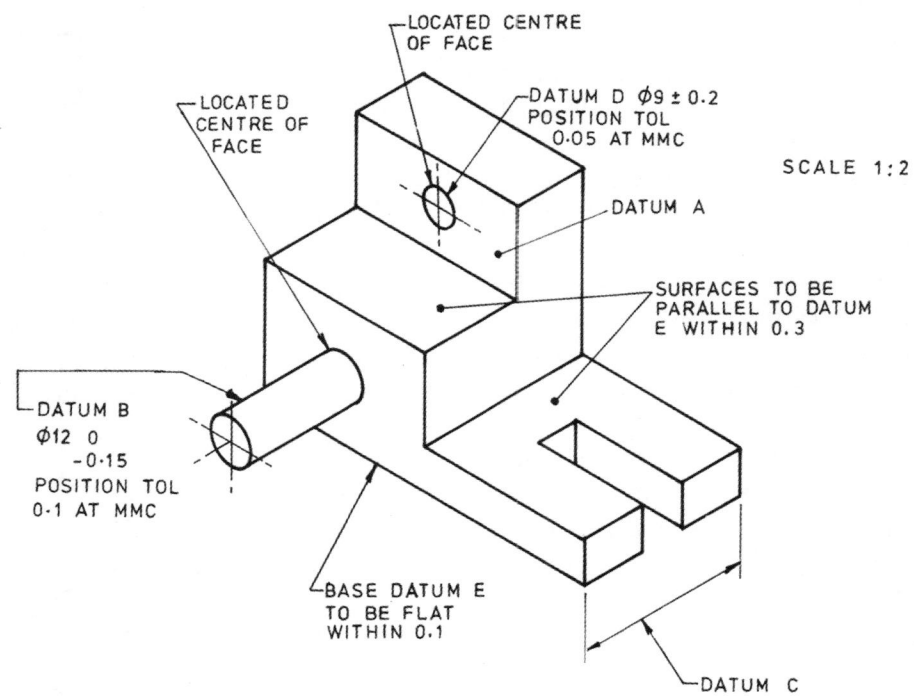

SCALE 1:2

(a) Show how this can be done in the following cases:
  (i) by co-ordinate tolerancing
  (ii) by positional tolerancing regardless of feature size (RFS)
  (iii) by positional tolerancing at maximum material condition
(b) What would be the maximum departure from the true position allowed if:
  (i) the hole was toleranced as stated in (a)(ii)?
  (ii) the hole was toleranced as stated in (a)(iii)?

**FIGURE 1.111**

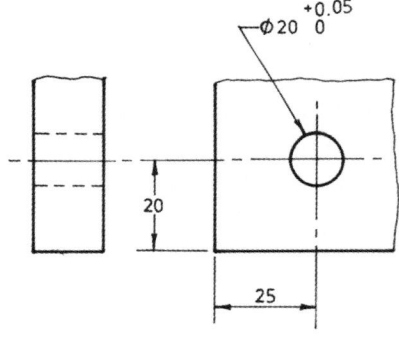

**FIGURE 1.112**

**1.19**

(a) Name three features to which a circularity tolerance can be applied.
(b) A circularity tolerance zone may cross the boundary of perfect form at the maximum or minimum material size. True or false?
(c) The centre axis of a circularity tolerance zone boundary always coincides with the centre axis of the feature. True or false?
(d) If a cylindrical part is mounted between centres and its surface checked for roundness using a dial indicator while the part is being revolved, the resulting readings are a true indication of circularity errors. True or false?
(e) Measurements are made at three cross-sections A–A, B–B and C–C along the shaft shown (Fig. 1.113). Each

section indicates that all points on the surface fall within the annular rings shown of the sections.

  (i) Would you accept the shaft on the criteria of circularity alone?

  (ii) Would you accept the shaft on both criteria, size and circularity?

**FIGURE 1.113**

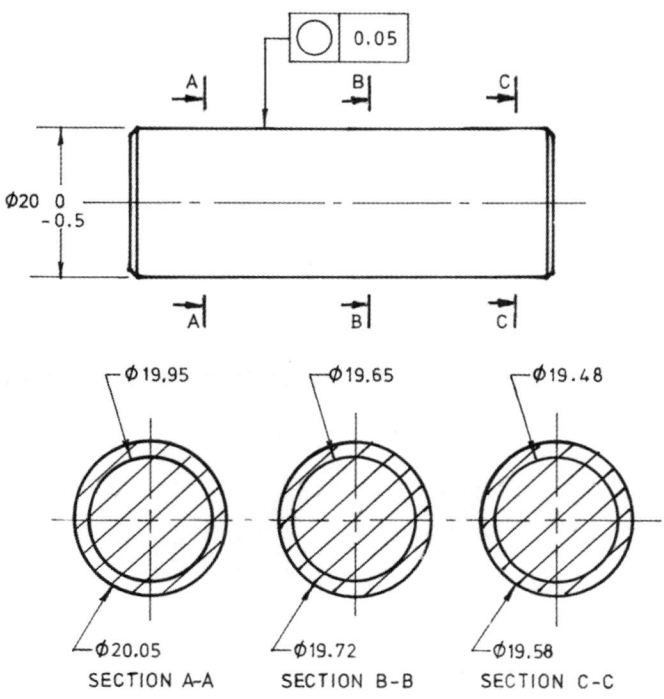

SECTION A-A    SECTION B-B    SECTION C-C

**1.20** In the sketch (Fig. 1.114) the pin must always fit into the hole without interference.

(a) Apply the largest cylindricity tolerance equally to each part to comply with this condition.

(b) What is the maximum permissible combined error in cylindricity allowed if each part is toleranced on a maximum material condition basis?

**FIGURE 1.114**

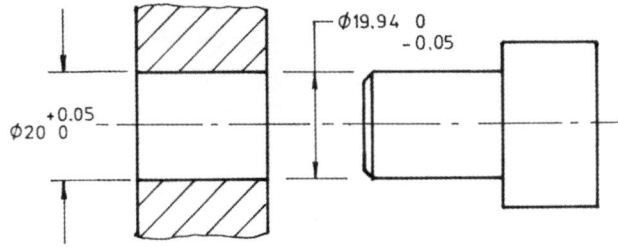

**1.21** The shaft end of the component shown (Fig. 1.115) is required to be concentric with each recess diameter on the other end to within 0.15 mm.

Sketch the part and show how this can be achieved using the appropriate geometry tolerancing methods.

**FIGURE 1.115**

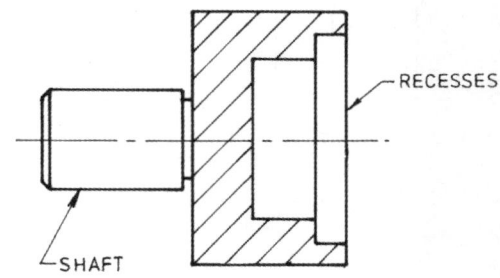

RECESSES

SHAFT

# 1.29 CAD corner

**1.22 Standard drawing sheet: visible requirements**

Either starting from new, or by editing a standard drawing sheet from your CAD program, create an A2 drawing sheet suitable for detail drawings with the following requirements:

(a) Create the sheet geometry from the details shown in pages 12–14.

(b) Include text in the appropriate areas of the title block that can easily be edited when required.

(c) The drawing sheet file is to contain the following elements to assist with drawing production:

  (i) To enable the drawing of outlines, centre lines, hidden detail, dimensions, hatching and annotations in the correct line types and at the correct line widths.

  (ii) Dimensions are to be set up to the details shown on page 8 and should be able to cater for different scaled drawings.

**Symbol library**

Create a library with the following symbols:

(a) 1st and 3rd angle projections as shown below

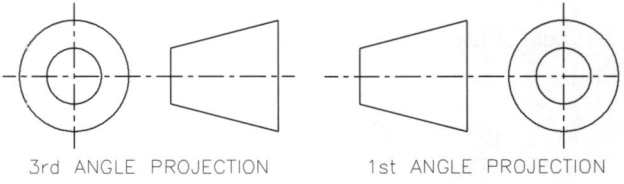

3rd ANGLE PROJECTION    1st ANGLE PROJECTION

(b) Hexagonal bolt, nut, washer and cap-screw to the details shown below (page 88).

Hints:

- The above fasteners are drawn to a unit scale of 1. This means that the scale of the symbols when placed on a drawing = the diameter of the fastener.

- Dimensions are for reference only and should not be included as part of the symbol.

**Challenge:** Paste a logo (company/College/your design) into the title block area of the drawing sheet.

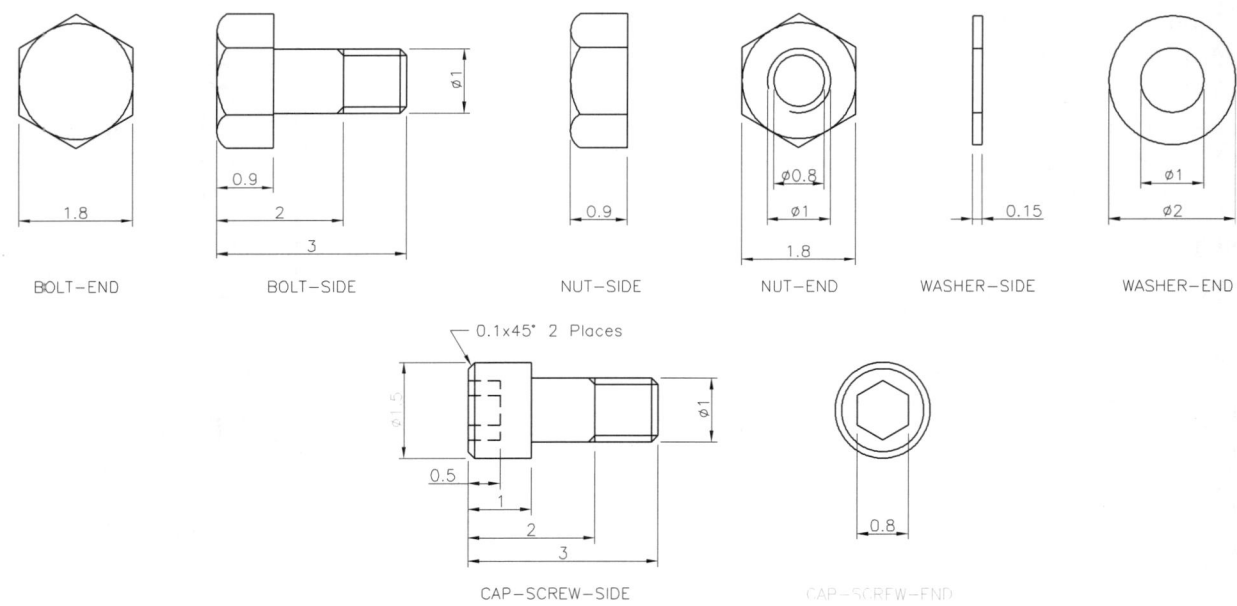

BOLT–END     BOLT–SIDE     NUT–SIDE     NUT–END     WASHER–SIDE     WASHER–END

CAP–SCREW–SIDE     CAP–SCREW–END

## 1.23 Graphical solutions of MMC (Maximum Material Condition) and LMC (Least Material Condition)

The figure below shows an assembly similar to the assembly shown in Figure 1.91 (page 67). This time the toleranced lengths of each of the components are shown.

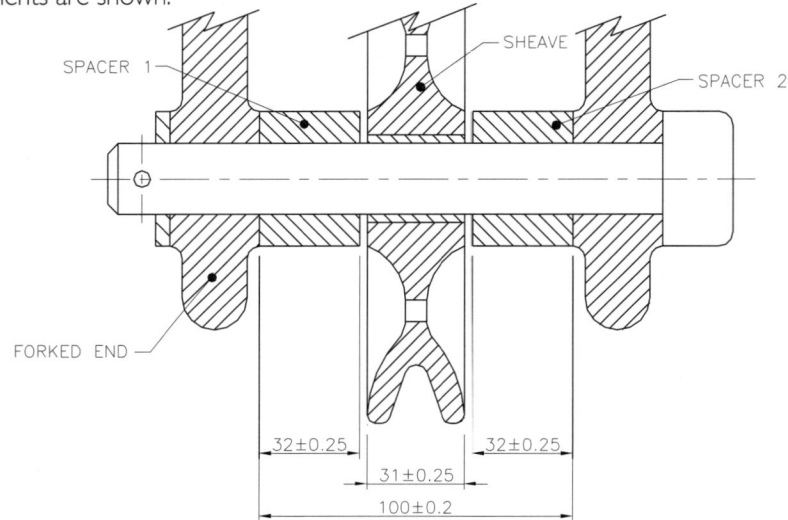

Use your CAD program to determine the maximum and minimum clearances in the assembly as shown in the figure below.

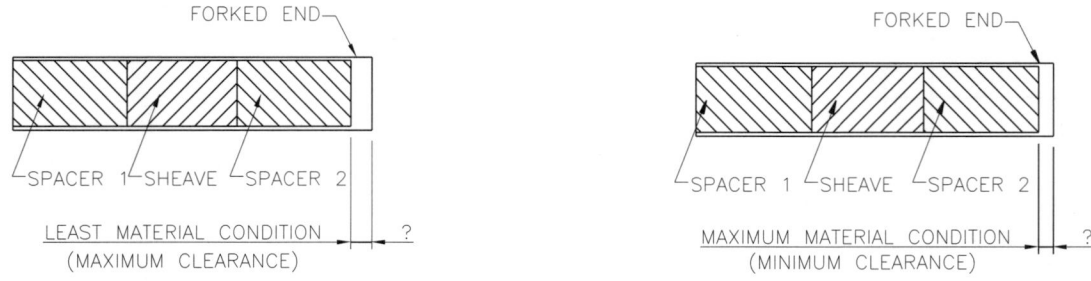

The rectangles represent the lengths of each component in the assembly. Your task is to determine graphically the value of '?' for the LMC and MMC conditions.

Hints:

- Set your CAD program to the appropriate number of decimal places to guarantee accurate results.
- The heights of the rectangles are only representative of each component, however, the lengths must be drawn accurately to reflect either the LMC or MMC condition.

# 1.30 Reference list for geometry tolerancing

## STANDARD PUBLICATIONS

AS1100.101—1992 *Technical Drawing, Part 101: General Principles* (Standards Australia, Sydney, 1992)

Farmer, Leonard E, SAA HB47—1993 *Handbook—Dimensioning and Tolerancing to AS1100.101—1992 and AS1100.201—1992* (Standards Australia, Sydney, 1993)

SAA HB7—1993 *Engineering Drawing Handbook* (Standards Australia, Sydney, 1993)

ISO 2768-1—1989 and ISO 2768-2—1989 *General Tolerances—Part 2 Geometrical tolerances for features without individual tolerance indications*

ISO/TR 5460—1985 *Technical drawings—Geometrical tolerancing—Tolerancing of form, orientation, location and run-out—Verification principles and methods—Guidelines*

ASME Y24.5M—1994 *Dimensioning and Tolerancing* (American Society of Mechanical Engineers, New York, 1994)

## BOOKS

Drake, Paul (ed.), *Dimensioning and Tolerancing Handbook* (McGraw-Hill, New York, 1999)

Farmer, Leonard E, *Dimensioning and Tolerancing for Function and Economic Manufacture* (Blue Print Publications, Sydney, 1999)

Henzold, G, *Handbook of Geometrical Tolerancing: Design, Manufacturing and Inspection* (John Wiley and Sons, England, 1995). Note: comprehensively covers ISO geometry tolerancing and also compares with American, British, German (DIN) and East European standards.

Krulikowski, Alex, *Geometric Dimensioning and Tolerancing, Self-Study Book* (Effective Training Incorporated, Westland, 1995). Note: this book uses only metric units.

Williams, Professor Roy A, *Fundamentals of Dimensioning and Tolerancing* (Edward Arnold, Melbourne 1991)

# Geometrical constructions

**2**

After studying and working on this chapter you should be able to do the following:

- competently use a range of drawing instruments
- understand the principle and construction of cams
- use basic geometric constructions
- lay out two-dimensional geometric shapes

# 2.1 Drawing instrument exercises

The following exercises are designed to give some degree of efficiency in using drawing instruments. These exercises could be used in the first practical sessions or as a first assignment.

**2.1**

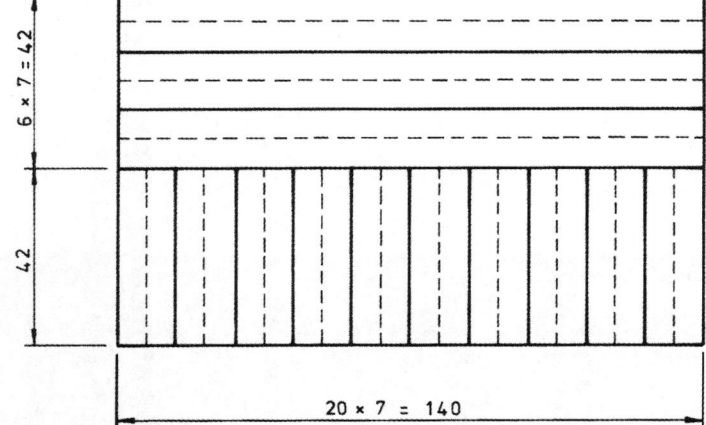

Full lines are outlines, type A. Dash lines are hidden detail, type E.

**2.2**

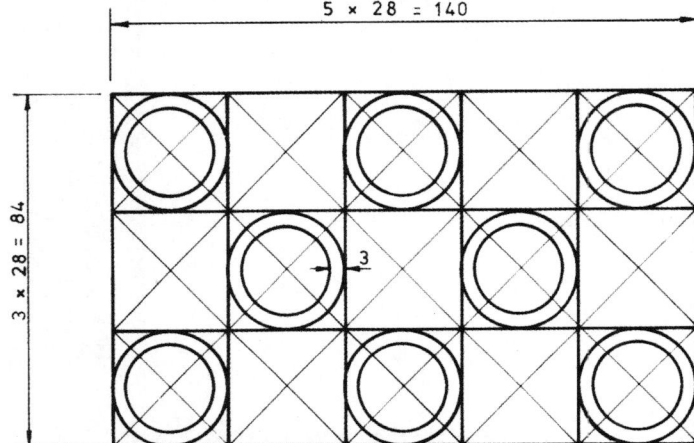

Mark off the 140 mm and the 84 mm sides into 28 mm segments with dividers to ensure perfect squares.

**2.3**

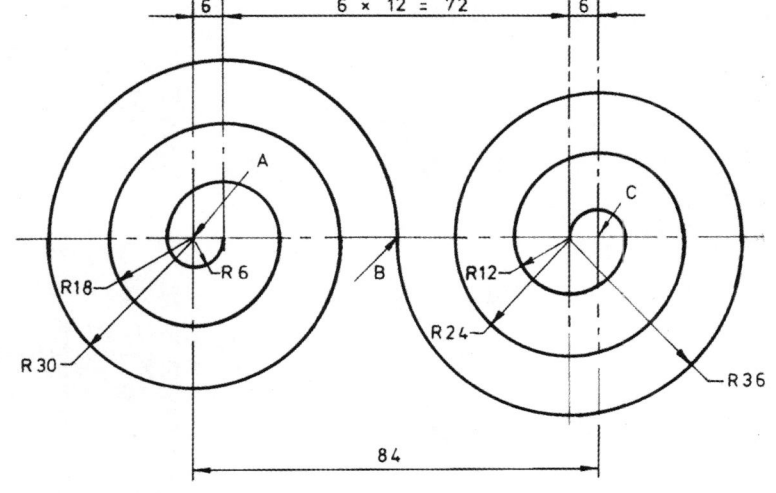

Commence at centre A with 6 mm radius, work out to B, then in to C.

ENGINEERING DRAWING

## 2.4

Locate centres of 70 mm diameter circles on the corners of a 70 mm square, then draw circles lightly. Construct triangles using the 30°–60° set square.

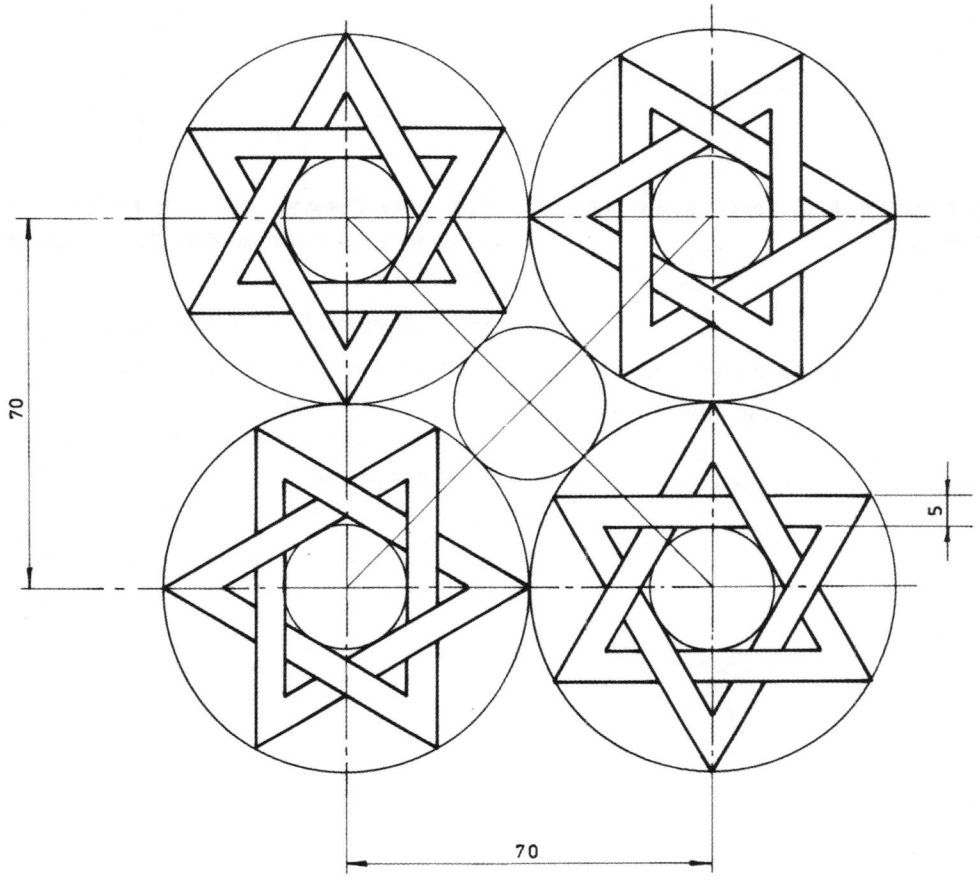

## 2.5

Divide the horizontal diameter into twelve 7 mm segments and use these points as centres for the semicircles.

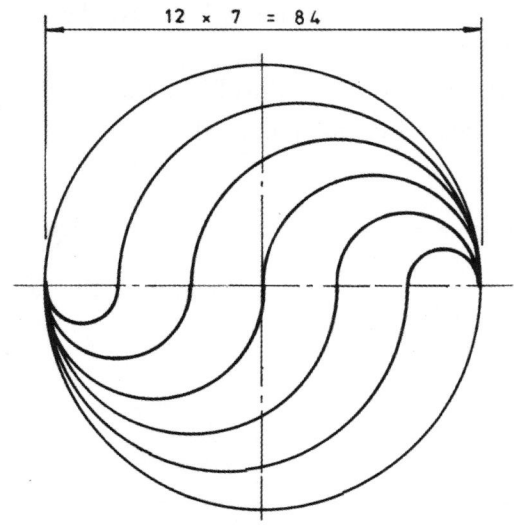

## 2.6

Divide the circle into 24 sectors using only the 45° and 30°–60° set squares, singly and in combination.

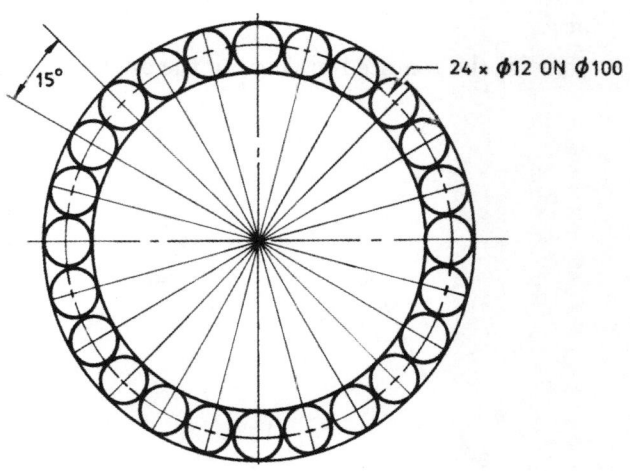

# 2.2 Geometrical constructions used in engineering drawing

In the course of engineering drawing, it is often necessary to make certain geometrical constructions in order to complete an outline.

The following basic constructions are given for reference.

### 2.7 TO CONSTRUCT AN ANGLE EQUAL TO A GIVEN ANGLE

ABC is the given angle.

1. Draw a line EF.
2. With centres B and E, describe two equal arcs.
3. Transfer the arc length from angle ABC to the arc drawn from E to give point D.
4. Join E to D, and produce to give the required angle.

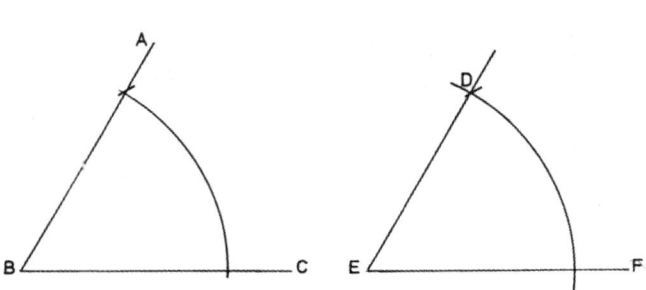

### 2.8 TO BISECT A GIVEN ANGLE

ABC is the given angle.

1. From B describe an arc to cut AB and BC at E and D respectively.
2. With centres E and D, draw equal arcs to intersect at F.
3. Join BF, the required bisector of the angle.

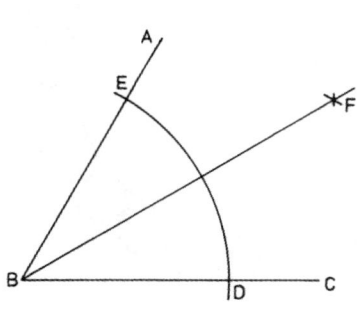

### 2.9 TO DRAW A LINE PARALLEL TO A GIVEN LINE AT A GIVEN DISTANCE FROM IT

AB is the given line, and c is the given distance.

1. From any two points well apart on AB, draw two arcs of radius equal to c.
2. Draw a line tangential to the two arcs to give the required line.

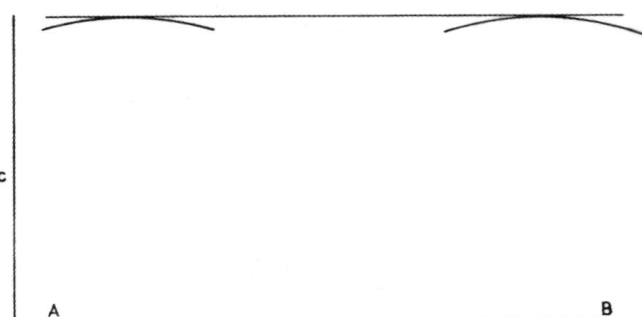

### 2.10 TO DRAW A LINE PARALLEL TO A GIVEN LINE THROUGH A GIVEN POINT

AB is the given line, and P is the point.

1. From P, and with radius a little less than AP, draw an arc to cut AB at D.
2. With centre D and radius DP, describe an arc to cut AB at E.
3. With centre D and radius EP, mark off DF.
4. Join FP to give the required line.

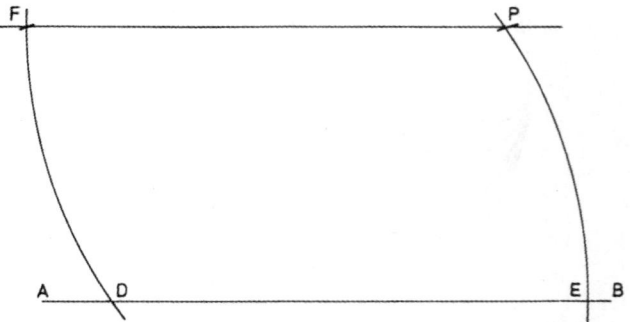

## 2.11 TO DRAW A PERPENDICULAR FROM THE END OF A LINE

AB is the given line.

1. With centre B and radius less than AB, describe an arc to intersect AB at C.
2. From C, and with the same radius, mark off D; from D, and with the same radius, mark off E.
3. From D and E describe any two equal arcs to intersect at F.
4. Join BF to give the required perpendicular.

## 2.13 TO DRAW A PERPENDICULAR FROM A POINT OUTSIDE A LINE (METHOD 1)

AB is the given line, and P is the point outside it.

1. From P draw a sloping line to intersect AB at C.
2. Bisect PC at O.
3. With O as centre, draw a semicircle with PC as diameter to intersect AB at D.
4. Join PD to give the required perpendicular.

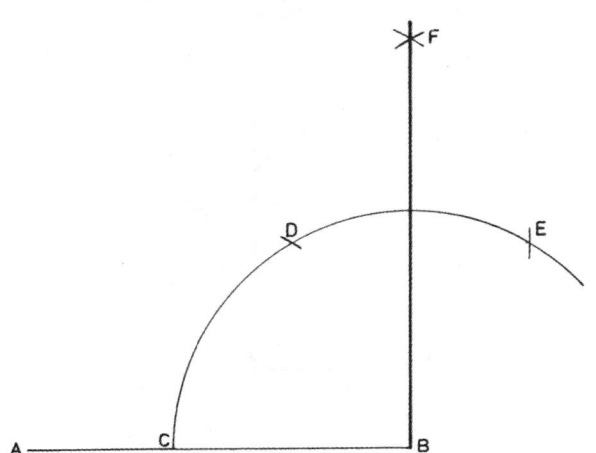

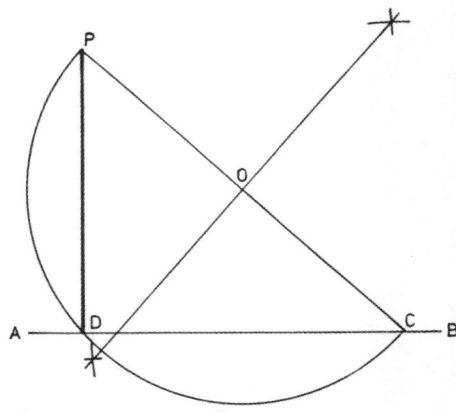

## 2.12 TO DRAW A PERPENDICULAR FROM A POINT IN A LINE

AB is the line, and C is the point on it.

1. With centre C and any radius, describe equal arcs to cut AB at E and F.
2. From E and F describe equal arcs to intersect at D.
3. Join CD to give the required perpendicular.

## 2.14 TO DRAW A PERPENDICULAR FROM A POINT OUTSIDE A LINE (METHOD 2)

AB is the given line, and P is the point outside it.

1. With centre P describe an arc to intersect AB at C and D, a fair distance apart.
2. From C and D, and with the same radius, describe two arcs to intersect at E.
3. Join PE to give the required perpendicular.

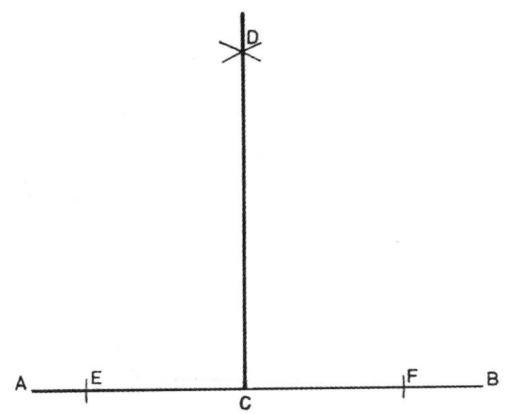

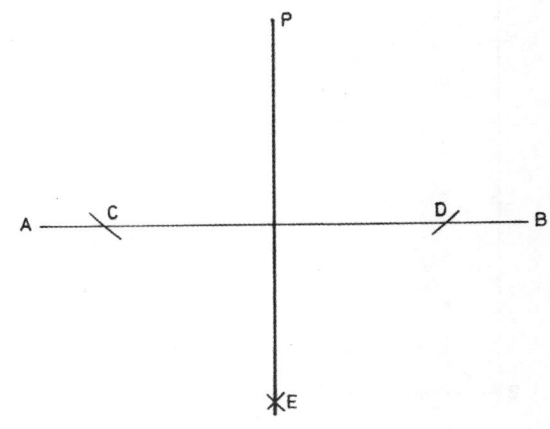

## 2.15 TO DIVIDE A LINE INTO ANY NUMBER OF EQUAL PARTS, SAY SIX

AB is the given line.

1. Draw AC at an angle of approximately 30° to AB.
2. Using dividers, mark along AC six equal lengths, each approximately equal to one-sixth of AB.
3. Join the sixth mark, D on AC, to B.
4. Using sliding set squares, draw lines parallel to DB from the points of division on AC to intersect AB and give the required points of division.

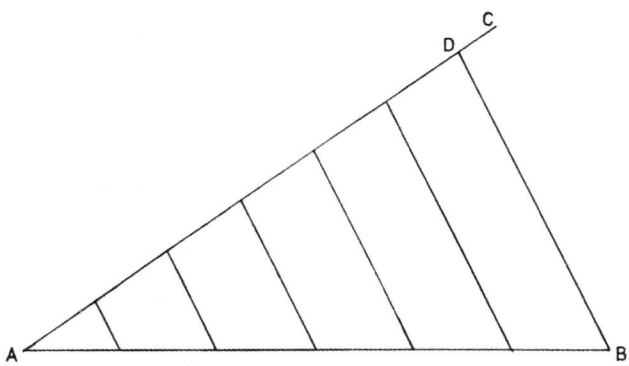

## 2.16 TO TRISECT A RIGHT ANGLE

ABC is the given right angle.

1. With centre B and any radius, describe arc DE.
2. From D and E and with the same radius, mark off F and G respectively on arc DE.
3. Join FB and GB to trisect the right angle.

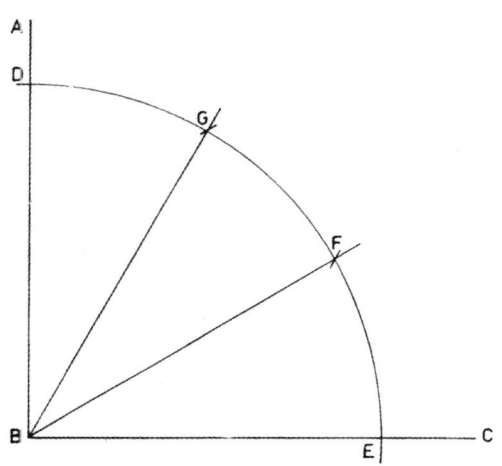

## 2.17 TO CONSTRUCT A REGULAR PENTAGON ON A GIVEN LINE

AB is the given line.

1. Bisect AB at C.
2. Erect a perpendicular at B (construction 5), and mark off BD equal to AB.
3. With C as centre and radius CD, describe an arc to intersect AB produced at E.
4. From A and B, and with radius AE, describe arcs to intersect at F.
5. With radius AB and centres A, B and F, describe arcs to intersect at G and H.
6. Join FG, GA, FH and HB to complete the pentagon.

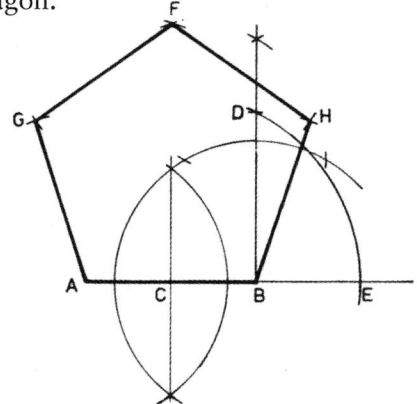

## 2.18 TO CONSTRUCT A REGULAR HEXAGON ON A GIVEN LINE

AB is the given line.

1. From A and B, and with radius AB, draw two equal arcs to intersect at O.
2. With radius OA or OB and centre O, draw a circle.
3. From A or B, using the same radius, step off arcs around the circle at C, D, E and F.
4. Join these points to complete the hexagon.

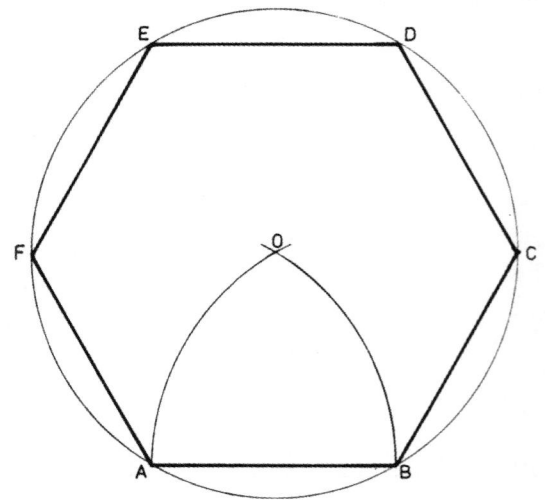

## 2.19 TO CONSTRUCT A REGULAR POLYGON, SAY A HEPTAGON, ON A GIVEN LINE

AB is the given line.

1. Bisect AB at C.
2. Along the bisector mark off C4 equal to AC.
3. With centre A and radius AB, describe an arc to intersect the bisector at 6.
4. Bisect distance 4–6 to give 5.
5. Add distance 4–5 to 6 to give 7 and so on. These points are the centres of circles around which AB will step that number of times.
6. With centre 7 and radius 7A, describe a circle. Step AB around seven times and join the points to give a heptagon.

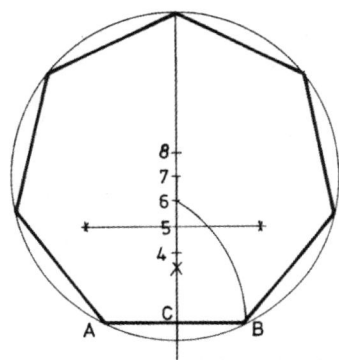

## 2.20 TO DRAW AN ARC TANGENTIAL TO A GIVEN ARC, CENTRES ON OPPOSITE SIDES

A is the centre of the given arc and a is its radius. Line b is the other arc radius.

1. From A describe an arc with radius a + b.
2. With centre B anywhere on this arc, describe an arc of radius b which will be tangential to the given arc.

The actual point of tangency of the two arcs is O, the point where the line joining the two centres intersects the arcs.

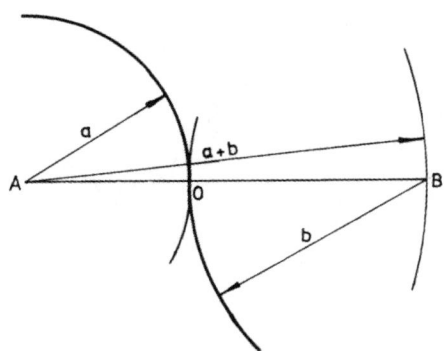

## 2.21 TO DRAW AN ARC TANGENTIAL TO TWO ARCS AND ENCLOSING ONE OF THEM

A and B are centres of two arcs of radius a and b respectively. Line c is the radius of the required arc.

1. With A and B as centres, describe arcs of radii a + c and c – b respectively to intersect at C.
2. With centre C and radius c, describe the required arc.
3. Join AC to intersect the curve at E, and produce CB to intersect the curve at F. Then E and F are the points of tangency of the three arcs.

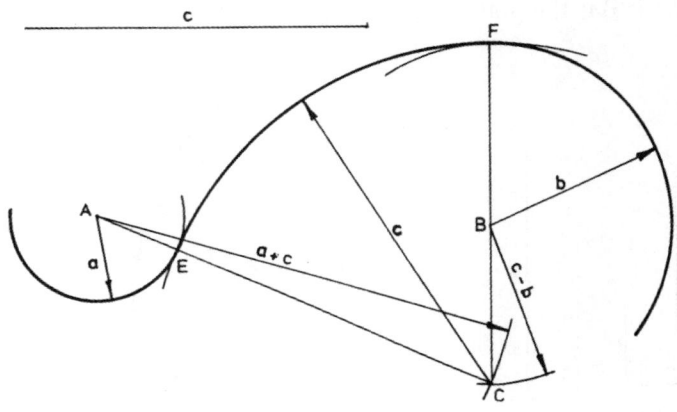

## 2.22 TO DRAW A CURVE TANGENTIAL TO THREE INTERSECTING LINES

AB, BC and CD are three intersecting straight lines, and P is the point of tangency of the required curve.

1. Make BE equal to BP and CF equal to CP.
2. Erect perpendiculars at E, P and F to intersect at O and R.
3. With centre O and radius OE, describe arc EP.
4. With centre R and radius RF, describe arc FP to complete the curve.

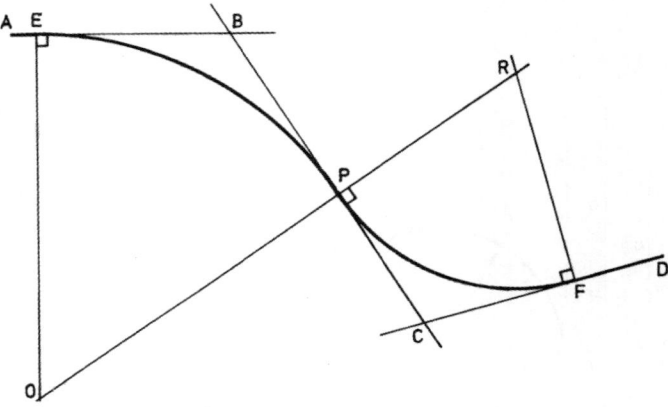

## 2.23 TO DRAW ARCS TANGENTIAL TO A GIVEN ARC, WITH CENTRES ON THE SAME SIDE

A is the centre of the given arc of radius a; b and b' are radii of arcs which are to be drawn tangential to the given arc.

1. With centre A describe two arcs of radii a – b and b' – a to intersect the line of centres at F and E respectively.
2. With E as centre, describe an arc of radius b', which will be tangential to the given arc externally.
3. With F as centre, describe an arc of radius b, which will be tangential to the given arc internally. B is the point of tangency of all three arcs.

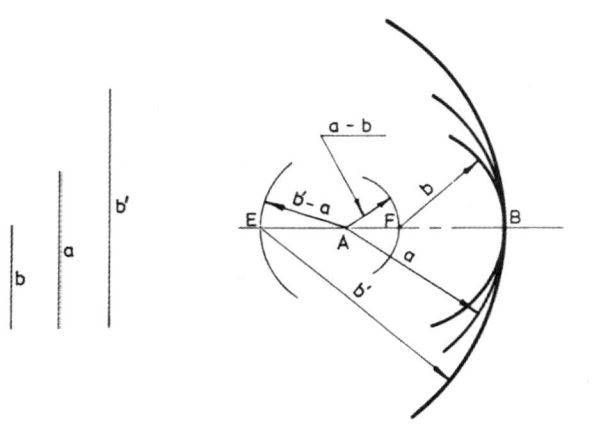

## 2.24 TO DRAW AN ARC TANGENTIAL TO TWO ARCS (EXTERNALLY)

A and B are the centres of the given arcs of radii a and b respectively; c is the external arc radius.

1. From centres A and B, describe two arcs of radii a + c and b + c respectively to intersect at C.
2. With centre C and radius c, describe an arc which will be tangential to the given arcs.

E and F are the points of tangency of the three arcs.

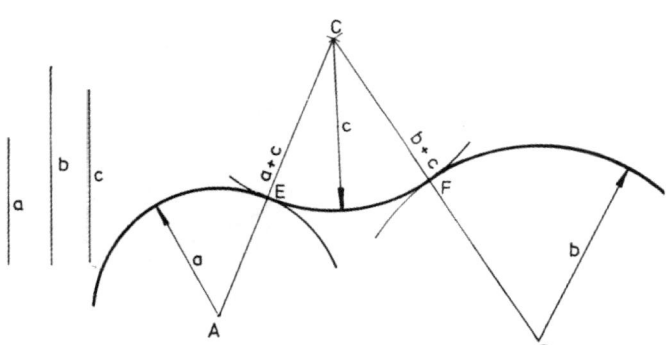

## 2.25 TO DRAW AN ARC TANGENTIAL TO TWO ARCS (INTERNALLY)

A and B are the centres of two given arcs of radii a and b respectively; c is the required tangential arc radius.

1. From A and B, describe arcs of radii c – a and c – b respectively to intersect at C.
2. With centre C and radius c, describe an arc which will be tangential to the given arcs.
3. Produce CA and CB to intersect the curve at E and F respectively. These are the points of tangency of the arcs.

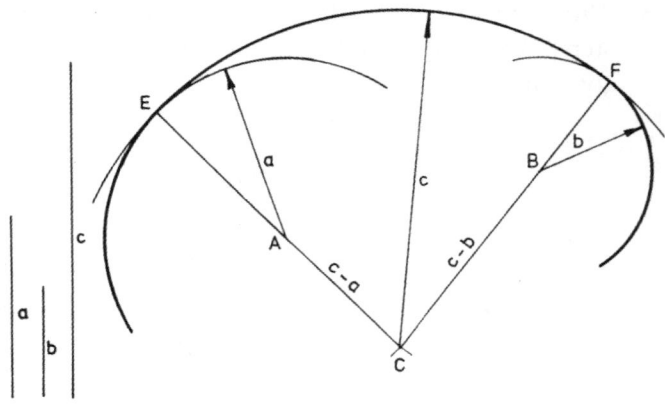

## 2.26 TO DRAW AN ARC TANGENTIAL TO TWO STRAIGHT LINES

AB and CB are the given lines, and c is the radius of the required arc.

1. Draw two lines parallel to the given lines at a distance c from them to intersect at D.
2. With centre D and radius c, draw an arc, which will be tangential to both given lines.
3. Erect perpendiculars at D to intersect AB and BC at E and F respectively. These are the points of tangency of the lines with the arc.

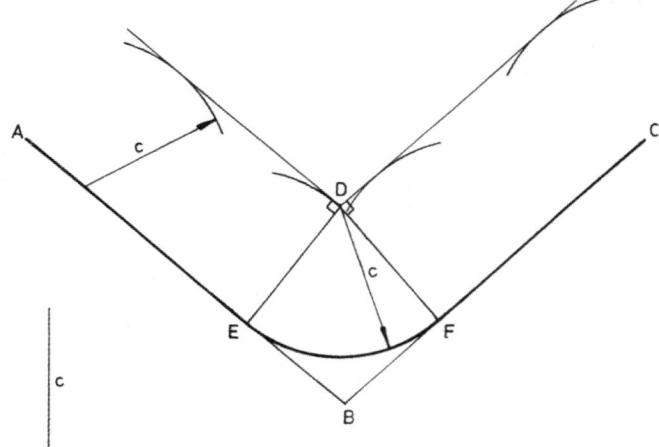

## 2.27 TO DRAW AN ARC TANGENTIAL TO TWO STRAIGHT LINES AT RIGHT ANGLES

AB and BC are the given lines, and c is the radius of the required arc.

1. From B, mark off BE and BF equal to c.
2. From E and F, draw arcs of radius c to intersect at D.
3. From D with radius c, describe an arc, which will be tangential to both AB and BC.

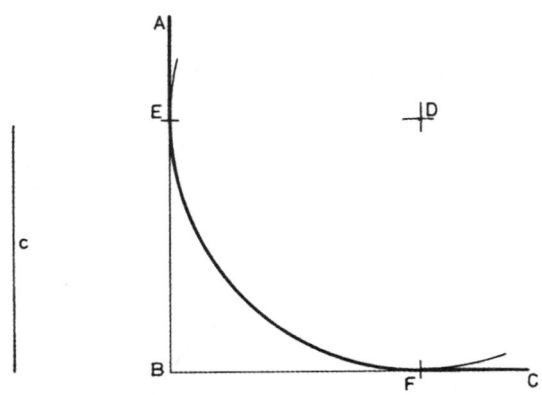

## 2.28 TO DRAW AN ARC TANGENTIAL TO A LINE AND ANOTHER ARC

A is the centre of the given arc of radius a. BC is the given line, and b is the radius of the required arc.

1. From A, describe an arc with radius a + b.
2. Draw a line parallel to BC and distant b from it to intersect the arc a + b at D.
3. From D, describe an arc of radius b, which will be tangential to the given line BC and the given arc a.

E and F are the points of tangency.

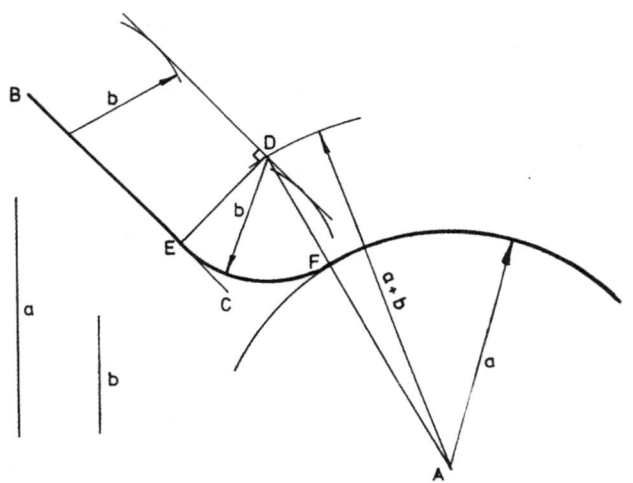

## 2.29 TO DRAW A TANGENT TO A CIRCLE FROM AN OUTSIDE POINT

O is the centre of the given circle, and P is the point outside it.

1. Join OP, and bisect OP at X.
2. Draw a semicircle on OP to intersect the given circle at T. Join PT to give the required tangent.

Note: Angle OTP is a right angle.

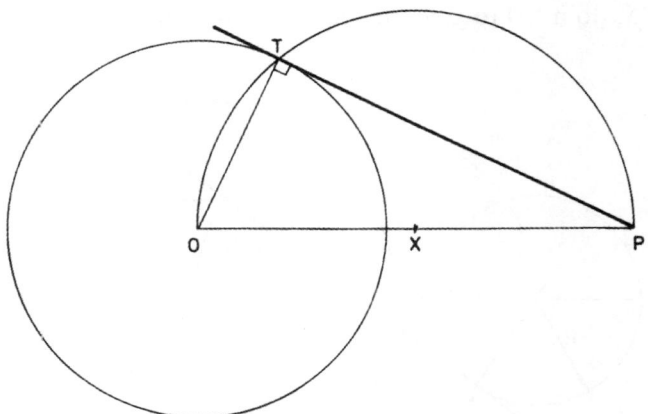

## 2.30 TO DRAW A DIRECT COMMON TANGENT TO TWO GIVEN CIRCLES

A and B are the centres of two given circles of radii r and R respectively.

1. With centre B and radius R − r, describe a circle.
2. Bisect AB at X, and draw a semicircle on AB to cut circle R − r at C.
3. Join BC, and produce it to cut the larger circle at D.
4. Draw AE parallel to BD.
5. Join ED to give the required tangent.

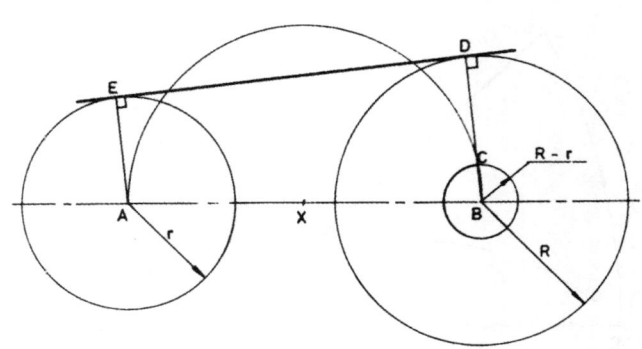

## 2.31 TO DRAW A TRANSVERSE COMMON TANGENT TO TWO GIVEN CIRCLES

A and B are the centres of two given circles of radii R and r respectively.

1. With centre B and radius R + r, describe a circle.
2. Bisect AB at X, and draw a semicircle on AB to intersect circle R + r at C.
3. Join BC to intersect the inside circle at D.
4. Draw AE parallel to BD.
5. Join ED to give the required tangent.

## 2.33 TO MARK ON A GIVEN CIRCLE AN ARC APPROXIMATELY EQUAL TO A GIVEN LENGTH

O is the centre of the given circle, and a is the given length.

1. Draw a tangent AC to the given circle, making AC equal to the given length a.
2. Mark off AD = $\frac{1}{4}$ AC along AC (construction 9).
3. With centre D and radius DC, describe an arc to intersect the given circle at B. Then AB is the required arc.

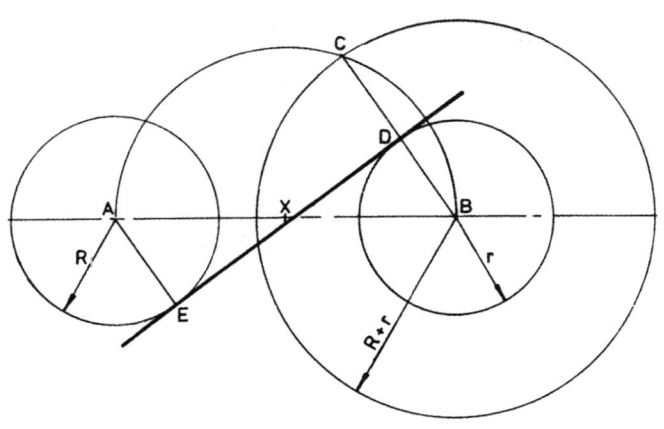

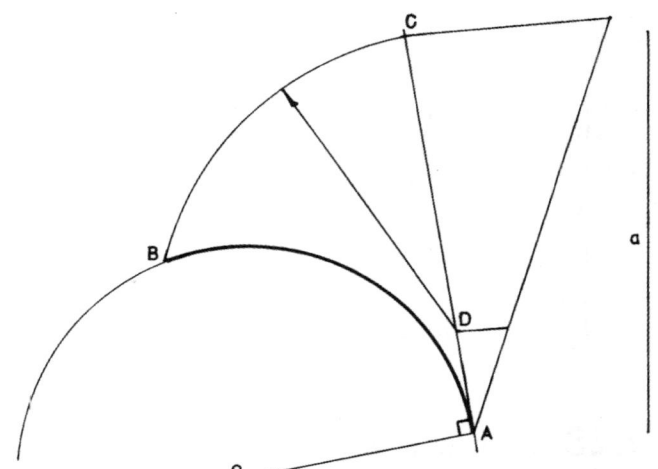

## 2.32 TO DRAW A STRAIGHT LINE APPROXIMATELY EQUAL IN LENGTH TO A GIVEN ARC

AB is the given arc.

1. Join AB and produce it to D, making AD = $\frac{1}{2}$ AB.
2. Draw the tangent to the given arc at A. (Note: The tangent at A makes a right angle with the radius.)
3. With centre D and radius DB, describe an arc to cut the tangent from A at C. AC is the required straight line.

## 2.34 TO DRAW A STRAIGHT LINE APPROXIMATELY EQUAL IN LENGTH TO THE CIRCUMFERENCE OF A GIVEN CIRCLE

O is the centre of the given circle, and AOB is its diameter.

1. From A, draw a tangent AC, making AC = 3AB.
2. Construct OD at 30° to the diameter (bisect angle BOF = 60°, construction 2).
3. Construct DE at right angles to AB (construction 7).
4. Join EC, which is the required line.

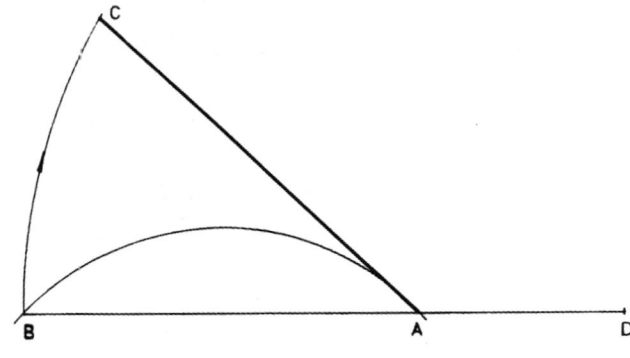

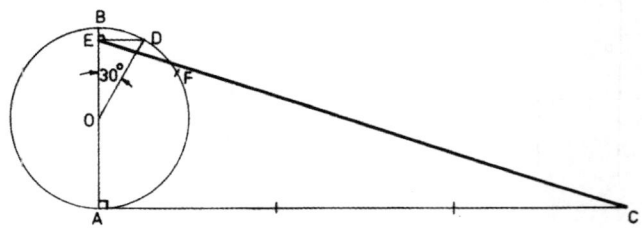

## 2.35 TO CONSTRUCT A REGULAR HEXAGON GIVEN THE DISTANCE ACROSS CORNERS

### First method

AB is the given distance and O its centre.

1. With centre O construct a circle with AB as diameter.
2. With the same radius and A and B as centre draw arcs intersecting the circle.
3. Connect the points.

### Second method

1. Draw lines with the 30°–60° set square in the order shown.

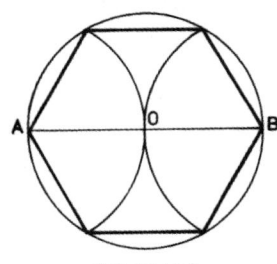

First method

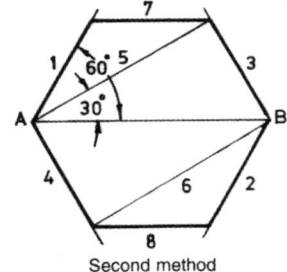

Second method

## 2.36 TO CONSTRUCT A REGULAR HEXAGON GIVEN THE DISTANCE ACROSS FLATS

1. Draw a circle equal to the distance across the flats. This circle is the inscribed circle of the required hexagon.
2. Draw external tangents to this circle ensuring that the hexagon is oriented according to requirements using the 60° or 30° angle as shown below.

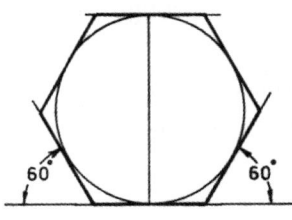

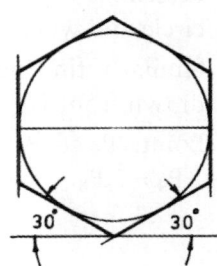

## 2.37 TO DRAW A CYCLOID, GIVEN THE DIAMETER OF THE ROLLING CIRCLE

1. Construct a tangent AB to the rolling circle so that AB is equal in length to the circumference of the circle.
2. Divide the tangent AB and the circumference of the circle into twelve equal parts so that when the circle rolls along the tangent, points of division on each will coincide.
3. Let P be a point on the circle coincident with A in the initial position.
4. As the circle rolls along the tangent and point 1 on the circle and tangent coincide, P moves up and across slightly to the position $P_1$ level with point 11.
5. With radius equal to the perpendicular from the line of centres to 1, describe an arc to intersect the horizontal line from point 11 at $P_1$.
6. Similarly find points $P_2$, $P_3$, etc., which are successive positions of point P as the circle rolls along the tangent.
7. Draw a smooth curve through points P to $P_{12}$ to give the required cycloid.

**Note:** The first half of the curve, P to $P_6$, may be constructed as above, and the second half, $P_6$ to $P_{12}$, obtained by marking off equal distances to the right of the centre line $P_6 6$, for example $XP_4 = XP_8$. A set of French curves is most helpful for lining in the cycloidal curve and other curves of this nature.

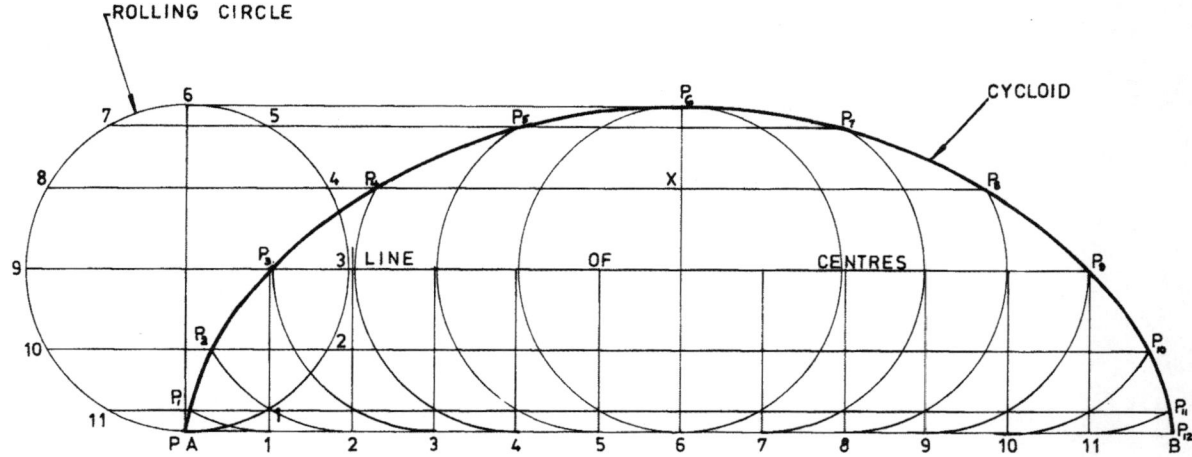

## 2.38 TO DRAW AN EPICYCLOID, GIVEN A ROLLING CIRCLE AND A BASE CIRCLE

1. Draw the rolling circle tangential to the base circle at A.

2. Divide the rolling circle into twelve equal parts, starting at A.

3. Let P be a point on the rolling circle coincident with A in the initial position.

4. Mark AB on the base circle equal to the twelve parts of the rolling circle. This means that as the rolling circle rolls on the base circle, points 1, 2, 3, 4, etc. on the rolling circle will coincide with points 1, 2, 3, 4, etc. on the base circle.

5. As the rolling circle rolls to point 1 on the base circle, P rises and moves across slightly to position $P_1$. To determine $P_1$, draw in that part of the rolling circle which intersects the arc whose centre is that of the base circle and whose radius passes through 11–1.

6. Similarly find points $P_2$, $P_3$, $P_4$, etc. by taking successive positions of the rolling circle to intersect arcs drawn from 10, 9, 8, 7 and 6 respectively.

7. Points $P_6$ to $P_{11}$ may be obtained by marking off equal distances to the right of $P_66$, for example arc $XP_4 = XP_8$.

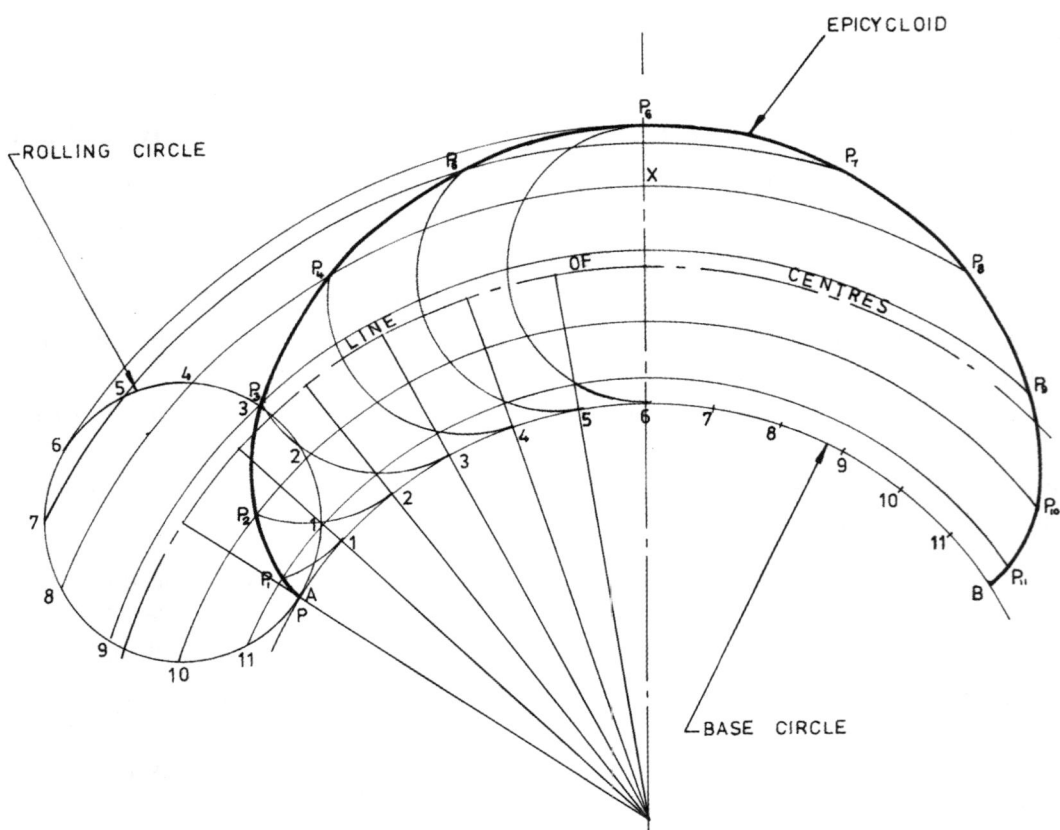

## 2.39 TO DRAW A HYPOCYCLOID, GIVEN A ROLLING CIRCLE AND A BASE CIRCLE

The hypocycloid is constructed in a manner similar to that shown for the epicycloid in the previous construction, except that the rolling circle rolls inside the base circle. It is interesting to note that when the rolling circle has a diameter equal to the radius of the base circle, the hypocycloid so formed is a diameter of the base circle.

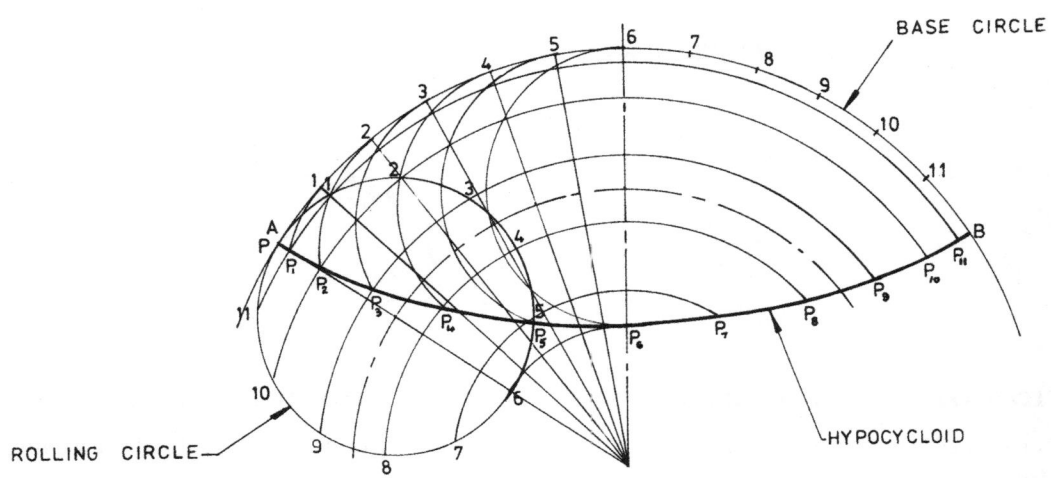

## 2.40 TO DRAW AN INVOLUTE TO A GIVEN CIRCLE

An involute can best be described as the path traced out by the end of a piece of string as it is unwound from the circumference of a circle called the base circle.

1. Draw a straight line AB tangential to the base circle and equal to its circumference.
2. Divide AB into twelve equal parts, and divide the base circle into twelve equal parts. This means that each part of the line AB is equal in length to each part of the circle.
3. From the points of division on the base circle, draw tangents using the 60°–30° set square.
4. Commencing at the next tangent to A, transfer length A1 to this tangent.
5. Transfer A2 to the next tangent, then A3 and so on until AB is reached.
6. Draw a smooth curve through the points A, 1, 2, . . ., 11, B, and this curve is the involute of the base circle.

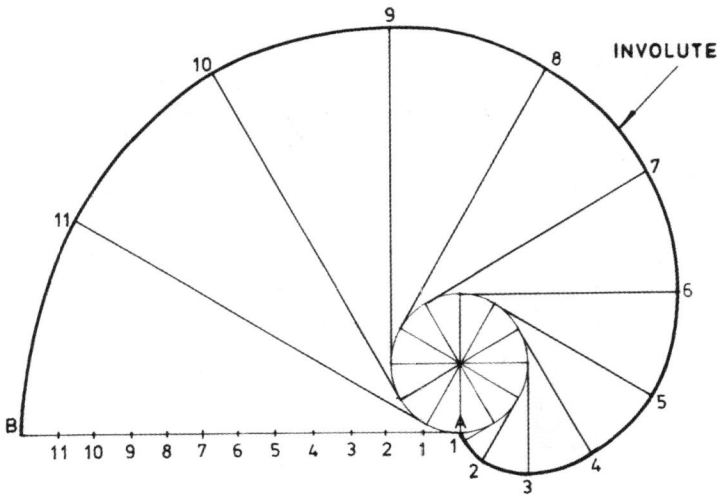

FIGURE 2.1 ▶ Involute spur gear tooth

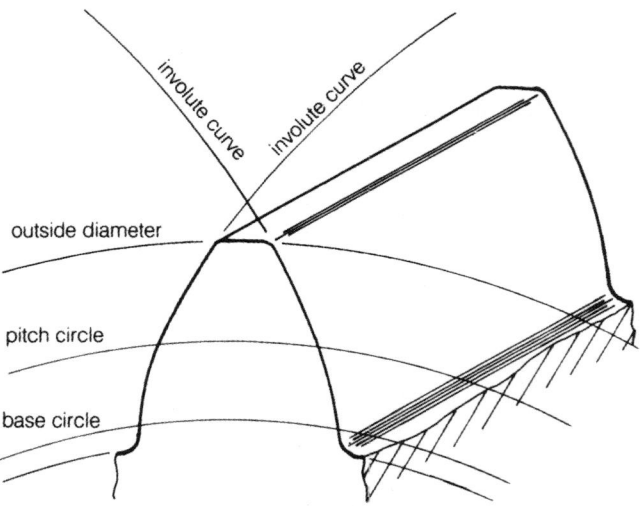

## APPLICATION OF THE INVOLUTE CURVE

One of the most useful applications of the involute curve in engineering is on the profile of gear teeth. Figure 2.1 illustrates an involute gear tooth in which that part of the tooth between the top and the base circle is of involute form.

The sides of the tooth are generated by two separate involutes from a common base circle and are spaced so that the tooth thickness at the pitch circle is a known value depending on the circumference of the gear and the number of teeth.

An involute may be generated from a straight line, polygon, circle or indeed any closed figure. It is simply the curve traced by the end of an imaginary piece of string unwound from the figure. Figure 2.2 illustrates involutes formed from various shapes.

**FIGURE 2.2** ▶ Involutes formed from various shapes

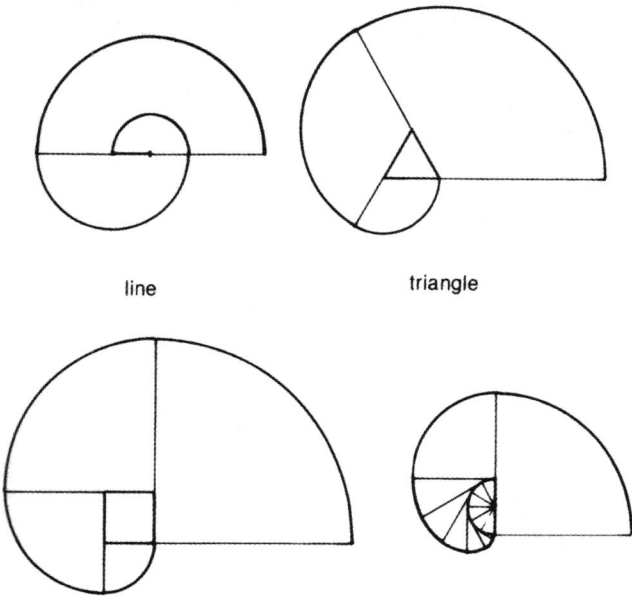

line                    triangle

## THE CYLINDRICAL HELIX

A helix is the path traced out by a point as it moves along and around the surface of a cylinder with uniform angular velocity and, for each circumference traversed, moves a constant length (called the lead) in a direction parallel to the axis.

The helix angle can be found by constructing a right-angled triangle, the base of which is the circumference of the cylinder and the vertical height of which is equal to the lead, as shown in Figure 2.3.

The helix finds many applications in industry: screw threads, springs and conveyors are typical examples of its use, and the drawing of these items is illustrated in Figures 2.4(a)–(e) (opposite).

The geometrical construction of the helix is illustrated in construction 2.41 (page 106).

**FIGURE 2.3** ▶ Right-hand helix

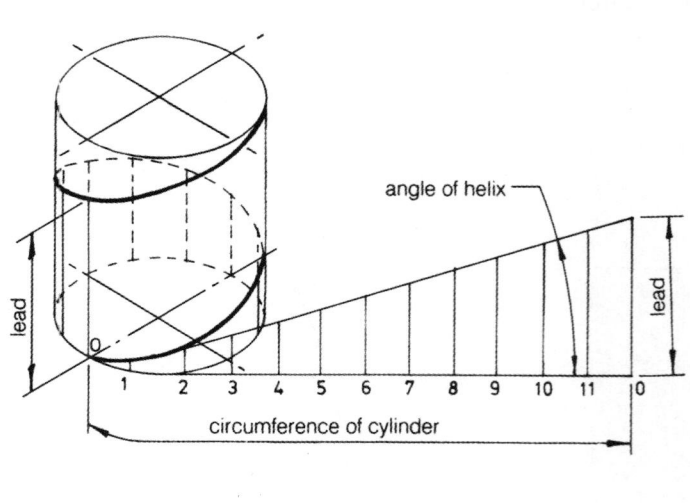

**FIGURE 2.4** ▶ Practical applications of the helix

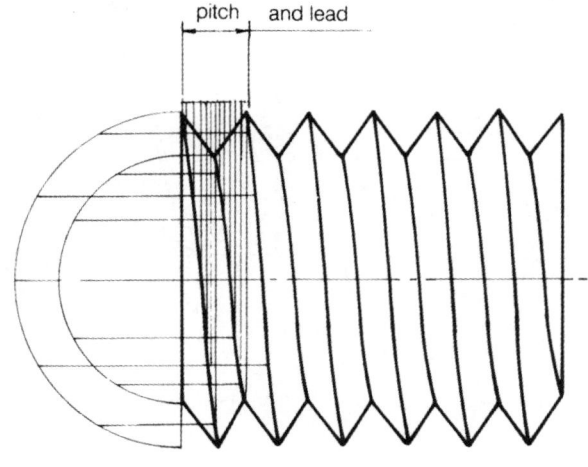

(a) right-hand V thread (single start)

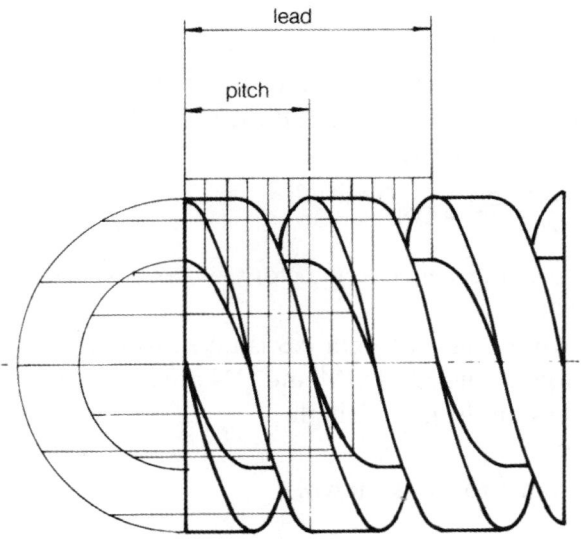

(b) right-hand square thread (double start)

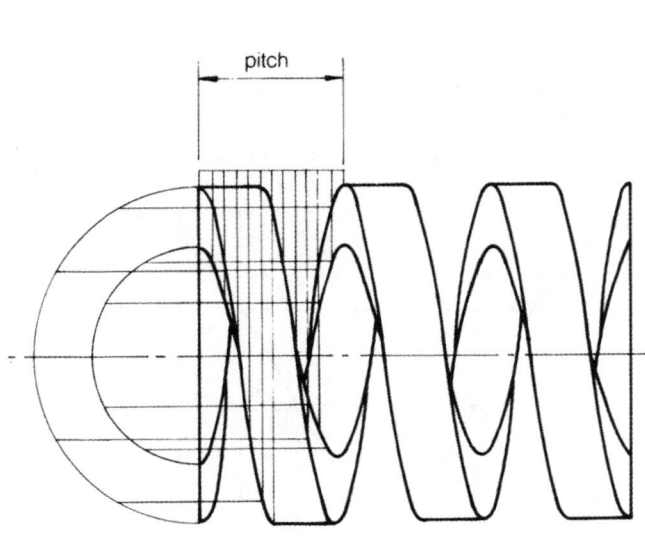

(c) right-hand square compression spring

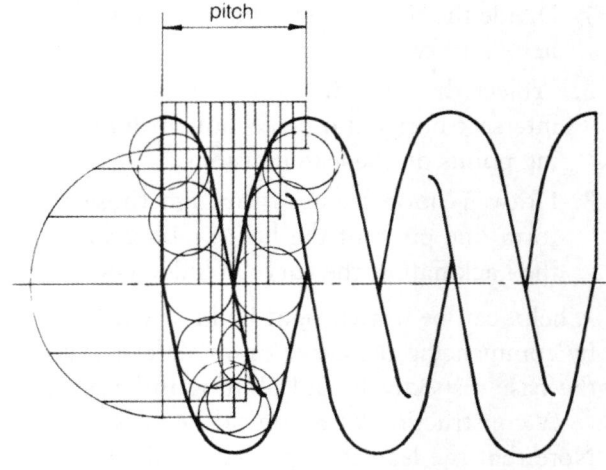

(d) right-hand round compression spring

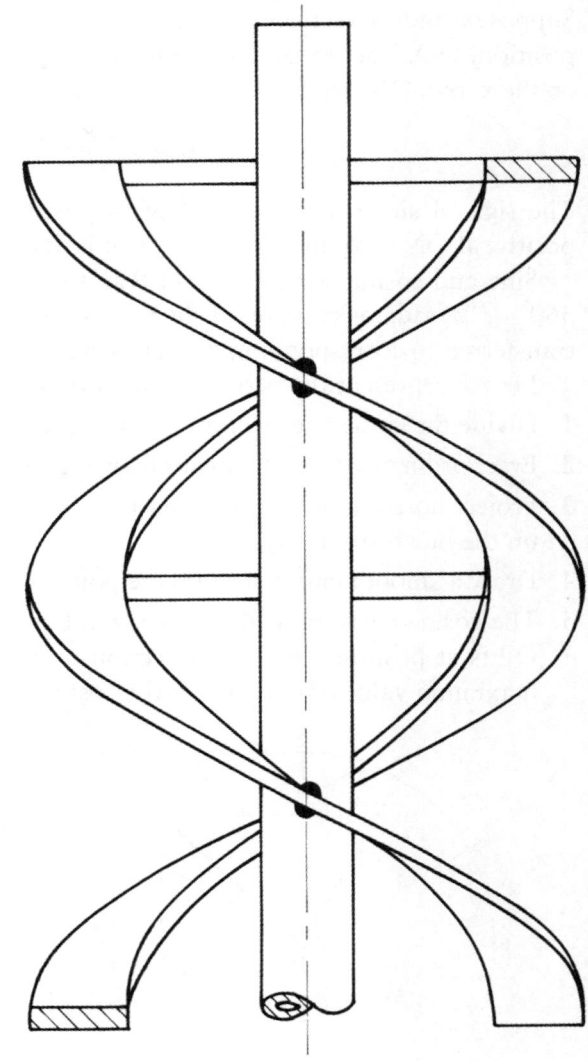

(e) left-hand double-flight ribbon-type screw conveyer

## 2.41 TO CONSTRUCT A CYLINDRICAL AND A CONICAL HELIX

1. Divide the base of the cylinder and the lead of the helix into twelve equal parts.
2. Project the base divisions parallel to the axis to intersect horizontal projections from corresponding points on the lead divisions.
3. Draw a smooth curve through these points to form one pitch of the helix, taking care to show the back half of the curve in hidden detail.

A helix can be drawn right-hand or left-hand simply by commencing the curve on the left or right side of the base respectively and plotting the near side first.

A construction of a conical helix is also shown. Note that the lead is measured axially and the slant height is divided, not the lead height.

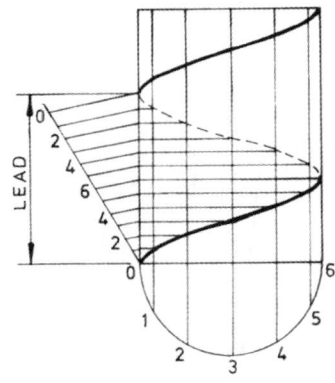
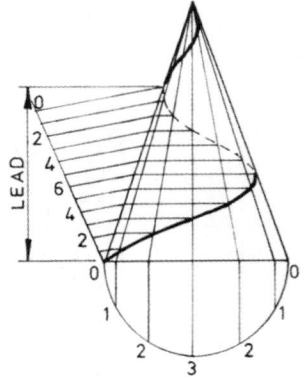

(a) right-hand helix          (b) right-hand conical helix

## 2.42 TO CONSTRUCT THE SINE AND COSINE CURVES

Suppose a radius vector OP rotates about O through an angle $\theta$ in an anticlockwise direction from its initial position, OO. The projection of OP on the y-axis, OM, represents the value of $\sin \theta$ and the projection of OP on the x-axis, ON, represents the value of $\cos \theta$ when the radius vector is taken as unity:

$$\sin \theta = \frac{OM}{OP} = \frac{OM}{1} = OM \text{ and } \cos \theta = \frac{ON}{OP} = \frac{ON}{1} = ON$$

The sign of $\sin \theta$ is such that OM is positive if above the x-axis and negative if below, whereas $\cos \theta$ is positive if ON is to the right of the y-axis and negative if to the left.

Sine and cosine curves ($y = \sin \theta$ and $y = \cos \theta$) may be drawn if a given length AB is taken to represent 360° or $2\pi$ radians rotation of the radius vector OP, and the projected lengths of OM and ON are respectively transferred to corresponding ordinates erected at points of divisions on the given length.

Let AB represent the period of one revolution of the radius vector.

1. Divide the circle and AB into twelve equal parts, numbering from 1 to 11 as shown.
2. Erect ordinates at the points of division on AB.
3. Project horizontally from points 1, 2, 3, . . ., 11 on the circle to intersect the ordinates from 1, 2, 3, . . ., 11 on the line respectively.
4. Draw a smooth curve through the points of intersection to give the sine curve.
5. The cosine curve may also be drawn, but lagging 90° or $\pi/2$ radians behind the sine curve, because when OP is at position OO, its projection on the y-axis OM ($\sin \theta$) is zero and on the x-axis ON ($\cos \theta$) is a maximum value of one, that is the cosine curve commences at the top when the sine curve is zero.

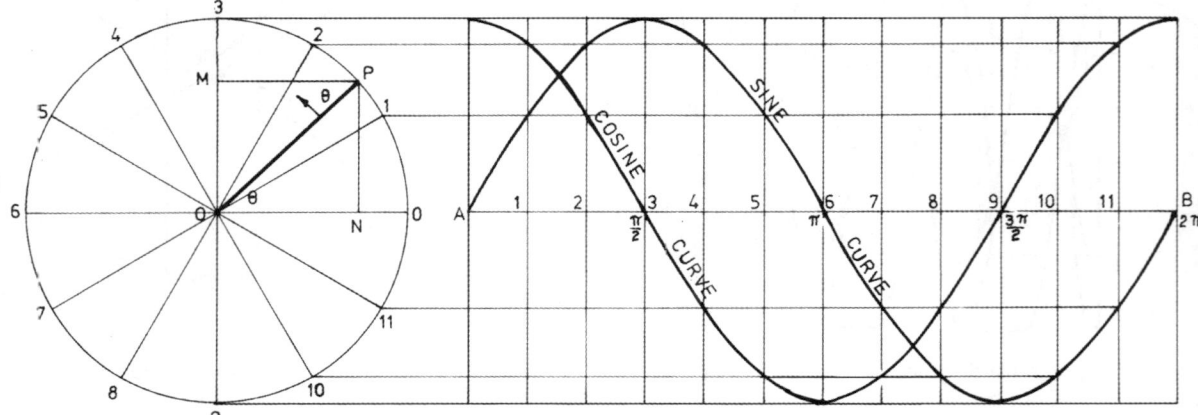

## 2.43 TO CONSTRUCT THE SPIRAL OF ARCHIMEDES

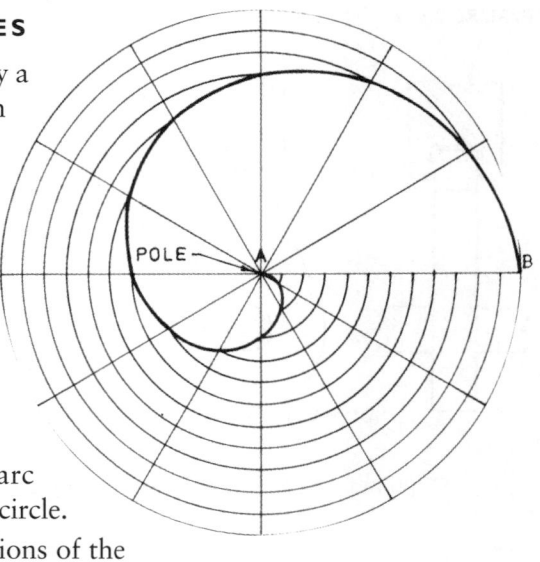

The spiral of Archimedes may be defined as the curve generated by a point which moves uniformly along a straight line as it rotates with uniform angular velocity about a fixed point.

The spiral is used in cam design to convert uniform rotary motion into uniform reciprocal motion. A further application is in the scroll plate of the universal lathe chuck where teeth on the jaws engage with the spiral thus synchronising the inwards and outwards movement of the three jaws.

AB is the given line.

1. Describe a circle with AB as radius.
2. Divide AB and the circle into twelve equal parts.
3. With centre A and radius one division of AB, describe an arc clockwise (or anticlockwise) to intersect the first division of the circle.
4. With centre A, continue describing arcs of 2, 3, 4, . . ., 11 divisions of the line to intersect successive clockwise (or anticlockwise) divisions of the circle.
5. Draw a smooth curve through the points of intersection to form one convolution of the spiral. Further convolutions may be obtained by extending AB and continuing the above procedure.

**Note:** A, the beginning of the spiral, need not commence at the pole, but at any specified distance from it.

# 2.3 Cams

A *cam* is a machine part which has a surface or groove specially formed to impart an unusual or irregular motion to another machine part called a *follower* which presses against and moves according to the rise and fall of the cam surface.

The follower is made to oscillate over a specific distance called the *stroke* or *displacement* with a pre-determined motion governed by the design of the cam profile.

### TYPES OF CAM

There are two general types of cam distinguished by the direction of motion of the follower in relation to the cam axis (refer to Fig. 2.5):

1. radial or disc cams in which the follower moves at right angles to the cam axis
2. cylindrical and end cams, in which the follower moves parallel to the cam axis.

### APPLICATIONS

Basically, cams are used to translate the rotary motion of a camshaft to the straight-line reciprocating motion of the follower. Cams are used as machine elements in a variety of applications including machine tools, motor cars, textile machinery and many other machines found in industry.

On the turret automatic lathe, for example, disc cams are used to move tool slides backwards and forwards in their guideways. In the motorcar engine, a well-known application is the camshaft on which a number of cams raise and lower the inlet and exhaust valves via a push rod and lever system.

Figure 2.5 illustrates various configurations of cam and follower combinations.

### DISPLACEMENT DIAGRAM

Since the motion of the cam follower is of primary importance, the follower's rate of speed and its various positions during one revolution of the cam must be carefully planned on a *displacement diagram* before the cam profile is constructed (see Fig. 2.6 (page 109)).

The displacement curve is plotted on the displacement diagram, which is essentially a rectangle, the base of which represents 360° or one revolution of the cam and the height of which represents the total displacement or stroke of the follower. It must be remembered that because the follower returns to its lowest position in every revolution of the cam, the displacement curve should begin and end at the lowest position of the stroke.

Three types of motion are commonly used in cam design:

1. constant velocity or straight-line motion
2. simple harmonic motion
3. constant acceleration–deceleration or parabolic motion

**FIGURE 2.5** ▶ Types of cam

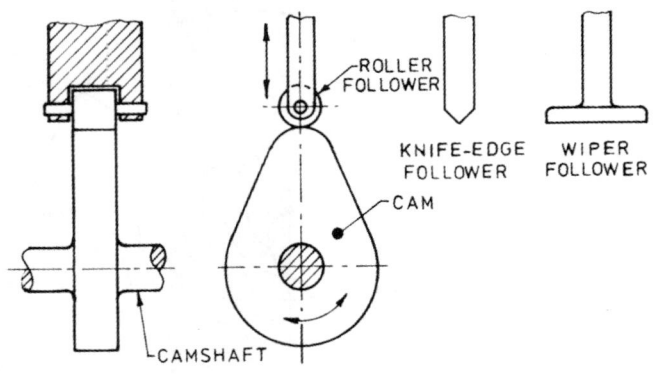

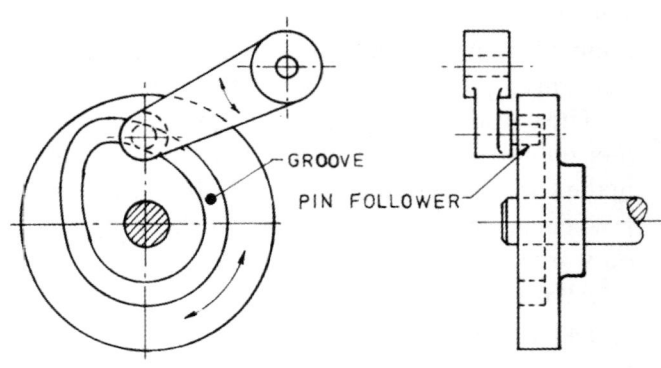

(a) radial disc cam and types of followers

(b) radial face cam with angular displacement

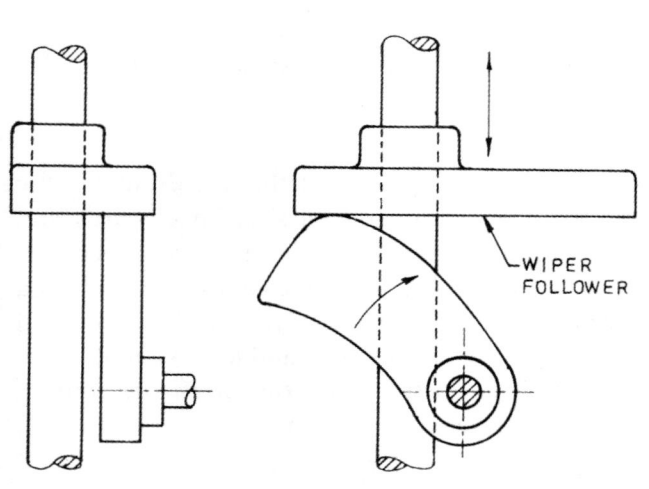

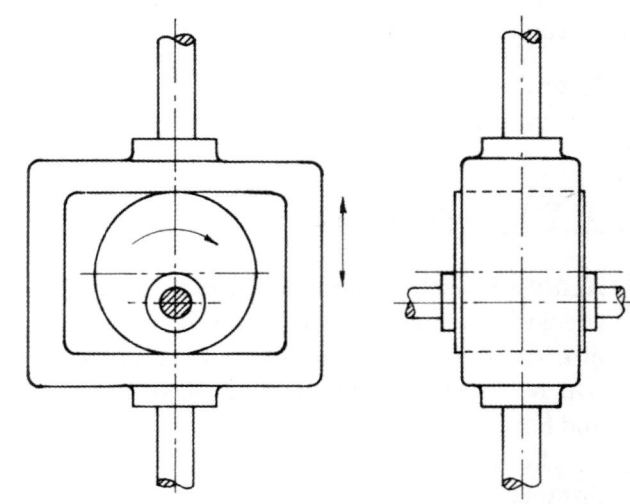

(c) radial toe and wiper cam

(d) radial yoke cam

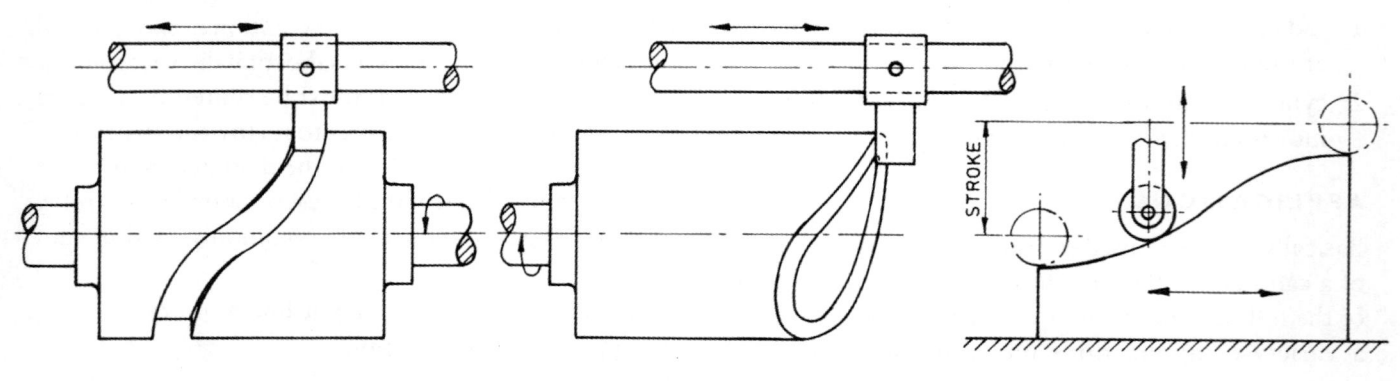

(e) axial cylindrical groove cam

(f) axial cylindrical end cam

(g) slider cam

**FIGURE 2.6** ▶ Types of cam and follower motions

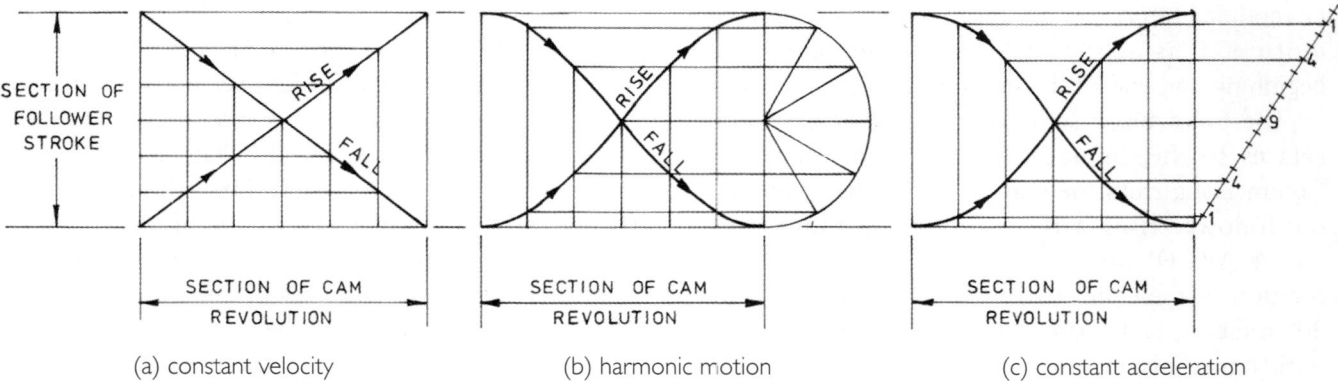

(a) constant velocity      (b) harmonic motion      (c) constant acceleration

*Uniform or constant velocity motion* is represented on the displacement diagram (Fig. 2.6(a)) by dividing the relevant section of the follower stroke and cam revolution into the same number of equal parts. This means that for each part of the cam revolution, the follower will rise or fall by equal amounts.

*Simple harmonic motion* is represented on the displacement diagram (Fig. 2.6(b)) by drawing a semicircle on the relevant section of the follower stroke, dividing the semicircle into six equal parts, and projecting them horizontally to intersect ordinates drawn from six equal divisions on that section of the cam revolution over which harmonic motion is required.

Harmonic motion imparts a movement to the follower that commences from zero, gradually builds up to a maximum speed halfway through the motion, and then slows down to zero during the second half of the motion.

*Constant acceleration–deceleration or parabolic motion* is represented on the displacement diagram (Fig. 2.6(c)) by dividing the relevant section of the follower stroke into parts proportional to $1^2$, $2^2$, $3^2$, etc. (1, 4, 9, etc.) and projecting them horizontally to intersect ordinates drawn from the same number of equal divisions on that section of the cam revolution over which parabolic motion is required.

As with harmonic motion, parabolic motion commences with zero follower movement, accelerates uniformly to a maximum velocity at halfway through the motion, then decelerates uniformly back to zero over the second half of the motion.

In each of the above cases, the motion may be applied to either the rise or fall of the follower, the curve beginning at the bottom or top of the displacement diagram respectively and progressing in the direction of the arrows.

Figure 2.7 illustrates a typical cam displacement diagram on which the three types of motion in Figure 2.6 are used. The use of dwell periods is also shown; for that section of the cam revolution the follower is stationary within its stroke. The cam profile for a dwell period is circular.

**FIGURE 2.7** ▶ Cam displacement diagram with typical curves

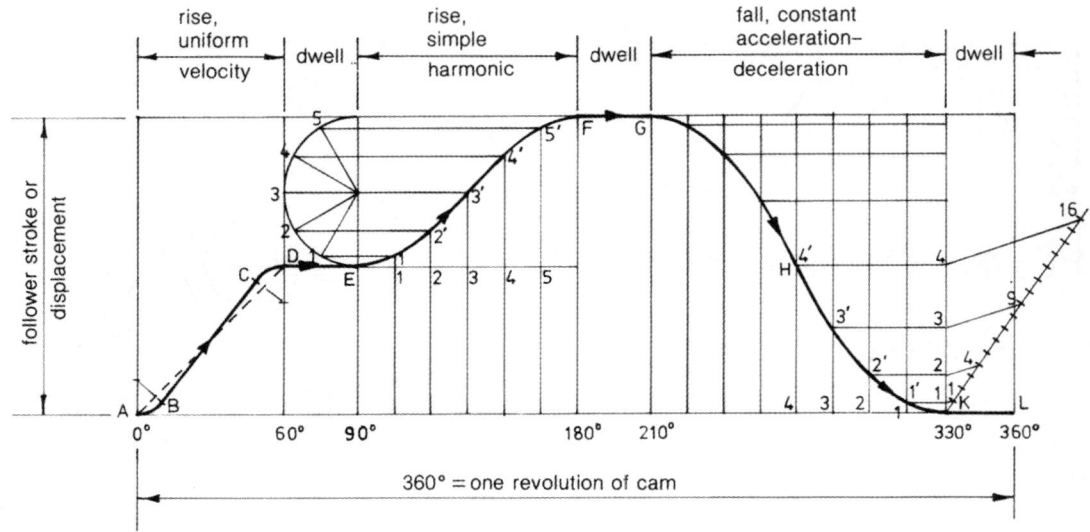

The constant velocity motion, dashed line AD, may be modified to prevent abrupt changes in the follower's motion. It is achieved by inserting radii at the beginning and end of the motion to give curve ABCD.

The following description of the follower motion relates to the displacement diagram Figure 2.7. Commencing from the bottom of the stroke (point A) the follower rises with modified constant velocity (curve ABCD) through half the stroke during 60° rotation of the cam. It then dwells (DE) for the next 30° rotation, and finally completes the rise (curve EF) with harmonic motion. The cam at this stage has completed half a revolution (180°), and the follower is at the top of its stroke. For the next 30° revolution, the follower dwells (FG), then falls with constant acceleration for half the stroke over 60° cam rotation to point H, where its speed is maximal, and finally decelerates back to zero speed in falling through the remainder of the stroke (curve HK), the cam having rotated through a further 60° to a total of 330° of the complete revolution. For the remaining 30° of cam rotation the follower dwells (KL) after which the next cam revolution commences. The motion of the follower is repeated according to the displacement diagram.

## 2.44 TO CONSTRUCT A RADIAL DISC CAM FOR A WEDGE-SHAPED FOLLOWER

Construct a cam for a wedge-shaped follower, imparting to it the following motion:

(a) upward stroke during 120° of cam rotation at constant velocity

(b) dwell for 60° of cam rotation

(c) fall to its original position through a further 120° rotation with simple harmonic motion

(d) dwell for the remainder of the cam's revolution

The stroke of the follower is 20 mm and in line with the vertical axis of the camshaft. The minimum radius of the cam is 10 mm, and it is to rotate in a clockwise direction. Draw a displacement diagram to a scale of 15 mm = 90° cam rotation.

1. Draw the centre lines of the camshaft and the highest and lowest positions of the follower.
2. Draw the displacement diagram comprising the required motions.
3. Project the constant velocity motion from the displacement diagram across to the stroke position of the follower.
4. Divide the first 120° rotation of the cam into six equal parts (20° each), and describe radii from the constant velocity points to intersect these divisions in sequence at points 1 to 6.
5. Describe an arc from 6 to 7 (60°) representing the dwell at the top of the stroke.
6. Project the harmonic motion from the displacement diagram across to the stroke position of the follower.
7. Divide the next 120° of cam rotation into six equal parts (20° each), and describe radii from the harmonic motion points to intersect these divisions in sequence at points 8 to 13.
8. Describe an arc from 13 to 0 (60°) representing the dwell at the bottom of the stroke.
9. Line in the outline of the cam through the plotted points.

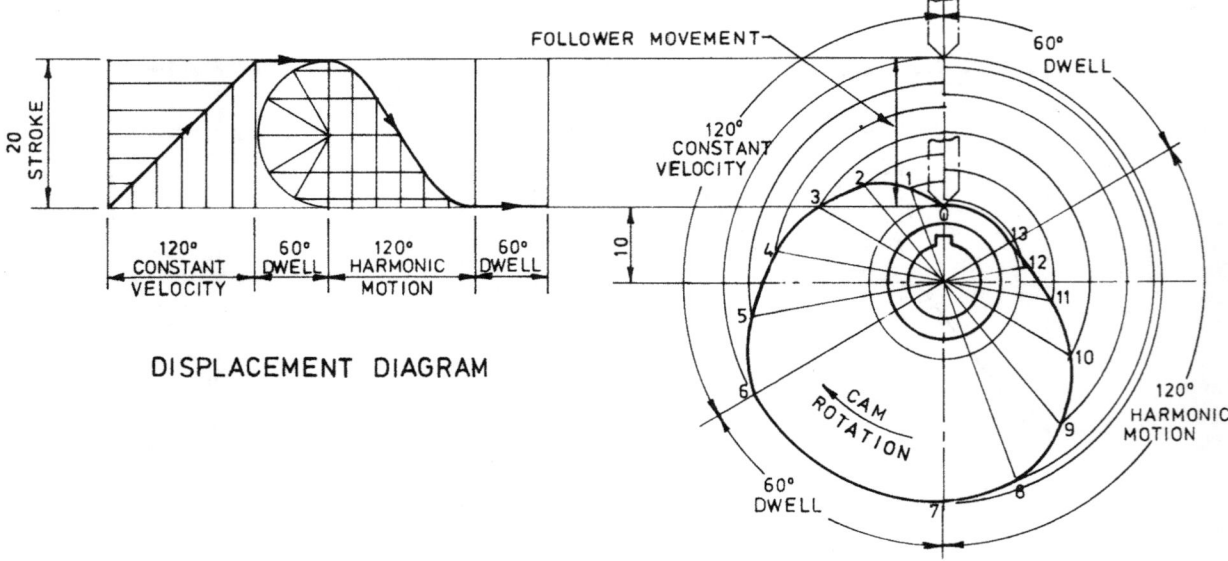

DISPLACEMENT DIAGRAM

## 2.45 TO CONSTRUCT A RADIAL DISC CAM PROFILE FOR A ROLLER FOLLOWER

Construct a radial cam profile for a roller follower, 4 mm in diameter, so that it rises and falls a distance of 16 mm with harmonic motion equally over one revolution of the cam. Consider the following two cases:

(a) when the follower axis is in line with the camshaft centre line

(b) when the follower axis is offset a distance of 6 mm to the left of the camshaft centre line

The least radius of the cam in each case is 10 mm, and the cam rotation is in a clockwise direction (scale: full size).

*Case (a)*

1. Draw the displacement diagram as shown.
2. Position the camshaft axis to the side of the displacement diagram and a distance equal to the radius of the roller (2 mm) plus the least radius of the cam (10 mm) below it.
3. Draw six radial lines through the camshaft axis 30° apart to give the equivalent number of divisions on the cam as on the displacement diagram.
4. Project points 0 to 6 from the harmonic curve to the follower axis.
5. With centre the camshaft axis and radius to the points of division on the follower axis, describe arcs to intersect the radial line through the camshaft axis at points 1, 2, 3, 4, 5 and 6.
6. Draw roller circles at the points of intersection found in the previous step.
7. Draw a tangential curve to the roller circles to give the required cam profile.

*Case (b)*

1. Draw the displacement diagram as shown.
2. Position the follower axis to the side of the displacement diagram, indicating the highest and lowest positions of the roller centre.
3. Draw the vertical centre line of the camshaft axis 6 mm to the right of the follower axis.
4. With centre the lowest roller centre position (O) and radius equal to the roller radius (2 mm) plus the least cam radius (10 mm), describe an arc to intersect the vertical centre line of the camshaft to give the camshaft axis.
5. With centre the camshaft axis and radius equal to the offset, describe a circle and divide it in to twelve equal parts.
6. Draw tangents to each of the twelve divisions on the offset circle.
7. Project points 0 to 6 from the harmonic curve to the follower axis.
8. With centre the camshaft axis and radii to the points of division on the follower axis, describe arcs to intersect the tangents drawn from the offset circle at points 1, 2, 3, 4, 5 and 6.
9. Draw roller circles at the points of intersection found in the previous step.
10. Draw a tangential curve to the roller circles to give the required cam profile.

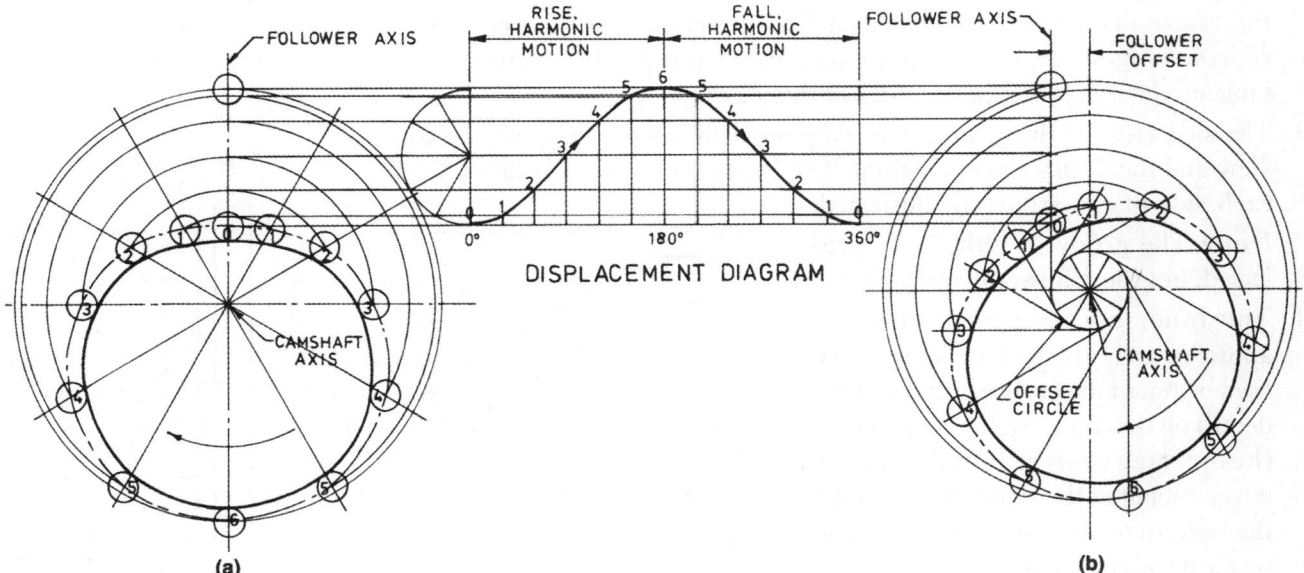

DISPLACEMENT DIAGRAM

(a)  (b)

## 2.46 TO CONSTRUCT A RADIAL DISC CAM PROFILE FOR A WIPER FOLLOWER

Construct a cam profile to raise and lower a wiper follower pivoted at one end through an angle of 30° with uniform velocity. The follower is horizontal in its lowest position, and the cam rotates anticlockwise.

1. Let OO' be the initial position of the follower surface and A be the camshaft axis.

2. Describe an arc of 30° from OO' and divide it into six equal segments numbered 1' to 6'.

3. With centre A and radius AO, describe a circle and divide it into twelve equal parts numbered 0 to 11.

4. With centre A and radii 1', 2', to 6', describe arcs.

5. With centres 1, 2, . . ., 11 and radius OO' successively cut the arcs drawn in the previous step at points a, b, . . ., k.

6. Join points 1a, 2b, . . ., 11k. These lines represent successive positions of the follower as it rises and falls through an arc of 30° while the cam makes one revolution.

7. Draw a smooth curve tangential to these successive positions of the follower, produced if necessary, to give the cam profile.

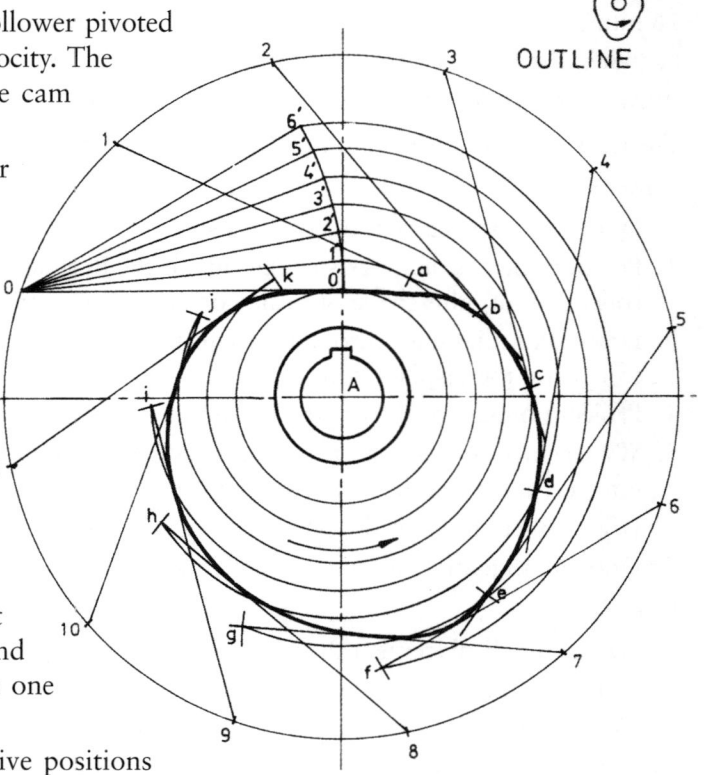

## 2.47 TO CONSTRUCT A CYLINDRICAL CAM FOR A ROD FOLLOWER

Draw a cylindrical cam of diameter D to lift a follower rod a distance AB with harmonic motion during 180° cam rotation and return it with the same motion in the remaining 180°.

1. Draw a displacement diagram opposite the side view of the cam (only half of this diagram is shown).

2. Draw the groove development on the displacement diagram, by plotting the centre line of the follower, then drawing circles on this centre line representing successive follower positions, and finally drawing two tangential curves on opposite sides of these circles.

3. The side view of the groove may be plotted by projecting from the top view and the displacement diagram. The location of one point (at 60°) on each side of the groove is illustrated.

   Note: The groove width, X, is projected, not the follower diameter.

4. The inner curves may be plotted approximately in a similar manner, using the bottom of the groove circle (hidden detail) on the top view. If a true view of the inner curves is required, then the development of the cylinder containing the bottom of the groove must be used on the displacement diagram.

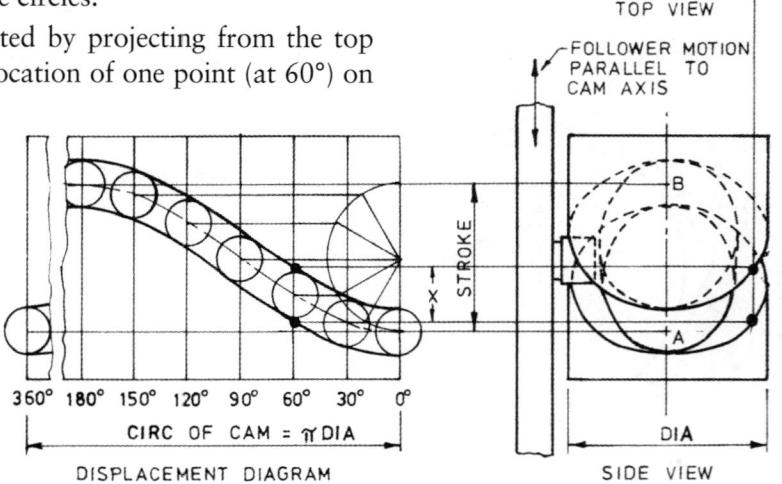

# 2.4 Conic sections

When a cone is intersected by a plane, one of four well-known geometrical curves is obtained, depending on the angle of intersection. Figure 2.8 shows the side view of a cone and the curves relevant to a given plane of intersection. When the intersecting plane:

1. is perpendicular to the axis, the section outline is a *circle*
2. makes a greater angle with the axis than does the sloping surface, the section outline is an *ellipse*
3. makes the same angle with the axis as does the sloping surface, the section outline is a *parabola*
4. makes a lesser angle with the axis than does the sloping surface, the section outline is a *hyperbola*

The true shape of these four sections can be found by projecting an auxiliary view from the edge view of the cutting plane (see construction 2.48).

The ellipse, parabola and hyperbola may also be constructed by considering the geometrical definition governing them, which is:

An ellipse, parabola or hyperbola is the locus of a point which moves so that its distance from a fixed point (called the *focus*) and its perpendicular distance from a fixed straight line (called the *directrix*) bear a constant ratio to each other (called the *eccentricity*). Figure 2.9 illustrates the three curves constructed in this manner.

AB and CD are two lines at right angles and are referred to as the directrix and axis respectively. A point F which can be positioned anywhere on the axis is called the focus. The shape of the three curves is governed by a ratio called the eccentricity (E) which is the ratio:

$$E = \frac{\text{distance from a point (P) on the curve to focus (F)}}{\text{perpendicular distance from point (P) on the curve to the directrix}}$$

Hence, if the distance of the focal point F from the directrix and the eccentricity of the curve are known, the curve may be constructed quite easily by plotting a series of points which conform to the eccentricity ratio, which is as follows for the three conics:

ellipse—any ratio less than unity, for example $\frac{1}{2}$

parabola—a ratio equal to unity

hyperbola—any ratio greater than unity, for example $\frac{2}{1}$

**FIGURE 2.8** ▶ Various conic sections

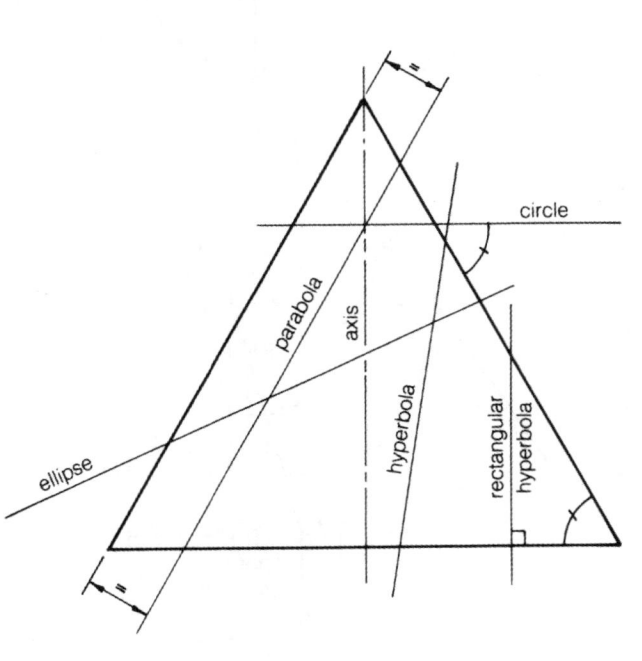

**FIGURE 2.9** ▶ Construction of conic sections by geometric ratios

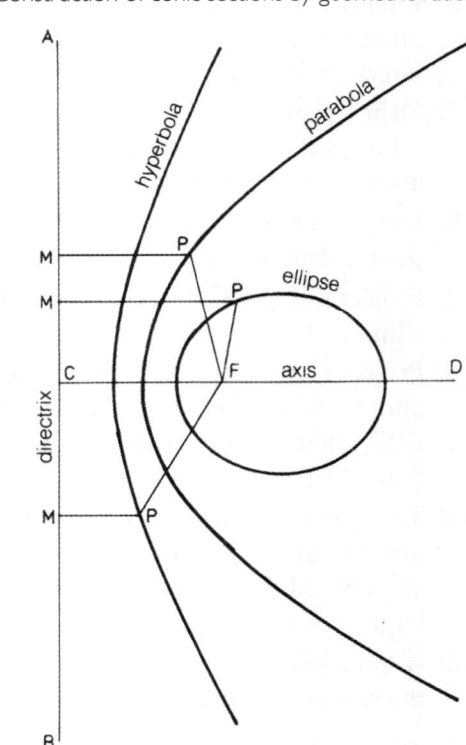

Ellipse

$E = \dfrac{FP}{PM} < 1$

Parabola

$E = \dfrac{FP}{PM} = 1$

Hyperbola

$E = \dfrac{FP}{PM} > 1$

## **2.48** TO DRAW THE CONIC SECTIONS BY AUXILIARY PROJECTION

The conic sections are four types of curves produced when a cone is intersected by a plane as shown in Figure 2.8, namely the circle, ellipse, parabola and hyperbola. The curves may be drawn by the method of auxiliary projection explained in Chapter 4.

The first curve, the *circle*, is drawn by determining the diameter of the cone at that section through which the intersecting plane passes.

The *ellipse* is drawn as shown in diagram (a) by projecting an auxiliary view perpendicular to the line of intersection AB.

This gives the true shape of the section which is elliptical.

1. Draw the side view of the cone and a half top view (semicircle) attached to the base.

2. Let AB be the edge view of the cutting plane.

3. To draw the half top view of this cutting plane, project A and B to intersect the base of the cone at A1 and B1, these two points being the extreme points on the half top view of the cutting plane A-B.

4. Draw any horizontal section X-X to intersect AB at C.

5. Draw the half top view of X-X, that is the semicircle $X_1$-$X_1$.

6. Project C on to $X_1$-$X_1$ to intersect at $C_1$, which is one point on the half top view of A-B. Let this projector intersect the base at $D_1$.

7. Other points on this view are similarly found by taking other horizontal sections through A-B. Two are shown on the diagram.

8. Draw a smooth curve $A_1C_1B_1$ passing through the points determined to give the half top view of A-B.

9. Project OO parallel to AB. OO is the long axis of the ellipse.

10. Project C at right angles to AB to intersect OO at $D_2$ and mark off $D_2C_2$ on either side of OO equal to $D_1C_1$; that is, $C_2C_2$ is the true width of the section A-B at point C.

11. The points $C_2$ are on the required ellipse. The four other points are similarly found to give eight points all told. More should be determined if the ellipse is larger.

12. A freehand curve is drawn through the points to give the required ellipse.

The *parabola*, diagram (b), and the *hyperbola*, diagram (c), are drawn in a similar manner except that these two curves have a flat base because the intersecting plane A-B cuts through the base of the cone.

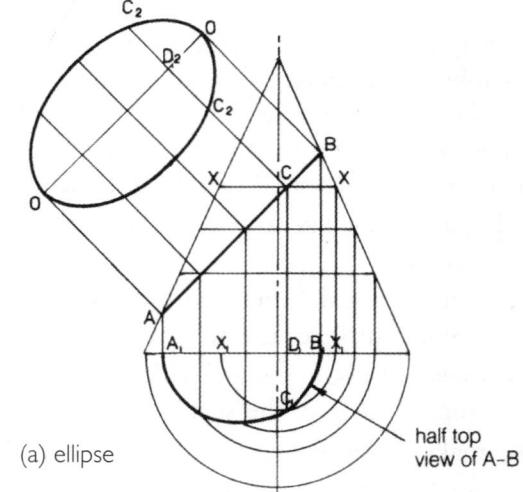

(a) ellipse

half top view of A-B

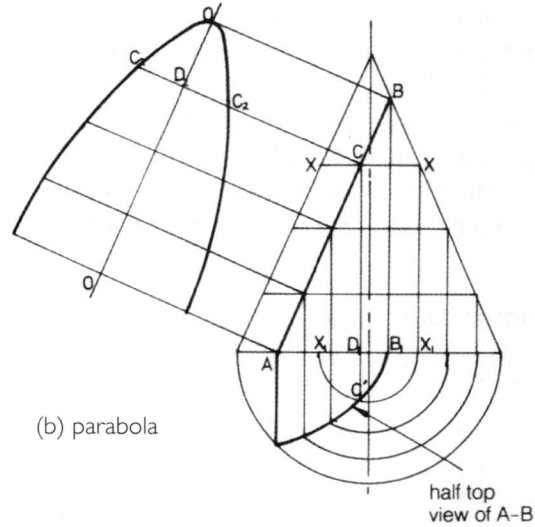

(b) parabola

half top view of A-B

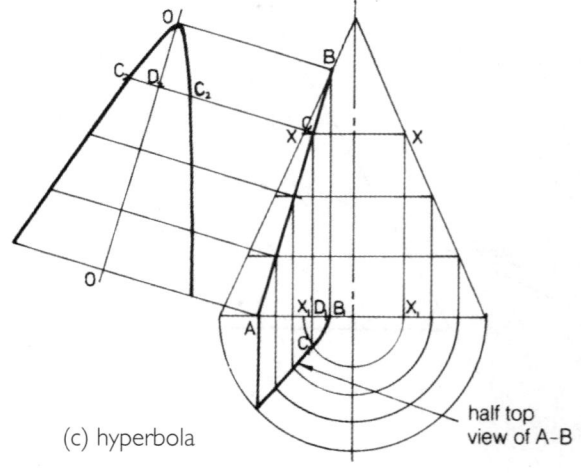

(c) hyperbola

half top view of A-B

## THE ELLIPSE

An ellipse is a closed symmetrical curve with a changing diameter which varies between a maximum and minimum length. These two lengths are known as the *major* axis and *minor* axis respectively. The lengths of the axes may vary greatly, and it is upon their relative sizes that the shape of the ellipse depends.

An ellipse may be defined geometrically as the curve traced out by a point (P) which moves so that the sum of its distances from two fixed points (F and F') is constant and equal in length to the major axis.

In Figure 2.10, AB is the major axis, CD is the minor axis, and F and F' are the *focal points*. From the definition of an ellipse,

$$FP + PF' = AB$$

The definition also leads to a construction for finding the focal points, F and F', when only the axes are given, because as C is a point on the curve,

$$CF + CF' = AB$$

Now CF and CF' are equal, and each is equal to half the major axis AB. Therefore by placing the major and minor axes so that they bisect each other at right angles and taking a radius equal to half AB from C or D, the focal points F and F' are obtained.

**FIGURE 2.10** ▶ Definition of the ellipse geometry

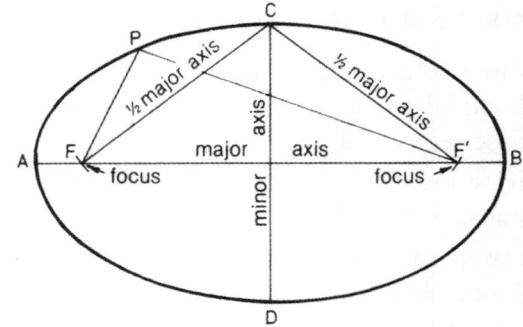

## 2.49 TO CONSTRUCT AN ELLIPSE (APPROXIMATE METHODS)

### The four-centre method

This is used when given the major axis AB and the minor axis CD.

1. Draw the major and minor axes to intersect at O. Join AC.

2. With centre O and radius OC, describe an arc to intersect AO at E.

3. With centre C and radius AE, describe an arc to intersect AC at F.

4. Draw the perpendicular bisector of AF to intersect AO at G and CD (produced) at H. G and H are the centres of two arcs for forming half of the ellipse.

5. Make OJ = OH and OK = OG to give two more centres for the other half of the ellipse.

6. Join JG, HK and JK, and produce all three.

7. With centres H, J, G and K, and radii HC, JD, GA and KB respectively, describe arcs to form the ellipse. The tangent points of the four arcs are at points 1, 2, 3 and 4.

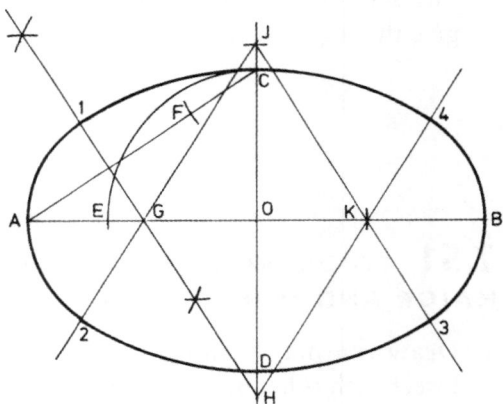

four-centre ellipse

### The isometric ellipse

This is used when given the centre O and the radius of the circle AO required for isometric representation.

1. Draw the isometric axes (any two lines 60° apart) to intersect at O.

2. With centre O and radius AO, describe a circle to intersect the isometric axes at A, D, B and C.

3. Through these four points draw lines parallel to the isometric axes to intersect at E, F, G and H.

4. Draw the long diagonal GE to pass through O.

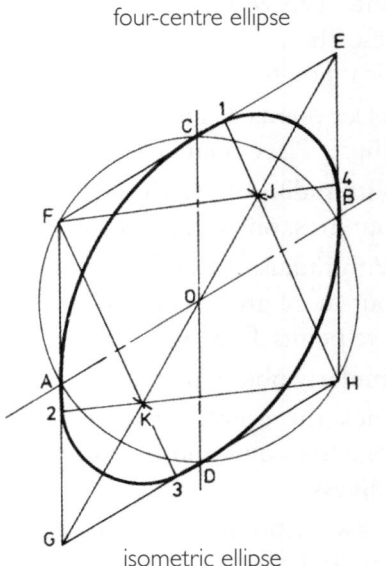

isometric ellipse

5. From E and G, mark off along this diagonal distances EJ and GK equal to AO.

6. Join FK, HK, FJ and HJ, and produce these to intersect EFGH at points 1, 2, 3 and 4.

7. With centres H and F, describe arcs 1-2, 3-4 respectively.

8. With centres K and J, describe arcs 2-3, 1-4 respectively to complete the ellipse.

## 2.50 TO DRAW AN ELLIPSE BY THE CONCENTRIC CIRCLES METHOD, GIVEN THE MAJOR AND MINOR AXES

1. Draw two concentric circles having diameters AB and CD equal to the major axis and minor axis of the ellipse respectively.

2. Divide the circles into twelve equal parts with either compasses or 60°–30° set squares.

3. Draw perpendicular lines from points of division on the outside circle to intersect horizontal lines drawn from corresponding points of division on the inside circle.

4. Draw a smooth curve through these points of intersection together with points A, B, C and D to give the required ellipse.

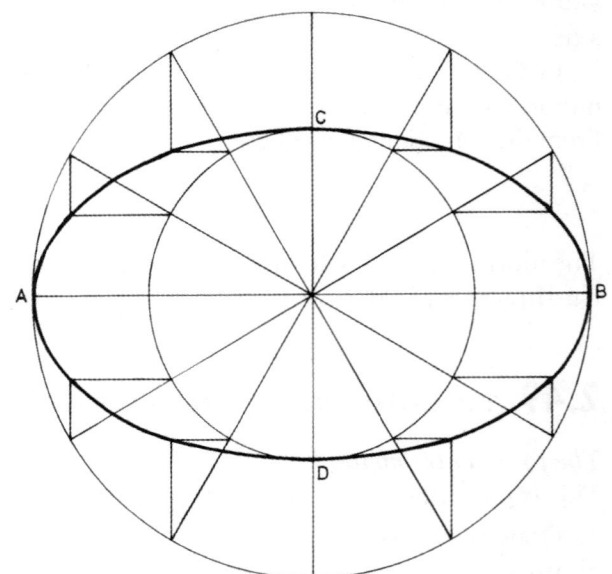

## 2.51 TO DRAW AN ELLIPSE BY THE INTERSECTING ARCS METHOD, GIVEN THE MAJOR AND MINOR AXES

1. Draw the major and minor axes, AB and CD, to bisect each other at right angles.

2. With centre C or D and radius equal to half AB, describe arcs to cut AB at F and F'. These are the focal points of the ellipse.

3. Select points 1 and 2 between F and the centre of the ellipse. Place point 1 fairly close to F.

4. With radius of length A1 and centre F and F', describe four arcs, one in each quadrant of the ellipse.

5. With radius of length B1 and centre F and F', describe four more arcs to intersect the arcs found in step 4 to give points 1' in each quadrant.

6. Similarly obtain points 2'.

7. Prick the points of intersection of the arcs with a compass point and then erase the arcs for the sake of tidiness.

8. Draw a smooth curve through these points to give the required ellipse.

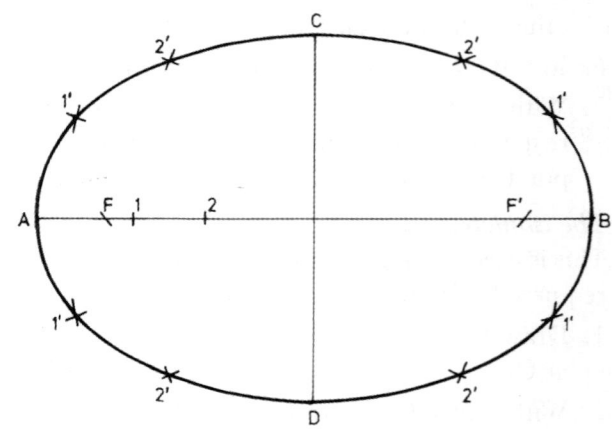

## 2.52 TO DRAW AN ELLIPSE BY THE RECTANGULAR METHOD

1. Draw the major and minor axes, AB and CD, to bisect each other at right angles at O.
2. Draw a rectangle EFGH of length AB and width CD.
3. Divide AO and AE into four equal parts.
4. Join C to the points of division on AE.
5. Join D to the points of division on AO, and produce these lines to meet C1, C2 and C3 to give three points on the ellipse.
6. Using horizontal and vertical ordinates from these points, obtain three points in each of the three quadrants.
7. Draw a freehand curve through these points to give the required ellipse.

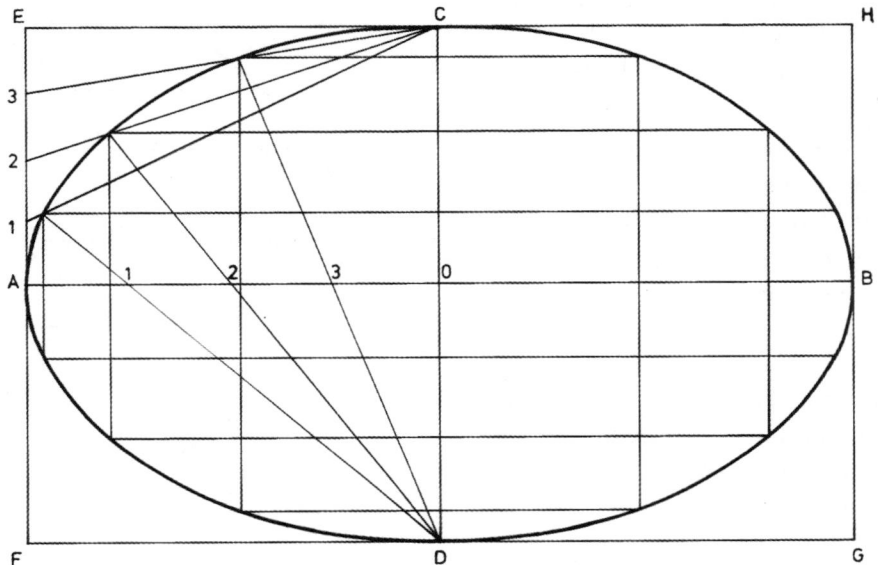

## 2.53 TO DRAW AN ELLIPSE WITH THE AID OF A TRAMMEL, GIVEN THE MAJOR AND MINOR AXES

1. Make a trammel out of a piece of paper as shown.
2. Draw the major and minor axes, AB and CD, to bisect each other at right angles.
3. Using the trammel as shown, and taking care to keep F on the minor axis and E on the major axis, plot a series of points using D as the marker.
4. Rotate the trammel through 360° to give a full range of points on the curve.
5. Draw a freehand curve through the points to give the required ellipse.

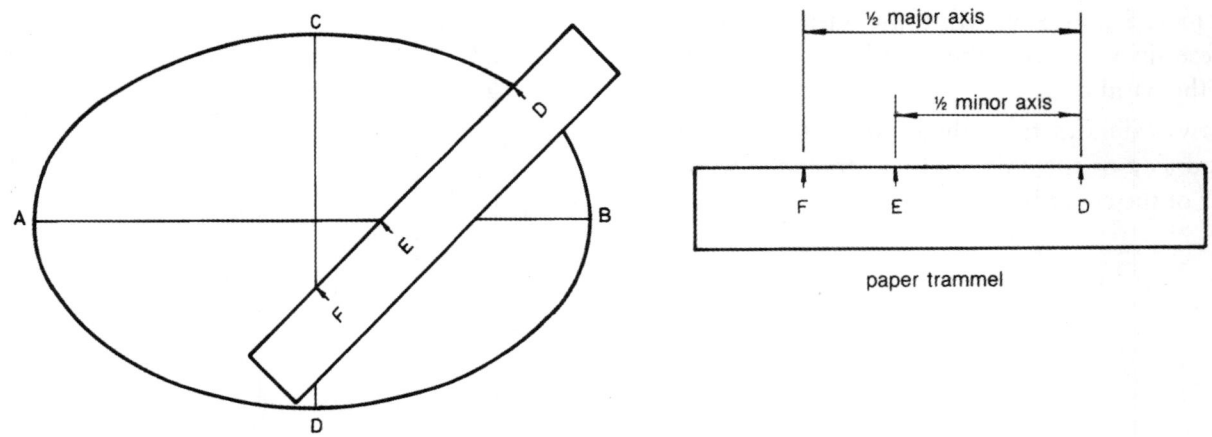

paper trammel

## THE PARABOLA

The parabolic curve has numerous uses throughout industry. Practical applications can be found in the reflection of light beams, for example searchlights. A property of the parabola enables light sources to be positioned relative to a parabolic reflector in such a way that the emerging light waves are parallel, which results in greater brilliance over a longer distance. Some loudspeakers also use the same principle. Civil engineering applications include vertical curves in highways, arch profiles, and cable curves on suspension bridges.

## 2.54 TO CONSTRUCT A PARABOLA, GIVEN THE DIRECTRIX AND THE FOCUS

This is the method outlined in Figure 2.9.

1. Draw the directrix AB, and CD perpendicular to it, to pass through the focus F.
2. Through any point G on CD, draw a line parallel to the directrix.
3. With centre F and radius CG, describe arcs to intersect the parallel line through G at H and J. Then H and J are two points on the parabola.
4. Plot sufficient points in this manner to enable a smooth curve to be drawn. E is the apex of the parabola and CE = EF.

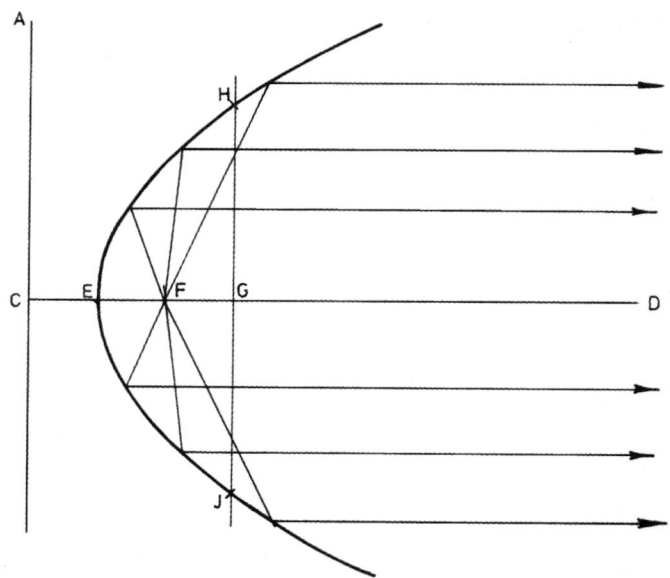

**Note:** If a light source is positioned at F, and the inside surface of the parabola is reflecting, rays of light from F will be reflected parallel to each other as shown by the arrows.

## 2.55 TO CONSTRUCT A PARABOLA, GIVEN THE AXIS AND THE BASE

1. Draw the base EF and the axis CD, the perpendicular bisector of EF.
2. Construct a rectangle ABFE on the base BF of length CD.
3. Divide AC and AE into the same number of equal parts (say 4) numbered 1, 2 and 3.
4. From 1, 2 and 3 on AC, draw lines parallel to CD.
5. From 1, 2 and 3 on AE, draw lines to point C. These lines intersect the parallel lines at points on the parabola.
6. Draw ordinates from these points to the right-hand side of the axis, and construct the second half of the curve by symmetry.

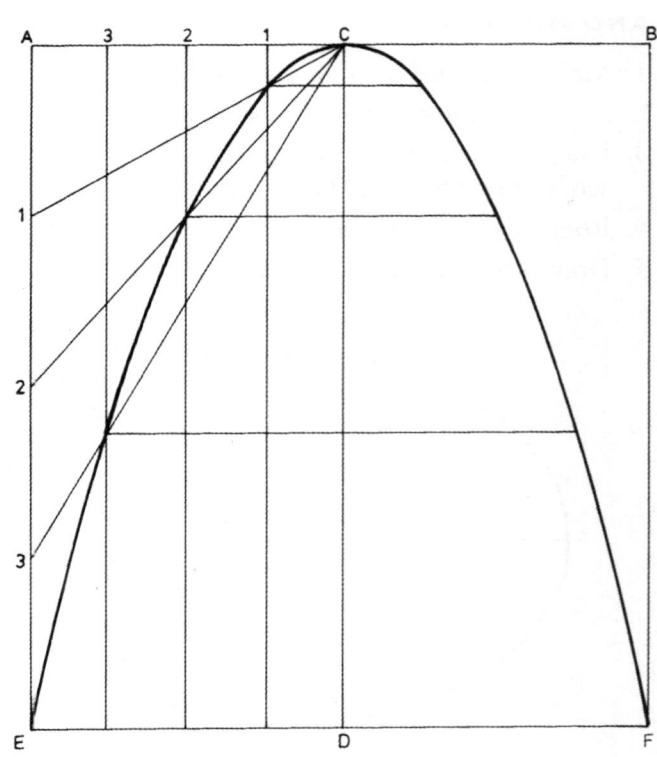

## 2.56 TO CONSTRUCT A PARABOLA, GIVEN THE RISE AND THE SPAN

This method is known as the *offset* method. Parabolic arches are usually drawn by this method.

1. Divide the half span AO into any number of equal parts (say 4).
2. Divide the rise AD into a number of parts equal to the square of the number of points in the half span (in this case 16).
3. Plot points on the parabola by drawing horizontal lines from points 1, 2 and 3 on AB to intersect vertical lines drawn from the square of that number on AD, for example 3 intersects line 9 (the square of 3). Join the points to form a curve.

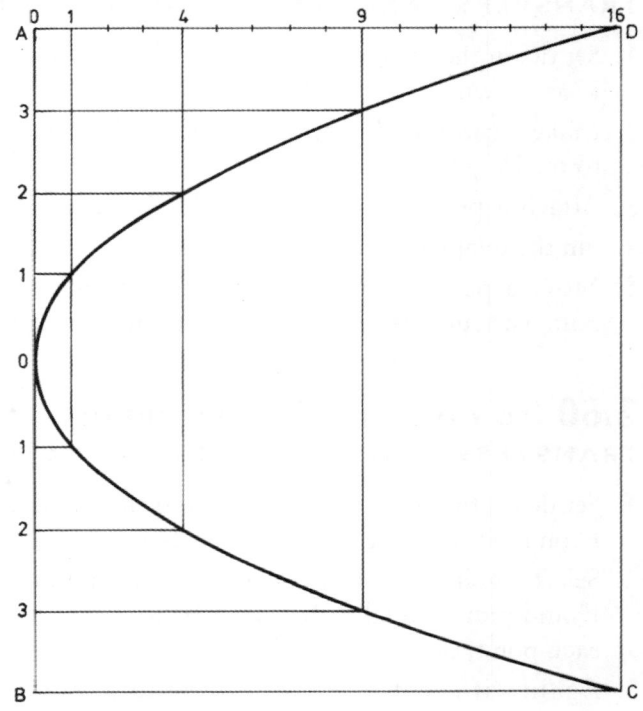

## 2.57 TO CONSTRUCT A PARABOLIC CURVE BETWEEN ANY TWO LINES WHICH MEET

1. Let OA and OB be any two lines which meet at O.
2. Divide OA and OB into the same number of equal parts, numbering the points of division as shown.
3. Join the correspondingly numbered points.
4. The resulting curve drawn tangential to these lines is a parabolic curve.

## 2.58 TO LOCATE THE FOCUS OF A PARABOLA, GIVEN TWO POINTS ON THE CURVE AND ITS VERTEX

1. Let A and B be two opposite points on the curve and C be the vertex.
2. Join AB, and from C draw the perpendicular CE to meet AB.
3. Mark off CD = CE on EC produced.
4. Join AD, which is a tangent to the parabola.
5. Draw the perpendicular bisector of AD to intersect CE at F, which is the focus of the parabola.

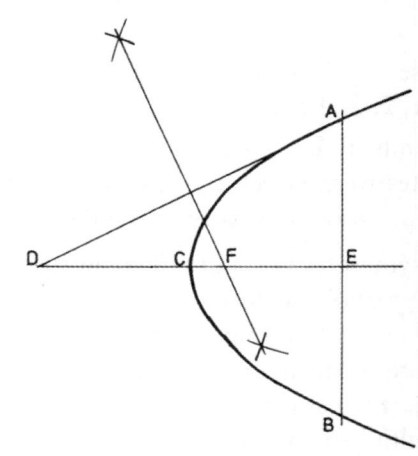

## 2.59 TO CONSTRUCT A HYPERBOLA, GIVEN THE FOCI AND THE TRANSVERSE AXIS (PRACTICAL METHOD)

1. Set down the transverse axis AB and the two foci F and F' as shown.
2. Make a cardboard template FC of any length according to the length of curve required.
3. Attach a piece of string of length FC–AB to F' and C.
4. Pin the template to F.
5. Move a pencil along the template, keeping the string taut. The curve traced out by the pencil is the hyperbola.

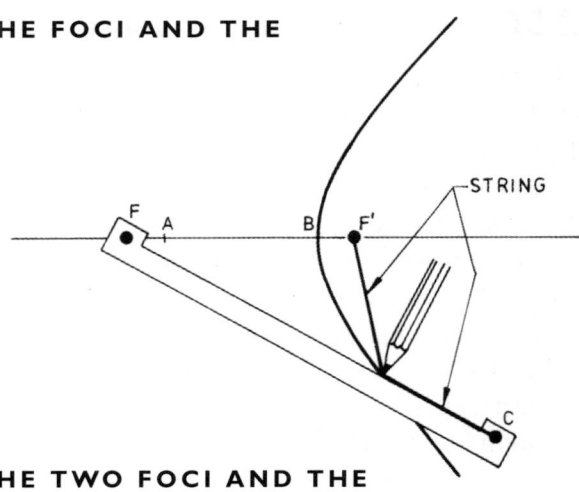

## 2.60 TO CONSTRUCT A HYPERBOLA, GIVEN THE TWO FOCI AND THE TRANSVERSE AXIS (GEOMETRICAL METHOD)

1. Set down the transverse axis AB and the two foci F and F' on the transverse axis extended as shown.
2. Select a point 1 to the right of F' and with centres F and F' and radius equal to B1, describe four arcs (two from each point, one on each side of the axis).
3. Again with centres F and F' and radius equal to A1, describe four arcs (two from each point, one on each side of the axis) to intersect the first four arcs at M. These are points on the required hyperbola.
4. Select two more points 2 and 3, and similarly plot points N and P.
5. Draw two curves through these points and A and B to give two identical hyperbolas.
6. To draw the asymptotes of the hyperbola, draw a circle with FF' as diameter, and erect perpendiculars at A and B to intersect the circle at XX. Join the lines XOX and produce them. These are the required asymptotes.

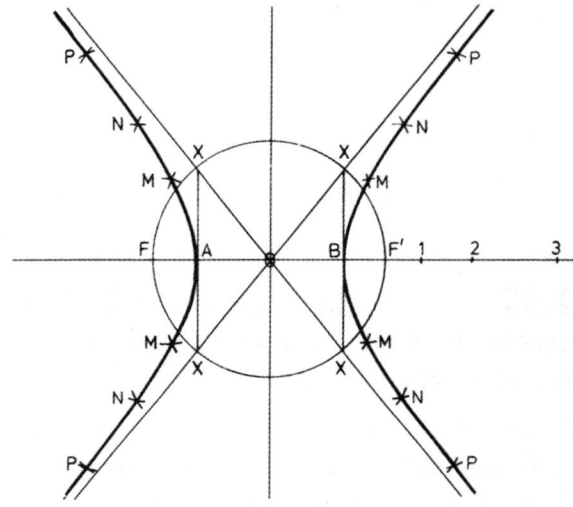

## 2.61 TO DRAW A HYPERBOLA, GIVEN THE ASYMPTOTES AND A POINT ON THE CURVE

1. Let OA and OB be the asymptotes and P be a point on the curve.
2. Through P draw lines parallel to the asymptotes OA and OB.
3. Draw a series of radial lines from O to cut these parallel lines at 1, 2, 3, 4 and 5.
4. From points 1, 2, 3, 4 and 5 draw lines parallel to the asymptotes to intersect and give points on the required hyperbola. Draw a smooth curve through these points.

When the asymptotes OA and OB are at right angles to each other, the hyperbola so formed can be used to express the well-known pressure–volume relationship of PV = k, that is if the pressure is on the horizontal axis and volume on the vertical axis, then for all points on the curve the volume ordinate multiplied by the pressure abscissa is a constant.

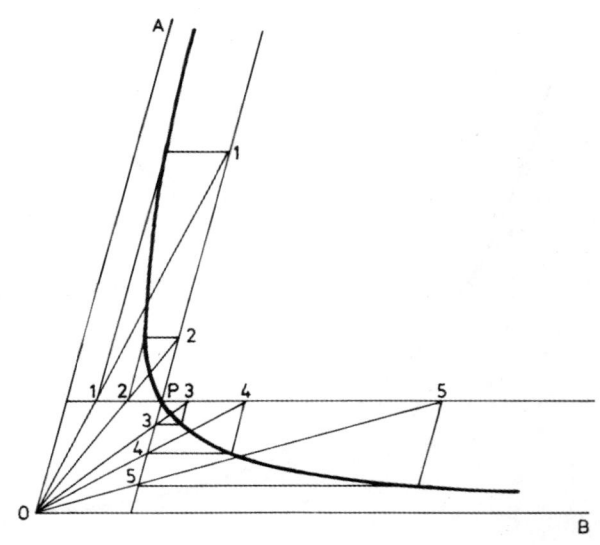

# 2.5 Problems

## CYCLOIDS, INVOLUTE, SPIRALS, CURVES

**2.62** Using a 50 mm diameter rolling circle and a 300 mm diameter base circle for (b) and (c), construct the following:

(a) a cycloid
(b) an epicycloid
(c) a hypocycloid

**2.63** Draw an involute to the following:

(a) a circle 40 mm diameter
(b) the shape shown in Figure 2.11

**FIGURE 2.11**

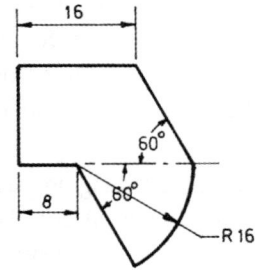

**2.64** Draw one convolution of an Archimedean spiral having its commencing point 10 mm from the pole and its finishing point 50 mm from the pole.

**2.65** Plot the sine and cosine curves from $\pi/2$ to $5\pi/2$ for a radius vector of 30 mm.

**2.66** Plot the locus of the point P as the circle ABC rolls from A to E in Figure 2.12.

**FIGURE 2.12**

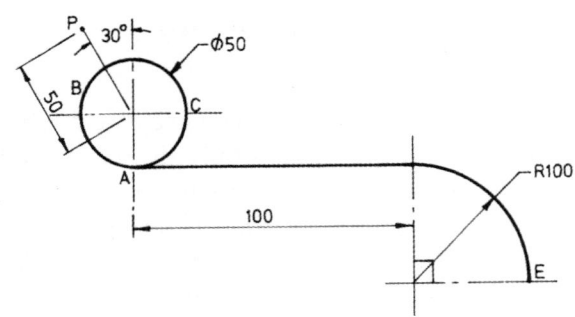

## HELIXES

**2.67** A right-hand helical compression spring is made from 10 mm square steel. Its external diameter is 75 mm. Project correctly two complete turns of the spring.
(pitch = 25 mm, scale 1:1)

**2.68** The profile of a single-start right-hand buttress thread is shown in Figure 2.13. Make a true projection of two complete threads. Omit hidden detail.
(scale 1:1)

**FIGURE 2.13**

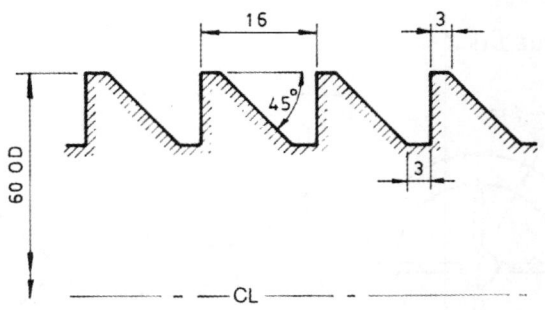

**2.69** Draw a true projection of two coils of a right-hand cylindrical compression spring whose mean diameter is 60 mm, sectional diameter 10 mm and pitch 25 mm.
(scale 1:1)

**2.70** A double-start 60° left-hand V thread has an outside diameter of 50 mm and a pitch of 12 mm. Draw four complete pitches of the thread.
(scale 1:1)

**2.71** The flights of a left-hand helical screw conveyor are formed from a continuous strip of MS bar and welded to the outside of a MS tube. The flight is 225 mm diameter with a 225 mm pitch, and is fitted to a 100 mm tube. The flight material is 5 mm thick. Figure 2.14 shows diagrammatically one pitch of the conveyor.

Draw, to scale of 1:2, a side view showing a true projection of one and a half pitches of the conveyor. The tube is to be shown, but hidden detail may be omitted.

**FIGURE 2.14**

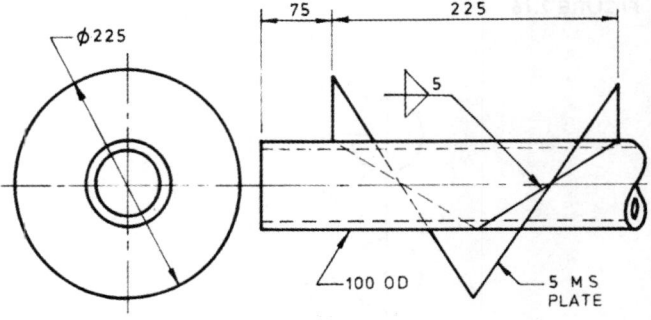

**2.72** A ribbon-type screw conveyor flight is to be made using a 65 mm OD tube and two 65 mm × 15 mm MS flat bars, twisted to form a double left-hand helix as shown in Figure 2.15. The flats are fastened to the tube with 15 mm diameter MS bars spaced every 90°.

Draw a length of 635 mm of the conveyor, including the 75 mm of tube protruding at the end, showing a true projection of the helixes and including the plate thickness. (scale 1:5)

**FIGURE 2.15**

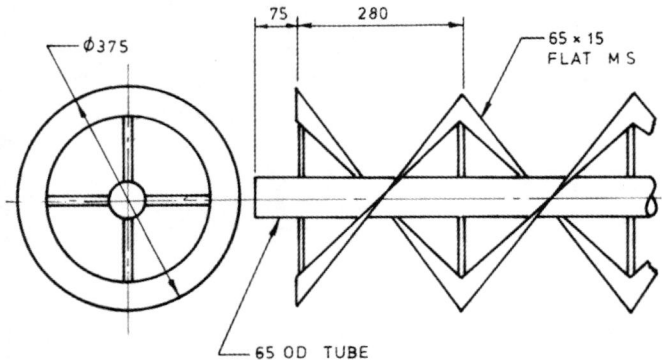

## CAMS

**2.73**  Draw the profile of a radial plate cam to give the following motion to a roller follower in one revolution of the camshaft in a clockwise direction:
(a) outward stroke during 120° of cam rotation at constant acceleration–deceleration
(b) dwell for 60° of cam rotation
(c) fall to the original level through a further 120° rotation with simple harmonic motion
(d) dwell for the remainder of the revolution

The stroke of the follower is 40 mm long and in line with the vertical axis of the camshaft. The diameter of the roller is 25 mm, and the minimum radius of the cam is 50 mm.

Draw a displacement diagram to a scale of 25 mm = 90°.

**2.74**  Figure 2.16 shows the position of a roller-ended follower in relation to the axis of a camshaft. The highest and lowest positions of the follower are shown.

**FIGURE 2.16**

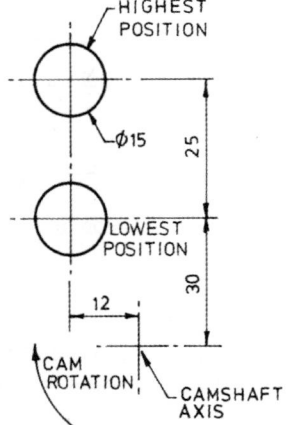

(a) Draw a displacement diagram to a base of 25 mm = 90°.
(b) Draw the profile of a radial disc cam which:
 (i) makes the roller rise with simple harmonic motion to its highest position during 90° rotation of the camshaft
 (ii) remains stationary for the next 90° rotation
 (iii) falls with constant velocity to its lowest position during the remainder of the revolution
 (scale 1:1)

**2.75**  A wedge-shaped cam follower is to have a rise of 36 mm and is offset 20 mm to the right of the camshaft axis, as shown in Figure 2.17. The least distance from the camshaft axis to the follower is 40 mm. The follower is to rise 18 mm with simple harmonic motion for half of a camshaft revolution, dwell for a quarter of a revolution, and rise the remaining 18 mm during the last quarter of the camshaft revolution with uniform acceleration–deceleration, then return instantaneously to the starting point.

Draw a displacement diagram (using a time scale of 30 mm = $1/4$ of a revolution) and the profile of the cam necessary to impart the above motion to the follower (using scale 1:1).

**FIGURE 2.17**

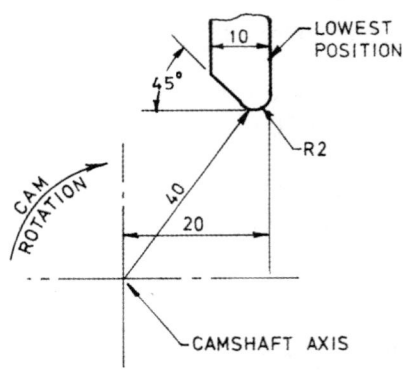

**2.76**  A cylindrical cam, 100 mm in diameter and 100 mm long, is mounted on a horizontal 25 mm diameter camshaft. A 12 mm roller follower is to be moved by the cam from its initial position 25 mm from the right-hand end to the left (a distance of 50 mm) with constant velocity during 180° rotation, dwell for 30° rotation and return to its initial position with uniform acceleration–deceleration.

Draw the side view of the cam showing the true profiles of the outside and bottom edges of the groove, together with a displacement diagram for 360° camshaft rotation. (scale 1:1)

**2.77**  Figure 2.18 shows an oscillating roller-ended follower, radius 125 mm.

Determine the profile of a cam, centre O, which in revolving once causes the follower to rise and fall through 30° about a mean horizontal position with uniform angular velocity.

**FIGURE 2.18**

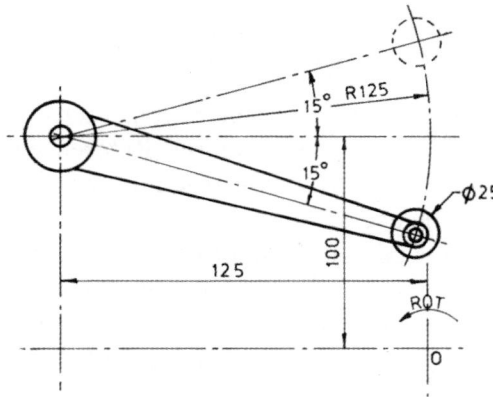

**2.78** A 12 mm diameter roller follower is constrained to move at an angle of 30° to the horizontal centre line of a camshaft, as shown in Figure 2.19. The extreme positions of the roller are also shown in relation to the camshaft axis.

(a) Draw the profile of a cam which would cause the follower to trace out the following motion during one revolution of the camshaft:
   (i) dwell for 30° at the initial position
   (ii) rise with uniform velocity for 180° rotation to the full extent of travel
   (iii) dwell for a further 30°
   (iv) return to the initial position during the remainder of the revolution with uniform acceleration–deceleration(scale 1:1)
(b) Draw a displacement diagram which is representative of the cam.

**FIGURE 2.19**

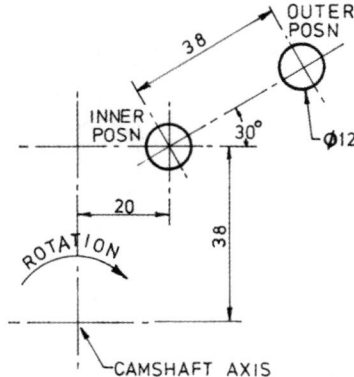

**2.79** A wiper follower has its axis in line with the camshaft axis, as shown in Figure 2.20, and the surface of the follower makes an angle of 60° with this axis. The least radius of the cam, AO = 38 mm.

Draw the profile of a radial plate cam which makes the follower rise 45 mm with uniform acceleration–deceleration over 180° rotation and fall back 45 mm with the same motion for the remaining 180° rotation.

Determine the least possible length of the follower surface.
(scale 1:1)

**FIGURE 2.20**

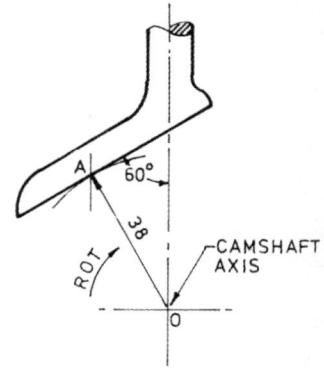

**2.80** Determine the profile of the cam in Figure 2.21 which lifts the tappet, T, 40 mm with simple harmonic motion, and then lowers it 40 mm with uniform velocity in successive half revolutions. In Figure 2.21, T is in its highest position, and the short arm CR of the lever crank is horizontal.
(scale 1:1)

**FIGURE 2.21**

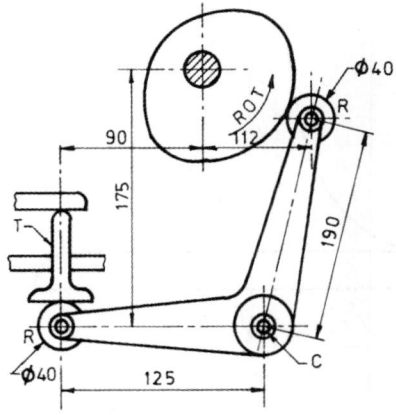

## CONIC SECTIONS

**2.81** Construct an ellipse having a major axis of 90 mm and a minor axis of 60 mm by the following methods:

(a) intersecting arcs
(b) concentric circles
(c) trammel
(c) four-centre

**2.82**  A cone, 60 mm base diameter and 75 mm high, is cut by a plane inclined 45° to the base halfway along the axis.

Determine and name the true shape of the section. (scale 1:1)

**2.83**  Draw a parabola with horizontal axis and distance from the directrix to the focus 16 mm.

**2.84**  A parabolic reflector for a motorcycle is designed so that the emerging beam consists of parallel rays of light. The diameter of the front of the reflector is 180 mm, and its vertex is 170 mm behind the front.

(a) Draw the parabolic profile of the reflector to a scale of 1:2.
(b) Determine the distance from the vertex to the point of location of the bulb element.
(c) Indicate approximately where the bulb must be located so that the beam is slightly converging.

**2.85**  The side view of a circular fan base is shown in Figure 2.22. A parabolic profile is used on the curved surface based on vertical lines through A and C and the horizontal line BD.

Draw the side view of the base, showing lightly the construction for the parabolic surface. (scale 1:1)

**FIGURE 2.22**

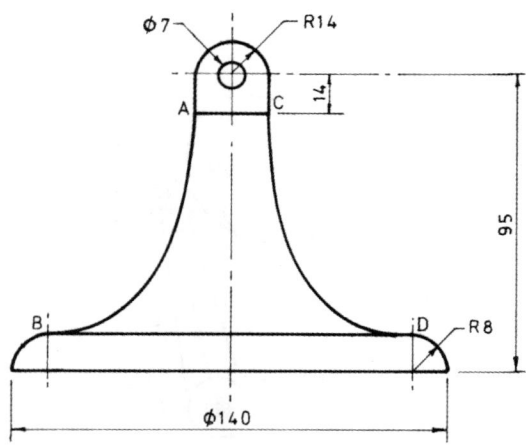

**2.86**  Draw a parabolic arch having a span of 150 mm and a rise of 65 mm using the offset method and dividing the half span into eight equal parts.

**2.87**  The asymptotes of a hyperbola intersect at 110° and at a distance of 30 mm from the vertices of the two branches which lie in the 110° angles.

Draw the two branches of the hyperbola.

# 2.6 Problems (*construction of geometrical shapes and templates*)

The following exercises should be constructed with the aid of compasses and set squares. Reference may be made to the geometrical constructions in this chapter to ensure the use of correct and accurate methods.

All construction lines to locate radii centres should be shown. Indicate all points of tangency with a neat cross. Do not dimension the shapes. A uniform thickness and darkness of outline is required throughout.

### 2.88  PLATE CAM

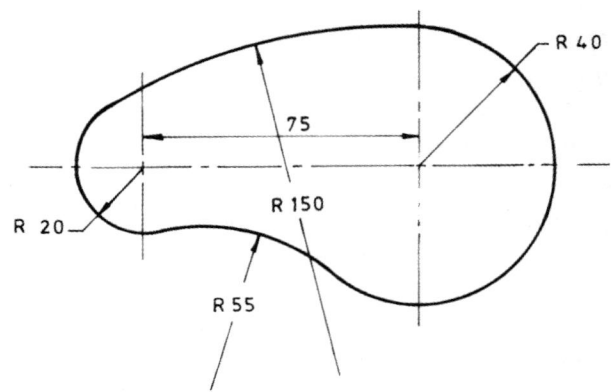

### 2.89  LOCKING PLATE

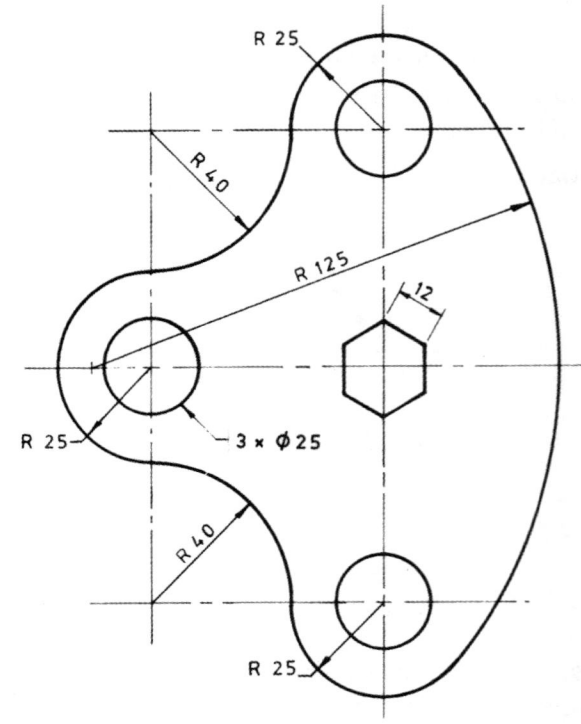

## 2.90 HANDRAIL SECTION

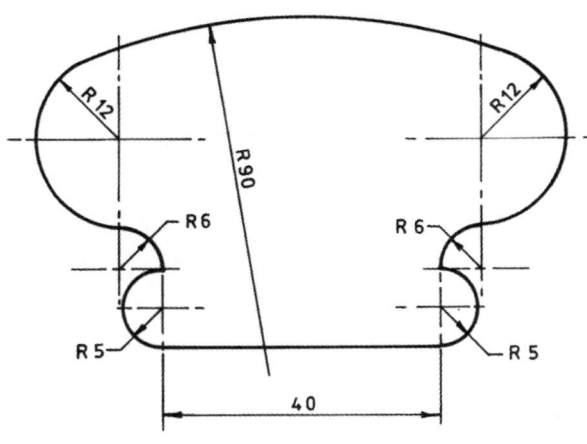

## 2.93 TRIP LEVER

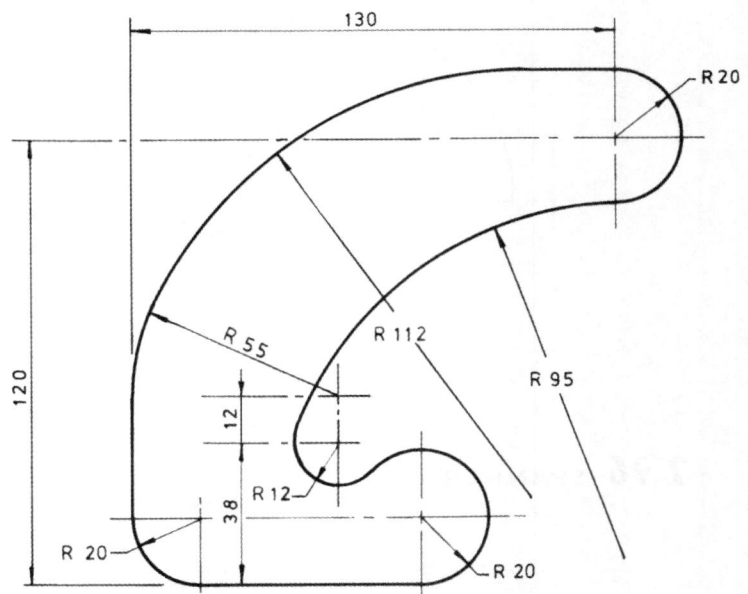

## 2.91 TRIP CATCH

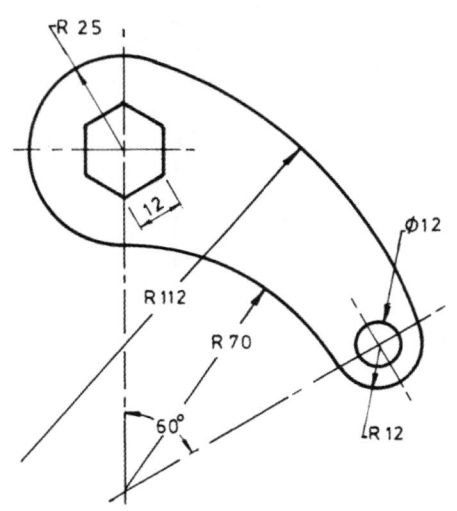

## 2.94 FAN BLADES

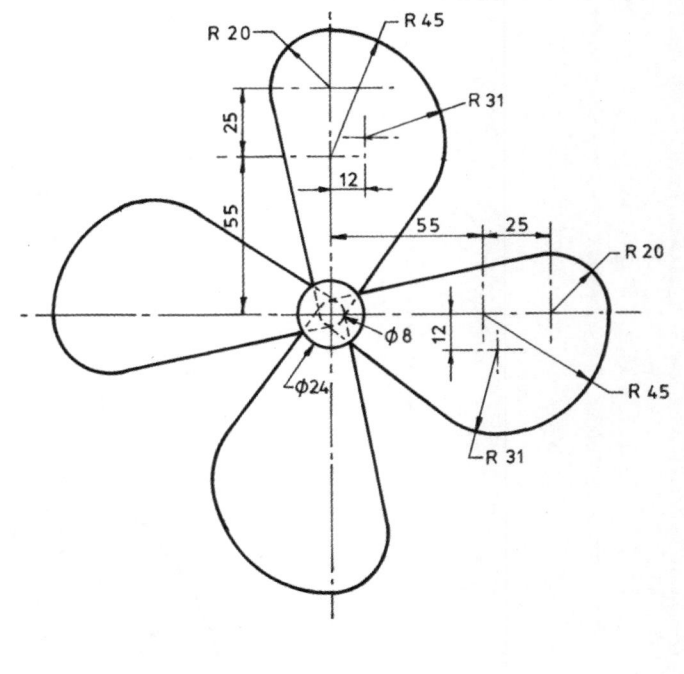

## 2.92 COVER PLATE

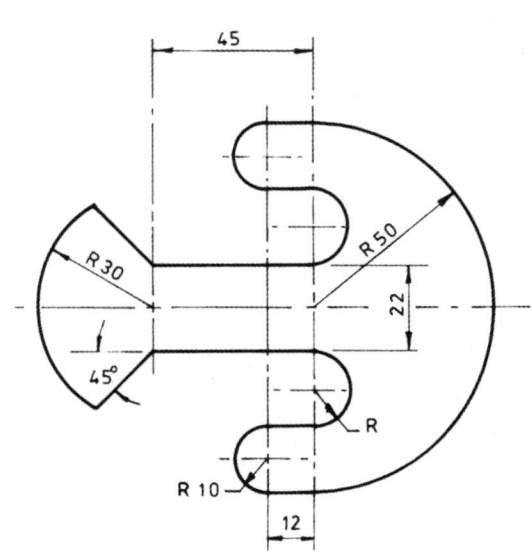

## 2.95 C WRENCH

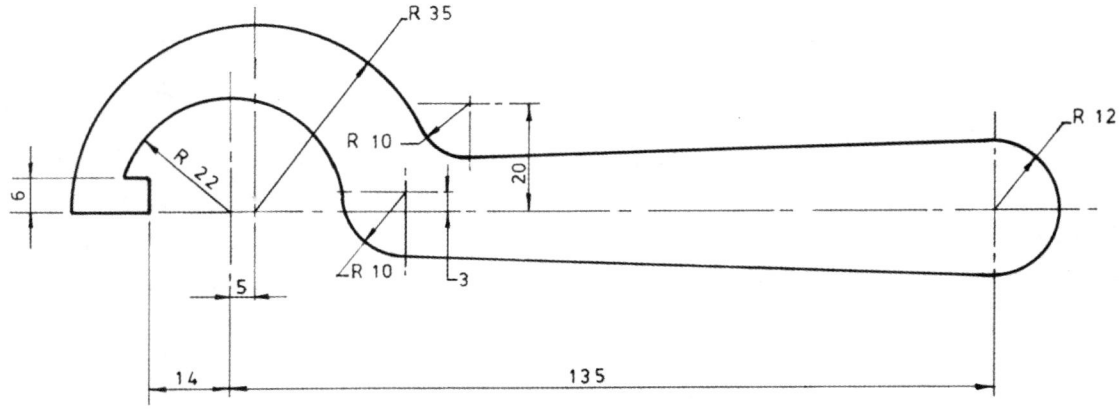

## 2.96 SPANNER

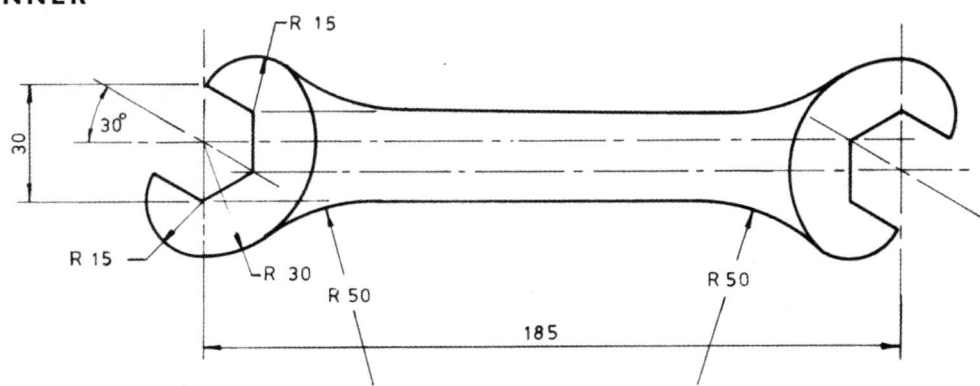

## 2.97 ROLLBAR LEVER

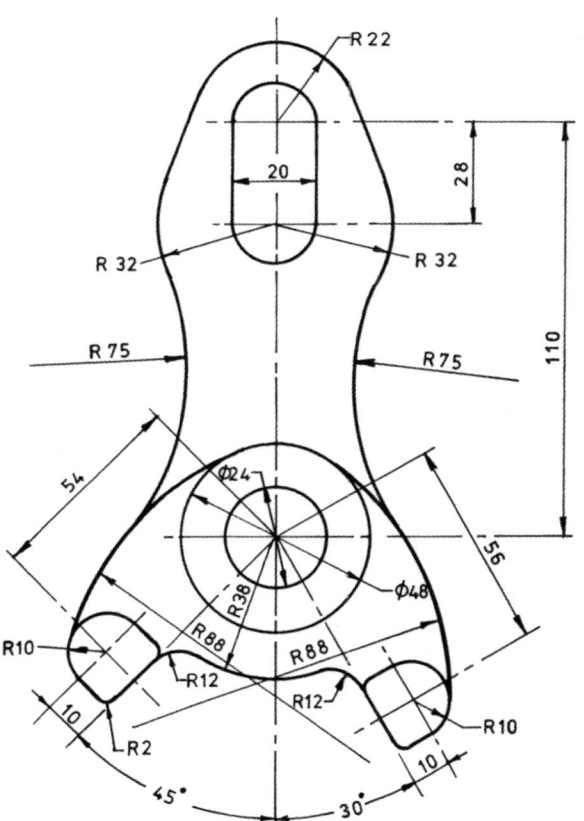

## 2.98 GASKET

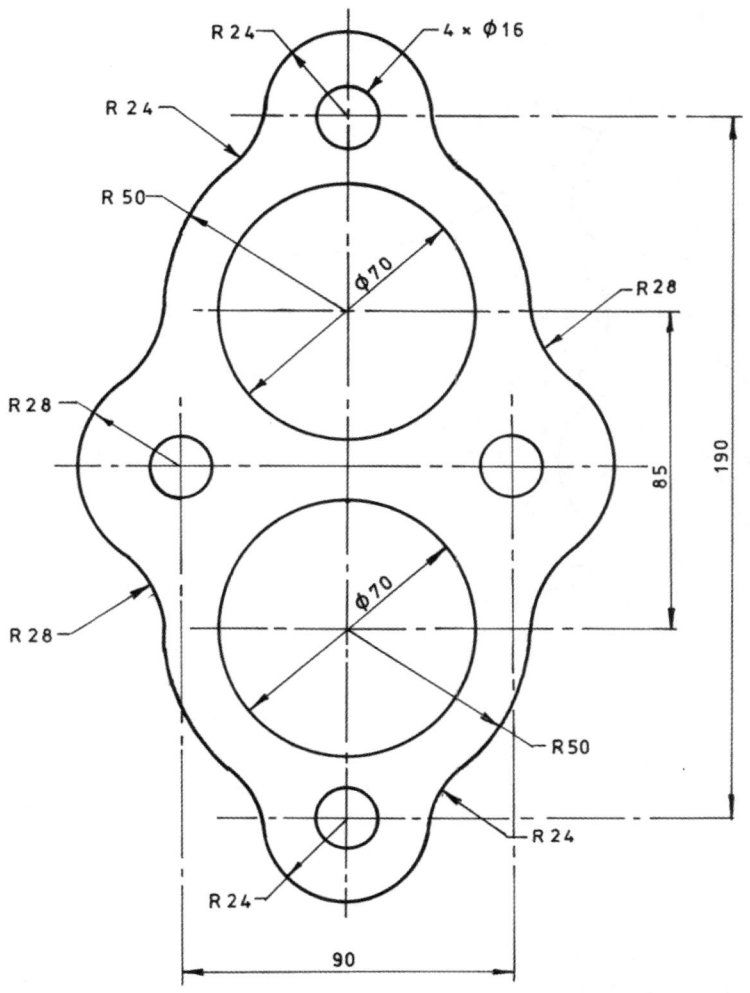

## 2.99 ROCKER ARM

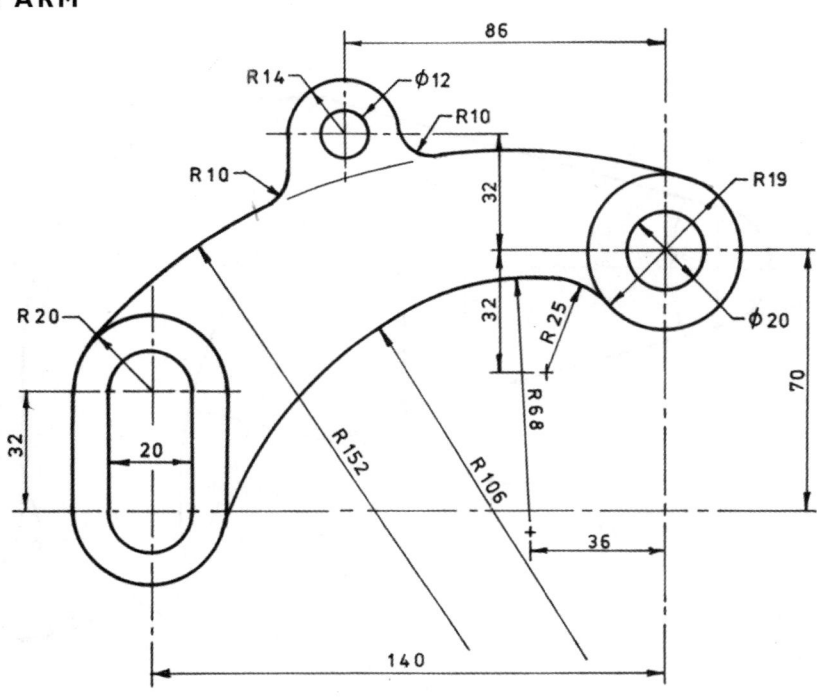

## 2.100 PLATE SPANNER

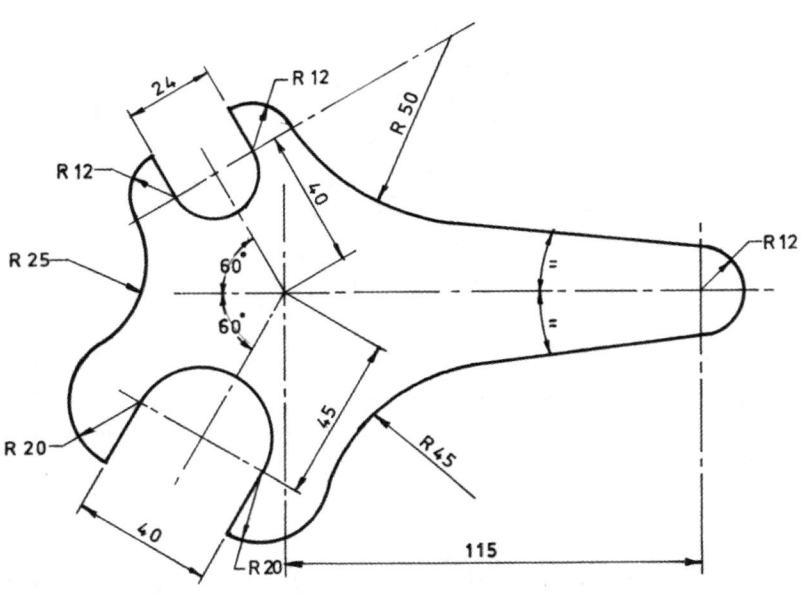

## 2.101 KEYHOLE SAW HANDLE

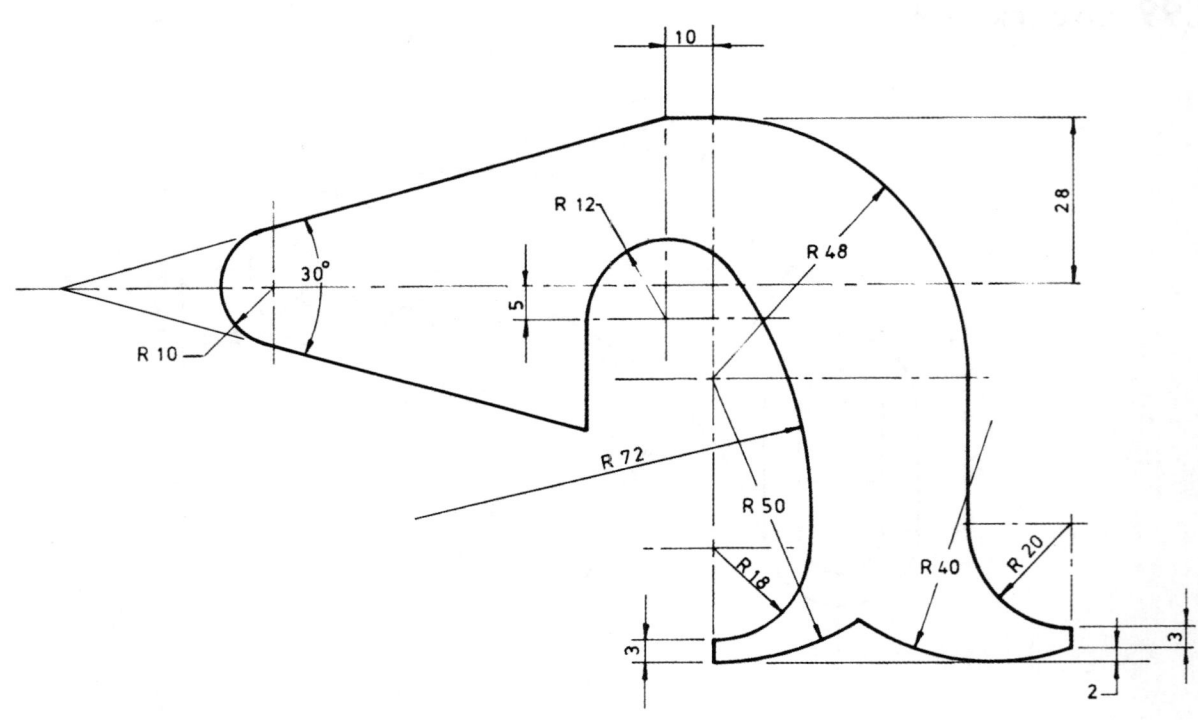

## 2.102 HACKSAW HANDLE

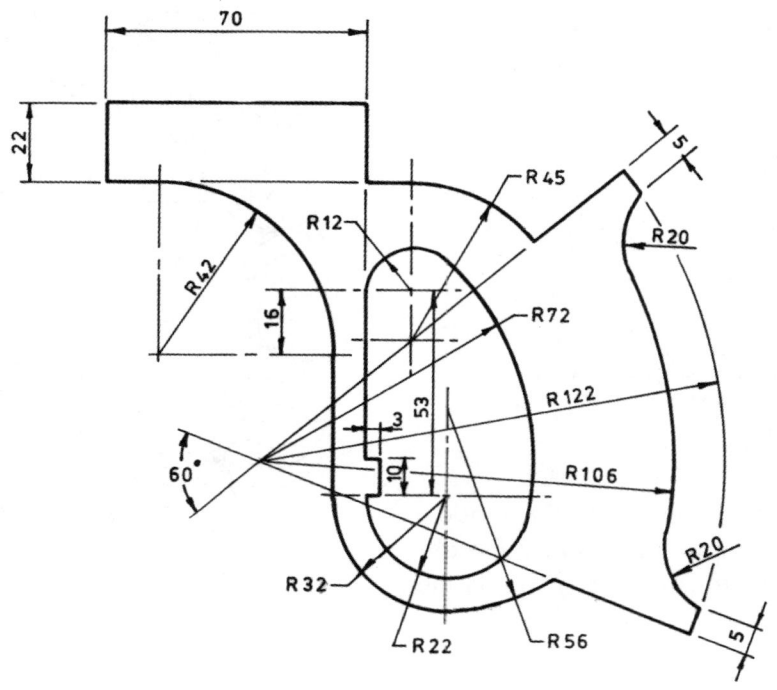

## 2.103 PLANE HANDLE

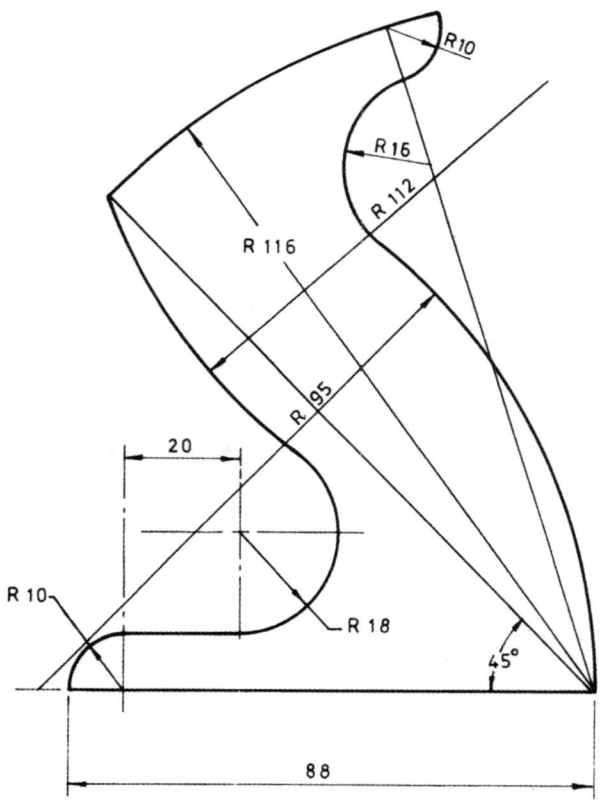

# 2.7 CAD corner

**2.104** Inside a construction circle, diameter 70 mm, complete the figure below to the dimensions shown in Exercise 2.4 on page 93.

Draw a circle, radius 12 mm as shown in the figure below and construct geometry similar to that shown in Exercise 2.5 on page 93, but using the dimensions shown in the figure below.

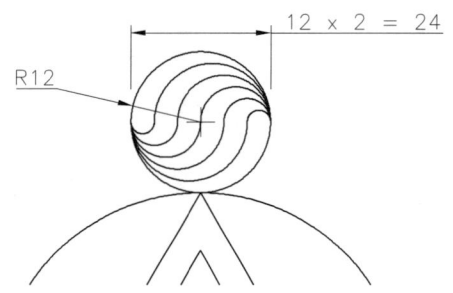

Pattern the circle and its contents to obtain the figure below.

Note: The construction circle is now deleted.

**Challenge:** Prior to the step above fill the areas with colour and add some zing to your pattern.

**2.105** Draw the keyhole saw handle to the dimensions shown in Exercise 2.101 on page 128.

Add the saw blade, complete with teeth, to the handle using the details shown in the figure below.

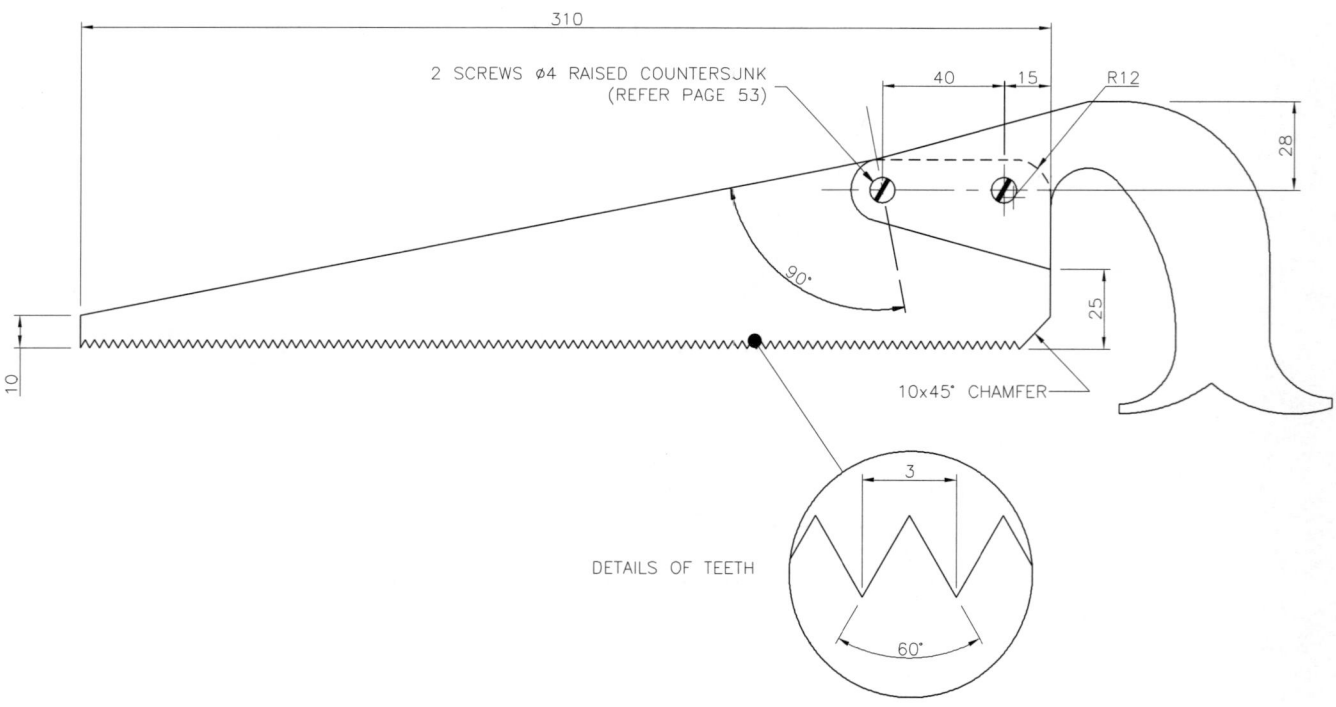

# Orthogonal projection

After studying this chapter and working on the problems you should be able to do the following:

- understand the principles of first and third angle orthogonal projection

- produce a detailed orthogonal drawing of a component including all information necessary for its manufacture

# 3.1 Orthogonal projection

## INTRODUCTION

Engineering drawings are normally intended to indicate the shape and size of an object. However, all objects have three dimensions, namely length, breadth and depth, and the problem of representing these on a two-dimensional drawing as well as conveying an impression of shape to the reader is overcome by the use of a technique called *orthogonal projection*.

## PRINCIPLES OF PROJECTION

Orthogonal projection is a method of viewing an object so that a number of plane views may be obtained, each of which includes two of the object's three dimensions of length, breadth and depth.

When a horizontal and vertical plane intersect at right angles, four right angles (known as *dihedral angles*) are formed and are numbered conventionally as shown in Figure 3.2(a). The first and third of these angles lend their name to the type of orthogonal projection commonly known as *first-angle projection* and *third-angle projection* respectively. *In the interests of standardisation, Standards Australia has recommended that the third-angle method of projection be used.*

# 3.2 Third-angle projection

Assume that the third dihedral angle forms the basis of a six-sided transparent box (Fig. 3.2(b)), in which the object is imagined to be placed so that two of its three principal dimensions (length, breadth and depth) are contained in each of six possible views reflected onto the sides of the box. (Note: Only four views are shown in Fig. 3.2(b); the views from underneath and the rear are omitted for clarity.) So that the views reflected onto the sides of the box may be represented on one plane, the box is unfolded as shown in Figure 3.2(c) and (d). Unfolding the box in this way positions the views in a unique manner with respect to each other. Such relative positioning applies universally to all views obtained by this method.

It is essential that drawings made by the third-angle projection method be identified, preferably by the use of the standard symbol illustrated in Figure 3.2 or by the words *third-angle projection* printed in a conspicuous place on the drawing, usually in the title block.

## DESIGNATION OF THIRD-ANGLE VIEWS

Figure 3.1 shows how the third-angle three-dimensional viewing box is unfolded to give the particular orientation of the six possible two-dimensional views. It will be noticed that the rear view can be positioned on either the extreme right or left of the elevation views, depending on which back corner of the viewing box the rear viewing plane is hinged when swung around to the front. Further consideration will also show that the rear view is orientated the same way irrespective of whether it is positioned on the right or left of the line of views. The alternative positioning of the rear view and its orientation is the same for first-angle projection (p. 135).

Figure 3.2(e) illustrates the six possible viewing positions associated with third-angle projection, and the preferred designation of each view. However, when the method of projection is indicated by the standard symbol, the principal views shown in Figure 3.2(f) require no further identification.

**FIGURE 3.1** ▶ Orientation of rear view

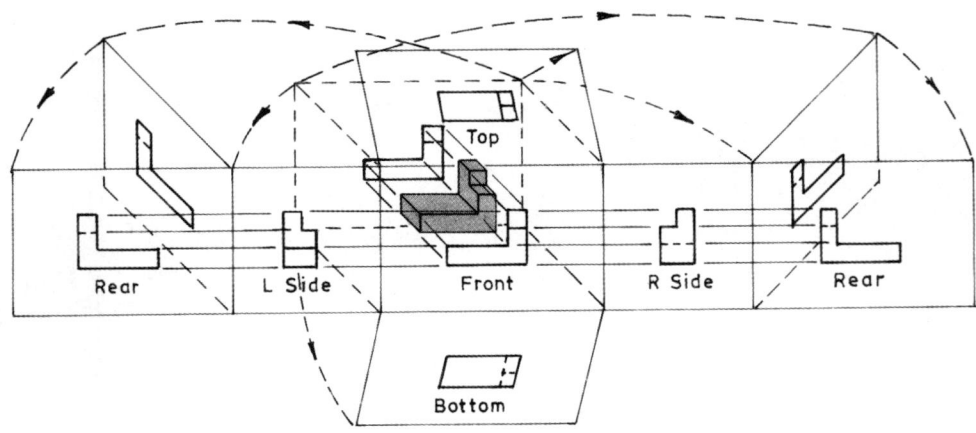

Figure 3.2(f) shows the relative placement of the six designated views, which include the rear view (F) and the bottom view (E) not illustrated on the previous figures. The rear view (F) may be positioned as shown or alternatively on the left of the left side view (C) as indicated by the fictitious outline.

**FIGURE 3.2** ▶ Third-angle projection

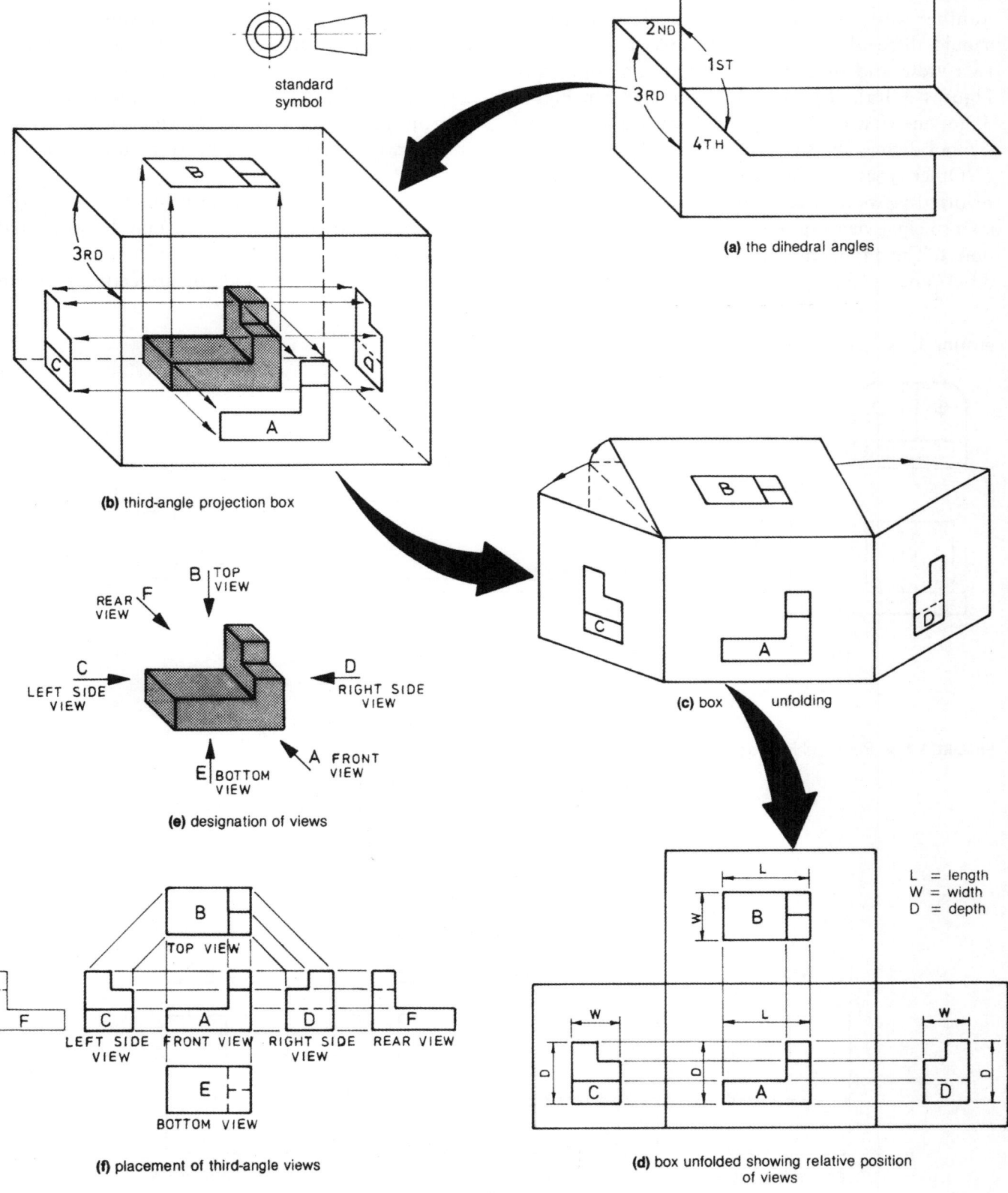

standard symbol

**(a)** the dihedral angles

**(b)** third-angle projection box

**(e)** designation of views

REAR VIEW F

B | TOP VIEW

C LEFT SIDE VIEW

D RIGHT SIDE VIEW

A FRONT VIEW

E | BOTTOM VIEW

**(c)** box unfolding

**(f)** placement of third-angle views

F — LEFT SIDE VIEW — C — FRONT VIEW — A — RIGHT SIDE VIEW — D — REAR VIEW — F

B TOP VIEW

E BOTTOM VIEW

**(d)** box unfolded showing relative position of views

L = length
W = width
D = depth

## NUMBER OF VIEWS

Although six possible views may be drawn, all six are very rarely required. The number used should be just sufficient to indicate the shape of the object and to enable a clear definition of size of all features. For most drawings, three views are adequate. However, the front view is always provided, and whatever number and combination are decided on, they should all be adjacent views. Examples of three-view, two-view and one-view drawings are shown in Figures 3.3(a), (b) and (c) respectively. In Figure 3.3(c) one view only is required because the diameter symbol defines the shape at right angles to the axis.

Other types such as section, auxiliary, partial and revolved views may be used in conjunction with the six principal views to more satisfactorily describe an object. These types are described in Chapter 1, pages 33–37 and in Chapter 4.

## PROJECTION OF ORTHOGONAL VIEWS

Because orthogonal views bear a standard relationship to each other according to the unfolding of the projection box, details such as edges, surfaces, holes, etc. which have been located on one view may be transferred to other views by projection methods. Projecting horizontally between the front, rear and side views with the aid of a tee-square enables height measurements to be transferred quickly and accurately from one view to another. The front view is normally drawn first, and from it detail may be projected horizontally to the side and rear views or vertically to the top and bottom views, and vice versa.

Figure 3.4 illustrates the principle for third-angle projection, showing how detail may be projected between the two side, front and top views.

There are three methods of projecting between

**FIGURE 3.3** ▶ Choosing number of views

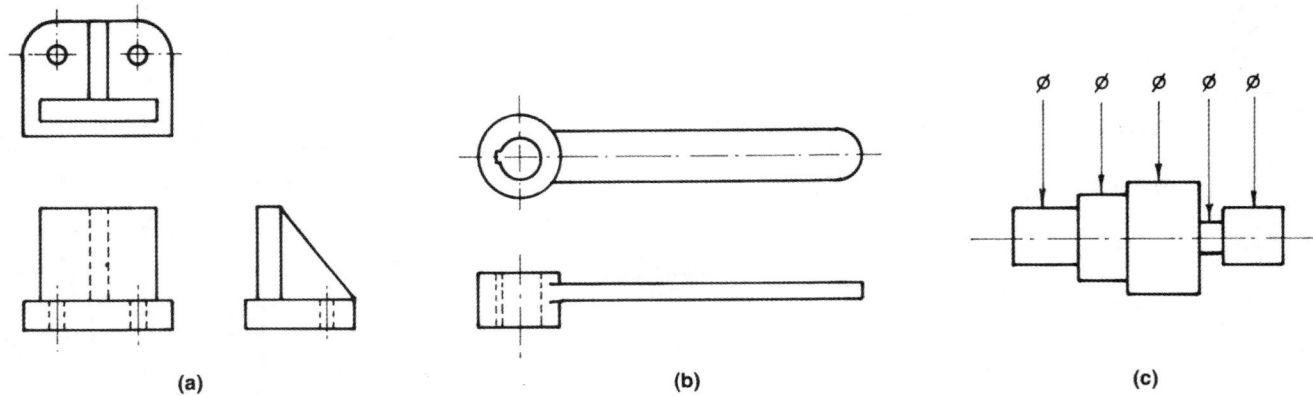

(a)        (b)        (c)

**FIGURE 3.4** ▶ Relationship of orthogonal views

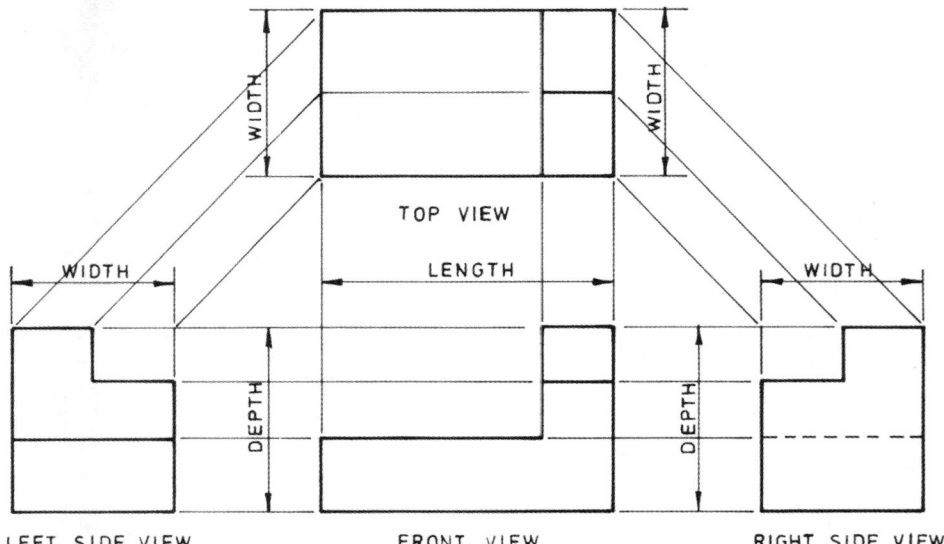

the top and side views: Figure 3.5(a) uses a 45° set square, Figure 3.5(b) compasses, and Figure 3.5(c) combines horizontal and vertical projection lines from a 45° line. In Figure 3.5(a), (b) and (c), the distances between views are the same; however, the distance may be varied by moving the projection quadrant to the side, as in Figure 3.5(d). The top view may be moved further from the front view without altering the side view in a similar manner. The ability to vary the distances between views at will is necessary for proper layout of the views on the drawing sheet.

## 3.3 First-angle projection

The second method of projecting plane views, known as first-angle projection, is illustrated in Figure 3.6 (page 136). *In the interests of standardisation, Standards Australia has recommended that this method not be used and that third-angle projection be the preferred method.*

However, first-angle projection is still used by many firms, and it is essential for the student of engineering drawing to understand the principles of both methods.

**FIGURE 3.5** ▶ Methods of projection between views

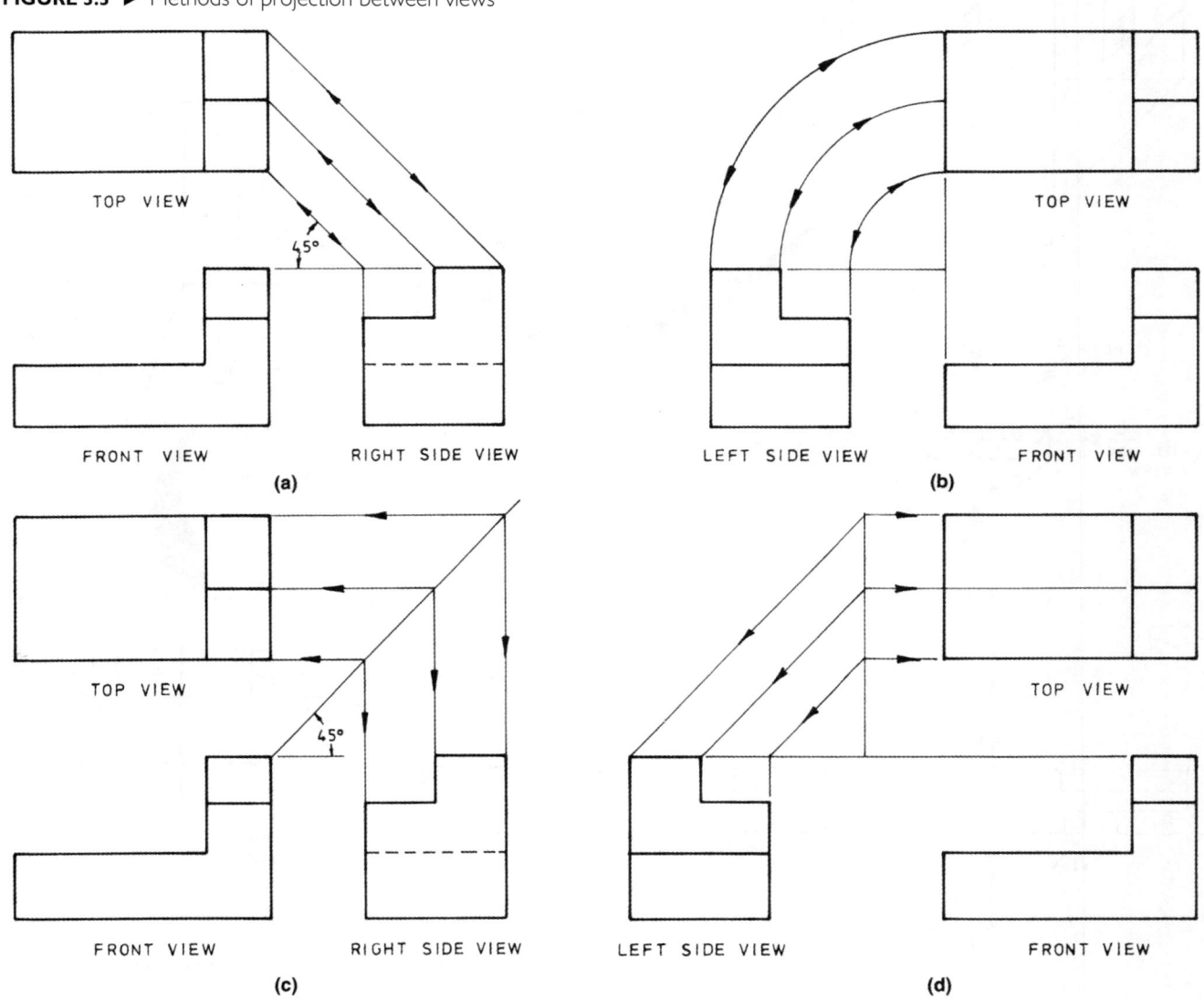

**FIGURE 3.6** ▶ First-angle projection

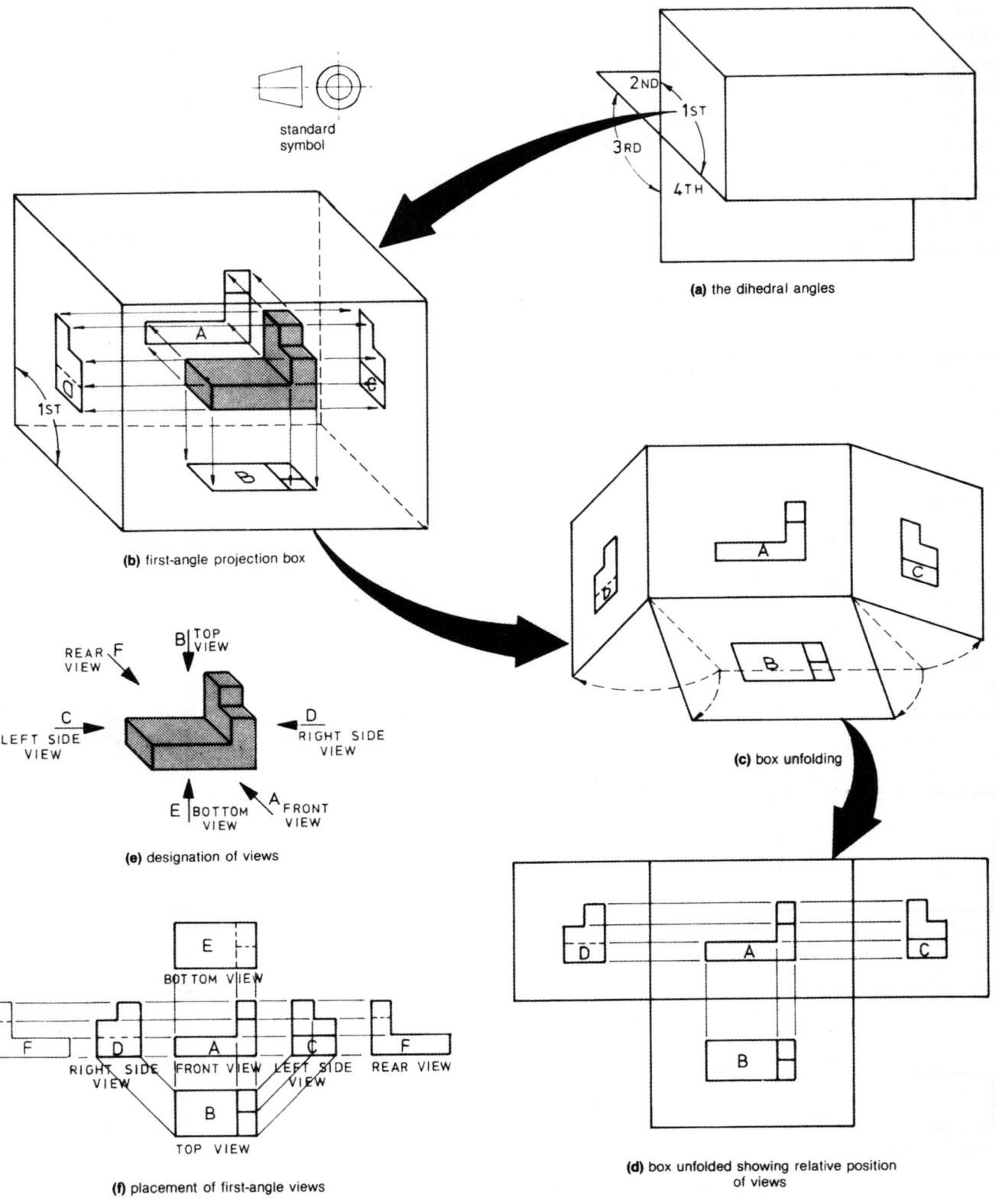

**(a)** the dihedral angles

standard
symbol

**(b)** first-angle projection box

**(c)** box unfolding

**(e)** designation of views

REAR F
VIEW

B TOP
VIEW

C
LEFT SIDE
VIEW

D
RIGHT SIDE
VIEW

E BOTTOM
VIEW

A FRONT
VIEW

**(f)** placement of first-angle views

E
BOTTOM VIEW

F

D

A

C

F

RIGHT SIDE
VIEW

FRONT VIEW

LEFT SIDE
VIEW

REAR VIEW

B
TOP VIEW

**(d)** box unfolded showing relative position
of views

D

A

C

B

# 3.4 Relationship between first-angle and third-angle views

As illustrated in Figure 3.6(e), the designation of views in first-angle projection is identical to that in third-angle projection (Fig. 3.2(e)).

However, a comparison between the two methods of unfolding the dihedral box will show that while the views are identical, their relative positions are different. The difference may be stated simply as follows:

A view in third-angle projection is placed so that it represents the side of the object nearest to it on the adjacent view (Fig. 3.2(f)).

A view in first-angle projection is placed so that it represents the side of the object farthest from it on the adjacent view (Fig. 3.6(f)).

In all other respects the rules of projection for the two methods are identical.

# 3.5 Production of a mechanical drawing

After deciding on a selection of views, the production of a mechanical drawing can be divided into five stages, as follows:

1. drawing of borderline and location of views on the drawing sheet
2. light construction of views
3. lining in of views
4. dimensioning and insertion of subtitles and notes
5. drawing of title block, parts list and revisions table

## I. DRAWING OF BORDERLINE AND LOCATION OF VIEWS

Consider the component shown by the isometric view in Figure 3.7, the orthogonal projection of which is to be drawn on an A2 size sheet (594 mm × 420 mm). The views to be drawn are indicated by the arrows.

It will be observed that the overall length of the front view is 150 mm long and its height is 75 mm. The top view is 150 mm long and its width is 100 mm, while the side view is 100 mm wide and 75 mm high.

**FIGURE 3.7**

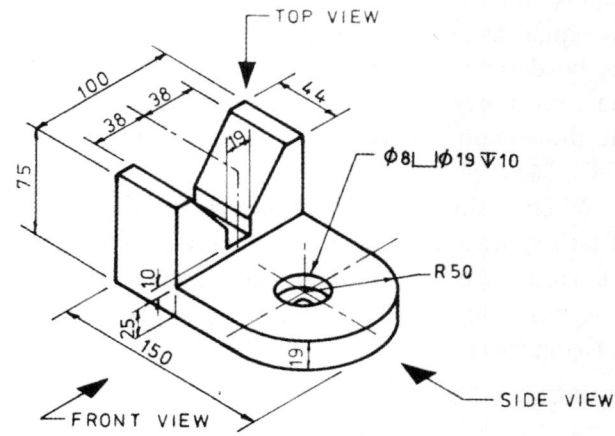

When the number and designation of views have been decided, their correct layout within the available working space is necessary to give the drawing an overall balanced and pleasing appearance. The available working space is that portion of the drawing sheet remaining after allowances have been made for the insertion of such items as the title block, parts list and revisions table. An indication of the dimensions of available working space on various types of drawing is shown in Figure 3.8.

Assuming that a title block 35 mm high is to be provided in the bottom right-hand corner of the drawing frame, the available working space is equal to 574 × (400 − 35) = 574 × 365. (Dimensions of

**FIGURE 3.8** ▶ Dimensions of working space

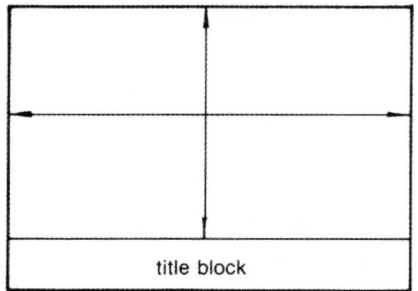
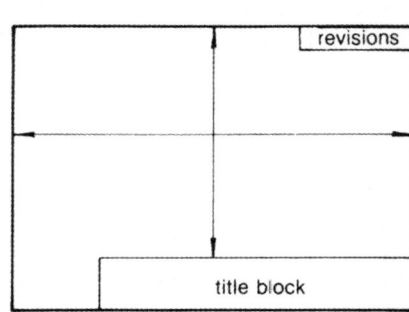
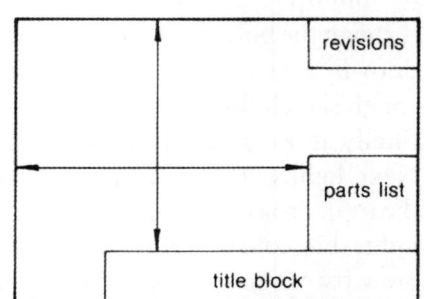

drawing frames and border widths for various sheet sizes are given in Table 1.5, page 12.)

On a piece of rough working paper, sketch a rectangle representing the working space (Fig. 3.9). Place the views in their relative positions on the sheet, and draw two lines, one horizontal and one vertical, along which dimensions are placed. Mark in the dimensions which represent the three views. Decide on distances to be allowed between views, and mark these as well. If possible at least 40 mm should be allowed between views to make room for the insertion of dimensions. Now mark on the sketch the equal distances remaining between the views and the borderline. Finally, add up the dimensions along the two dimension lines and make sure they total the dimensions of the working space, for example 574 × 365.

When making the rough sketch for the location of views, always see that the rectangles representing the views are in their correct relative positions, otherwise the views will be wrongly placed on the drawing sheet.

**FIGURE 3.9** ▶ Positioning of views

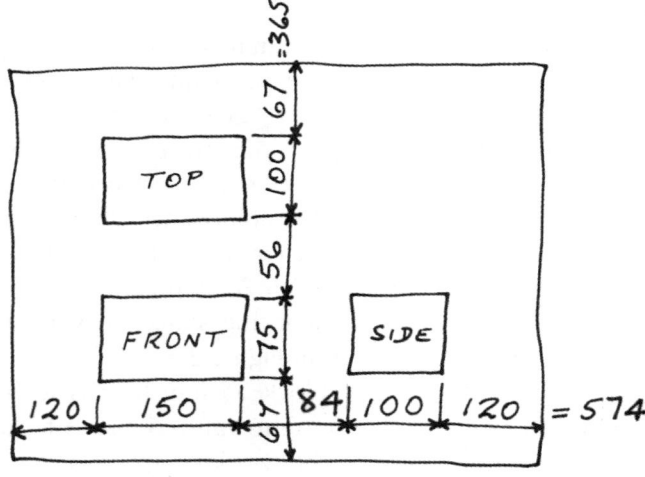

## 2. LIGHT CONSTRUCTION OF VIEWS

Refer to Figure 3.10. Draw a light horizontal line 67 mm up from the title block. This line will pass through the bottom of the front and side views. Lightly draw in the front view, taking measurements from the rough sketch (Fig. 3.9). Hidden detail lines are drawn finally at this stage, as lining them in later is difficult. Next, lightly construct either the top or side view (say the top). Project the top view from the front view using light, thin construction lines. Finally, project the side view from the front and top views using light, thin construction lines. Note that arcs and circles should be lined in at this stage.

It is important that construction and projection lines are ruled lightly, as they may have to be erased or altered as the drawing progresses.

## 3. LINING IN OF VIEWS

Lining in should be done systematically for the three views. Commencing with horizontal lines and at the top of the top view, line in progressively working down the page with the tee-square. Starting at the left-hand side and working across the page, line in all vertical lines using a tee-square and set square combined. Projection lines may be left on the drawing, provided they are very light. If any projection lines need to be erased, this should be done before lining in commences. The views, when lined in, are shown in Figure 3.11.

**FIGURE 3.10** ▶ Construction of views

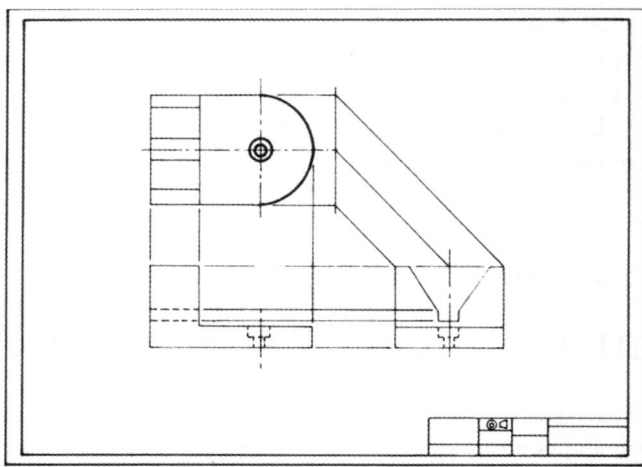

**FIGURE 3.11** ▶ Completed views

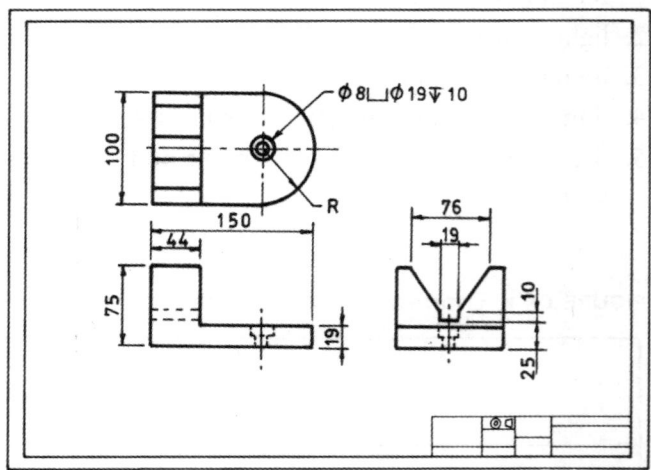

## 4. DIMENSIONING AND INSERTION OF SUBTITLES AND NOTES

At this stage it is necessary to explain the main principles involved when dimensioning a mechanical drawing. There are two rules which students must remember.

### Rule 1

Each dimension necessary to describe a component should be given, and it should not be necessary to deduce a functional dimension from other dimensions on the drawing.

### Rule 2

There should be no more dimensions than are necessary to define a component. That is, it should not be possible to deduce a dimension from other dimensions if that dimension has already been given directly.

Three correct methods of dimensioning lengths which conform to the above rules are given in Figures 3.12(a), (b) and (c). Referring to Figure 3.12, the following points are noted:

1. Where two dimensions together give the length of an object as in Figure 3.12(a), the overall dimension is omitted.
2. When an overall length is shown, as in Figures 3.12(b) and (c), a non-functional intermediate dimension is omitted.

An exception to rule 2 is permitted sometimes when it is useful to show an overall dimension, even though all intermediate dimensions are supplied. The overall dimension is shown as an auxiliary dimension, and this is indicated by placing it in brackets as shown in Figure 3.12(d). An auxiliary dimension is in no way binding as far as machining operations are concerned.

Dimensions which govern the working of a component are called *functional dimensions*. All other dimensions are called *non-functional dimensions*.

The above rules show that a drafter must fully understand the working of a component to be able to indicate functional dimensions; by correct dimensioning, the drafter ensures that features are correctly located on the finished product.

Referring to the completed example (Fig. 3.15, page 141), the following features are regarded as functional:

1. The axis of the bored hole is vertical, centrally located and is a toleranced distance from the back surface, which is machined.
2. The top surface of the boss must be correctly located in relation to the three 16 mm diameter fixing holes.
3. The bore of the boss is a toleranced size.

**FIGURE 3.12** ▶ Principles of linear dimensioning

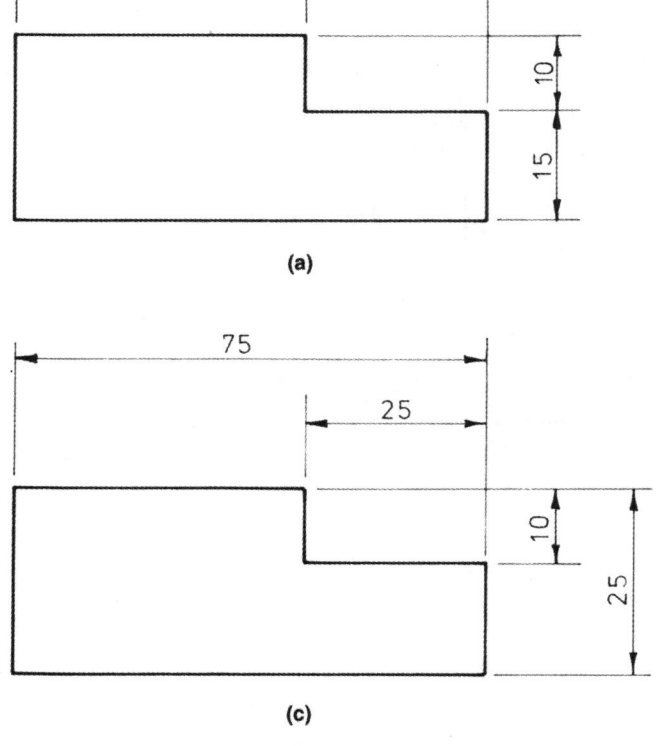

(a)

(c)

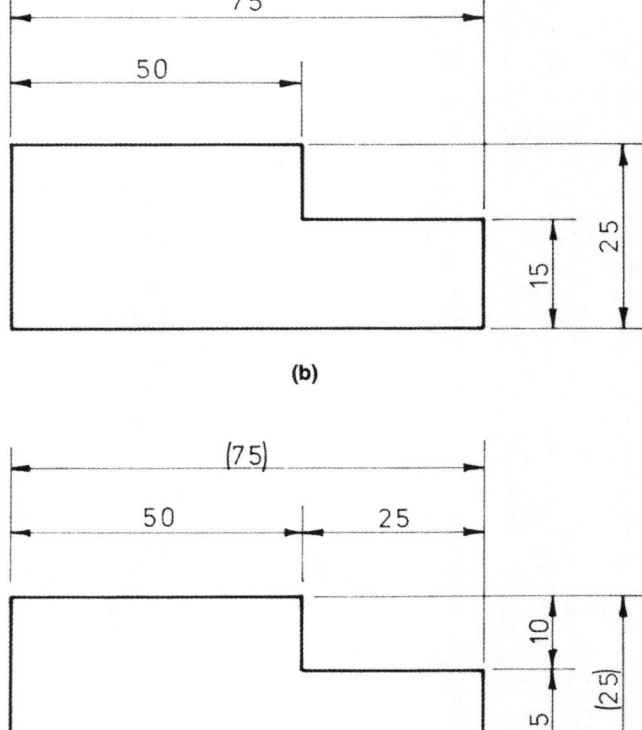

(b)

(d)

The above features are functional and must be dimensioned accordingly. Hence the centre line of the bored hole is dimensioned directly off the back surface. The axis of the boss is located centrally between the fixing holes, and the top of the boss is spotfaced and located 10 mm above the horizontal centre line of the two top fixing holes. The bottom fixing hole is dimensioned from the centre line of the top holes as well.

It is necessary when dimensioning a drawing to decide on one or more base or datum lines from which functional dimensions are taken. The datum lines for the above example are the back surface lines on the top and side views, the vertical centre line on the front view, and the horizontal centre line through the two fixing holes on the front view. Dimensioning of the three views (Fig. 3.15) follows the above rules as well as those given in Chapter 1.

## 5. DRAWING OF TITLE BLOCK, PARTS LIST AND REVISIONS TABLE

A suitable layout for these three items is given in Figure 1.17, and a general description on pages 11–15. For this exercise a title block only is required, and it is inserted in the bottom right-hand corner of the sheet as shown in Figures 3.10 and 3.11.

The following is an example of a typical exercise which involves the drawing of a simple mechanical component in third-angle orthogonal projection.

### Exercise
Figure 3.13 shows an isometric view of a cast-steel wall bracket. Draw the following views in third-angle orthogonal projection:
1. a front view in direction A
2. a side view in direction B
3. a top view

**FIGURE 3.13** ▶ Isometric view of a cast steel wall bracket

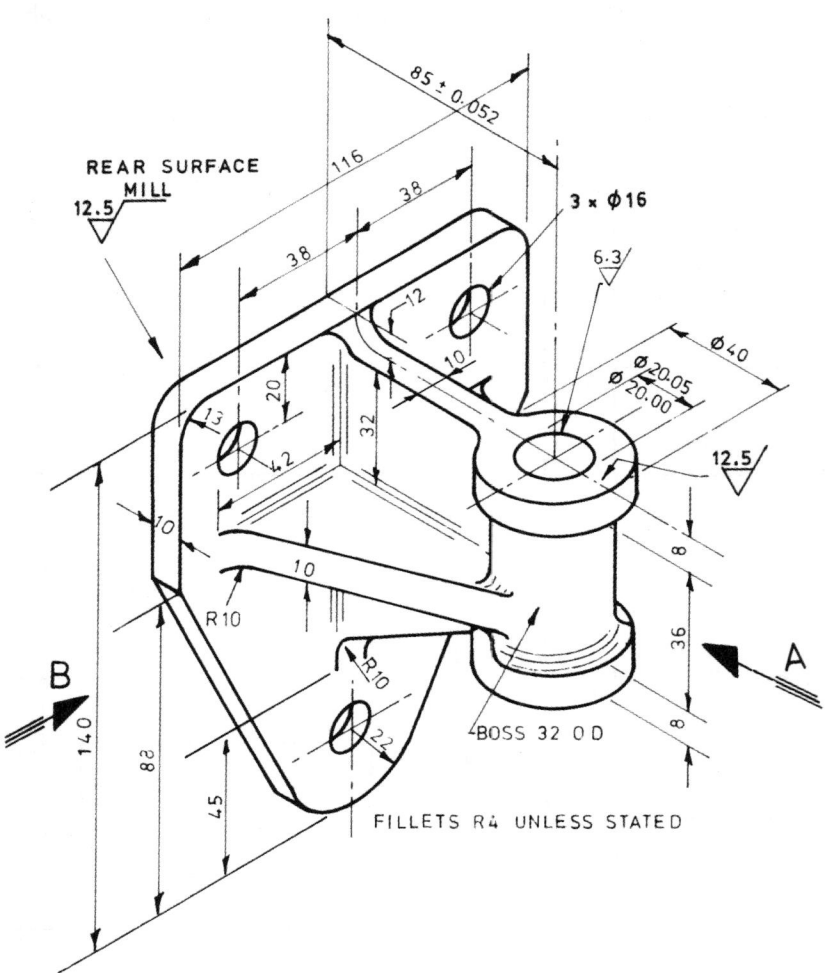

Fully dimension the drawing, and supply a suitable title block.

(scale: full size)

Figure 3.14 shows the rough sketch for the calculation of the positions of the three views on the drawing sheet. Notice the space between the top and front views is 40 mm compared with 75 mm between the front and side views. This is because the bracket is higher than it is wide, and if these two spaces were made the same, the drawing would appear cramped on the paper.

The completed orthogonal projection is shown in Figure 3.15. This should be studied carefully to ensure full understanding of the relationship between the detail on the views. Attempt to do the drawing within an A2 size drawing frame using the measurements given in Figures 3.13 and 3.14. Try not to refer to Figure 3.15.

**FIGURE 3.14** ▶ Calculations for view positions

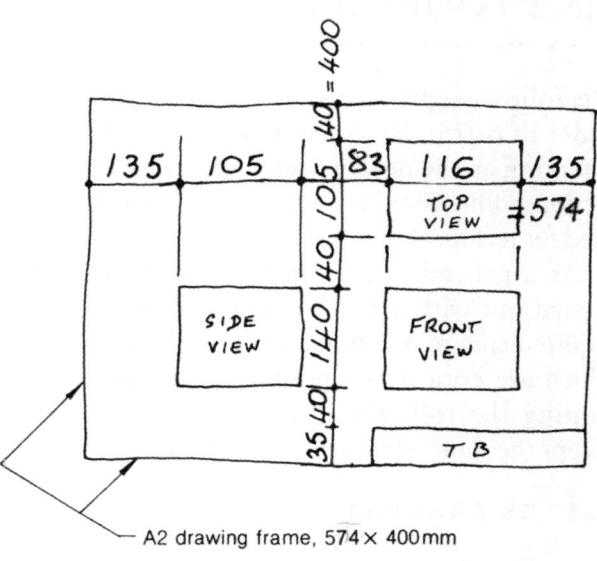

A2 drawing frame, 574 × 400mm

**FIGURE 3.15** ▶ Complete orthogonal projection

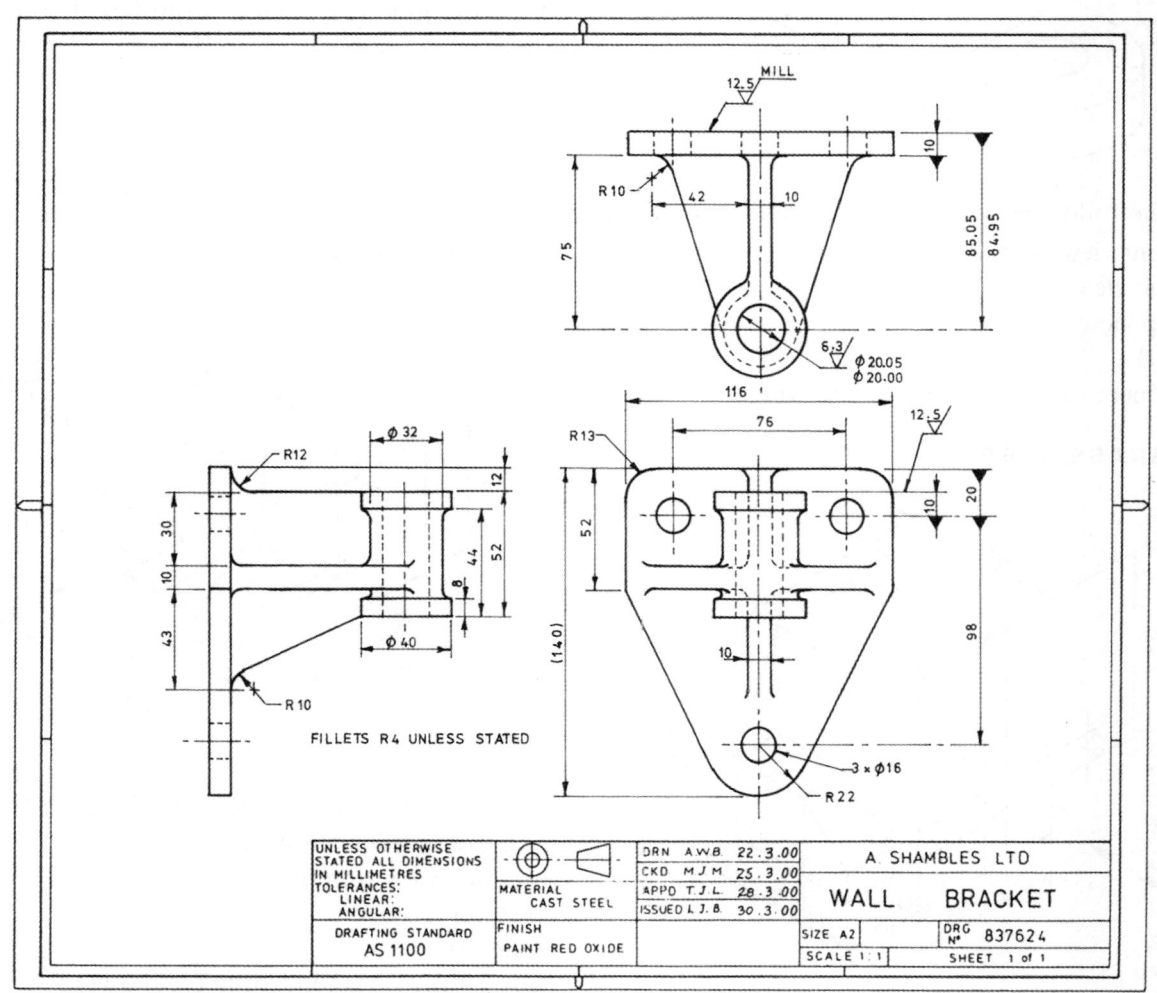

# 3.6 Problems

The following problems are graded in approximate order of difficulty. Start at 1 and work through, referring as the need arises to the relevant text on sections, dimensioning, etc. The exercises may also be used for technical sketching on squared or plain paper.

As a general rule, dimensioning of drawings is carried out with a full knowledge of the functional requirements of a component, and those dimensions which are critical are inserted. However, in dimensioning the following exercises students should accept the dimensions given as critical.

## 3.1 CS BRACKET

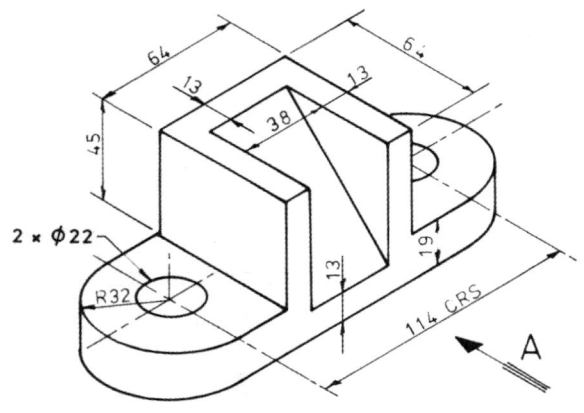

Draw the following views in third-angle projection:
(a) a front view from A
(b) a side view
(c) a top view
(scale 1:1)
Fully dimension and identify the drawing.

## 3.2 BRASS STEP

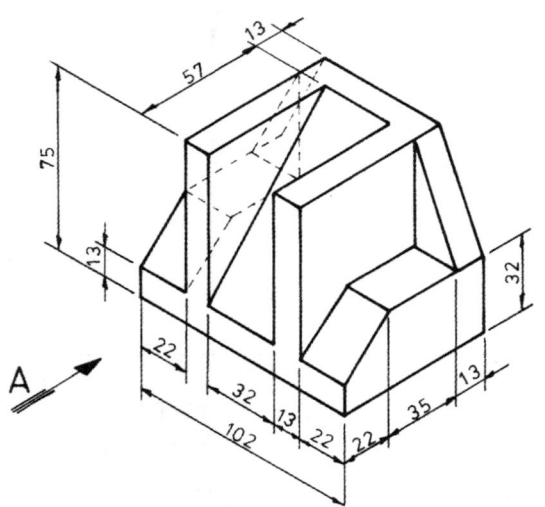

Draw the following views in third-angle projection:
(a) a front view from A
(b) a side view
(c) a top view
(scale 1:1)
Fully dimension and identify the drawing.

## 3.3 CI BENCH BLOCK

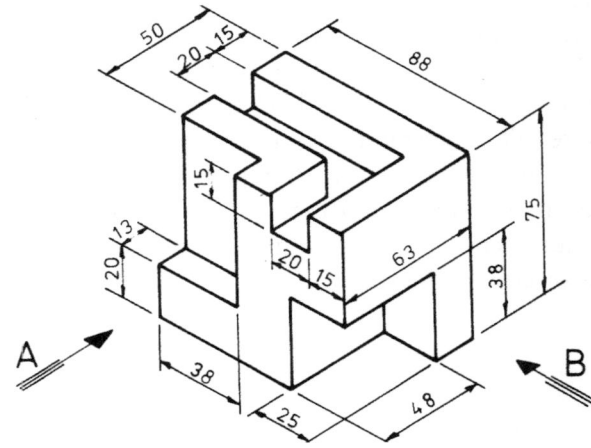

Draw the following views in third-angle projection:
(a) a front view from A
(b) a side view from B
(c) a top view
(scale 1:1)
Fully dimension and identify the drawing.

## 3.4 CI BRACKET

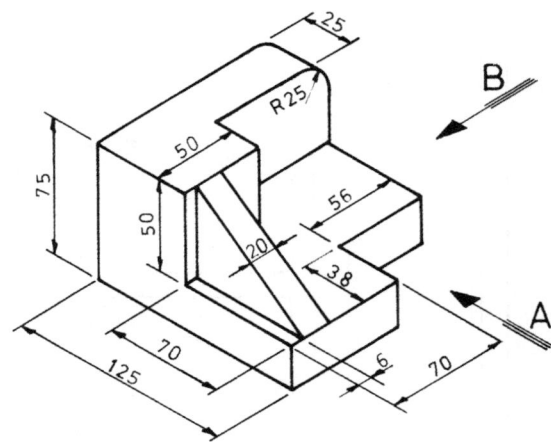

Draw the following views in third-angle projection:
(a) a front view from A
(b) a side view from B
(c) a top view
(scale 1:1)
Fully dimension and identify the drawing.

## 3.5 MS BRACKET

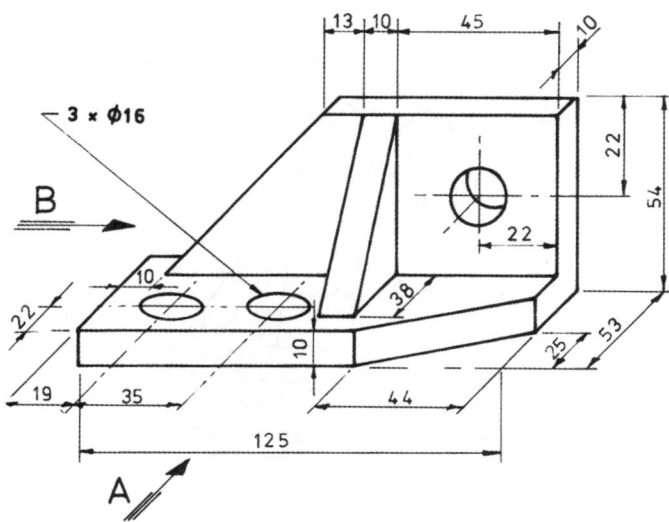

Draw the following views in third-angle projection:
(a) a front view from A
(b) a side view from B
(c) a top view
(scale 1:1)
Fully dimension, provide a title block, and identify the drawing.

## 3.7 JIG FRAME

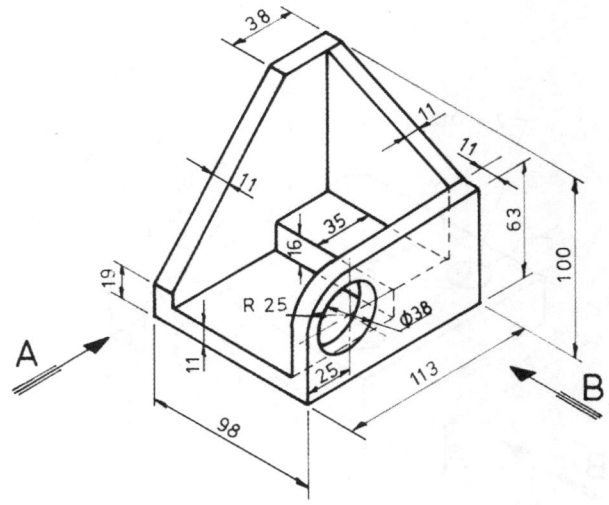

Draw the following views in third-angle projection:
(a) a front view from A
(b) a side view from B
(c) a top view
(scale 1:1)
Fully dimension, provide a title block, and identify the drawing.

## 3.6 CI SUPPORT

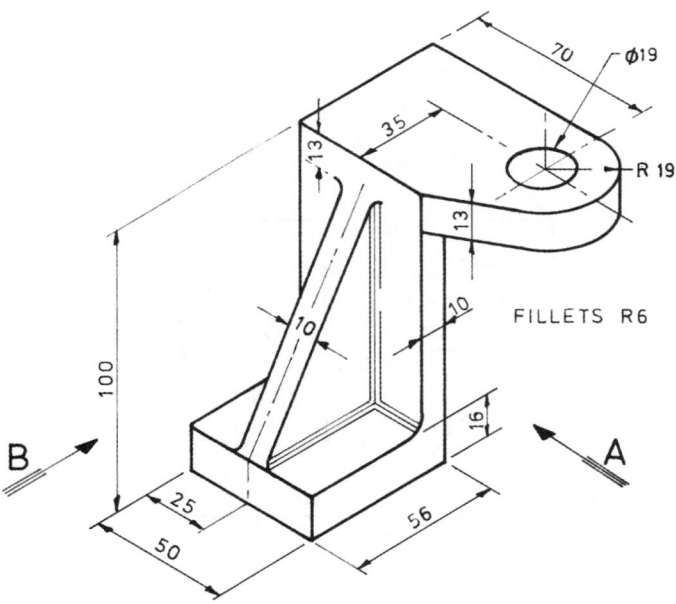

FILLETS R6

Draw the following views in first-angle projection:
(a) a front view from A
(b) a side view from B
(c) a top view
(scale 1:1)
Fully dimension, provide a title block, and identify the drawing.

## 3.8 GRINDER REST

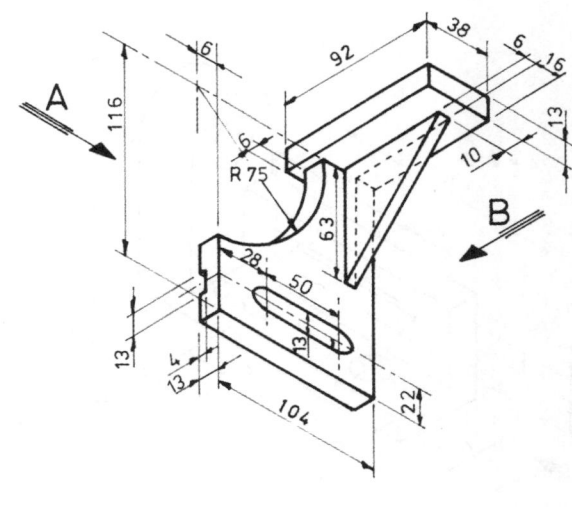

Draw the following views in third-angle projection:
(a) a front view from A
(b) a side view from B
(c) a top view
(scale 1:1)
Fully dimension, provide a title block, and identify the drawing.

## 3.9 MS ROD BRACKET

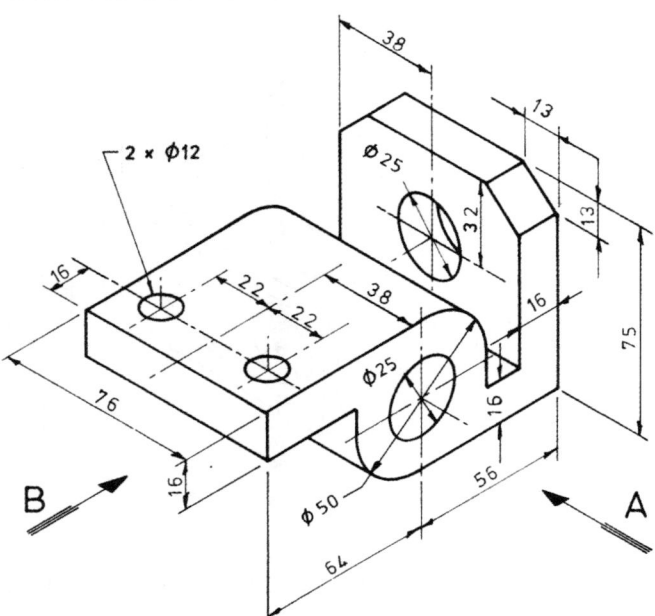

Draw the following views in third-angle projection:
(a) a front view from A
(b) a side view from B
(c) a top view
(scale 1:1)
Fully dimension, provide a title block, and identify the drawing.

## 3.10 SAFETY BRACKET

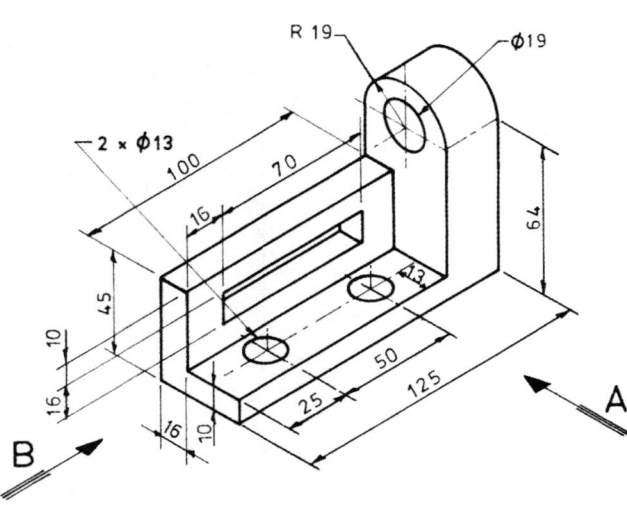

Draw the following views in third-angle projection:
(a) a front view from A
(b) a side view from B
(c) a top view
(scale 1:1)
Fully dimension, provide a title block, and identify the drawing.

## 3.11 CI BRACKET

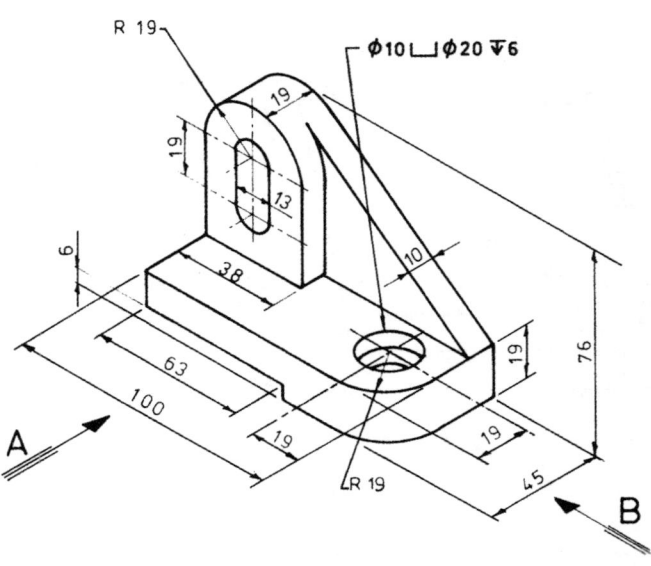

Draw the following views in third-angle projection:
(a) a front view from A
(b) a side view from B
(c) a top view
(scale 1:1)
Fully dimension, provide a title block, and identify the drawing.

## 3.12 SUPPORT BRACKET

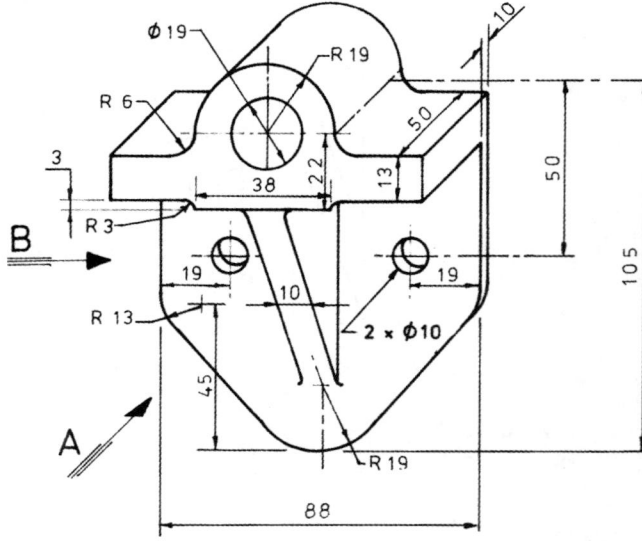

Draw the following views in third-angle projection:
(a) a front view from A
(b) a side view from B
(c) a top view
(scale 1:1)
Fully dimension, provide a title block, and identify the drawing.

# 3.13 CS SOCKET

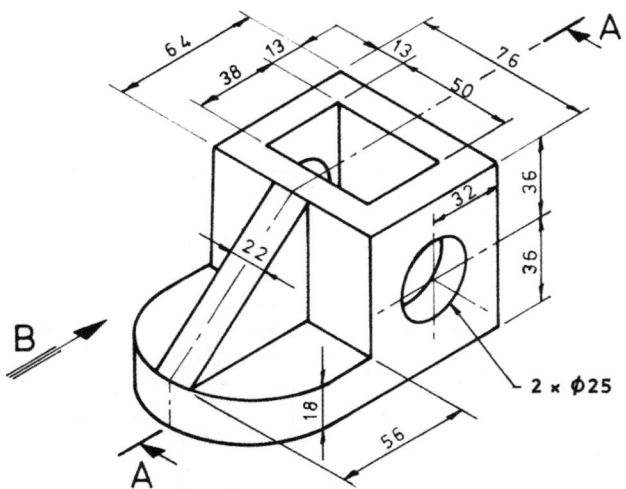

Draw the following views in third-angle projection:
(a) a sectional view on A-A
(b) a side view from B
(c) a top view
(scale 1:1)
Fully dimension, provide a title block, and identify the drawing.

# 3.15 CS BRACKET

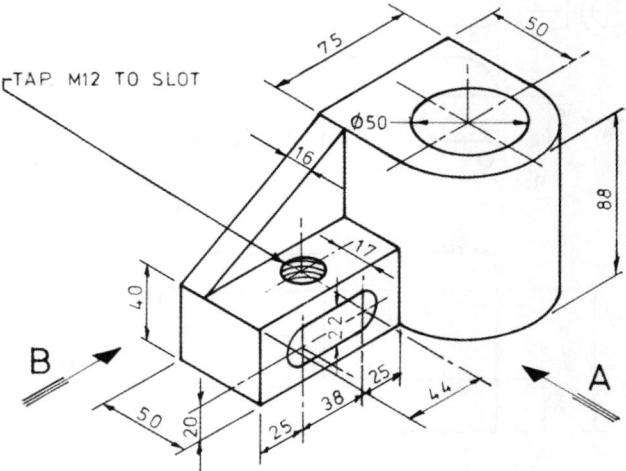

Draw the following views in third-angle projection:
(a) a front view from A
(b) a side view from B
(c) a top view
(scale 1:1)
Fully dimension, provide a title block, and identify the drawing.

# 3.14 MS CLAMP

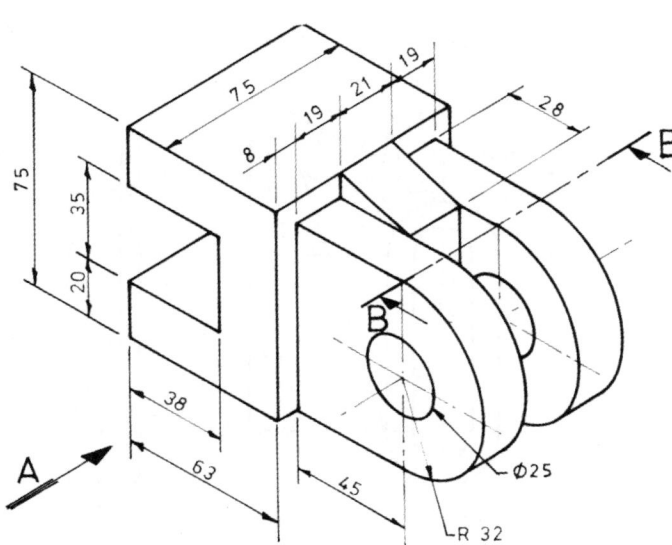

Draw the following views in third-angle projection:
(a) a front view from A
(b) a sectional view on B-B
(c) a top view
(scale 1:1)
Fully dimension, provide a title block, and identify the drawing.

# 3.16 MS SLEEVE BRACKET

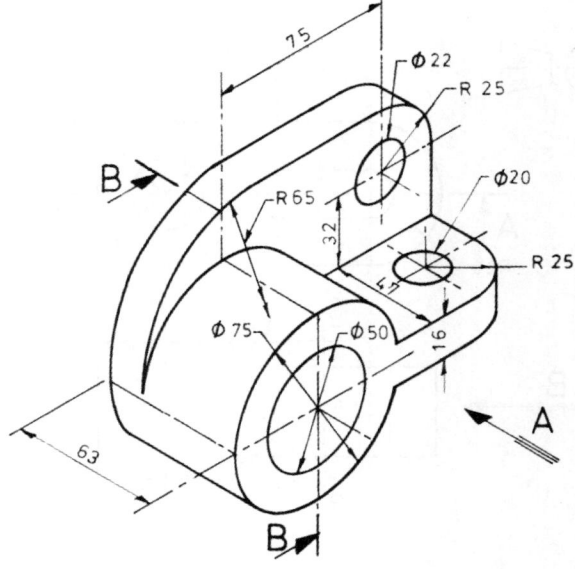

Draw the following views in third-angle projection:
(a) a front view from A
(b) a sectional side view on B-B
(c) a top view
(scale 1:1)
Fully dimension, provide a title block, and identify the drawing.

## 3.17 FORKED BRACKET

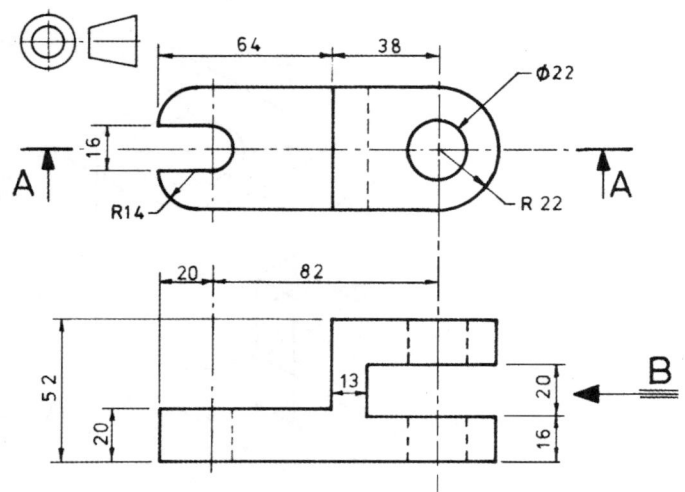

Draw the following views in third-angle projection:

(a) a sectional front view on A-A
(b) a side view from B
(c) the given top view
(scale 1:1)

Fully dimension, provide a title block, and identify the drawing.

## 3.18 MS SLOTTED LINK

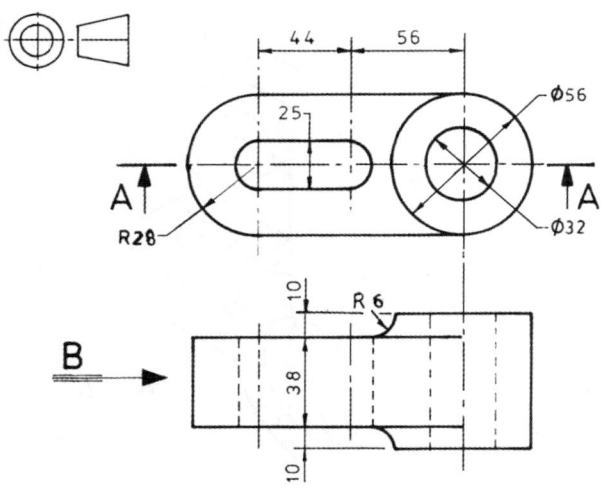

Draw the following views in first-angle projection:

(a) a sectional front view on A-A
(b) a side view from B
(c) the given top view
(scale 1:1)

Fully dimension, provide a title block, and identify the drawing.

## 3.19 MS HALF CLIP

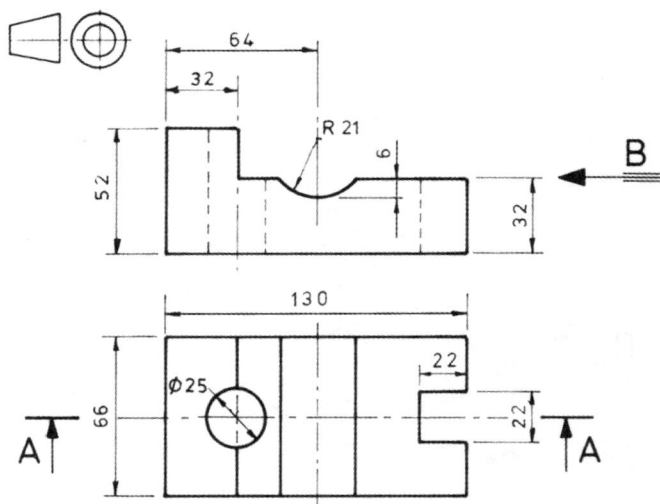

Draw the following views in third-angle projection:

(a) a sectional front view on A-A
(b) a side view from B
(c) the given top view
(scale 1:1)

Fully dimension, provide a title block, and identify the drawing.

## 3.20 FAN BRACKET

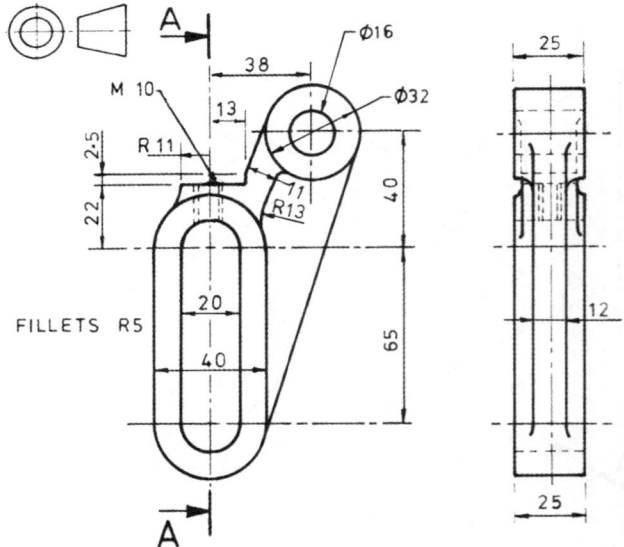

Draw the following views in third-angle projection:

(a) a front view as given
(b) a sectional side view on A-A
(c) a top view
(scale 1:1)

Fully dimension, provide a title block, and identify the drawing.

## 3.21 CI FORKED END

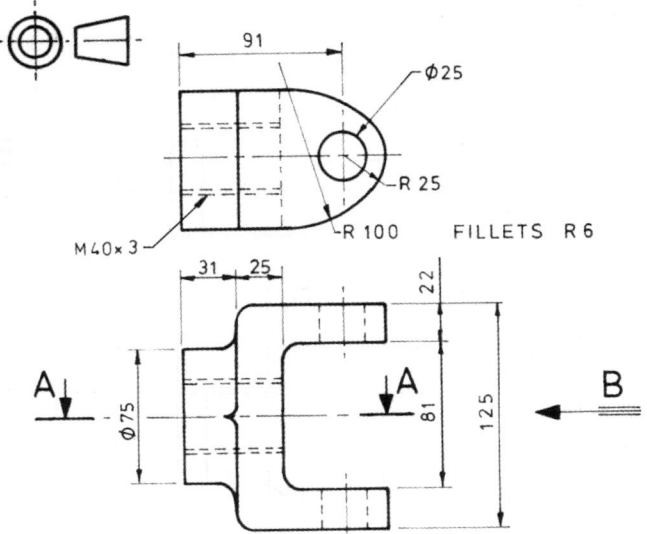

Draw the following views in third-angle projection:

(a) a front view as given

(b) a side view from B

(c) a sectional top view on A-A

(scale 1:1)

Fully dimension, provide a title block, and identify the drawing.

## 3.22 SHAFT BRACKET

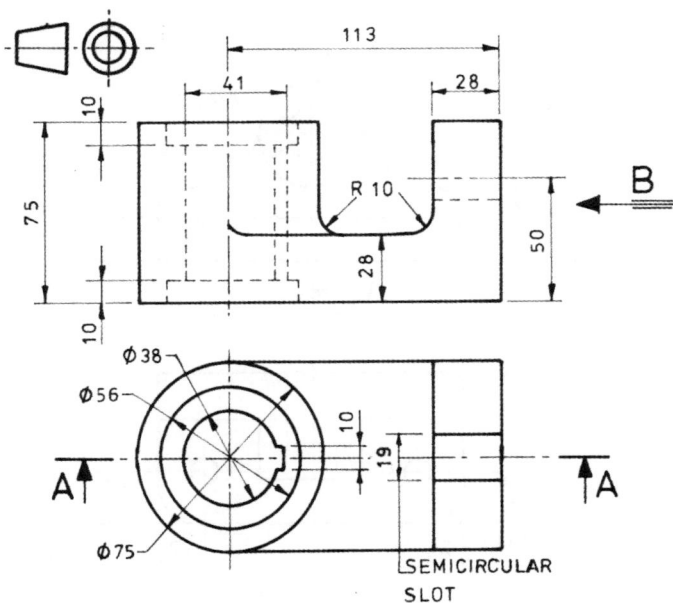

Draw the following views in third-angle projection:

(a) a sectional front view on A-A

(b) a side view from B

(c) the given top view

(scale 1:1)

Fully dimension, provide a title block, and identify the drawing.

## 3.23 CI CRANK ARM

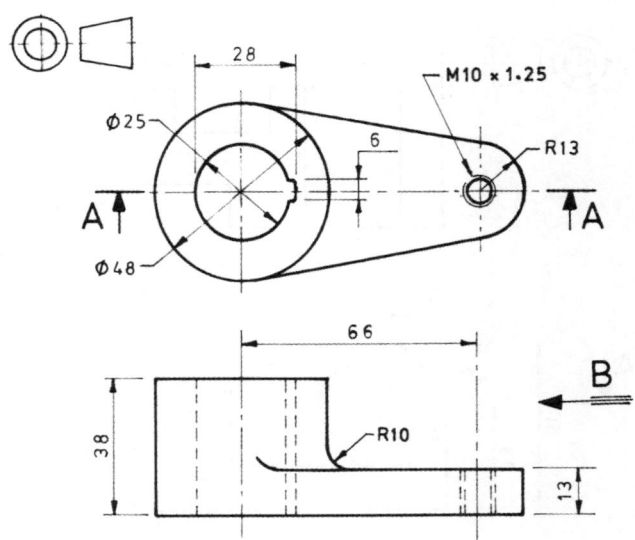

Draw the following views in third-angle projection:

(a) a sectional front view on A-A

(b) a side view from B

(c) the given top view

(scale 1:1)

Fully dimension, provide a title block, and identify the drawing.

## 3.24 SLIDING SUPPORT

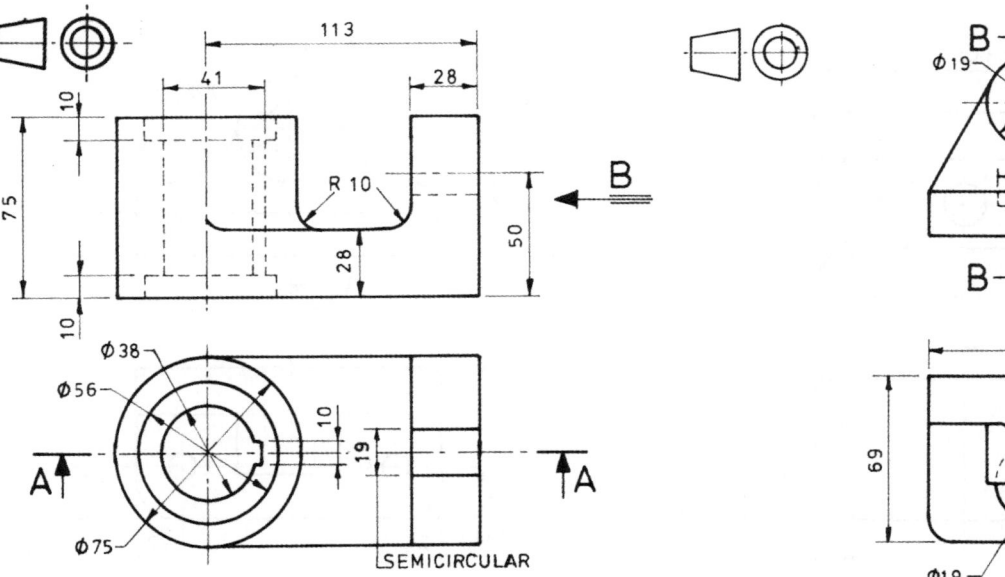
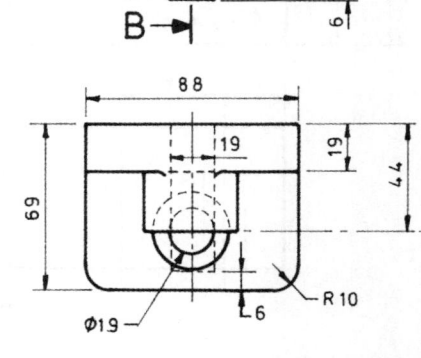

Draw the following views in first-angle projection:

(a) the given front view

(b) a sectional side view on B-B

(c) a bottom view

(scale 1:1)

Fully dimension, provide a title block, and identify the drawing.

## 3.25 GUIDE BRACKET

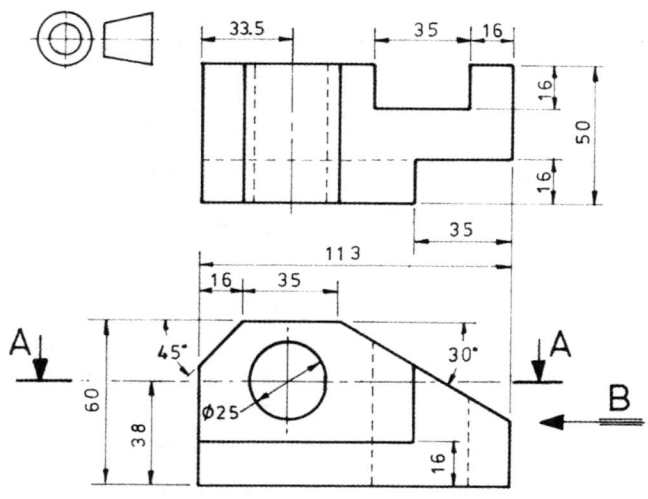

Draw the following views in third-angle projection:
(a) the front view as given
(b) a side view from B
(c) a sectional top view on A-A
(scale 1:1)
Fully dimension, provide a title block, and identify the drawing.

## 3.26 V STOP

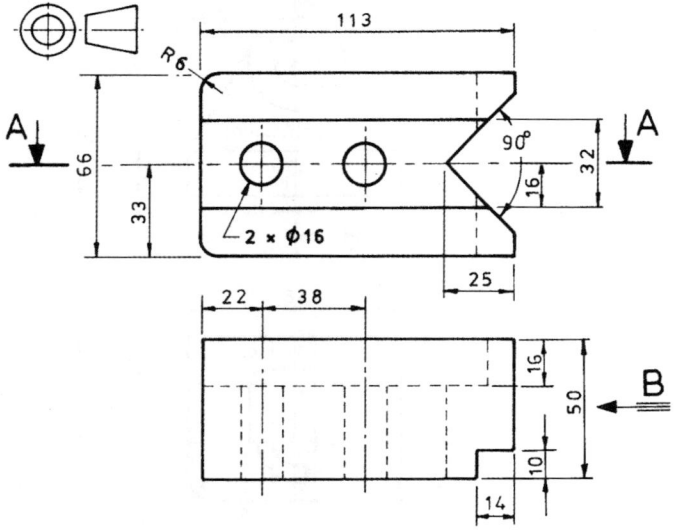

Draw the following views in third-angle projection:
(a) a sectional front view on A-A
(b) a side view from B
(c) a top view
(scale 1:1)
Fully dimension, provide a title block, and identify the drawing.

## 3.27 SIDE BRACKET

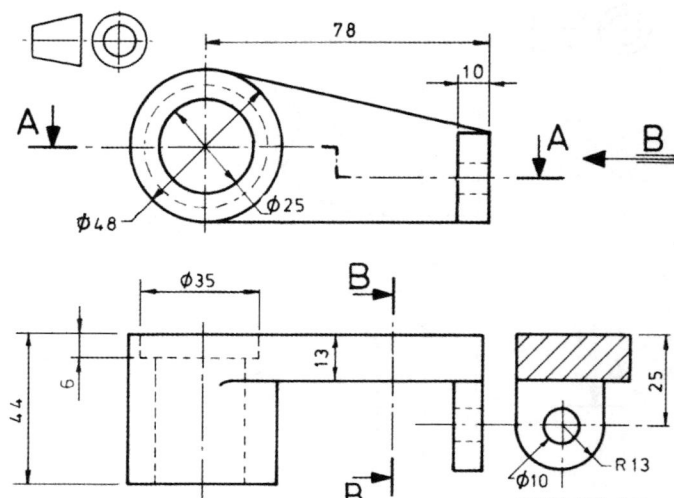

Draw the following views in third-angle projection:
(a) the front view as given
(b) a side view from B
(c) an offset sectional view on A-A
(scale 1:1)
Fully dimension, provide a title block, and identify the drawing.

## 3.28 CI SUPPORT

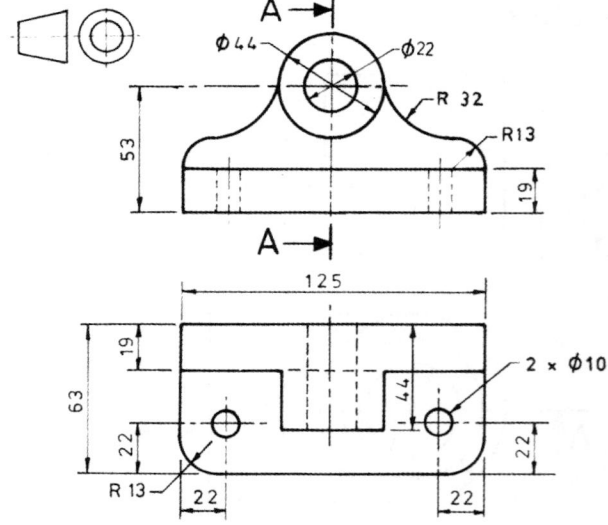

Draw the following views in first-angle projection:
(a) the front view as given
(b) a sectional side view on A-A
(c) the given top view
(scale 1:1)
Fully dimension, provide a title block, and identify the drawing.

## 3.29 CI GUIDE BLOCK

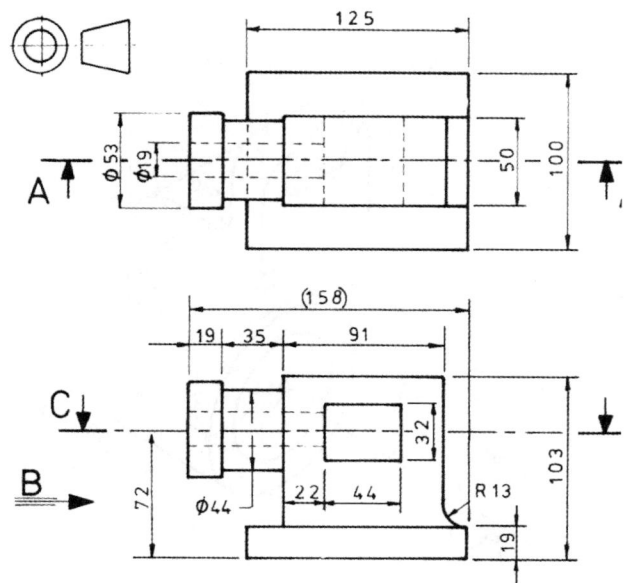

Draw the following views in third-angle projection:
(a) a sectional front view on A-A
(b) a side view from B
(c) a sectional top view on C-C
(scale 1:1)
Fully dimension, provide a title block, and identify the drawing.

## 3.30 BRASS SUPPORT

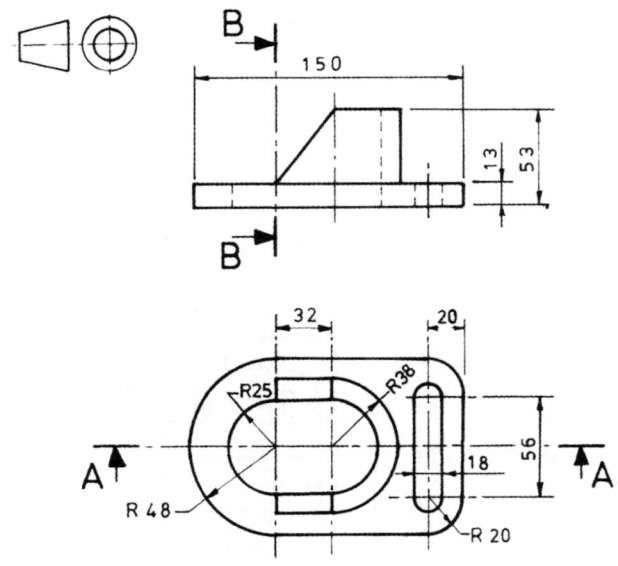

Draw the following views in third-angle projection:
(a) a sectional front view on A-A
(b) a sectional side view on B-B
(c) the given top view
(scale 1:1)
Fully dimension, provide a title block, and identify the drawing.

## 3.31 CI ROD GUIDE

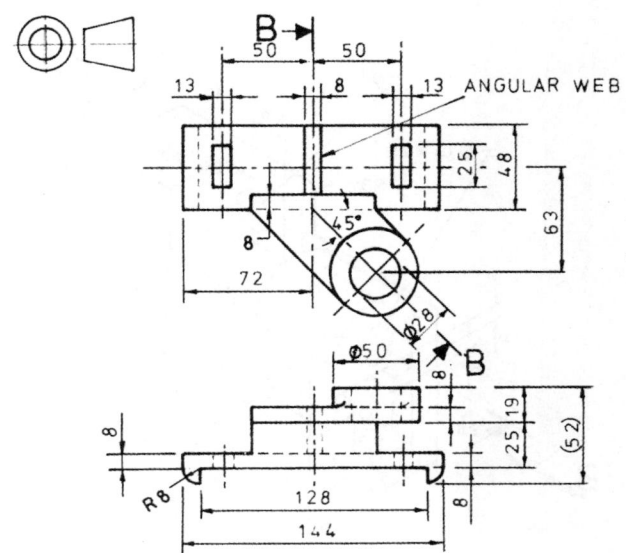

Draw the following views in third-angle projection:
(a) a rear view
(b) an aligned sectional side view on B-B
(c) a bottom view
(scale 1:1)
Fully dimension, provide a title block, and identify the drawing.

## 3.32 CI JIG

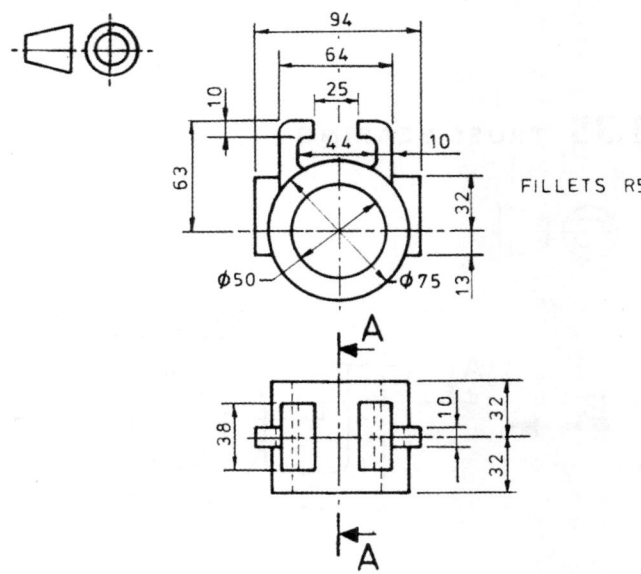

Draw the following views in third-angle projection:
(a) the front view as given
(b) a sectional side view on A-A
(c) the given top view
(scale 1:1)
Fully dimension, provide a title block, and identify the drawing.

## 3.33 CI JAW SUPPORT

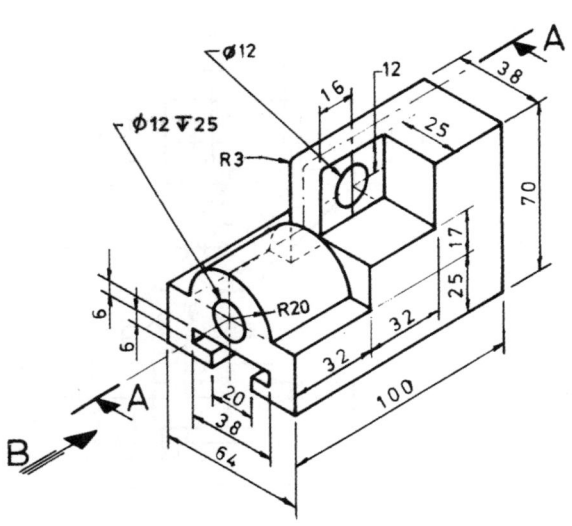

## 3.34 OFFSET CRANK

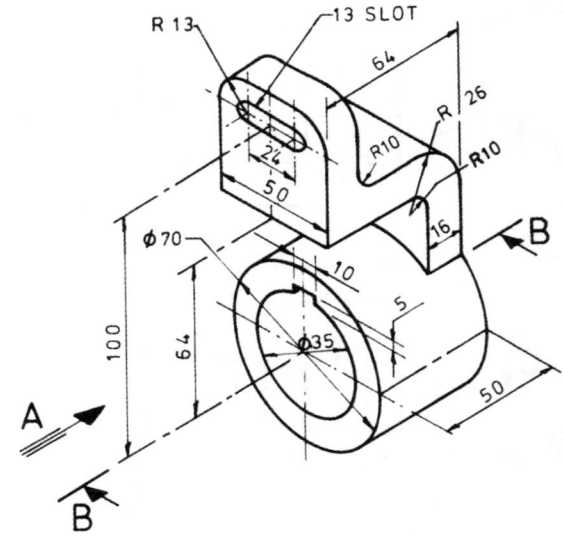

Draw the following views in third-angle projection:

(a) a sectional front view on A-A

(b) a side view from B

(c) a top view

(scale 1:1)

Fully dimension, provide a title block, and identify the drawing.

Draw the following views in third-angle projection:

(a) a front view from A

(b) a sectional side view on B-B

(c) a top view

(scale 1:1)

Fully dimension, provide a title block, and identify the drawing.

## 3.35 TRUSS BEARING

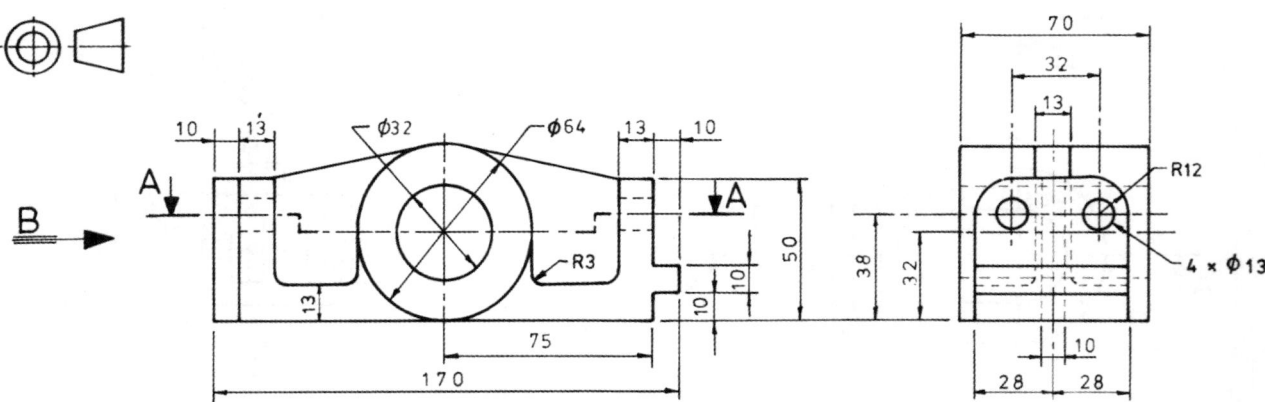

Draw the following views in third-angle projection:

(a) the front view as given

(b) a half sectional side view from B

(c) an offset section on A-A

(scale 1:1)

Fully dimension, provide a title block, and identify the drawing.

## 3.36 CI BEARING BRACKET

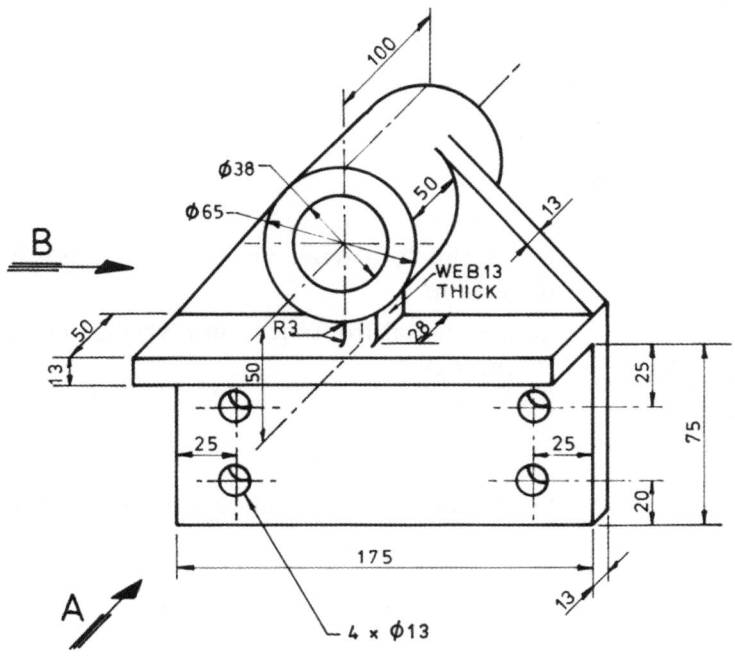

Draw the following views in third-angle projection:

(a) a front view from A

(b) a sectional side view on the centre line from B

(c) a top view

(scale 1:1)

Fully dimension, provide a title block, and identify the drawing.

## 3.37 BEARING RETAINER

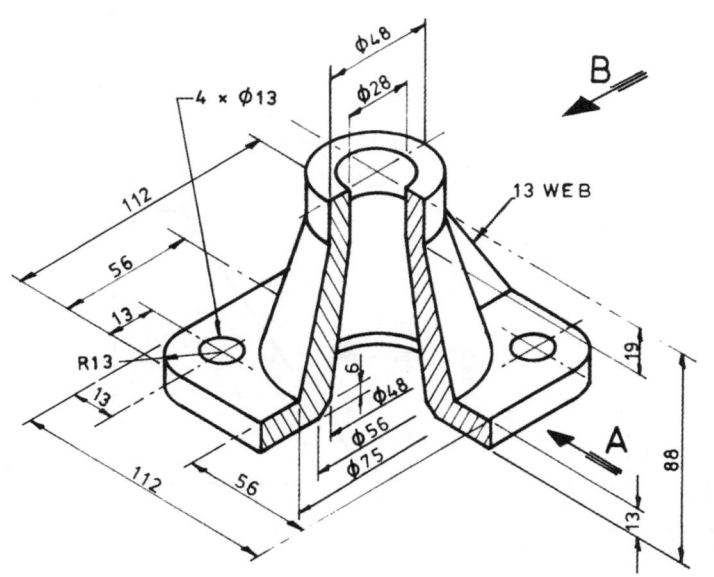

Draw the following views in third-angle projection:

(a) a sectional front view on the centre line viewed from A

(b) a side view from B

(c) a top view

(scale 1:1)

Fully dimension, provide a title block, and identify the drawing.

## 3.38 MAIN BEARING CAP

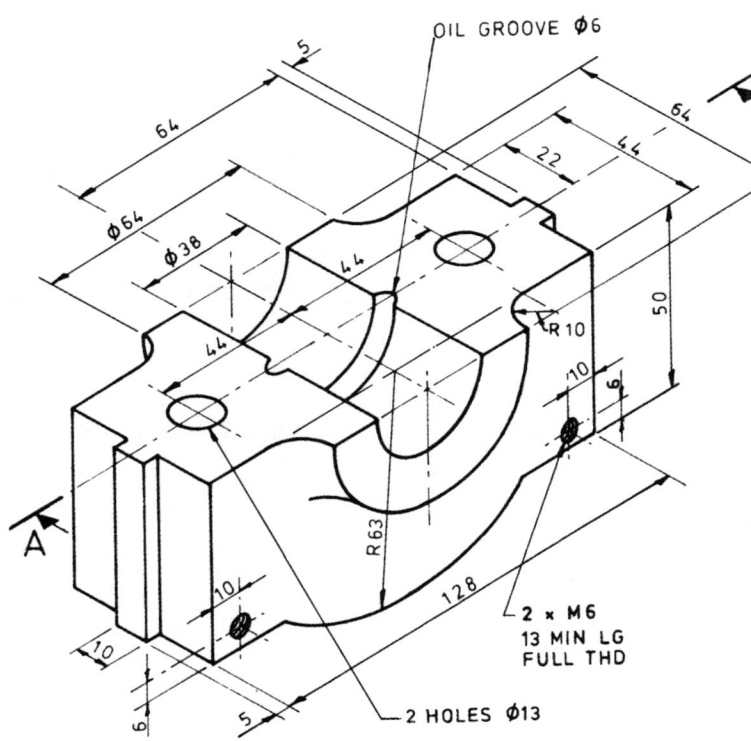

Draw the following views in third-angle projection:

(a) a sectional front view on a vertical plane through A-A

(b) a side view

(c) a top view

(scale 1:1)

Fully dimension, provide a title block, and identify the drawing.

## 3.39 CI PULLEY GUIDE

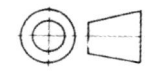

Draw the following views in third-angle projection:

(a) a sectional front view on A-A

(b) a side view from B

(c) a top view

(scale 1:1)

Fully dimension, provide a title block, and identify the drawing.

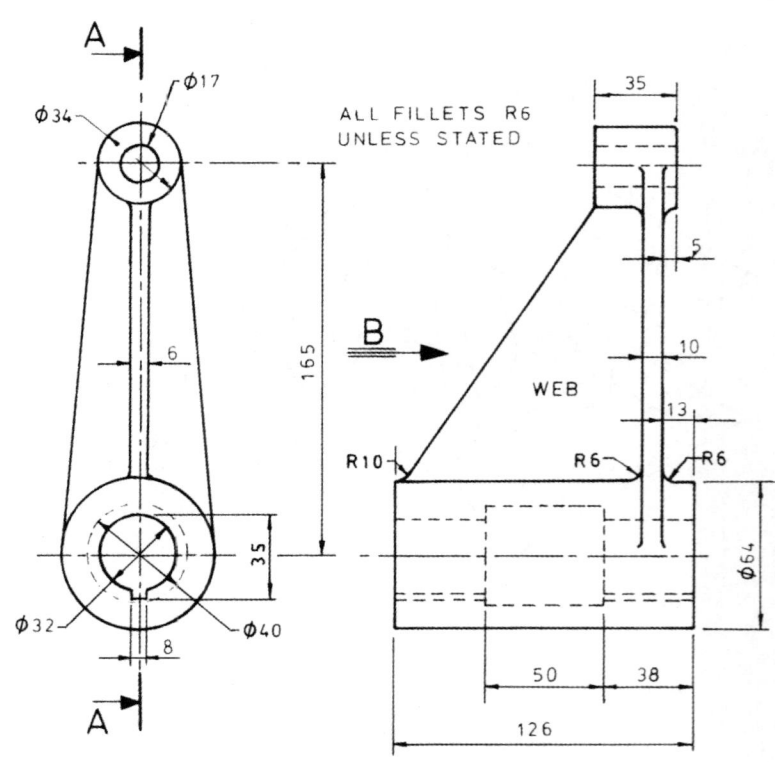

## 3.40 BELT-FORK BOSS

Draw the following views in third-angle projection:

(a) a sectional front view in place of the given front view, taken through the centre of the boss
(b) a sectional side view on A-A
(c) a top view
(scale 1:2)
Fully dimension, provide a title block, and identify the drawing.

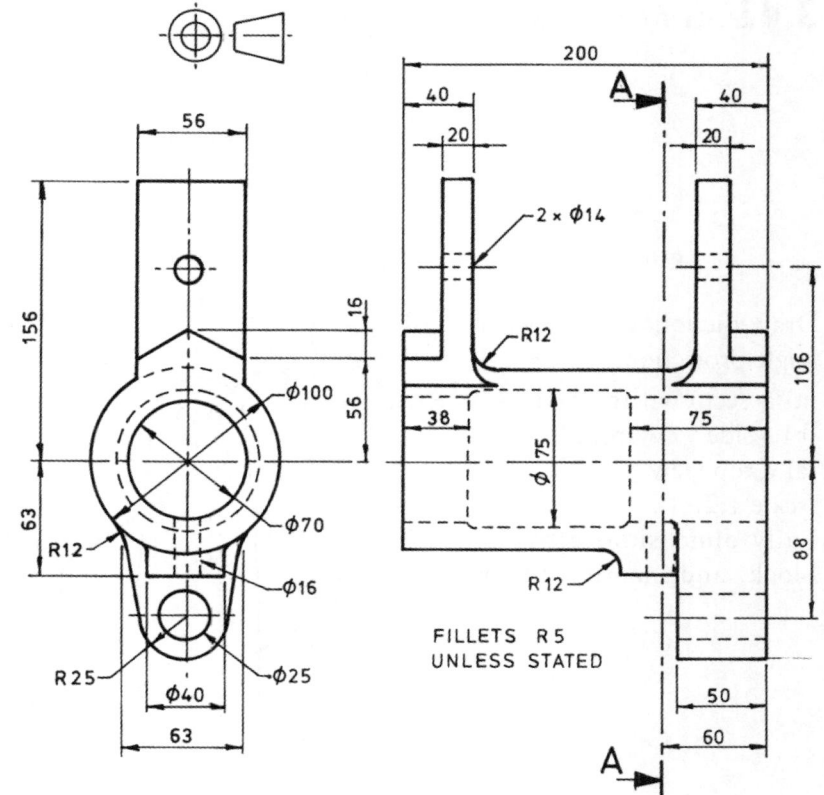

## 3.41 CI SUPPORT BRACKET

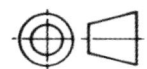

Draw the following views in third-angle projection:

(a) the given front view
(b) a sectional side view on B-B
(c) a top view
(scale 1:1)
Fully dimension, provide a title block, and identify the drawing.

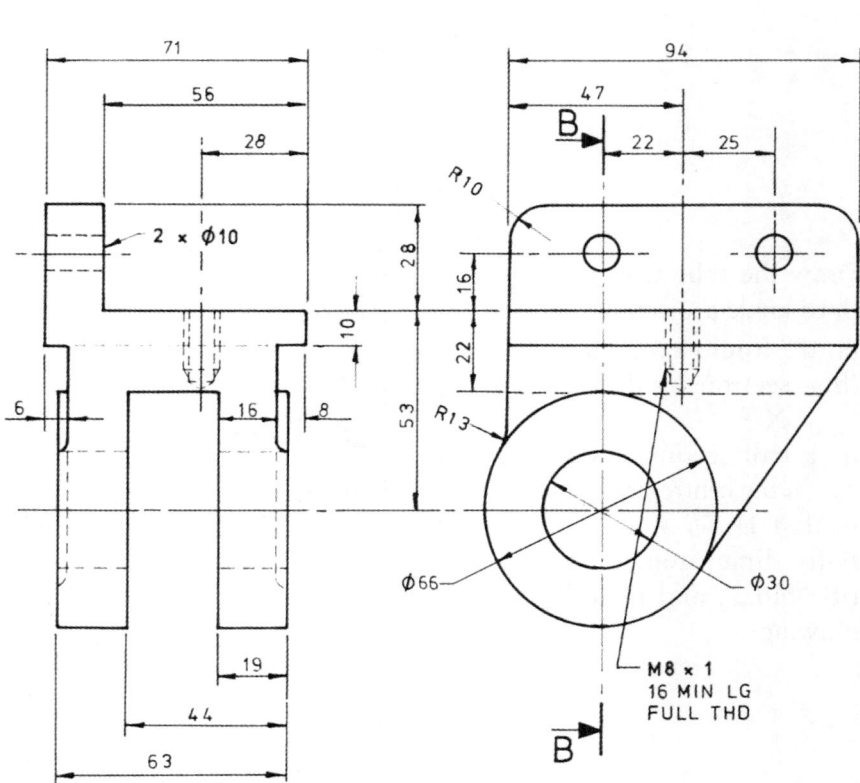

## 3.42 CI BRACKET

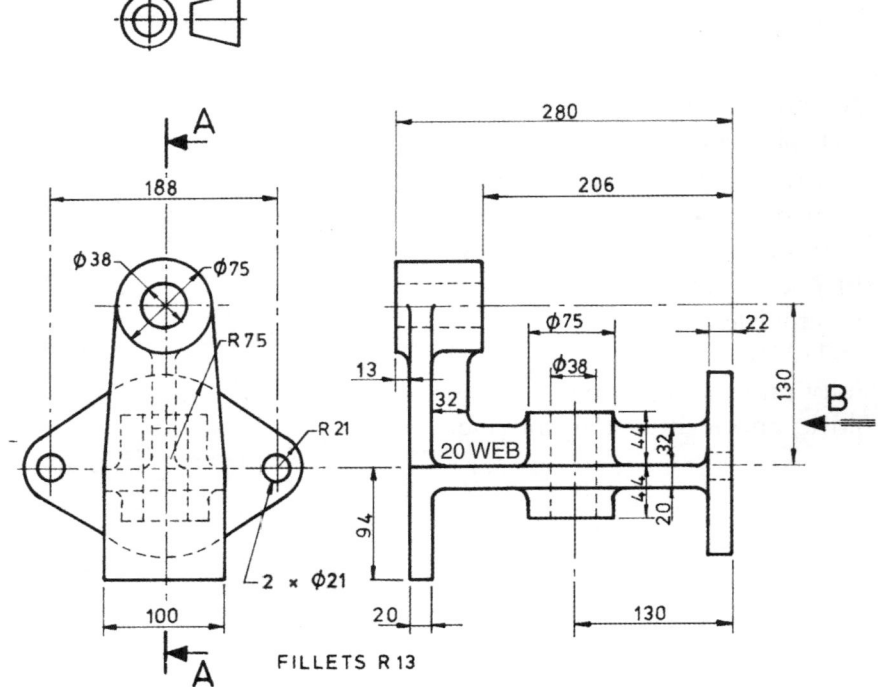

Draw the following views in third-angle projection:

(a) a sectional front view on A-A
(b) a side view from B
(c) a top view
(scale 1:2)
Fully dimension, provide a title block, and identify the drawing.

## 3.43 CI END PLATE

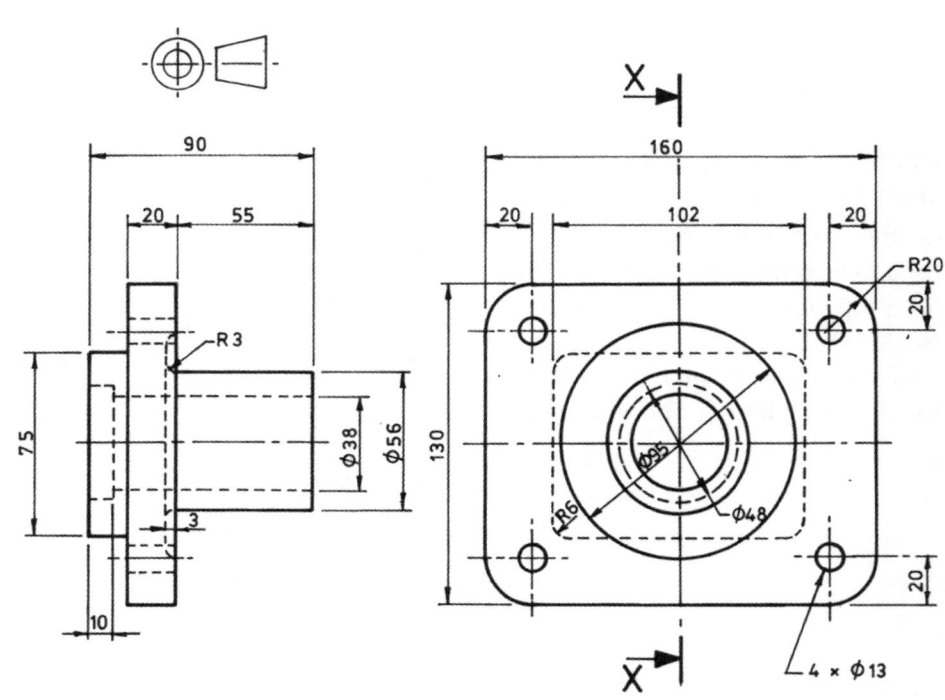

Draw the following views in third-angle projection:

(a) the front view as given
(b) a sectional side view on X-X
(c) a half sectional top view on the centre line
(scale 1:1)
Fully dimension, provide a title block, and identify the drawing.

## 3.44 TAIL ROD GUIDE

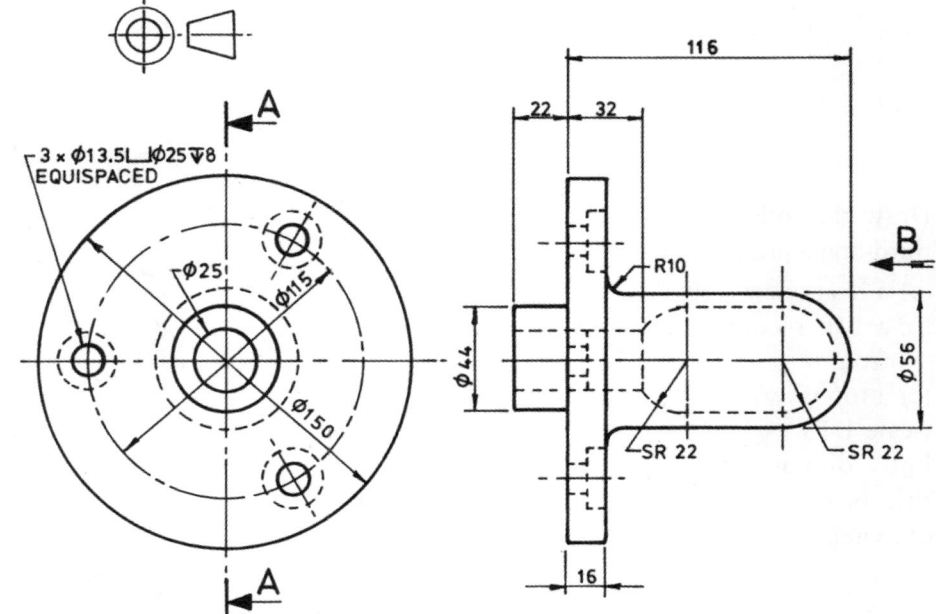

Draw the following views in third-angle projection:

(a) a sectional front view on A-A

(b) a side view from B

(scale 1:1)

Fully dimension, provide a title block, and identify the drawing.

## 3.45 CI PISTON

Draw the following views in third-angle projection:

(a) a sectional front view on A-A

(b) the side view as shown

(c) a sectional top view on C-C

(scale 1:1)

Fully dimension, provide a title block, and identify the drawing.

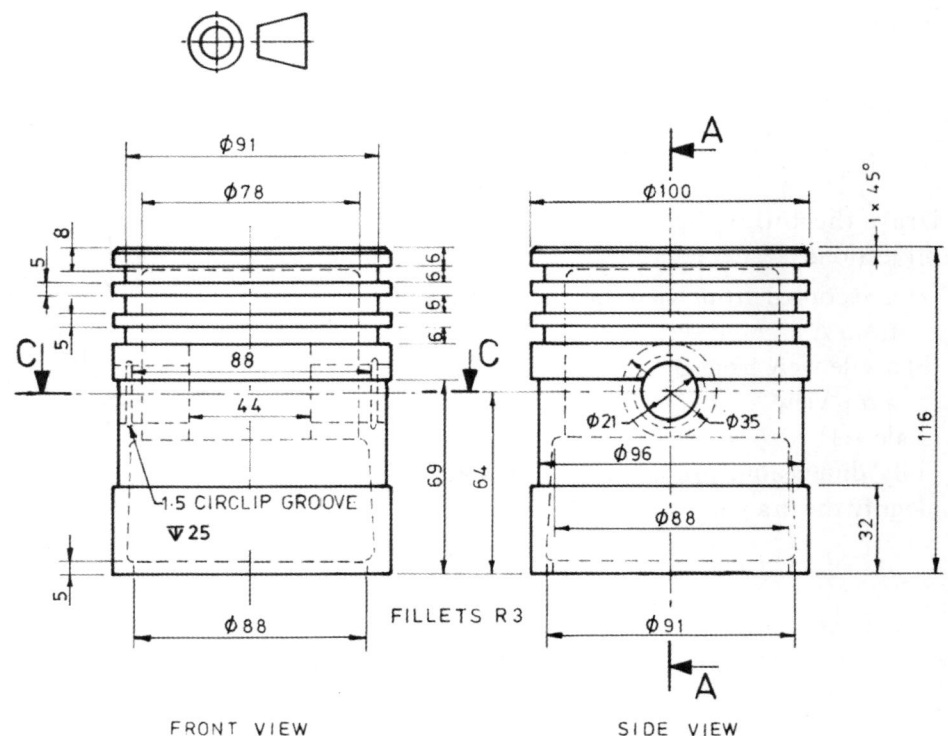

FRONT VIEW          SIDE VIEW

## 3.46 BEARING PEDESTAL

Draw the following views in third-angle projection:

(a) a front view from A

(b) a half sectional side view from B

(c) a top view

(scale 1:1)

Fully dimension, provide a title block, and identify the drawing.

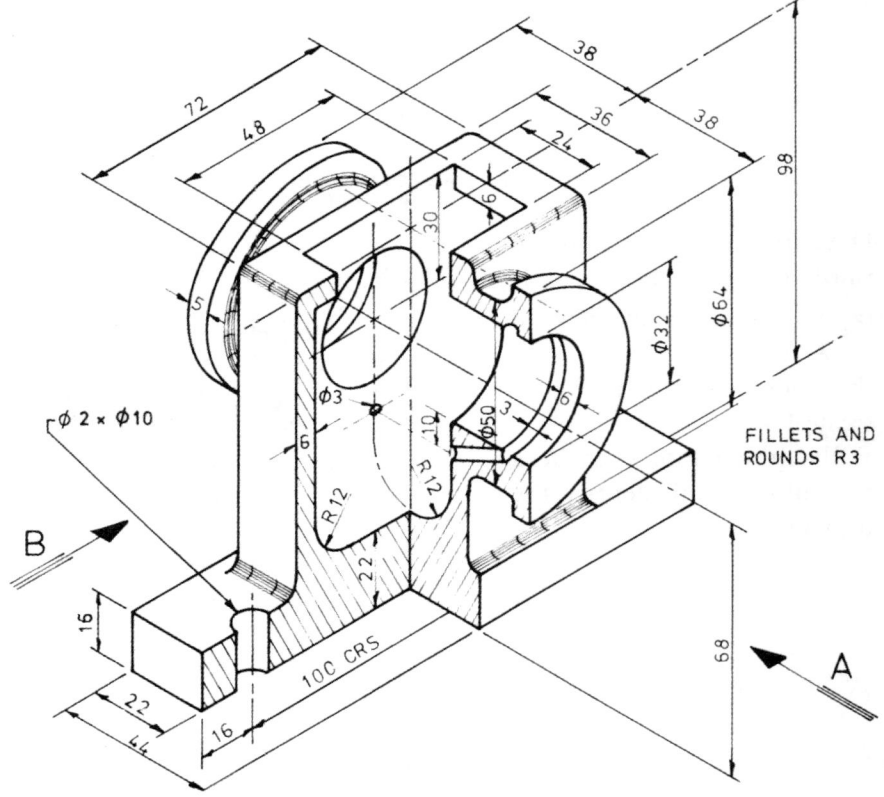

## 3.47 PUMP BRACKET

Draw the following views in third-angle projection:

(a) a sectional front view on the centre line from A

(b) a side view from B

(c) a top view

(scale 1:1)

Fully dimension, provide a title block, and identify the drawing.

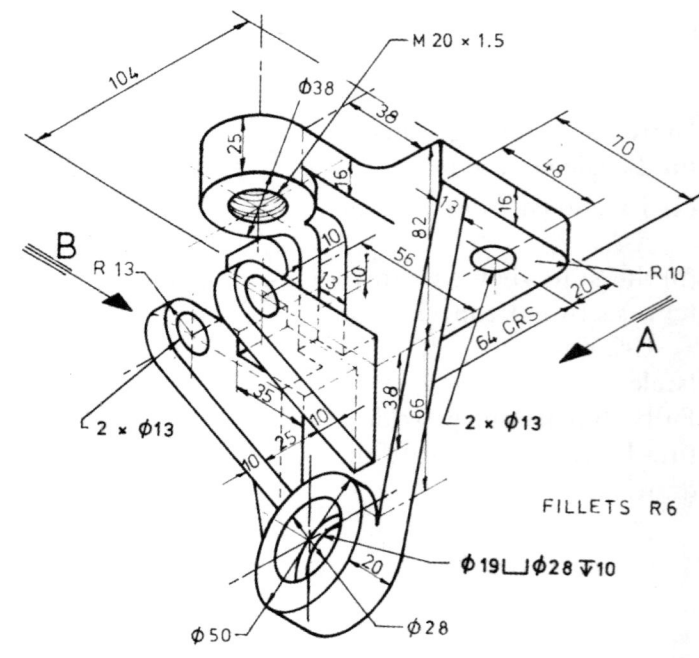

## 3.48 SPIGOT BEARING

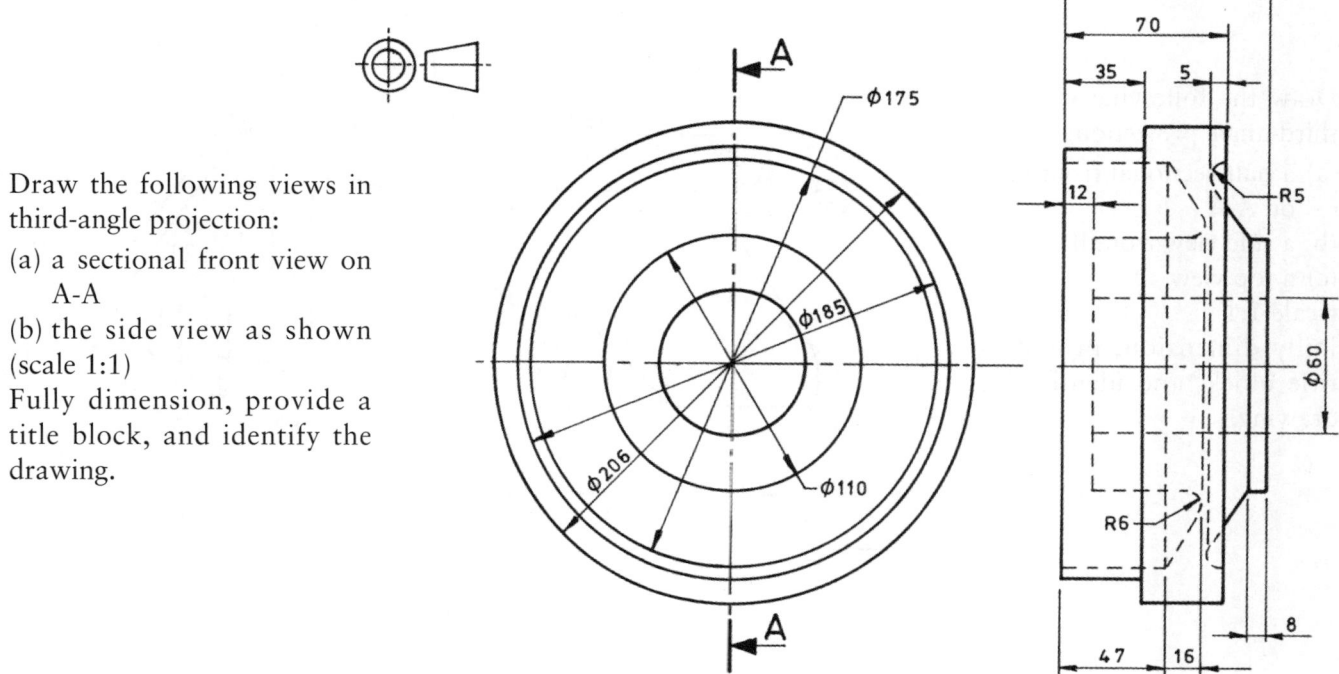

Draw the following views in third-angle projection:

(a) a sectional front view on A-A

(b) the side view as shown (scale 1:1)

Fully dimension, provide a title block, and identify the drawing.

## 3.49 WALL BRACKET

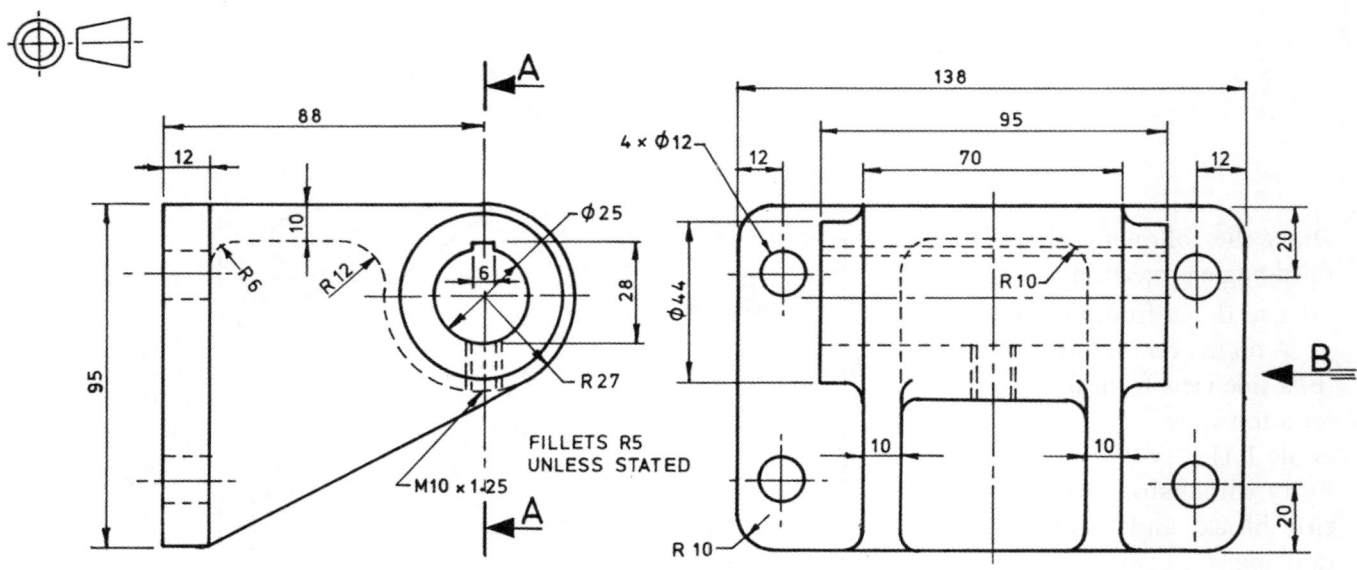

Draw the following views in third-angle projection:

(a) a sectional front view on A-A

(b) a side view from B

(c) a top view

(scale 1:1)

Fully dimension, provide a title block, and identify the drawing.

## 3.50 CS GUIDE BRACKET

Draw the following views in third-angle projection:

(a) a half sectional front view on A-A
(b) a side view from B
(c) a top view
(scale 1:1)
Fully dimension, provide a title block, and identify the drawing.

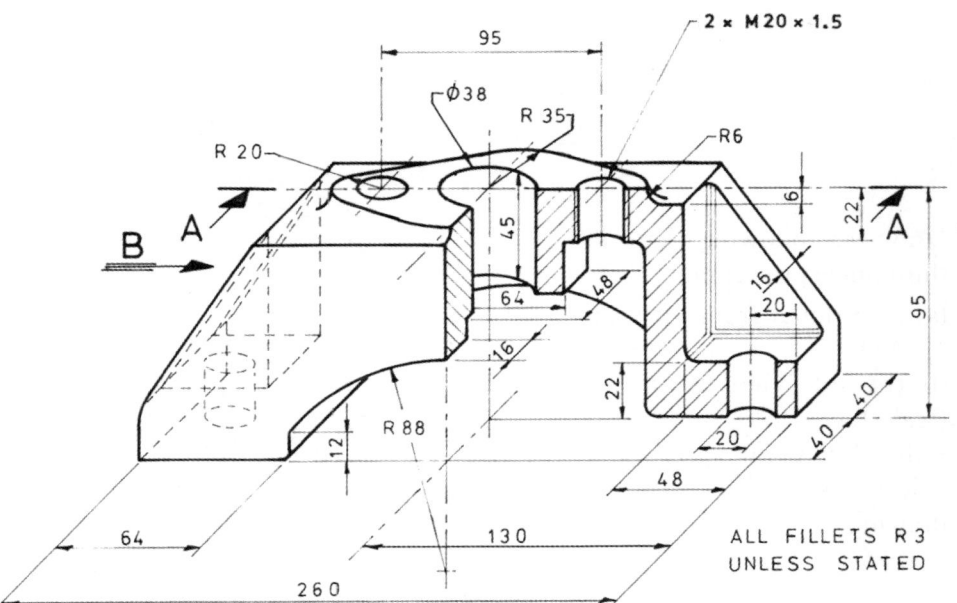

## 3.51 CS JIG

Draw the following views in third-angle projection:

(a) a sectional front view from A on the centre line
(b) a side view from B
(c) a top view
(scale 1:1)
Fully dimension, provide a title block, and identify the drawing.

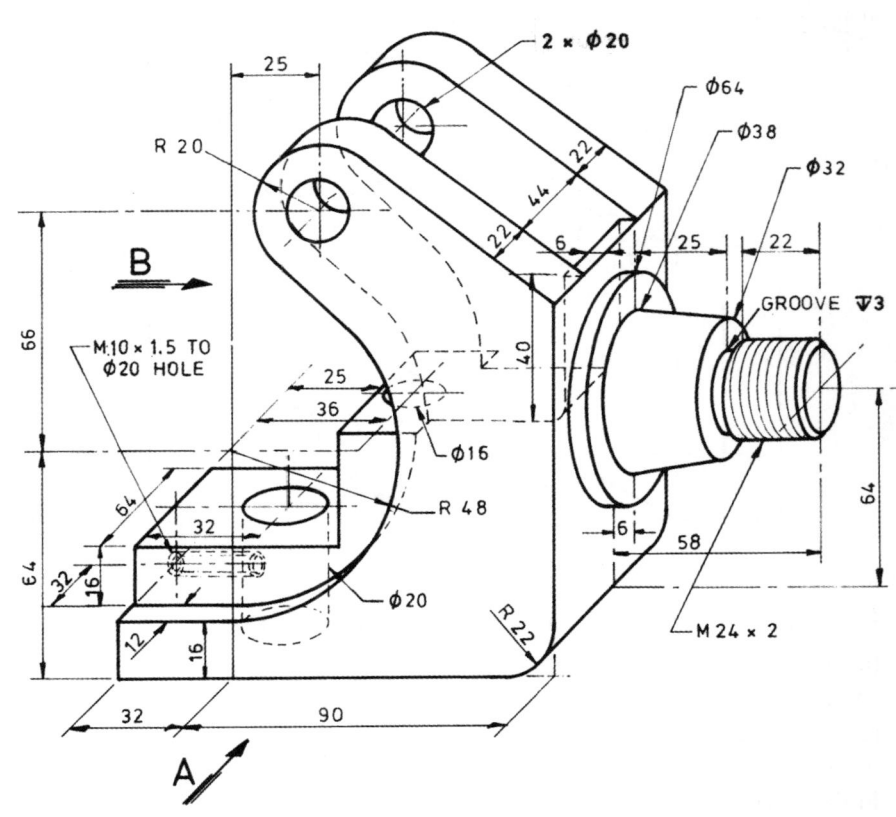

## 3.52 SHAFT BRACKET

Draw the following views in third-angle projection:

(a) a front view from A
(b) a sectional side view from B on the centre line
(c) a top view
(scale 1:1)
Fully dimension, provide a title block, and identify the drawing.

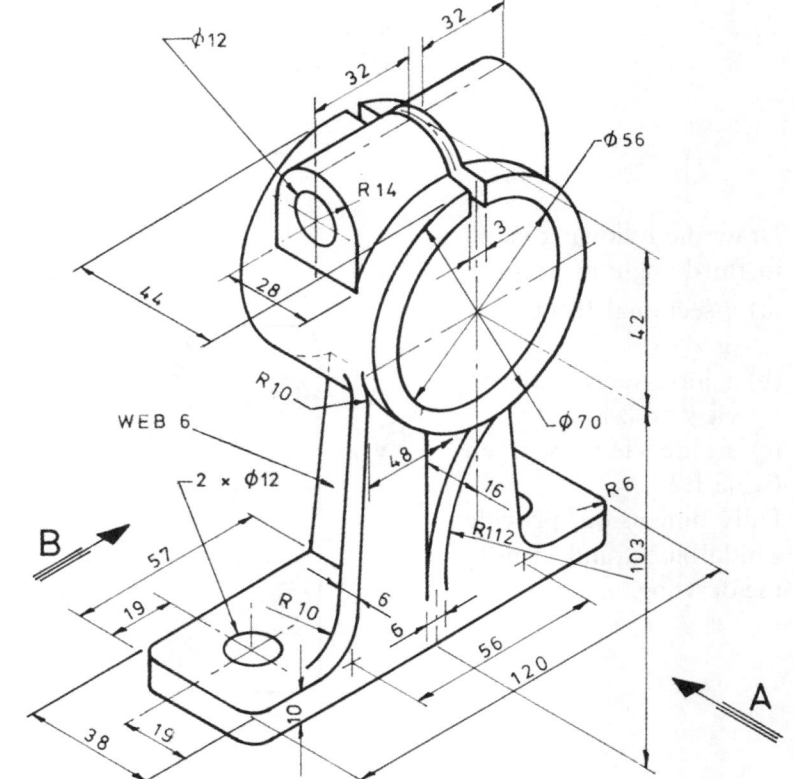

## 3.53 BEVEL GEAR SUPPORT

Draw the following views in third-angle projection:

(a) a sectional front view on a plane through A-A
(b) a side view from B
(c) a top view
(scale 1:1)
Fully dimension, provide a title block, and identify the drawing.

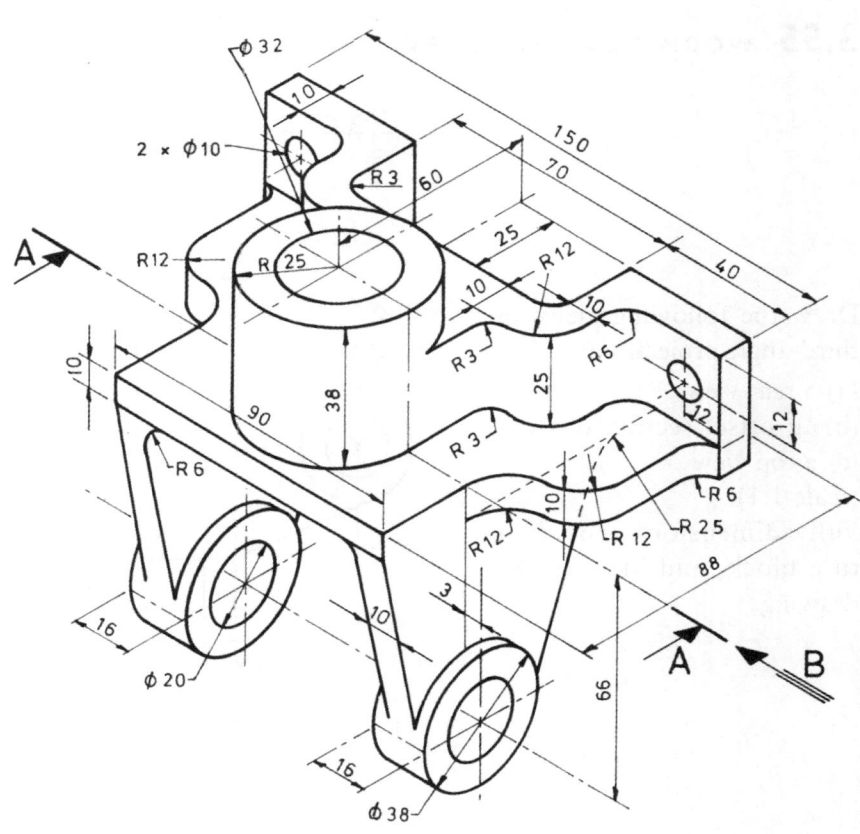

## 3.54 RACK AND GEAR BRACKET

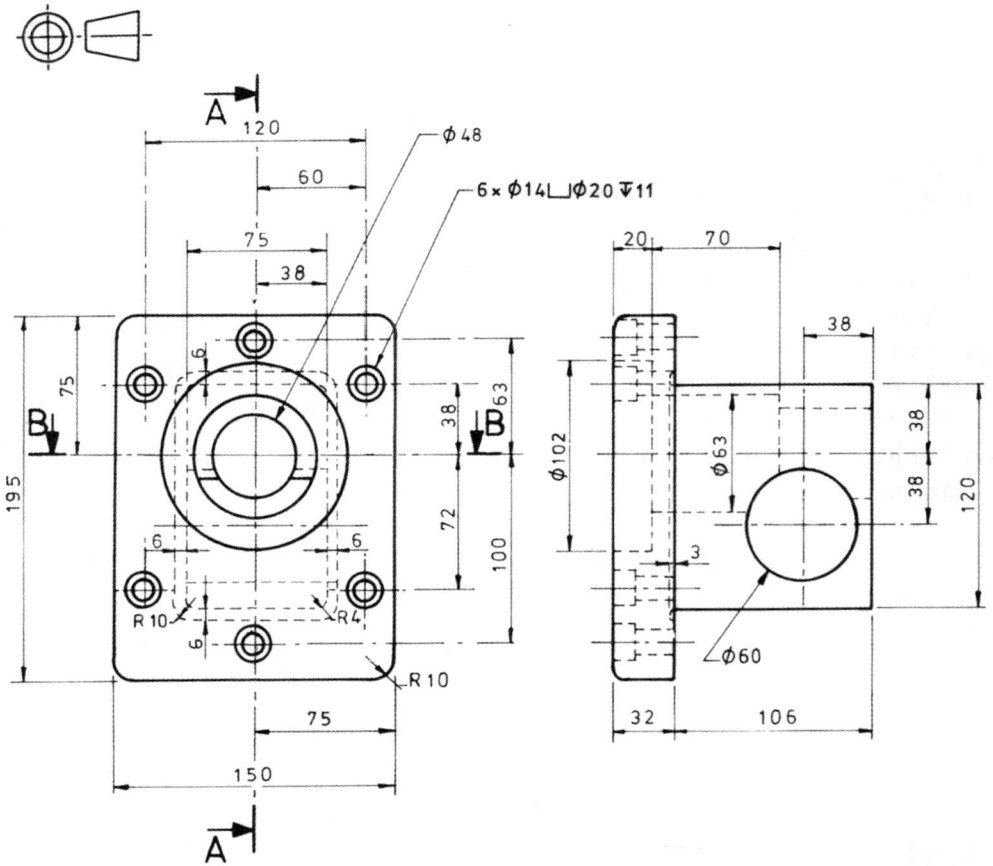

Draw the following views in third-angle projection:

(a) a sectional front view on A-A

(b) a half sectional top view on B-B

(c) a side view as given (scale 1:2)

Fully dimension, provide a title block, and identify the drawing.

## 3.55 WORM GEAR BRACKET

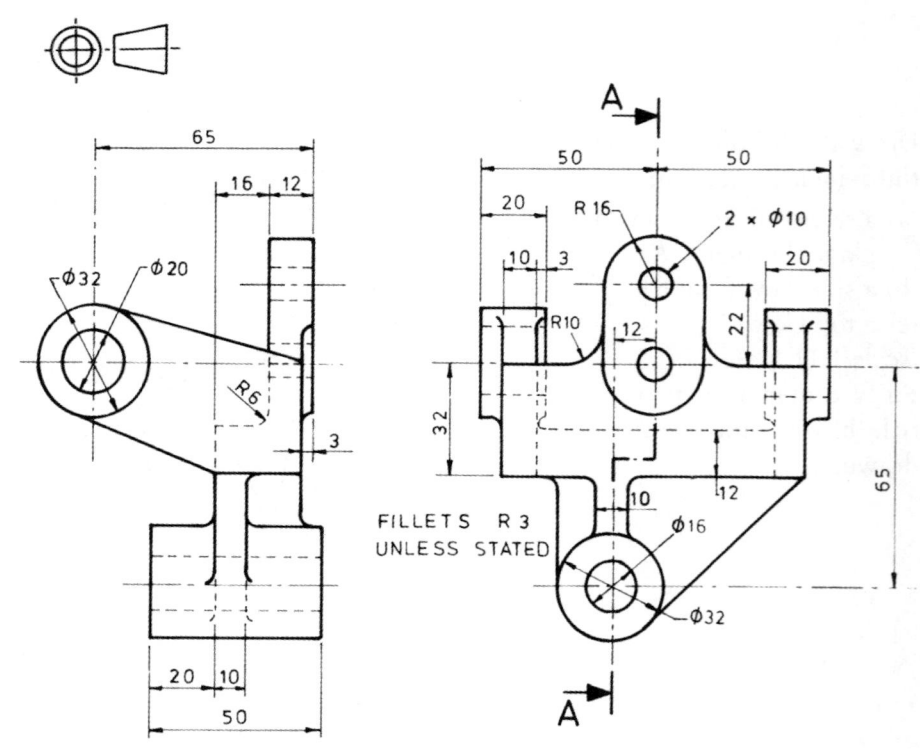

Draw the following views in third-angle projection:

(a) a rear view

(b) an offset section on A-A

(c) a top view

(scale 1:1)

Fully dimension, provide a title block, and identify the drawing.

## 3.56 CI HOUSING

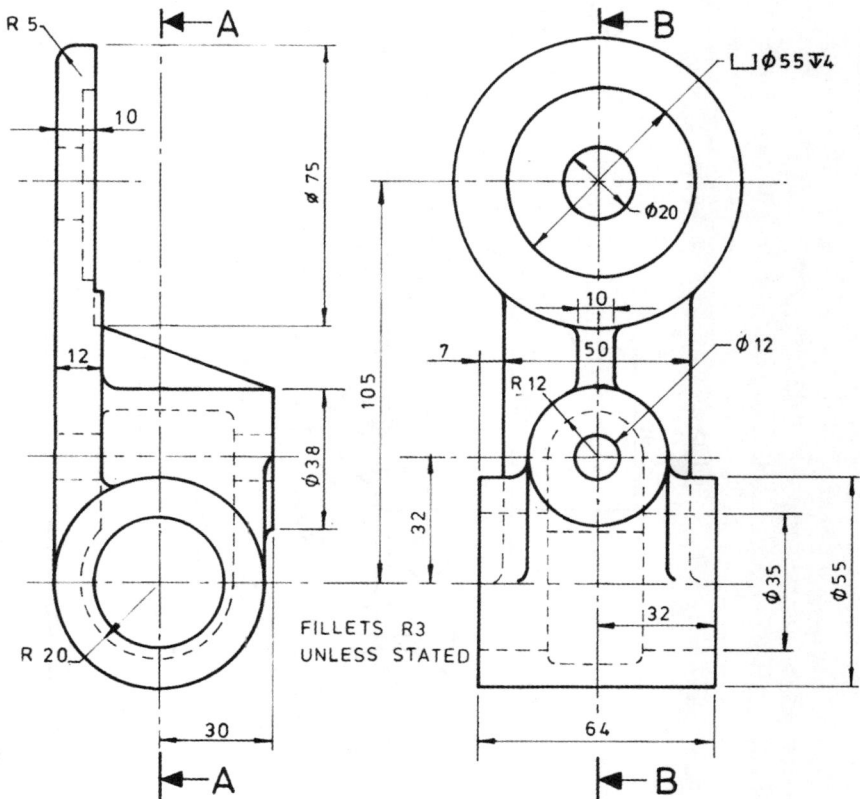

Draw the following views in third-angle projection:

(a) a sectional front view on A-A

(b) a sectional side view on B-B

(c) a top view
(scale 1:1)

Fully dimension, provide a title block, and identify the drawing.

## 3.57 CI SUSPENSION BRACKET

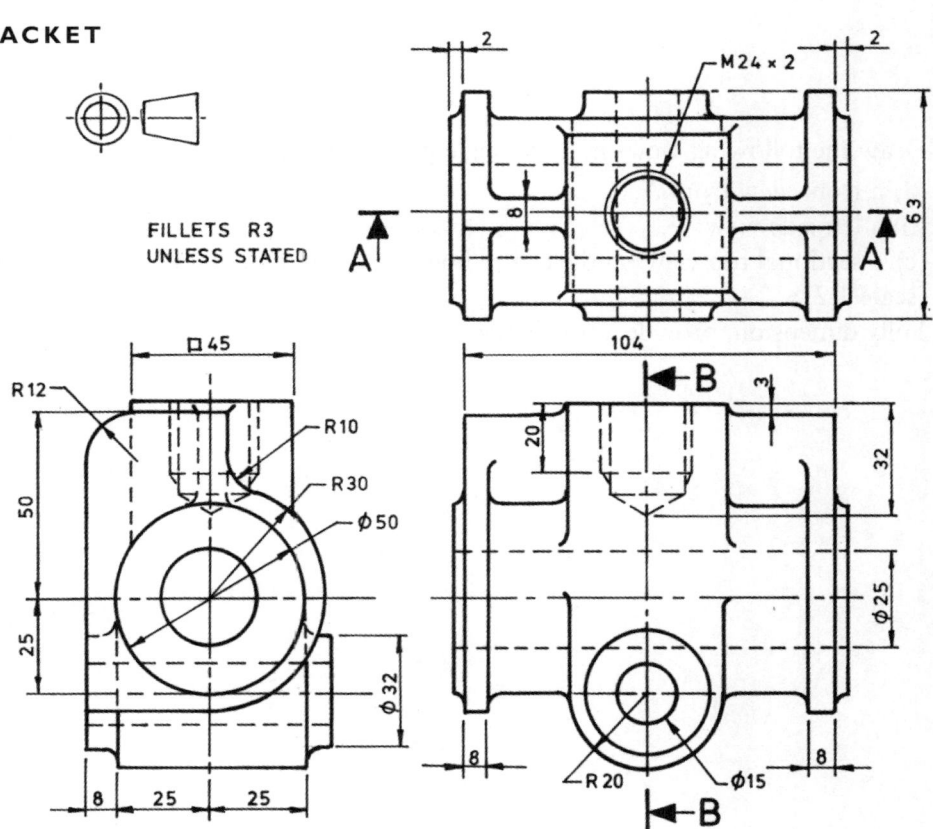

Draw the following views in third-angle projection:

(a) a sectional front view on A-A

(b) a sectional side view on B-B

(c) a top view
(scale 1:1)

Fully dimension, provide a title block, and identify the drawing.

## 3.58 CS FIXTURE BASE

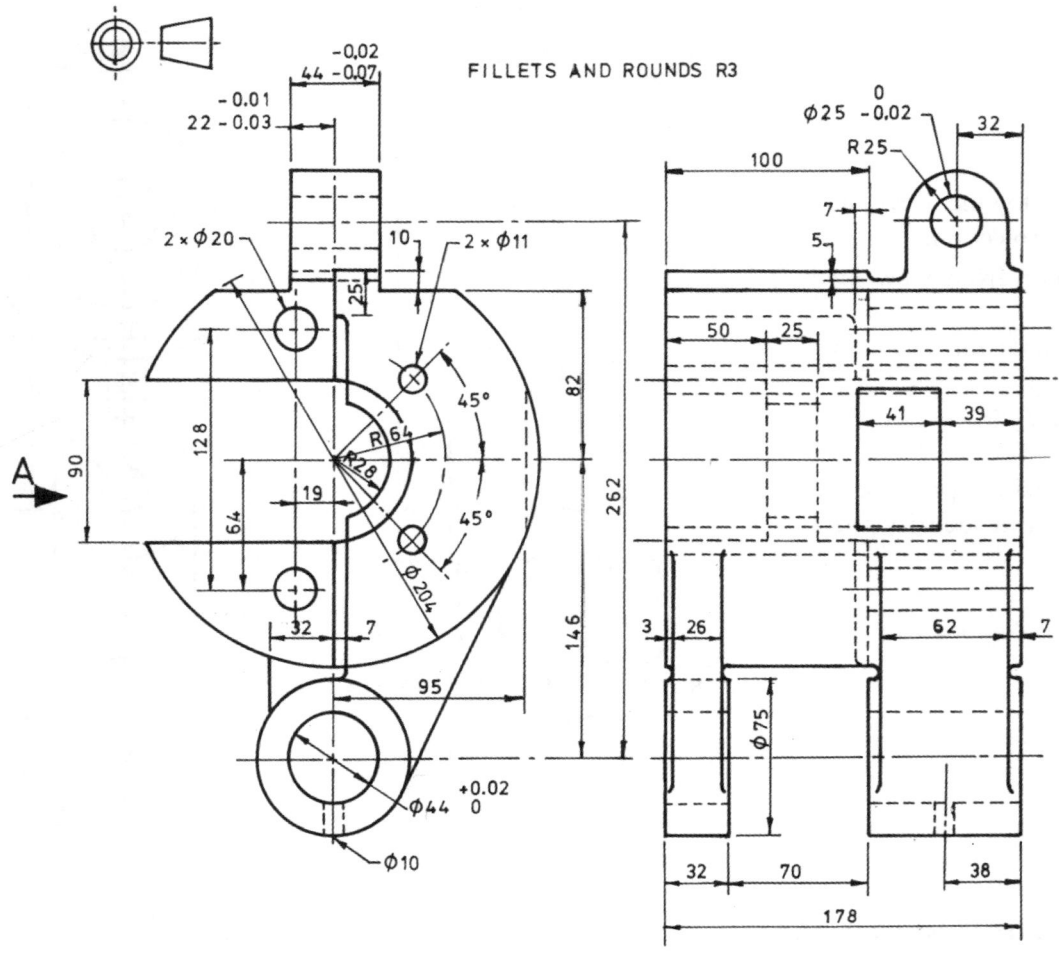

Draw the following views in third-angle projection:

(a) a front view from A

(b) a left side view

(c) a sectional top view on the centre line of the 204 mm diameter

(scale 1:2)

Fully dimension, provide a title block, and identify the drawing.

# 3.59 GEAR-BOX COVER

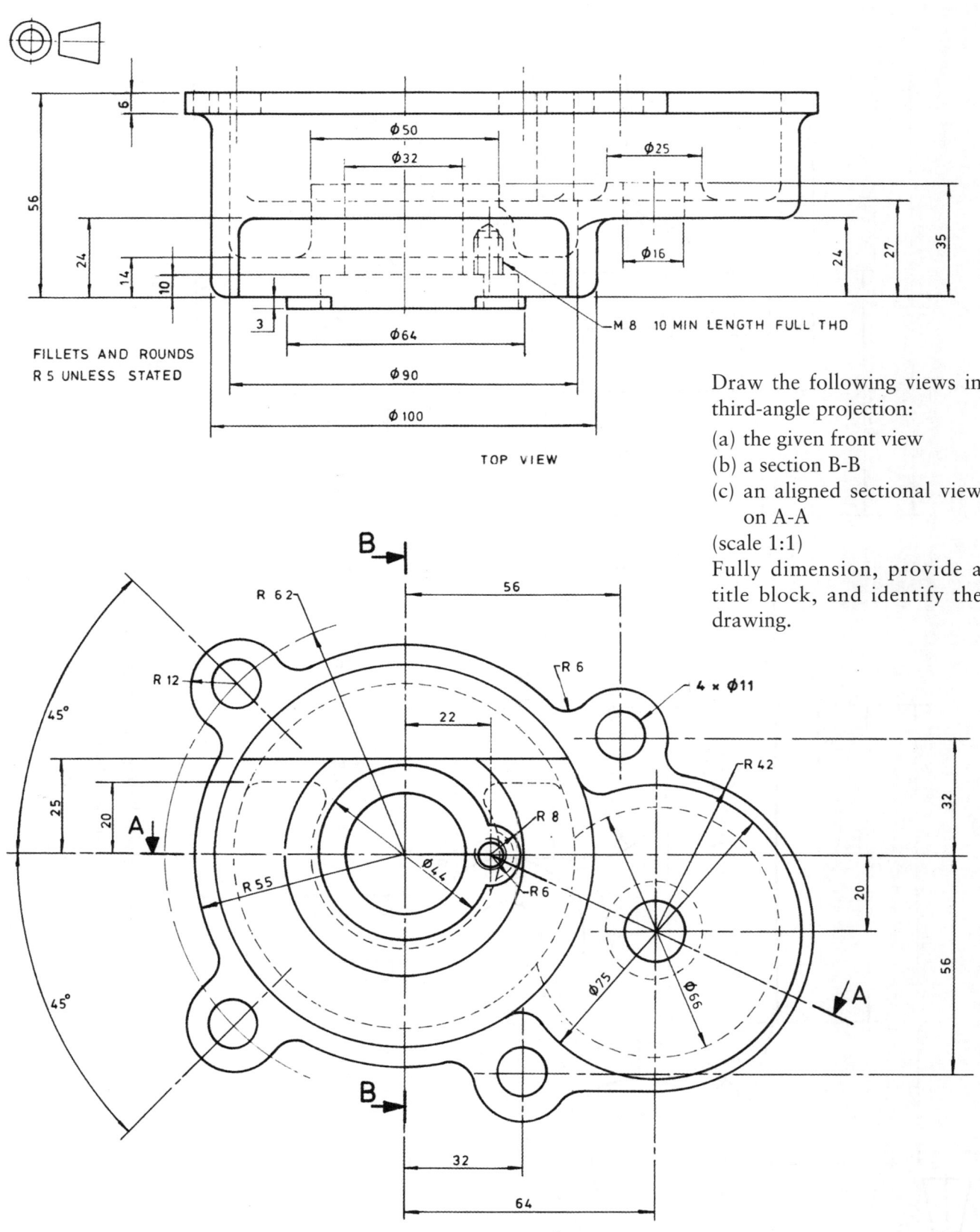

FILLETS AND ROUNDS
R 5 UNLESS STATED

Φ50
Φ32
Φ25
Φ16
Φ64
Φ90
Φ100

M 8   10 MIN LENGTH FULL THD

TOP VIEW

Draw the following views in third-angle projection:
(a) the given front view
(b) a section B-B
(c) an aligned sectional view on A-A
(scale 1:1)
Fully dimension, provide a title block, and identify the drawing.

R 62
R 12
45°
45°
56
R 6
4 × Φ11
22
R 42
25
20
A
R 8
Φ44
R 55
R 6
32
20
Φ75
Φ66
A
56
B
32
64

FRONT VIEW

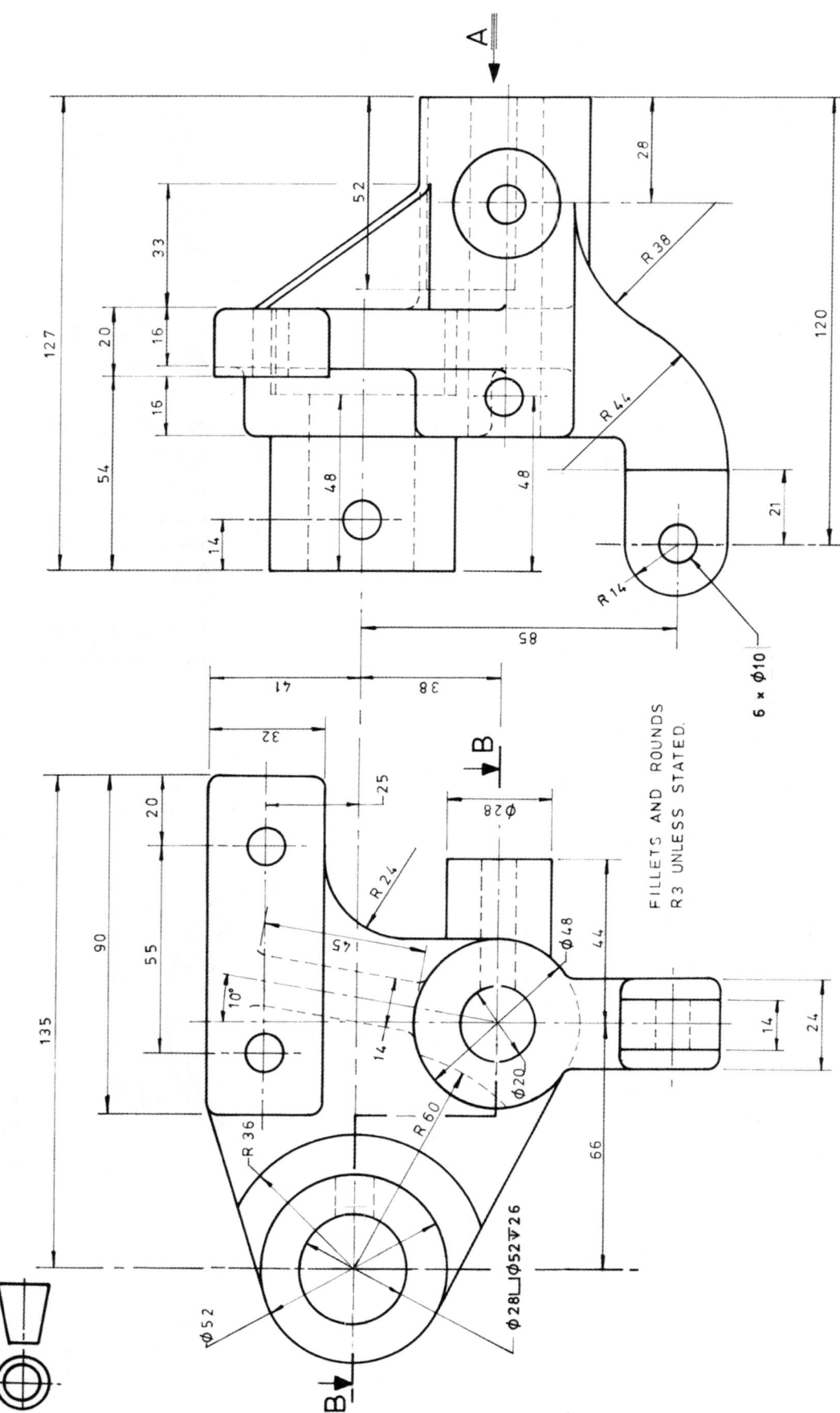

Draw the following views in third-angle projection:

(a) a front view from A
(b) a right side view
(c) an offset sectional view on B-B
(scale 1:1)
Fully dimension, provide a title block, and identify the drawing.

FILLETS AND ROUNDS
R 3 UNLESS STATED.

# 3.7 CAD corner

**3.61** Using your A2 drawing sheet from the first chapter, draw fully detailed component drawings of the adjustable bearing shown on page 227. The components are to be completed to the following requirements:

## General

- Multiple views of a component are to be 3rd angle projection.
- Finished plot scale 1:2.
- All components fully dimensioned.
- All centre lines, and notes.
- Completed title block.

## Base plate

- Top view.
- Fully sectioned front view through the horizontal centre line of the top view.
- Fully sectioned side view through the centre line of the right hand slots.

## Spur wheel

- Fully sectioned side view as shown.
- Front view.

## Bearing body

- Fully sectioned top view through the horizontal centre line of the front view.
- Front view.

## Bearing

- Front and top view.

## Post

- Front and side view.

## Shaft

- Front and side view.

**3.62** This question is suitable for people using a CAD system with 3D modelling capabilities.

(a) Model the clamp block shown below and, if possible, dimension the block similarly.

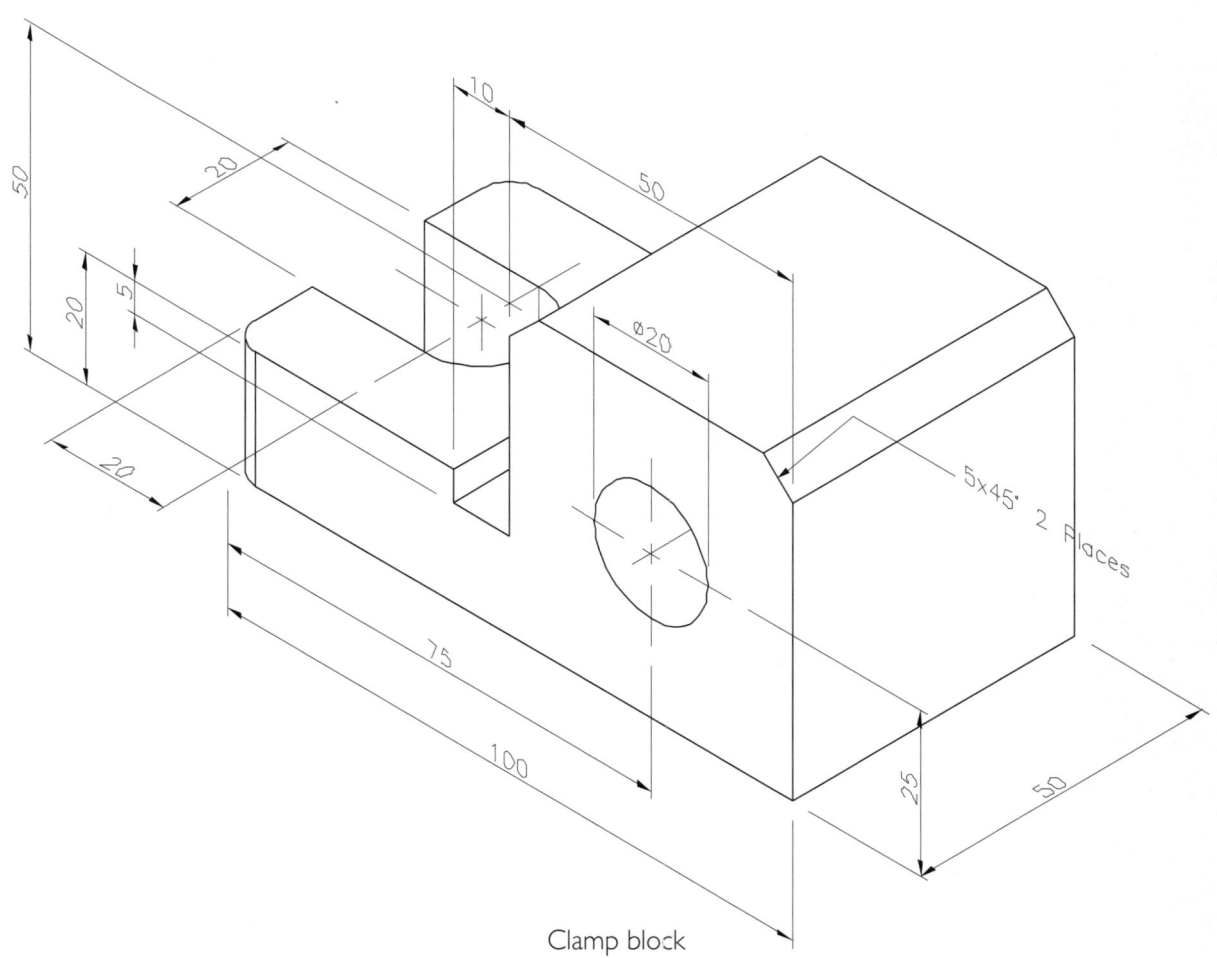

Clamp block

(b) Using your A2 sheet from Chapter 1, generate the drawing layout shown below.

Note: The four views shown below have been generated without any further drafting input, however how you complete this exercise will depend on the capabilities of the CAD package being used.

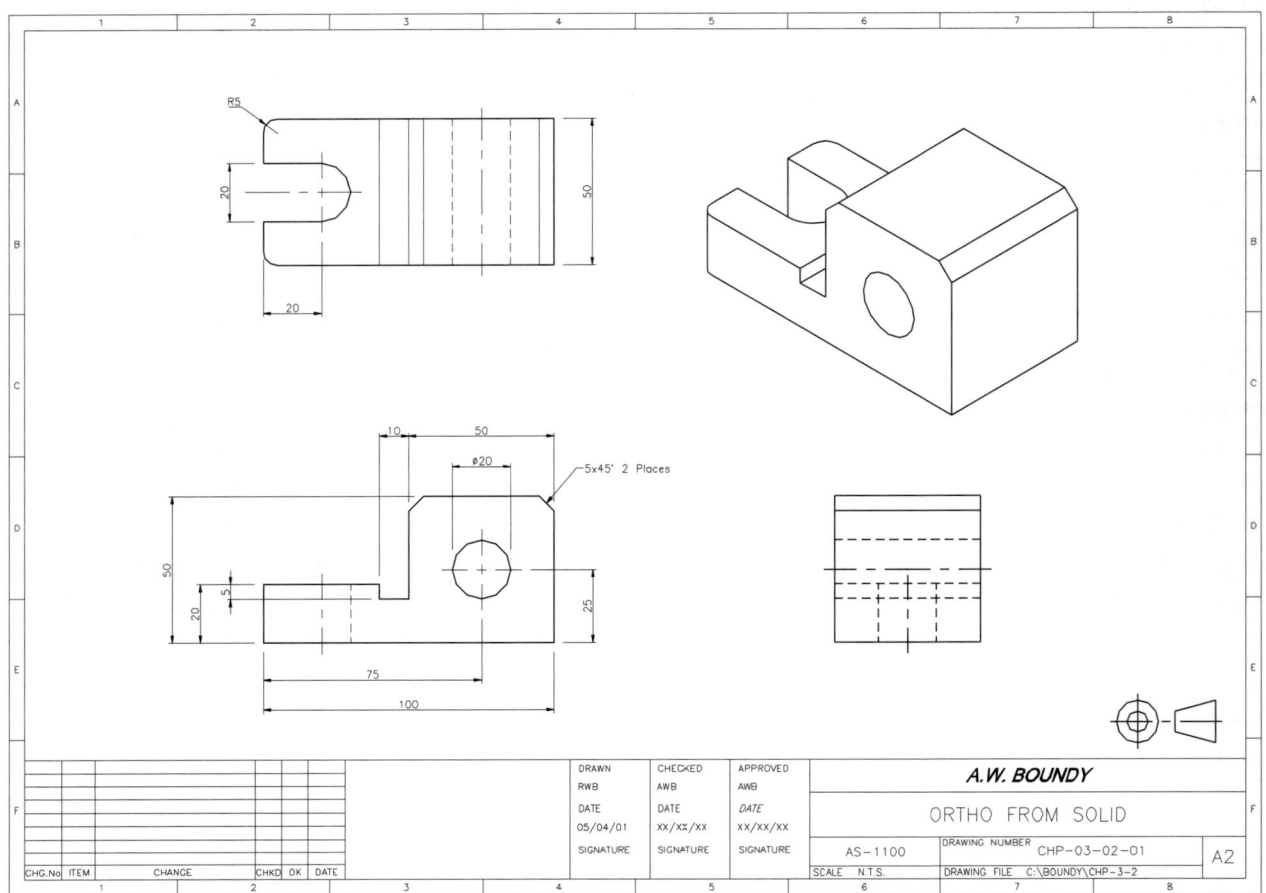

# Descriptive geometry:
## Auxiliary views

After studying this chapter and working through the problems you should be able to do the following:

- understand the spatial relationship between points, lines, plane figures and solids when these features are represented on orthogonal and auxiliary views

- draw primary and secondary auxiliary views

- draw full and partial auxiliary views

- use auxiliary views to determine the true shape of an inclined or oblique surface

# 4.1 Descriptive geometry

Descriptive geometry is a branch of drawing which involves the study of spatial relationships of points, lines, plane figures and solids. The understanding of these relationships is important in architectural and engineering drawing where it is necessary to represent three-dimensional objects on two-dimensional drawings.

In orthogonal drawing the projections of a point onto the principle planes are points also. Irrespective of which direction a point is viewed from, its projection is always a point identical to the original point. However, its position with respect to the reference planes may vary, and such variations can be measured and plotted. The plotting of points is also the basis of plotting lines, surfaces and solids, the latter formed by joining a series of outline points in a certain sequence to represent the particular shape of the surface or solid in question. Figure 4.1 shows three cases of the projection of a point, (a) where the point P is placed a particular distance from the three reference planes, (b) where point P is placed relative to two reference planes and an auxiliary inclined plane, and (c) where P is placed relative to two reference planes and an auxiliary vertical plane. In each case the left hand view shows the point P positioned inside the viewing box, with projectors onto the reference planes. The centre view shows the box opened out with the projections of the point only in relation to the reference planes and finally the right side view shows how the projections of each point are obtained using plane fold lines and traces.

**FIGURE 4.1** ▶

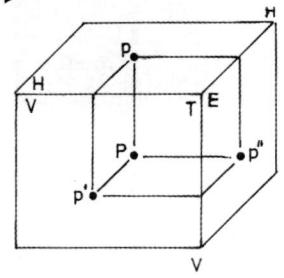 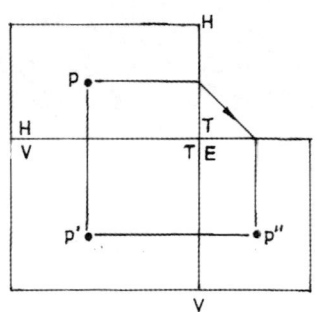 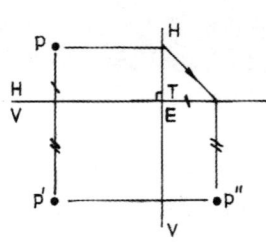

(a) Projections of a point onto HP, VP and EVP

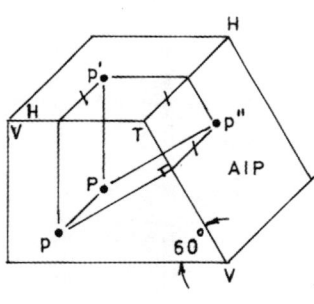 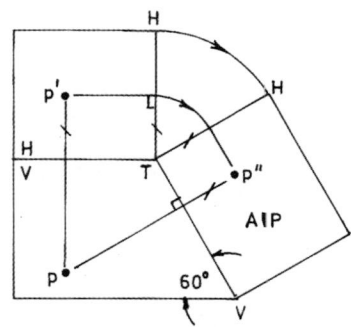 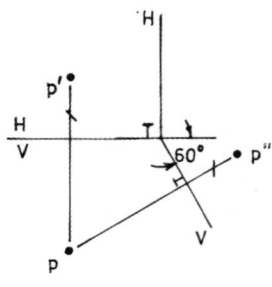

(b) Projections of a point onto H, V and AIP

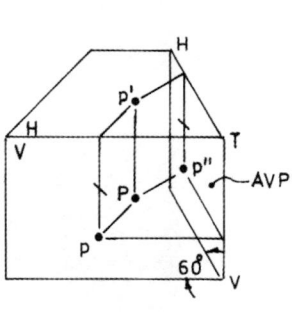 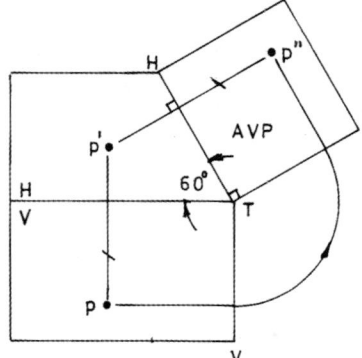 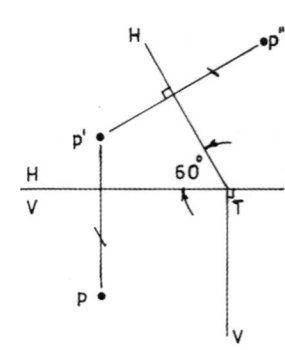

(c) Projections of a point onto H, V and AVP

A line can be represented in any direction of viewing simply by plotting the two ends of the line and joining them. If a line is parallel to any of the principle planes of projection, its true length and angle of inclination to the horizontal and vertical planes can be found by projection onto one or two of the principle planes of projection. A number of possible line positions relative to the principle planes of projection are shown both pictorially and orthogonally in Figures 4.2 (a)–(f). In most cases the true length (TL) of the line AB and angles of inclination (α,β) are readily available from the orthogonal views of the line, for example Figures

4.2 (a)–(e). However Figure 4.2(f) represents a line which is inclined to all reference planes (an oblique line) and the orthogonal views do not reveal its true length (TL) or angles of inclination (α,β). The determination of true length (TL) and angles of inclination are dealt with on pages 170–173.

The determination of the true shape of inclined surfaces may at times seem to be quite hard to visualise, let alone construct. However, if the inclination of the surface to the horizontal or vertical planes is known, we can proceed to find the true shape of the surface using either of the two methods

**FIGURE 4.2** ▶ Projections of a line onto H, V and E

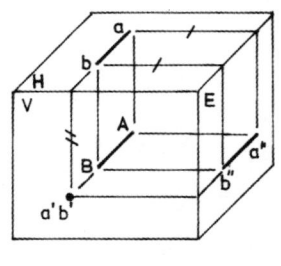 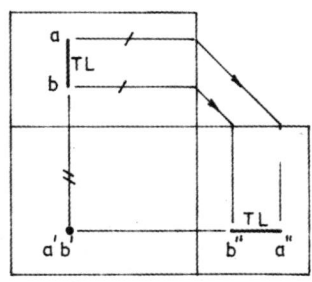

(a) Line—parallel to H and E, α = 0°, β = 90°

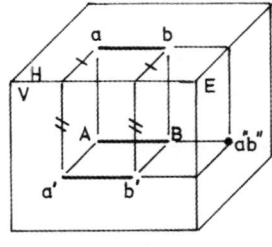 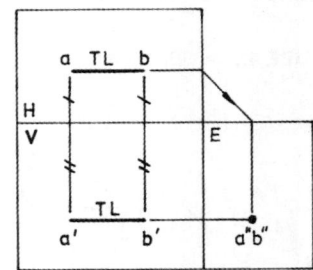

(b) Line—parallel to V and H, α = 0°, β = 0°

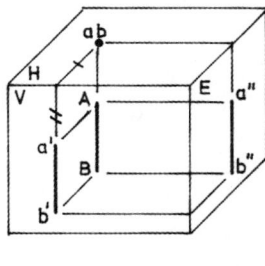 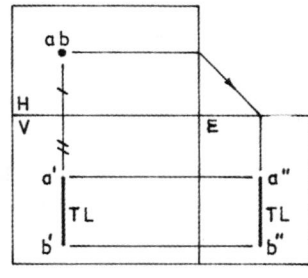

(c) Line—parallel to V and E, α = 90°, β = 0°

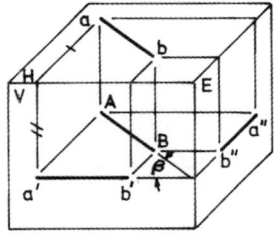 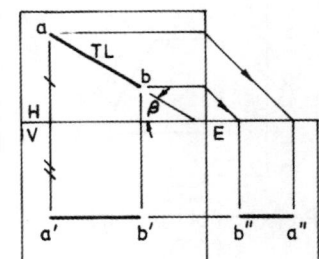

(d) Line—parallel to H, α = 0°, β = inclination to V

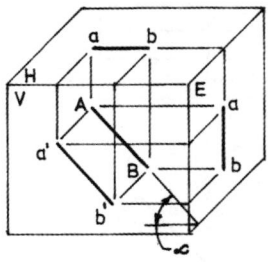 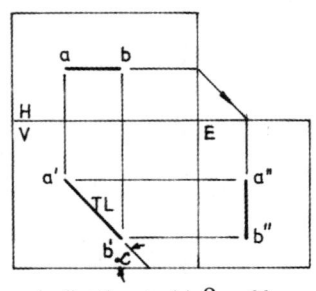

(e) Line—parallel to V, α = inclination to H, β = 0°

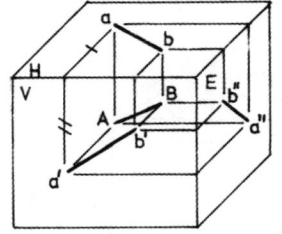 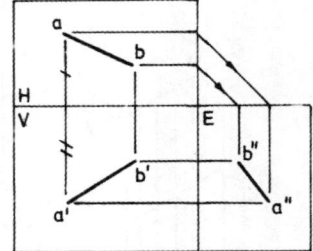

(f) Line—inclined to all reference planes (oblique)

Note:  H = horizontal plane
     V = vertical plane
     E = end vertical plane
     α = angle of inclination to H
     β = angle of inclination V

shown in Figure 4.3. Figure 4.3(a) shows an object inside the viewing box with its front and top views projected onto the horizontal and vertical planes respectively, as well as the true-shape view of the inclined edges projected onto an auxiliary inclined plane (AIP) which is parallel to the plane of the two inclined edges. The right hand view illustrates how the true shape can be constructed using projectors and fold lines similar to that shown on the pictorial view.

Figure 4.3(b) shows an object within the viewing box and having a vertically inclined surface which is not parallel to any of the main reference planes. In order to project the true shape of the vertically inclined surface, an auxiliary vertical plane is placed in front of and parallel to it. The pictorial view shows how the three principle views and the true shape are projected from the solid and how they relate to each other. For example, the depth distances D and D' apply to the front, end and true-shape views. When setting up the orthogonal views on the right it will be noticed that the HV, HE and HA fold lines are positioned so that the front, end and true-shape views can be located at the same distance d and d' below the fold line. These values do not have to be the same as D and D' but they must be all identical.

## TRUE LENGTH AND INCLINATION OF LINES

It is necessary at this stage to introduce a very important topic which has a bearing on the subject of development, namely the relationship of the front and top views of a line to its true length.

**FIGURE 4.3** ▶ Determination of true shape of surfaces

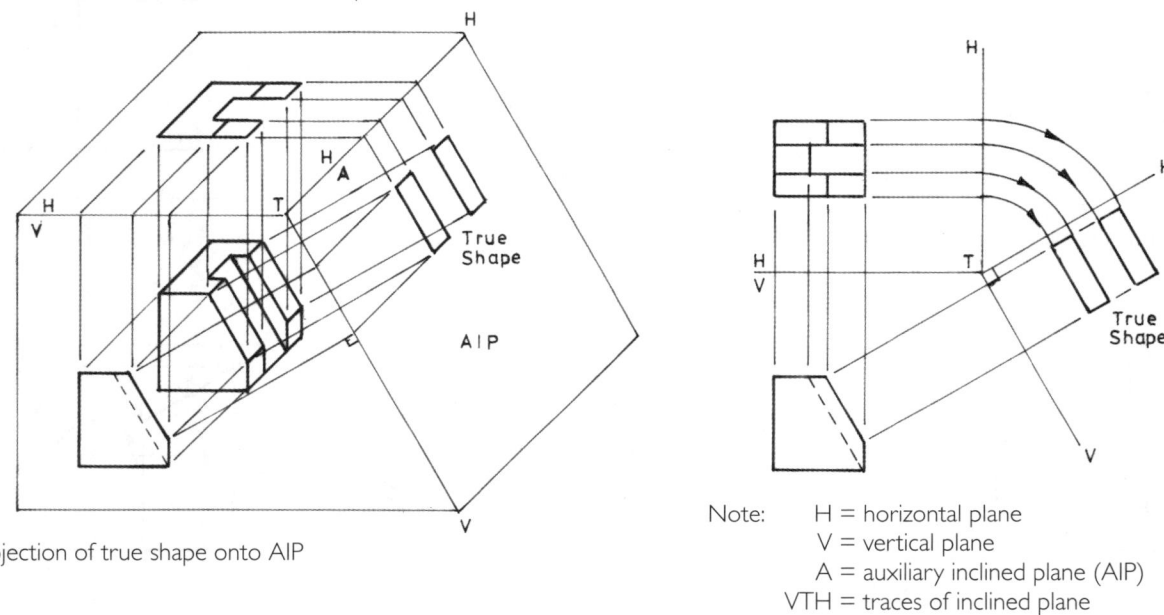

(a) Projection of true shape onto AIP

Note:     H = horizontal plane
V = vertical plane
A = auxiliary inclined plane (AIP)
VTH = traces of inclined plane

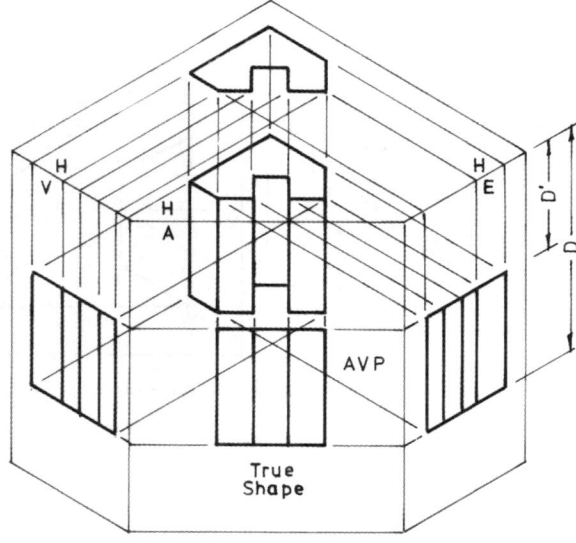

(b) Projection of true shape onto AVP

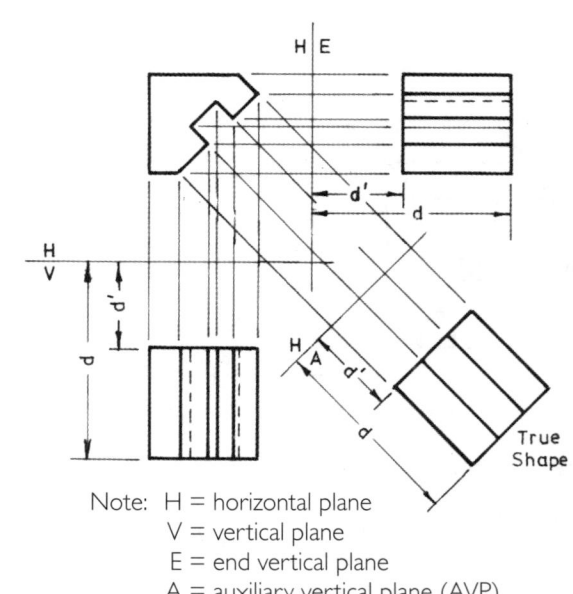

Note:  H = horizontal plane
V = vertical plane
E = end vertical plane
A = auxiliary vertical plane (AVP)

Consider the pictorial view of a line AB situated in space in the third dihedral angle, as illustrated in Figure 4.4. This figure shows the projection of the top view, ab, and the front view, a'b'. It can also be seen that the line, projected on, penetrates the vertical and horizontal planes at two points called the vertical trace (vt) and the horizontal trace (ht) of the line. The line AB is inclined at an angle α to the horizontal plane and β to the vertical plane as shown. Note the formation of two right-angled triangles which have the actual line AB as a common hypotenuse; these are triangle ABD and triangle ABC. They are formed as follows:

1. BD is drawn parallel to the front view, and is therefore equal in length to it; it also makes an angle β with AB.

2. AC is drawn parallel to the top view, and therefore is equal in length to it; it makes an angle α with AB.

There are seven important facts about a line and its position in the dihedral angle which enable it to be fully described in orthogonal projection:

1. its true length
2. its front view length
3. its top view length
4. its angle of inclination to the horizontal plane (α)
5. its angle of inclination to the vertical plane (β)
6. the vertical difference in height of the ends of the line below the horizontal plane
7. the horizontal difference in the distances of the ends of the line from the vertical plane

In the following problems, some of the above facts are given, and it is necessary to find the others. In development work, the front and top views of a line are generally given, and it is necessary to find its true length in order to use it on the development.

A knowledge of the composition of the two right-angled triangles ABC and ABD will enable all of the above seven facts about the line to be solved. These two triangles are now described in detail. Figure 4.5(a) represents triangle ABC and four of the above seven facts about the line are represented on it. These are:

1. AB, the true length
3. AC, the top view length
4. α, the angle of inclination of the line to the horizontal plane
6. BC, the vertical difference

An important property about this right-angled triangle is that it can be solved geometrically by knowing any two of the four facts represented on it.

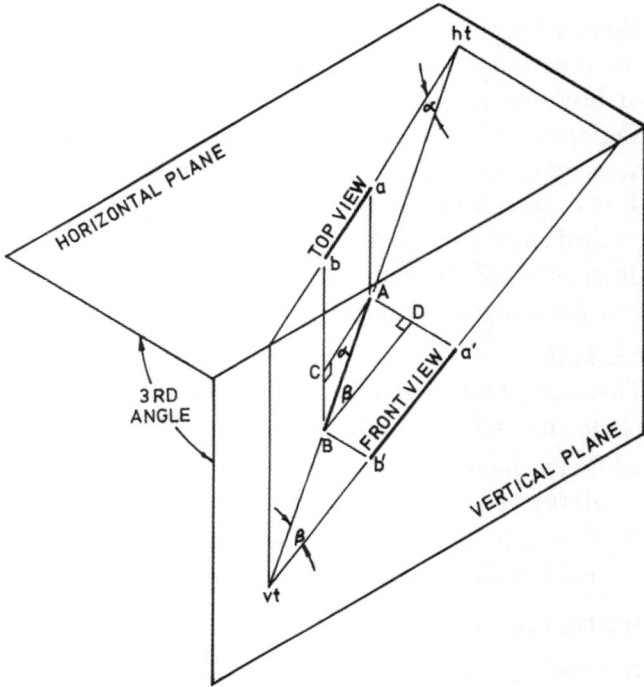

Note:  AB = actual line in space
       ab = projection of line onto horizontal plane (top view)
       a'b' = projection of line onto vertical plane (front view)
       α = angle of inclination of line to horizontal plane
       β = angle of inclination of line to vertical plane

Similarly, the right-angled triangle ABD (Figure 4.5(b)) can be solved geometrically by knowing any two of the following four facts which are represented on it:

1. AB, the true length
2. BD, the front view length
5. β, the angle of inclination to the vertical plane
7. AD, the horizontal difference

FIGURE 4.5 ▶ True length triangles

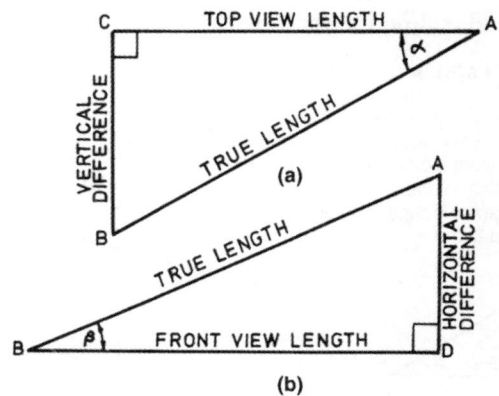

If one can remember and understand the origin of these two triangles, and be able to construct them, there will be very little difficulty in solving problems involving true length and inclinations of lines.

## METHODS OF DETERMINING TRUE LENGTH

Figures 4.6(a)–(f) illustrate six methods of determining the true length of a line, given the front and top orthogonal views a'b' and ab respectively in third-angle projection. Various methods are shown, but each determines one of the two triangles shown in Figure 4.5. In development work, it is usually necessary to find the true length of the line only, but the full description of the true length triangle is given in each case for recognition purposes.

### Method 1

This is used when given a line parallel to the vertical plane and inclined to the horizontal plane.

1. Draw the front and top views of the lines a'b' and ab respectively.
2. Then a'b' is also the true length of AB. Note the true length triangle a'b'c.

**FIGURE 4.6(a)** ▶ Determining true length—method 1

TL = true length of AB
TVL= top view length
VD = vertical difference in ends of line from horizontal plane
α = angle of inclination to horizontal plane
AB = actual line

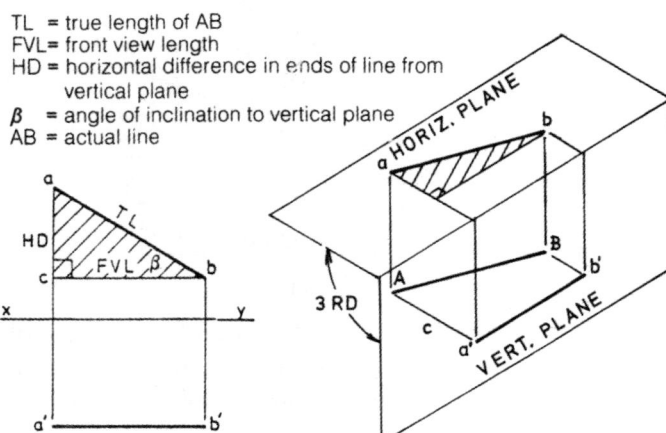

### Method 2

This is used when given a line parallel to the horizontal plane and inclined to the vertical plane.

**FIGURE 4.6(b)** ▶ Determining true length—method 2

TL  = true length of AB
FVL= front view length
HD = horizontal difference in ends of line from vertical plane
β = angle of inclination to vertical plane
AB = actual line

1. Draw the front and top views of the lines a'b' and ab respectively.
2. Then ab is also the true length of AB. Note the true length triangle abc.

### Method 3

This may be used when given a line inclined to both the horizontal and vertical planes. The aim is to construct one of the true length triangles, and this is possible in a number of ways, one of which follows.

1. Draw the front and top views of the lines a'b' and ab respectively.
2. Rotate the top view ab parallel to the vertical plane.
3. Project the line down level to b', that is to cb'.
4. From a', project horizontally to meet the projector through c at d. Triangle cdb' is the true length triangle, and b'd the true length of AB.

**FIGURE 4.6(c)** ▶ Determining true length—method 3

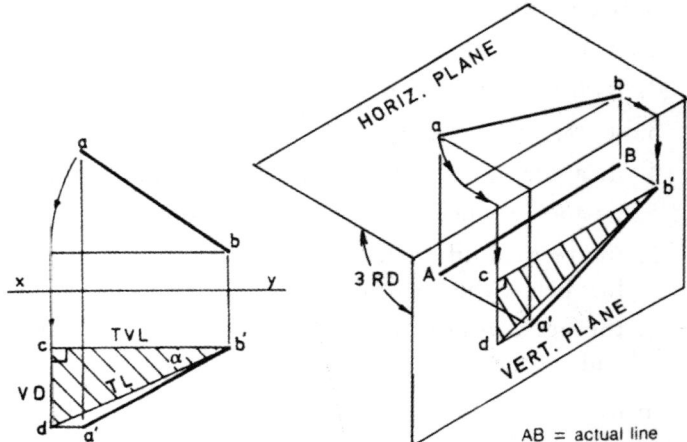

AB = actual line

### Method 4

This may be used when given a line inclined to both the horizontal and vertical planes. It is similar to method 3 except that the other true length triangle is constructed.

**FIGURE 4.6(d)** ▶ Determining true length—method 4

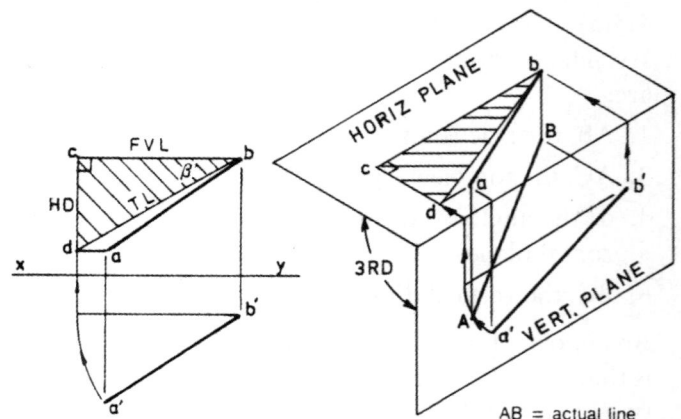

AB = actual line

172

1. Draw the front and top views of the lines a'b' and ab respectively.
2. Rotate the front view a'b' into the horizontal position.
3. Project the line up level to b, that is, to cb.
4. From a, project horizontally to meet the projector to c at d. Triangle bcd is the true length triangle, and bd is the true length of AB.

## Method 5

This may be used when given a line inclined to both the horizontal and vertical planes. This method uses an auxiliary view which determines the true shape of the triangle cross-hatched in the pictorial view.

1. Draw the front and top views of the lines a'b' and ab respectively.
2. Project at right angles from a' and b' the horizontal distances of a and b respectively from the vertical plane (these distances are shown bracketed).
3. Join the ends of these projectors c and d to give the true length of AB. Note the true length triangle (hatched).

**FIGURE 4.6(e)** ▶ Determining true length—method 5

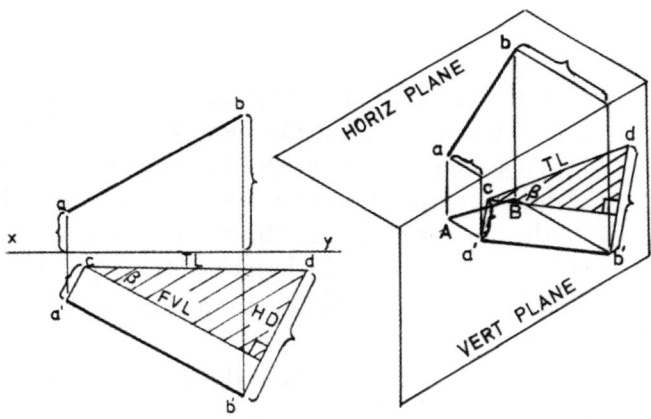

## Method 6

This may be used when given a line inclined to both the horizontal and vertical planes. This method uses a different auxiliary view from method 5, as shown in the pictorial view.

1. Draw the front and top views of the lines a'b' and ab respectively.
2. Project at right angles from a and b the vertical distances of a' and b' respectively below the horizontal plane (these distances shown bracketed).
3. Join the ends of these projectors c and d to give the true length of AB. Note the true length triangle (hatched).

**FIGURE 4.6(f)** ▶ Determining true length—method 6

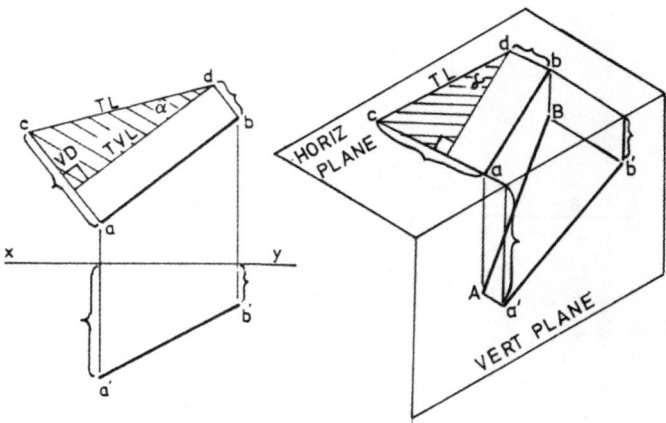

## TRUE LENGTH—WORKED EXAMPLES

Together with its use in development work, true length theory is used in many other instances of engineering construction. The basic principle holds true for all uses and involves the construction of one or two of the true length triangles described on page 171. A study of this section should be made before progressing on with the following exercises and problems, and further reference can be made to it as the need arises.

In summary, if two orthogonal views of an oblique line are given then it is possible to construct the third view and determine its true length and angles of inclination to either the vertical or horizontal planes of projection.

### Worked example 1

A TV mast located on a gable roof has a 15° pitch. The mast is 1 m from the ridge and stands vertically, held in position by three guy wires attached to the mast 5 m above the base. Each wire is tied to the roof 120° apart as shown in the top view of Figure 4.7(a).

Determine the true length of each wire and its true angle of inclination to the roof.

### Solution

1. Draw the front and top views of the roof line, mast and wires to a scale of 1:100.
2. The required wire lengths and true angles are contained in the three vertical triangles shown shaded in Figure 4.7(b).
3. DAE is represented in true shape on the front view, Figure 4.7(b), because the top view of DAE is at right angles to the direction of viewing. Therefore the true length of guy wire AD is 8.5 m and it makes an angle of 30° to the roof.
4. To obtain the true length and angle of inclination of wires DB and DC (which are the same) it is necessary to find the true shape of triangle DEB or DEC.

**FIGURE 4.7** ▶ View of a TV mast

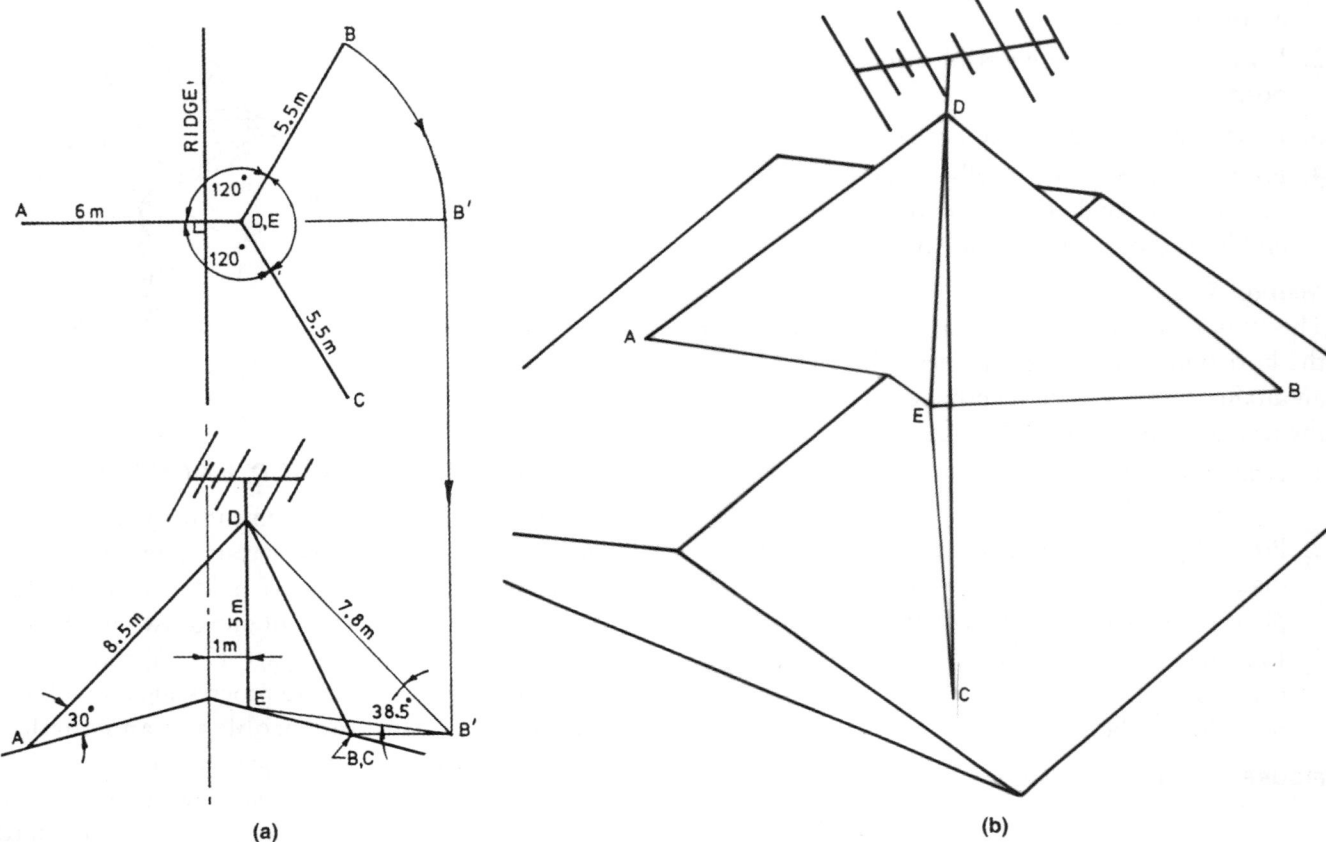

(a)                                                                                      (b)

5. Rotate the top view of triangle DEB until it is normal to the direction of viewing at DEB'. Project B' down to the front view level with B to give the true shape of triangle DEB'. DB' (7.8 m) is the true length of the guy wires DB and DC and angle DB'E (38.5°) is the true angle between wires DB, DC and the roof.

### Worked example 2

A sketch of two bulkheads connected by a pipe whose axis intersects the bulkheads at A and B is shown in Figure 4.8(a).

(a) Determine the true length of the pipe axis between the bulkheads, and the angles of inclination to the horizontal and vertical planes of projection, given that the front view is taken from direction X.

(b) Determine the true angle of the pipe cut so that it sits flush on each bulkhead.

### Solution

1. Draw the front and top views of the pipe axis including the edge view of the surface of the two bulkheads.

2. Construct the two true length triangles CDb' and aEF which provide the answer to part (a), i.e. true length = 2175 mm, $\alpha$ = 33.5°, $\beta$ = 27.5° (Fig. 4.8(b)).

3. The true angle of cut of the pipe at A is included in a right-angled triangle, the top view of which is baH, bH being a line drawn perpendicular to the plane of the upper bulkhead. Rotate this triangle about bH into the horizontal plane. This is done by marking off bG equal to the true length of the pipe axis. Angle bGH ($\Delta$ = 53.5°) is the true angle of cut of the pipe at A.

4. The lower bulkhead at B is in the horizontal plane and therefore the true angle of cut of the pipe at B is equal to the angle of inclination $\alpha$ = 33.5°.

# 4.2 Auxiliary orthogonal views

### INTRODUCTION

Sometimes an object has a face which is not parallel to the normal planes of projection. In this case a true shape view of the irregular face can only be obtained by projecting it onto an auxiliary plane which is parallel to the face. A view so obtained is called an *auxiliary view*. There are two kinds of auxiliary views: *primary and secondary*.

FIGURE 4.8 ▶ Bulkhead pipe connection

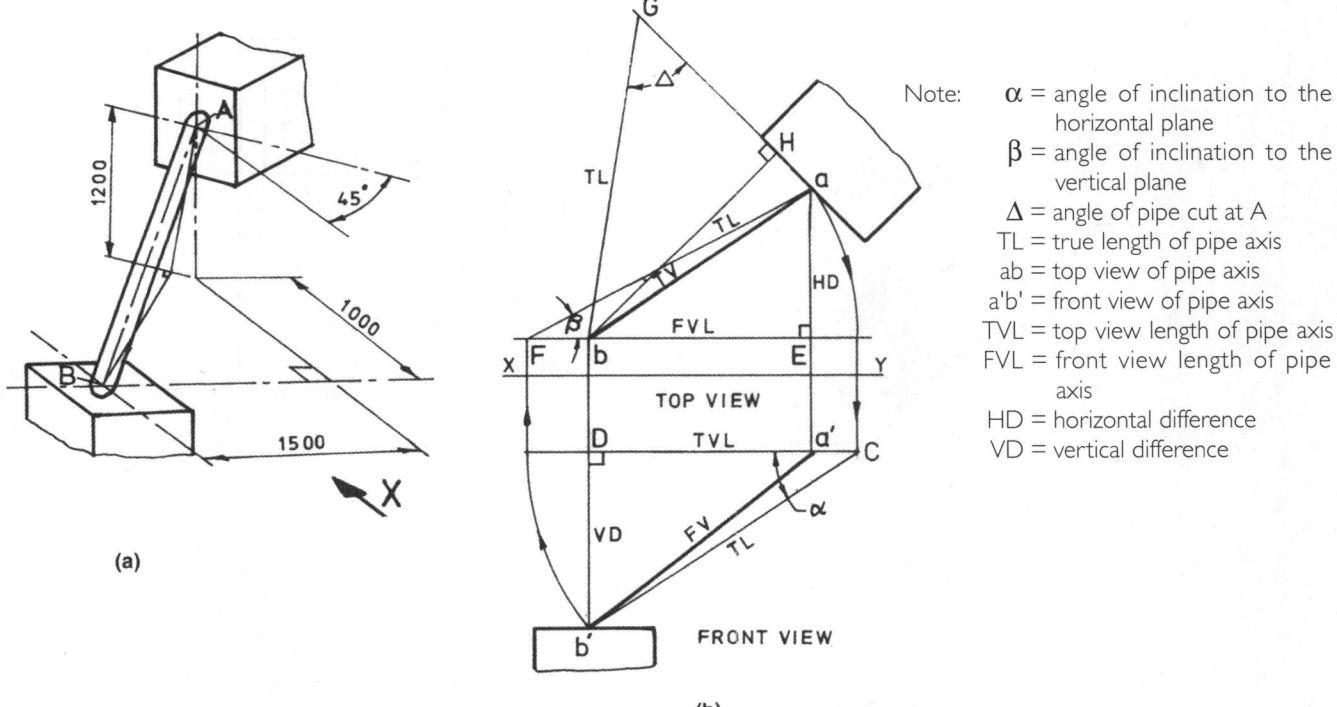

**(a)**

**(b)**

Note: α = angle of inclination to the horizontal plane

β = angle of inclination to the vertical plane

Δ = angle of pipe cut at A

TL = true length of pipe axis

ab = top view of pipe axis

a'b' = front view of pipe axis

TVL = top view length of pipe axis

FVL = front view length of pipe axis

HD = horizontal difference

VD = vertical difference

# 4.3 Primary auxiliary views

When the auxiliary plane is inclined to four of the six principal planes of projection and is square with the other two, the view so obtained is termed a *primary auxiliary view*. Such is the case illustrated by the pictorial views of Figures 4.9(a), (b) and (c).

### TYPES OF PRIMARY AUXILIARY VIEWS

An auxiliary view must be projected from the principal view which provides the edge view of the inclined irregular face. Hence there are basically three types of primary auxiliary views:

1. The view obtained when the edge view of the inclined irregular face is shown on the front (or rear) view (Fig. 4.9(a)).
2. The view obtained when the edge view of the inclined irregular face is shown on the top (or bottom) view (Fig. 4.9(b)).
3. The view obtained when the edge view of the inclined irregular face is shown on the side (right or left) view (Fig. 4.9(c)).

It will be noticed in each of the three figures that measurements of length are projected from the edge view of the inclined irregular face. Measurements of width required to complete the auxiliary view are transferred from the other principal view, and are normally measured from a reference plane (RP). (Note D1, D2 and D3.) The reference plane (RP) in this case is the plane on which the object is resting.

The position of the reference plane can be established by considering its edge representation on the orthogonal views and then relating it to the pictorial view in each case.

### PARTIAL AUXILIARY VIEWS

If a component has an irregular inclined face which must be detailed, it is often only necessary to draw an auxiliary view of the irregular face and not a complete auxiliary view of the whole object.

In many cases a complete auxiliary view distorts another part of the view, making it of little value.

The use of partial views saves drawing time, simplifies the drawing, and makes it easier to read. Figure 4.10(a) illustrates a complete auxiliary view, part of which is completely distorted due to the angle of viewing and is of little descriptive value. Figure 4.10(b) shows the use of the partial auxiliary view which omits the distorted portion.

**FIGURE 4.9** ▶ Types of primary auxiliary views in third-angle projection

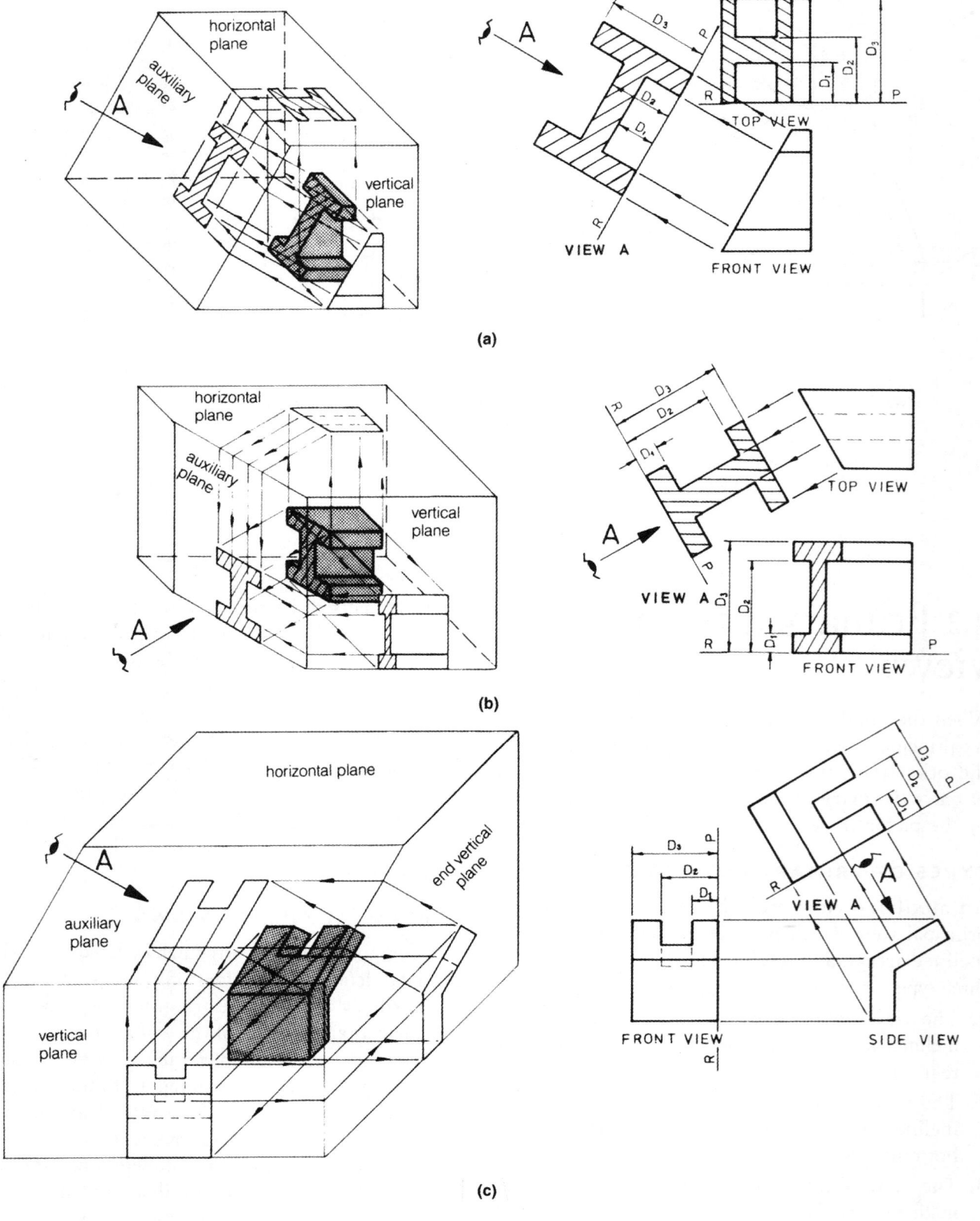

(a)

(b)

(c)

**FIGURE 4.10 ▶** Comparison of full and partial auxiliary views

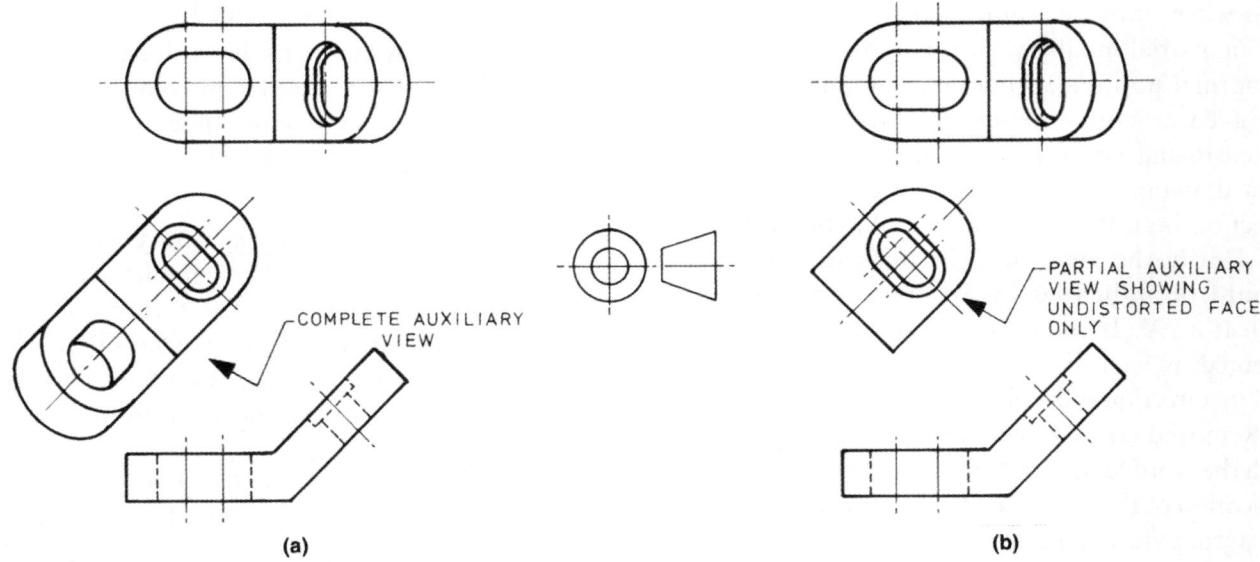

COMPLETE AUXILIARY VIEW

PARTIAL AUXILIARY VIEW SHOWING UNDISTORTED FACE ONLY

**(a)**

**(b)**

## ORIENTATION OF AUXILIARY VIEWS

*Auxiliary views should always be drawn using third-angle projection* irrespective of whether the method of projection used on the other views is first-angle or third-angle projection. The type of auxiliary view most commonly used is the normal view obtained by looking perpendicularly at the inclined face and projecting the true shape onto an auxiliary plane perpendicular to the line of viewing (Fig. 4.11(a)).

**FIGURE 4.11 ▶** Orientation of auxiliary views

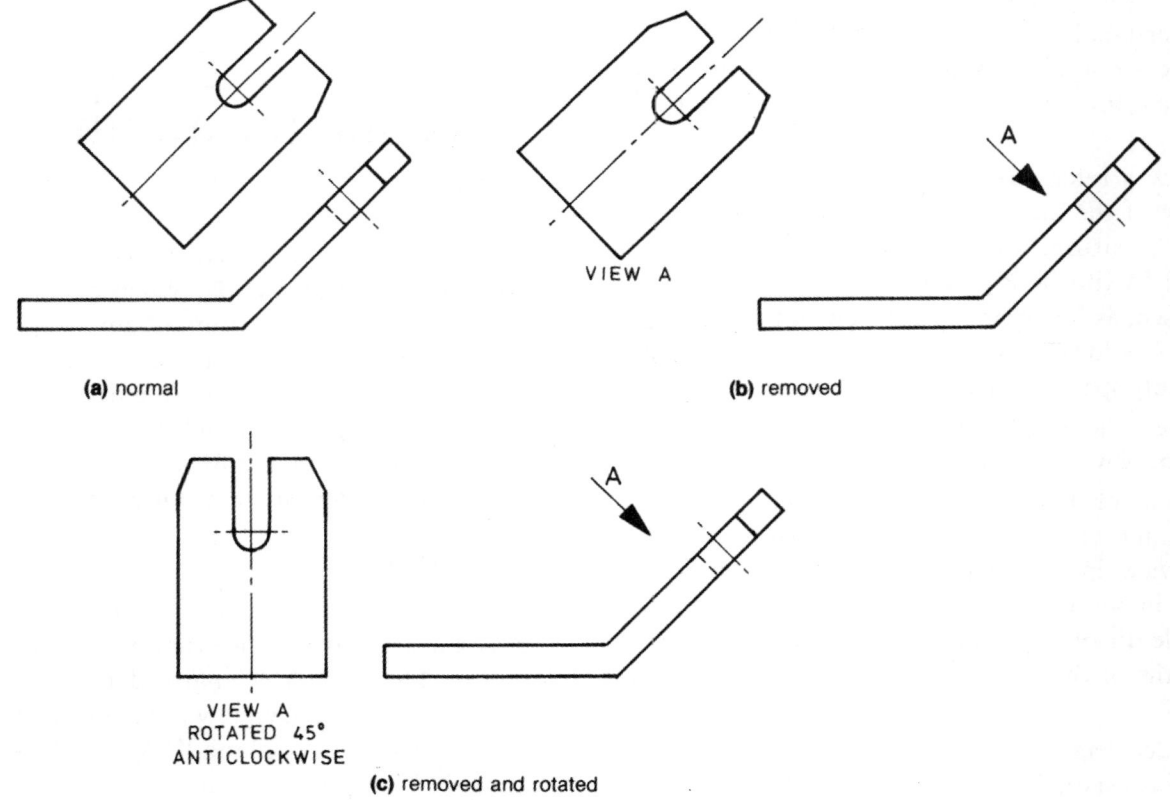

**(a)** normal

VIEW A

**(b)** removed

VIEW A
ROTATED 45°
ANTICLOCKWISE

**(c)** removed and rotated

The principle of auxiliary projection may also be used when drawing removed views (Fig. 4.11(b)). A full or partial auxiliary view may be removed from its normal position without changing its orientation for greater convenience or clarity, such as for dimensioning purposes or to achieve a better layout on a drawing sheet. The word 'view' followed by a direction indicator, for example 'A', should be used to identify the view, and the direction of viewing should be indicated by an arrow together with indicator 'A'. If the removed view needs to be re-oriented as well, the number of degrees of rotation and its direction must be stated (Fig. 4.11(c)).

Removed views drawn to a larger scale are labelled with the word 'detail' followed by a letter as well as an indication of the scale used (Fig. 4.12). The portion of the actual view removed is enclosed in a circle or a rectangle drawn with a thin type B line. If the removed view is close to the detail on the actual view, the circle or rectangle may be joined to the 'detail' by a leader.

### Example of primary auxiliary view

Figure 4.13 is an example of the application of a primary auxiliary view. It is one of the three types referred to on page 176 and illustrated in Figure 4.9(a). Other types are drawn in a similar manner to this example.

1. Draw the front view showing the edge of the inclined 30° face.
2. The right side view is included to enable an understanding of how the widths of this normal view are applicable to the auxiliary view. Note the aid for constructing the elliptical shape of the top.
3. Draw a reference line (RP) parallel to the edge view of the inclined face. RP may be selected in any position; an alternative is shown by the dashed line R'P'. The centre line position is chosen, as in a symmetrical view of this type most detail is located on the centre line and the rest is equally spaced on either side of it.
4. Project the lengths of the auxiliary view from the front view on to RP and beyond where necessary.
5. Widths of detail on the auxiliary view as well as arcs and circles may now be drawn to complete the view. In this example widths may be measured. It is important to realise, however, that the widths of detail on the auxiliary view are identical to widths of the same detail on the right side view.

Notes:

1. Hidden detail is not shown on this auxiliary view as little is to be gained by including it. However, where

a better shape description would be obtained, hidden detail should be included.
2. Auxiliary views should be dimensioned, but in this case dimensions are omitted in favour of showing projection lines for a better understanding of the method. Projection lines are always removed from the final drawing in practice.

### Example of complex primary auxiliary view

Figure 4.14 shows two rather complicated normal views of a box tool holder for a turret lathe. A primary auxiliary view shows the true shape of the face ABCDEFGHIJK. All projection lines have been left on the drawing so that detail from one view to another may be traced.

It may be seen that the auxiliary view is a combination of the normal views. The lengths are projected directly from the front view and the widths transferred from the side view. In drawing an auxiliary view of this complexity, it is difficult to visualise completely the whole view, and it is best to plot one edge at a time. For example, edge XY is hard to visualise on the auxiliary view, but it can be plotted quite easily by projecting from the front view; knowing the surface from which it starts, one can then transfer its width from the side view.

It should be pointed out that, in most cases, the whole auxiliary view would not be required, and that only the true shape of face ABCDEFGHIJK is necessary.

# 4.4 Secondary auxiliary views

Sometimes an object will have a face inclined to all principal planes of projection. When this is so, it is necessary to draw first a primary auxiliary view to obtain an edge view of the inclined face, and then a *secondary auxiliary view* to give the true shape.

Figure 4.15 shows front and top views of a block having an oblique face ABF on one corner. It is required to draw the true shape of this face.

**PROCEDURE**

1. Project a primary auxiliary front view in such a direction as to give an edge view of face ABF, that is looking along the true length edge FB on the top view. In this primary view the heights of points above the reference plane X'-Y' are the same as above X-Y on the front view.

**FIGURE 4.12** ▶ Removed and enlarged views

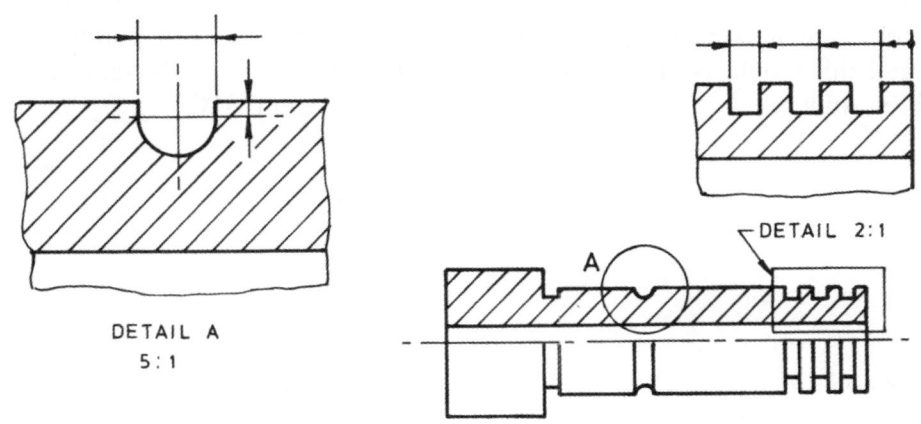

DETAIL A
5 : 1

DETAIL 2:1

A

**FIGURE 4.13** ▶ A primary auxiliary view

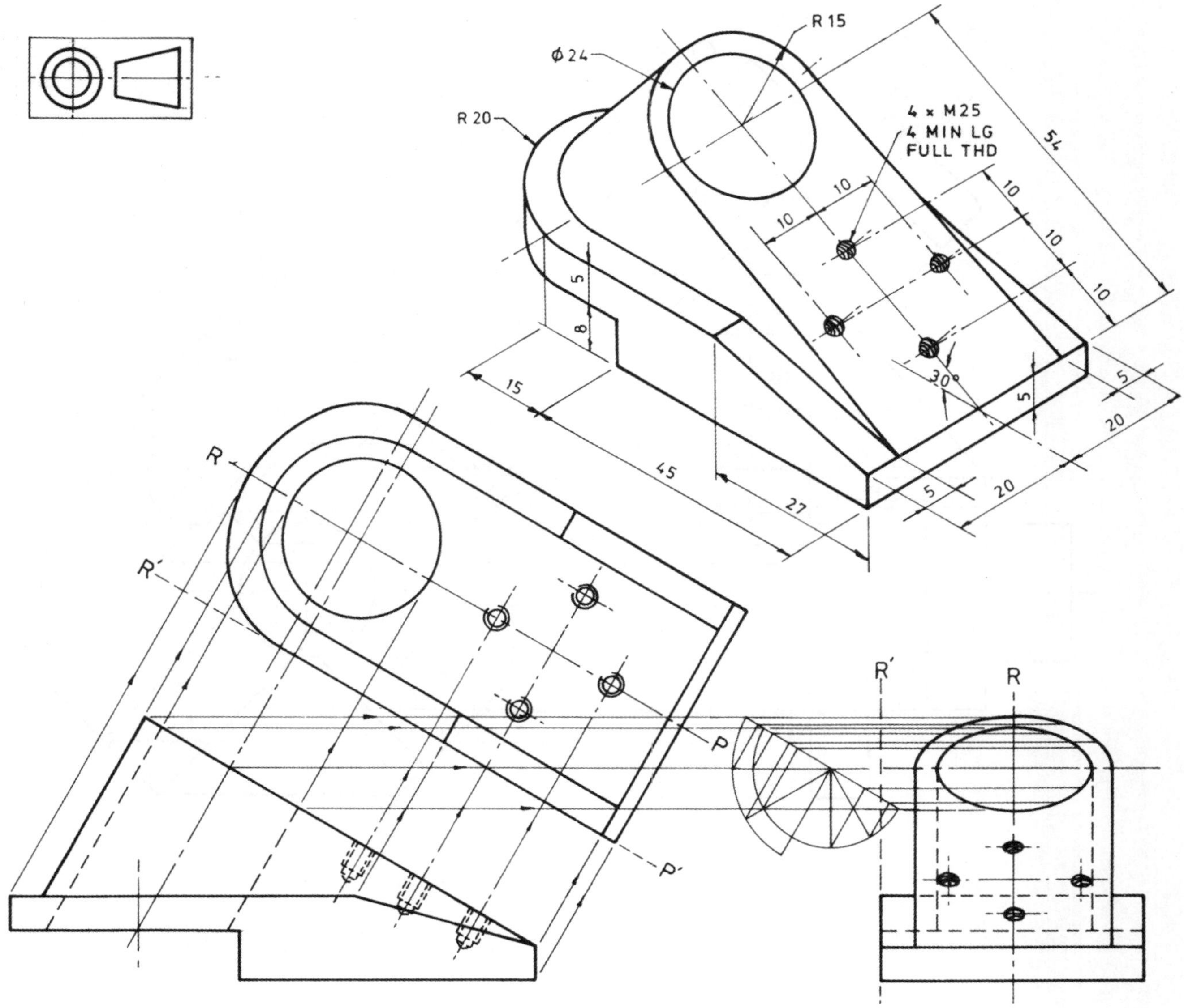

R 15

Ø 24

R 20

4 × M25
4 MIN LG
FULL THD

54

10

10

10

10

10

5

8

30°

5

5

20

15

45

27

5

20

R

R'

R'

R

P

P'

2. Project a secondary auxiliary view at right angles to the edge view of face ABF. Similarly on this view, the distances of points on the block from reference V'-W' are the same as for points on the top view from V-W.

**FIGURE 4.14** ▶ A complete auxiliary view

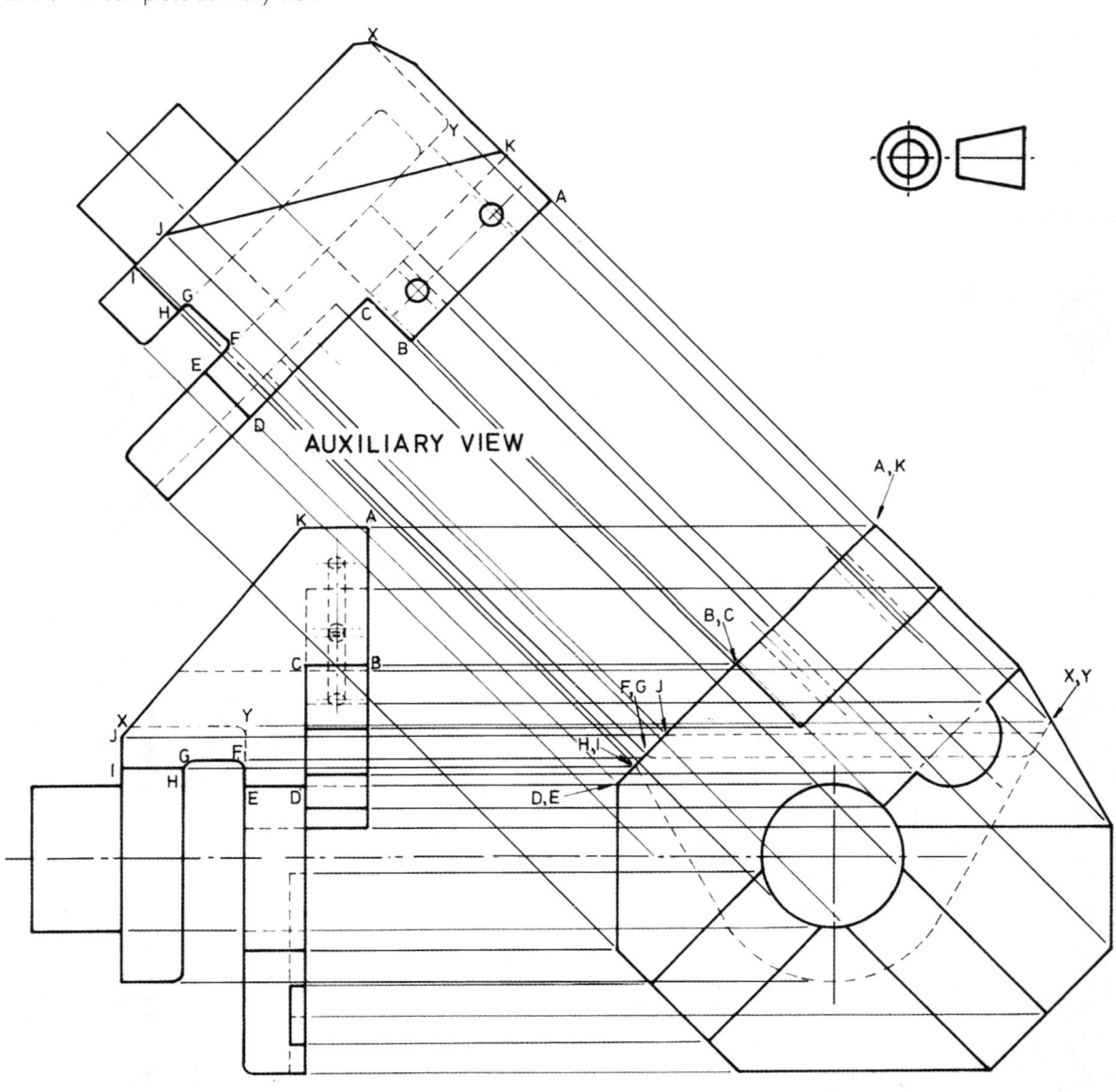

AUXILIARY VIEW

**FIGURE 4.15** ▶ A secondary auxiliary view

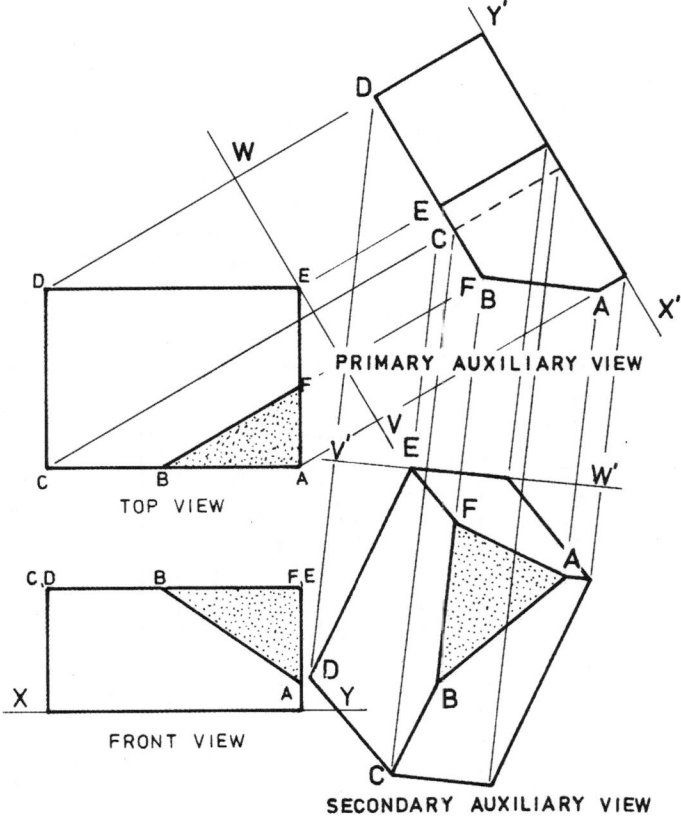

## USE OF A SECONDARY AUXILIARY VIEW TO CONSTRUCT NORMAL VIEWS

Figure 4.16(b) (page 182) illustrates the application of a secondary auxiliary view to enable the construction of an oblique face of a component on the normal front and top views. Figure 4.16(a) is a pictorial view of the component.

### Stage 1
Draw the front and top views of the undistorted portion of the bracket.

### Stage 2
Project primary and secondary auxiliary views showing the edge view and true shape of the oblique face respectively.

### Stage 3
Complete the construction of the oblique face on the top view by projecting points back from the true shape, for example points indicated by distances X.

### Stage 4
Project the points located on the top view down to the front view, and locate them by measuring their heights above the reference line, for example distances Y, Z and W.

# 4.5 General rules

To be able to draw an auxiliary view successfully, one must form a mental picture of how the object will look from the direction of viewing. The following rules may help the student to understand the use of the auxiliary view technique more clearly.

### Rule 1
An auxiliary view is normally used to detail an inclined face of an object which would be distorted on a principal orthogonal view.

### Rule 2
An auxiliary view is projected at right angles to the edge view of the inclined face contained in a principal orthogonal view.

### Rule 3
In third-angle projection, the auxiliary view is placed on the same side of the normal view as the position of viewing.

### Rule 4
In first-angle projection, the auxiliary view is placed on the same side of the normal view as the position of viewing. That is, the auxiliary view is treated as a third-angle view.

**FIGURE 4.16** ▶ Use of secondary auxiliary view to construct normal views

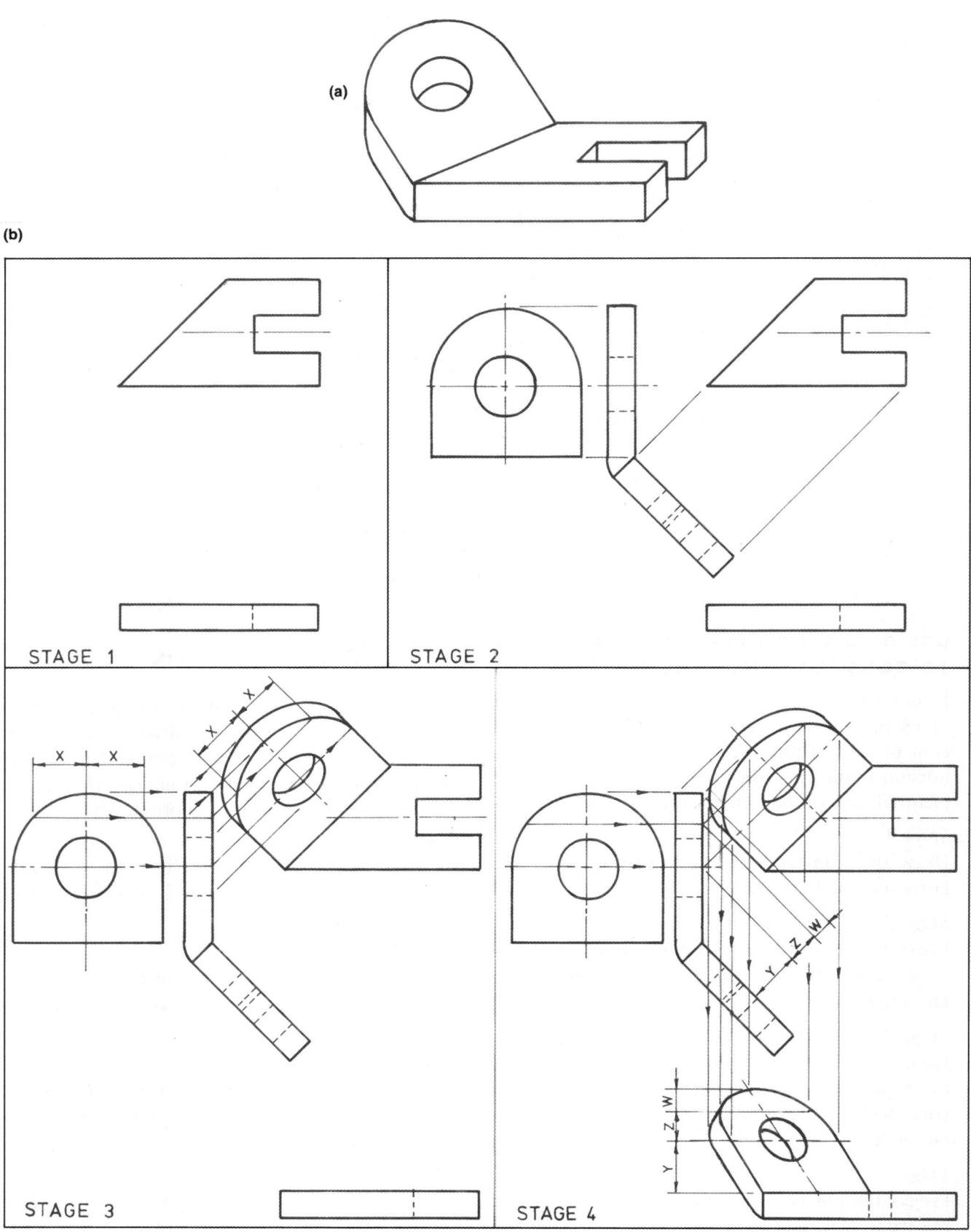

(a)

(b)

STAGE 1

STAGE 2

STAGE 3

STAGE 4

# 4.6 Problems

**Note:** From a practical point of view, most of the exercises in this section would be best drawn as 'partial' auxiliary views, but, in order to make them more challenging, 'complete' views are requested.

## 4.1 DRILL PRESS BRACKET

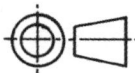

Draw the following views of the drill press bracket in third-angle projection:

(a) the front and top views

(b) a complete auxiliary view from A showing the true shape of the inclined vertical face

(scale 1:1)

Subtitle all views, but omit dimensions.

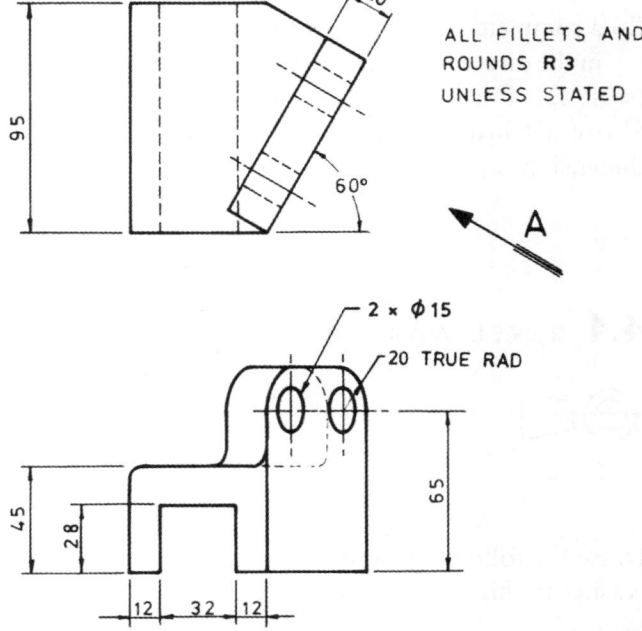

ALL FILLETS AND ROUNDS **R 3** UNLESS STATED

## 4.2 ANGLE SUPPORT

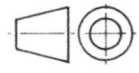

Draw the following views of the angle support in third-angle projection:

(a) the given top and a complete front view showing all hidden detail

(b) a complete auxiliary view showing the true shape of the inclined vertical face

Also show welded joint symbols as follows:

(a) 6 mm fillet all round

(b) 6 mm fillet both sides

(scale 1:1)

Fully dimension and subtitle the views.

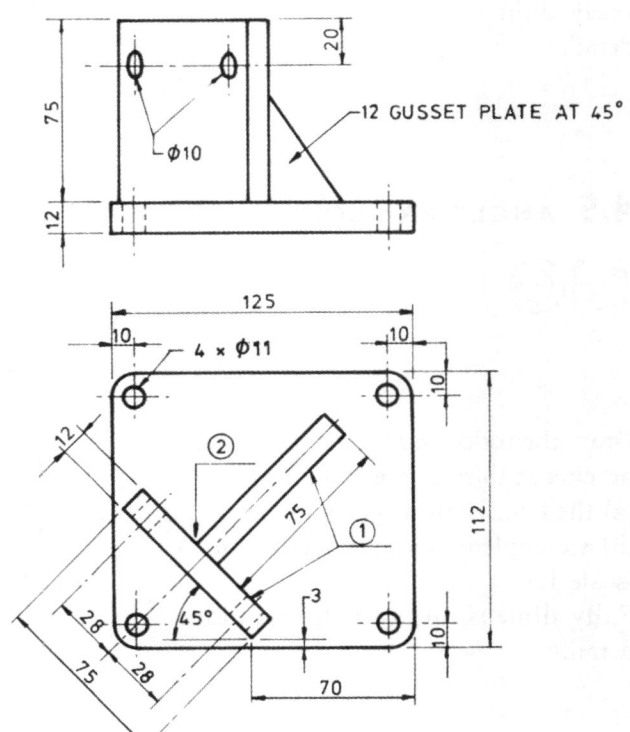

## 4.3 BOX SPANNER

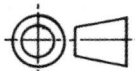

Draw the following views of the box spanner in third-angle projection:

(a) the given front and top views

(b) a complete auxiliary view projected in the direction of arrow B

(scale 1:1)

Show all hidden detail, but omit dimensions.

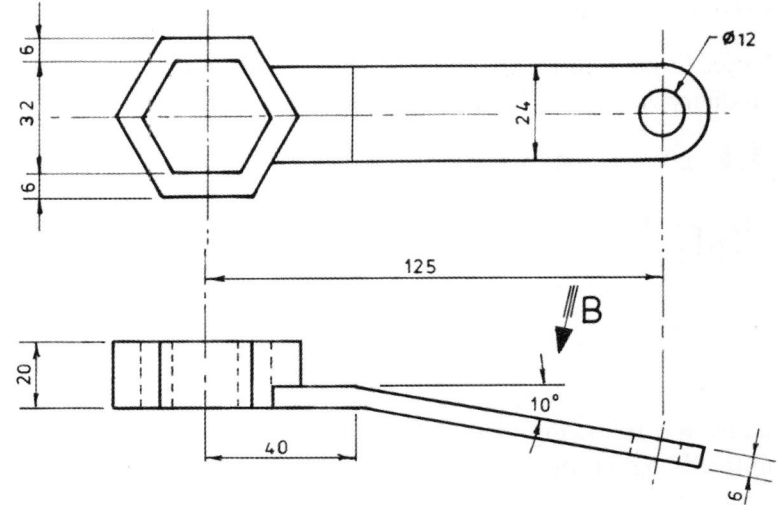

## 4.4 BEVEL WASHER

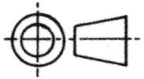

Draw the following views of the bevel washer in third-angle projection:

(a) a sectional view on A-A

(b) a top view

(c) a partial auxiliary view from B

(scale 1:1)

Fully dimension, and show hidden detail.

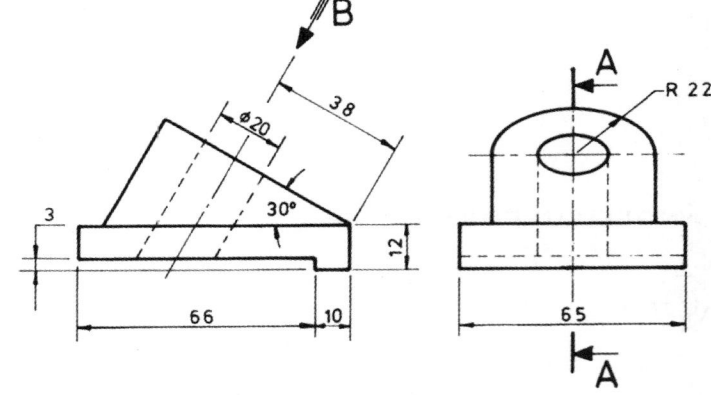

## 4.5 ANGLE BRACKET

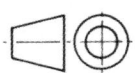

Draw the following views of the angle bracket in third-angle projection:

(a) the front view as given

(b) a complete auxiliary view from A

(scale 1:1)

Fully dimension, and show hidden detail.

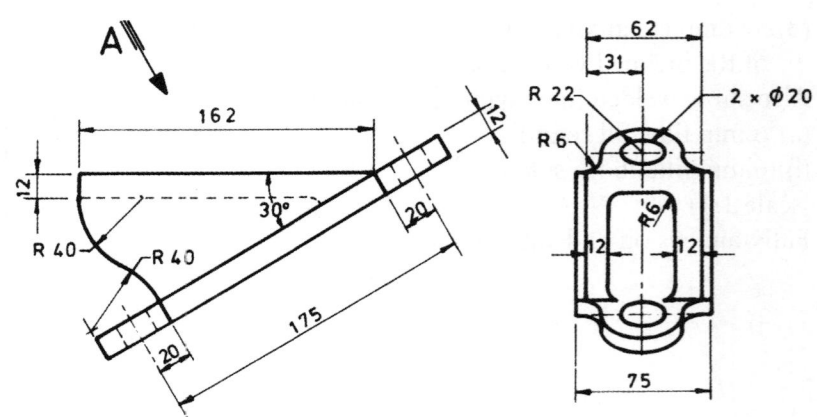

## 4.6 MS BRACKET

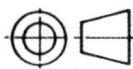

Draw the following views of the bracket in third-angle projection:
(a) the given front and top views
(b) a left side view
(c) a partial auxiliary top view from A (scale 1:1)
Fully dimension, and show hidden detail.

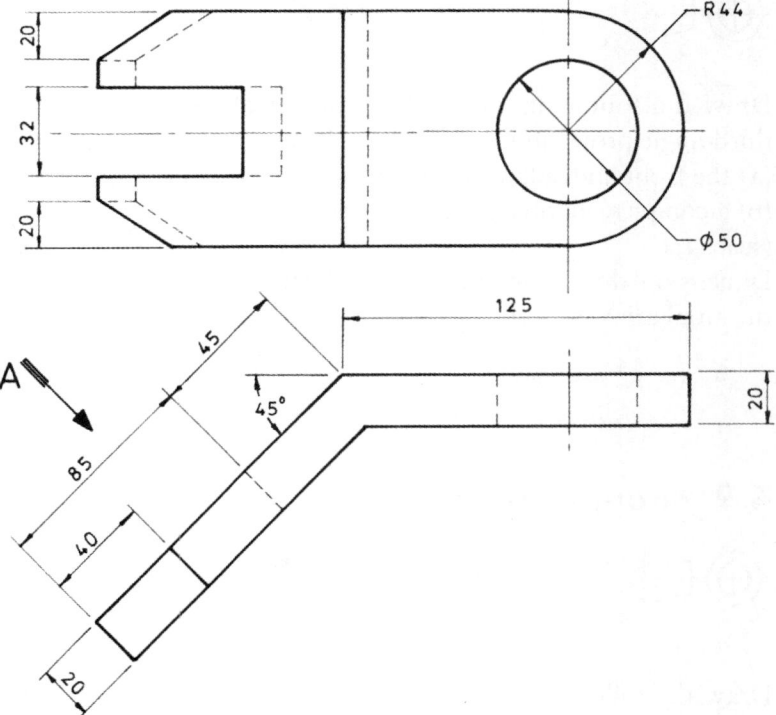

## 4.7 TIE-ROD SUPPORT

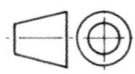

Draw the following views of the tie-rod support in third-angle projection:
(a) the given front and top views
(b) a complete auxiliary view from A (scale 1:1)
Fully dimension, and show hidden detail.

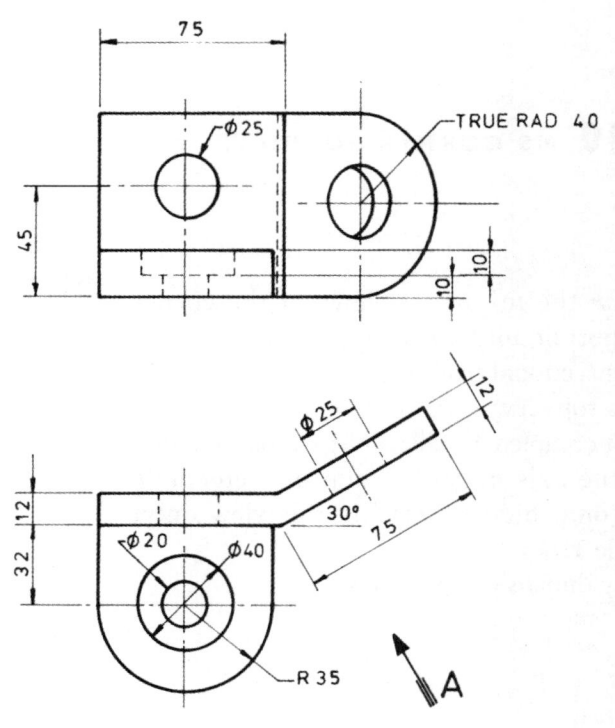

## 4.8 UNIVERSAL BASE

Draw the following views of the universal base in third-angle projection:
(a) the front and side views as shown
(b) a complete auxiliary view from A
(scale 1:1)
Dimension the views, and omit hidden detail from the auxiliary view only.

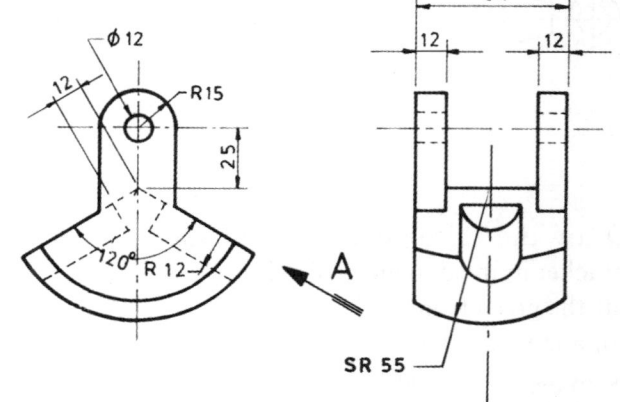

## 4.9 CI GUIDE BLOCK

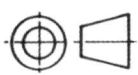

Draw the following views of the guide block in third-angle projection:
(a) a sectional front view on A-A
(b) the side view as shown, complete with hidden detail
(c) a complete auxiliary view from B, omitting hidden detail
(scale 1:1)
Fully dimension the views.

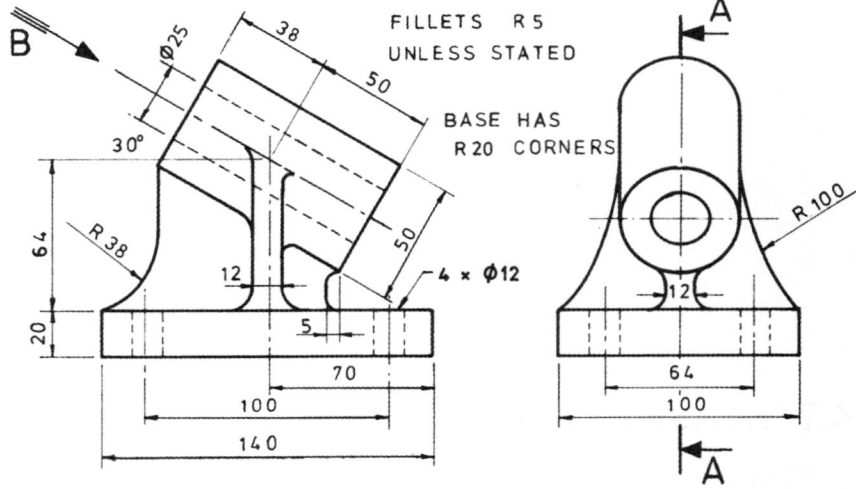

## 4.10 MS CORNER SUPPORT

Draw the following views of the corner support in third-angle projection:
(a) a sectional front view on A-A
(b) a top view
(c) a complete auxiliary view looking along the axis of the 22 mm diameter hole (omit hidden detail in this view only)
(scale 1:1)
Fully dimension the views.

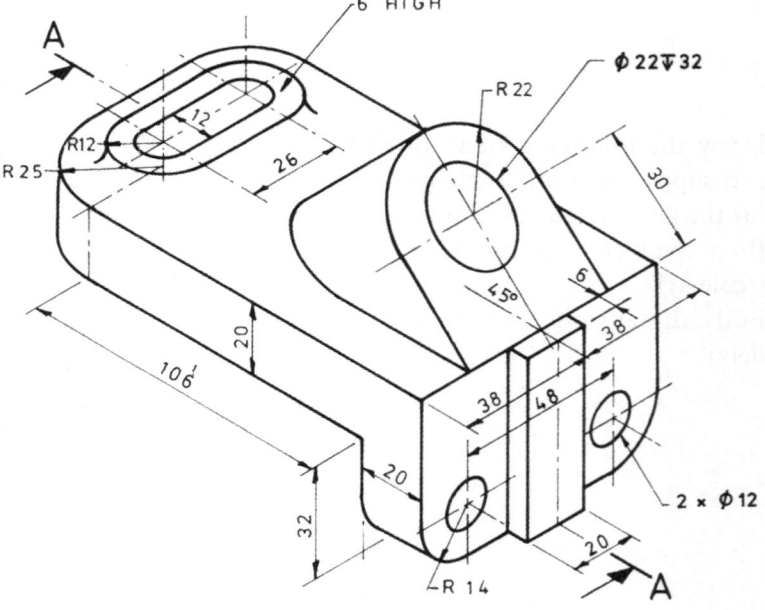

## 4.11 CI TRANSFER PIECE

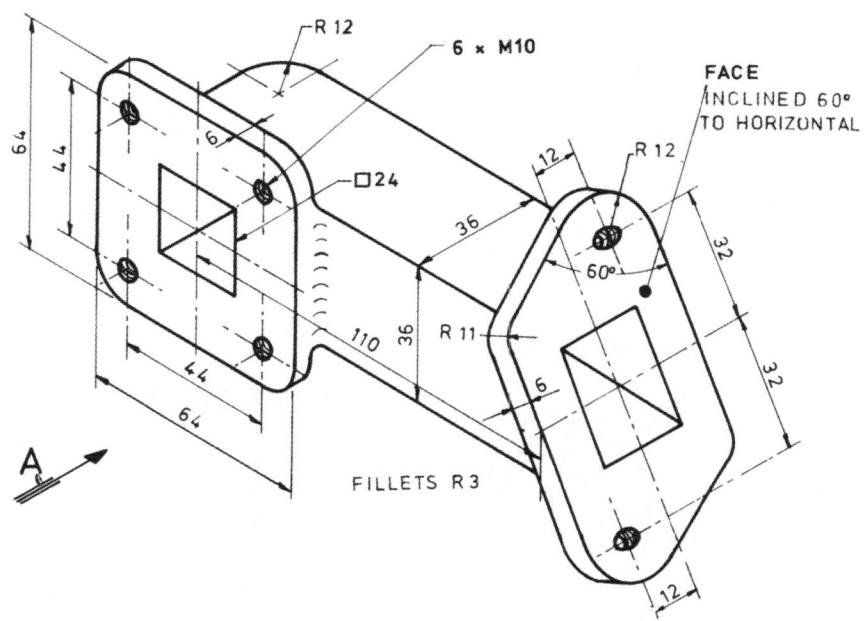

Draw the following views of the transfer piece in third-angle projection:
(a) a front view from A
(b) a top view
(c) a partial auxiliary view of the inclined face
(scale 1:1)
Fully dimension the views and show hidden detail.

## 4.12 CONTROL LEVER

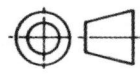

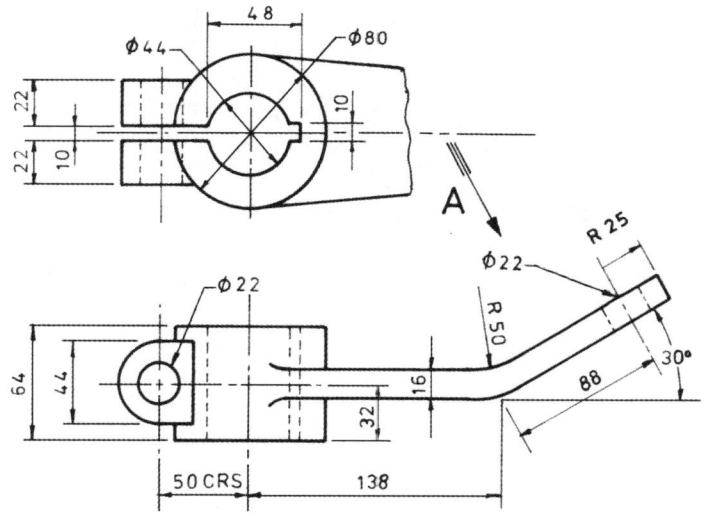

Draw the following views of the control lever in third-angle projection:
(a) the front view as given
(b) the completed top view
(c) a partial auxiliary view from A
(scale 1:1)
Fully dimension the views, and show hidden detail.

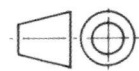

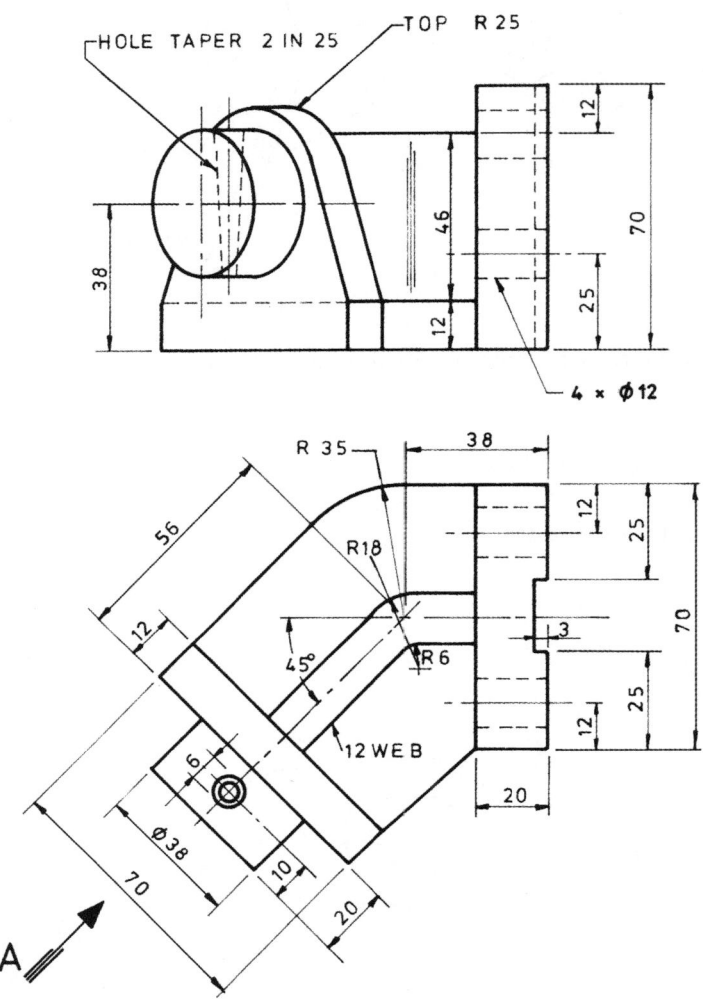

Draw the following views of the angle support bracket in third-angle projection:
(a) the given front and top views
(b) a complete auxiliary view from A
(scale 1:1)
Fully dimension the views, and show hidden detail.

## 4.14 CI PUSH PLATE

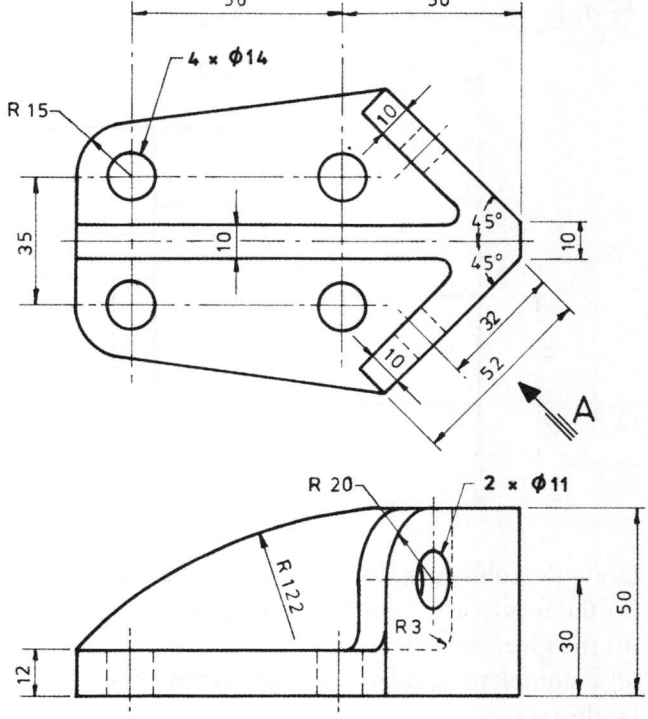

Draw the following views of the push plate in third-angle projection:
(a) the front and top views
(b) a complete auxiliary view from A
(scale 1:1)
Fully dimension the views, and show hidden detail.

## 4.15 MS JIG

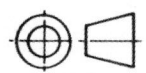

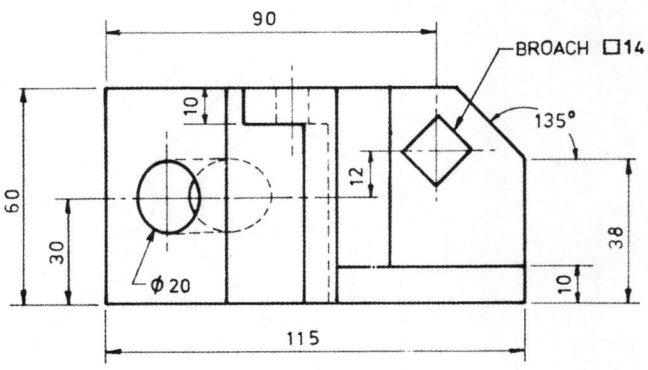

Draw the following views of the jig in third-angle projection:
(a) the front view as given
(b) a partial auxiliary view from A
(scale 1:1)
Fully dimension the views, and show hidden detail.

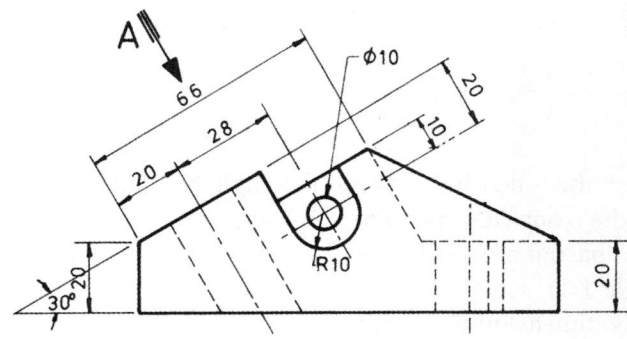

## 4.16 STEP BASE

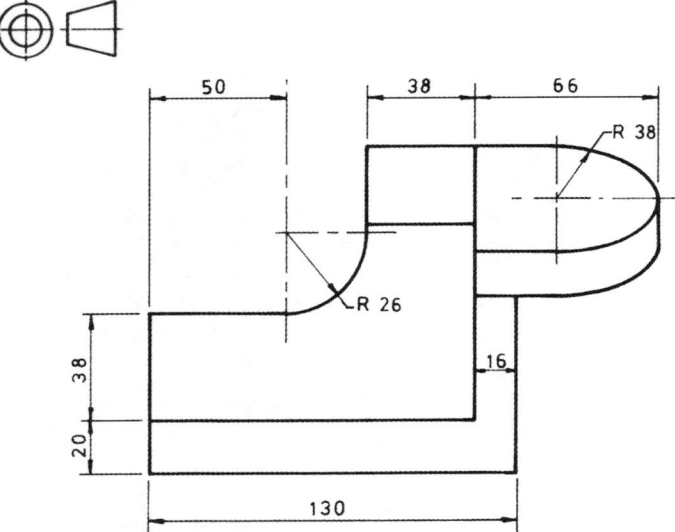

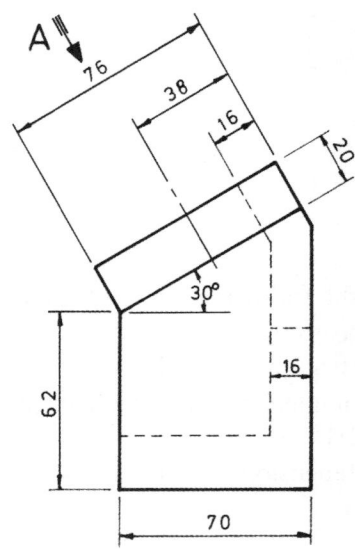

Draw the following views of the step base in third-angle projection:
(a) the front view, complete with hidden detail
(b) the given side view
(c) a complete auxiliary view projected from A
(scale 1:1)
Fully dimension the views.

## 4.17 ANGLE BEARING BRACKET

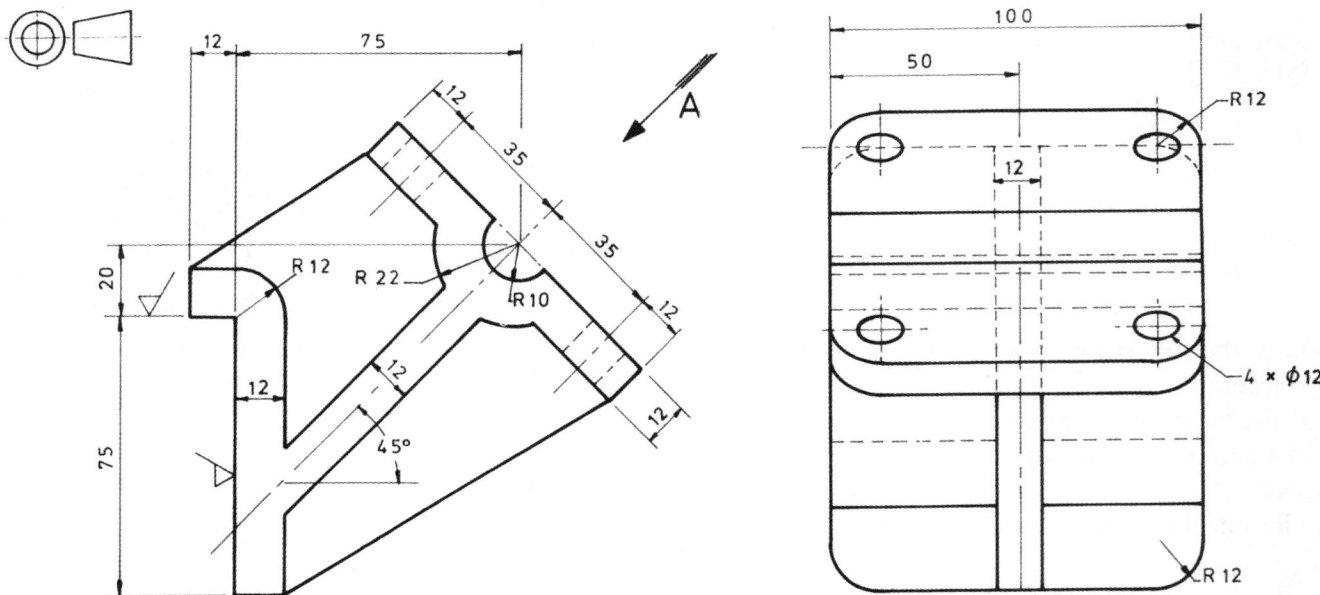

Draw the following views of the angle bearing bracket in third-angle projection:
(a) the front view as given
(b) a partial auxiliary view from A
(scale 1:1)
Fully dimension the views.

## 4.18 MACHINE GUIDE BLOCK

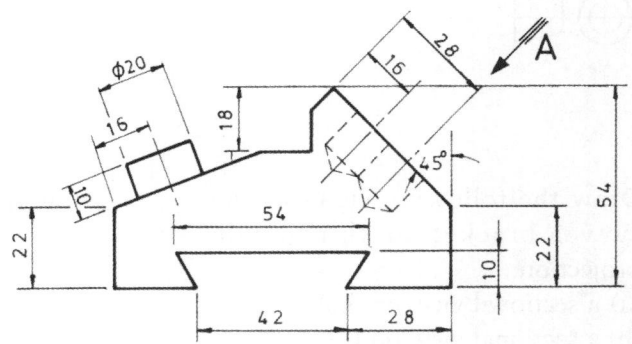

Draw the following views of the machine guide block in third-angle projection:
(a) the given front and top views
(b) a partial auxiliary view from A
(scale 1:1)
Dimensions and hidden detail are not required.

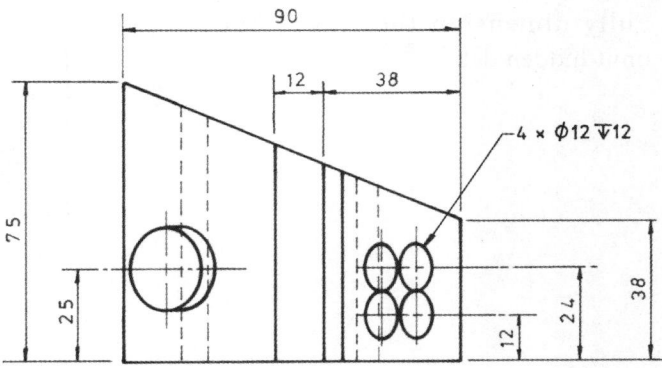

## 4.19 SUPPORT BRACKET

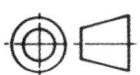

Draw the following views of the support bracket in third-angle projection:
(a) the given front view
(b) the top view
(c) a partial auxiliary view from A
(scale 1:1)
Fully dimension the views, and show hidden detail

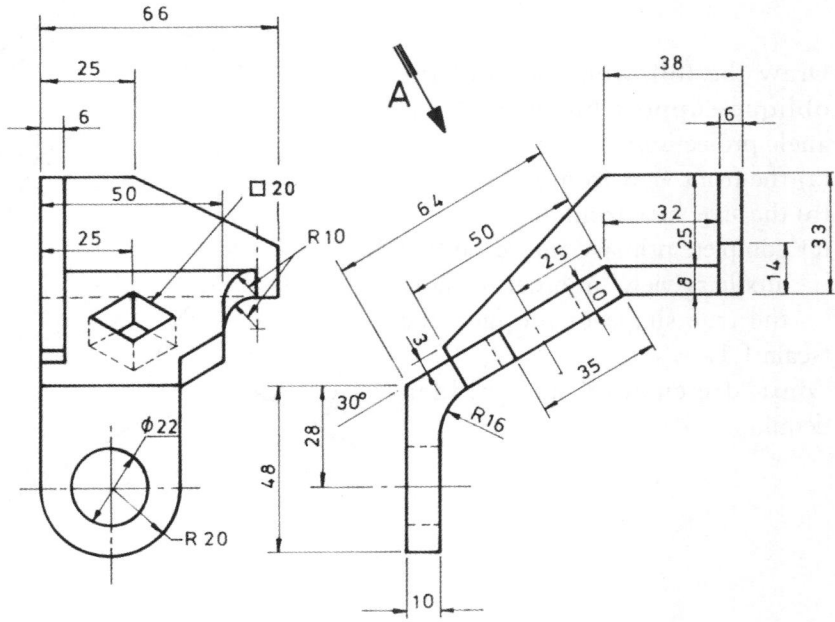

## 4.20 CI SWIVEL BRACKET

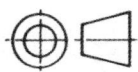

Draw the following views of the swivel bracket in third-angle projection:

(a) a sectional view on A-A
(b) a sectional view on B-B
(c) a partial auxiliary view from C (scale 1:1)

Fully dimension the views, but omit hidden detail.

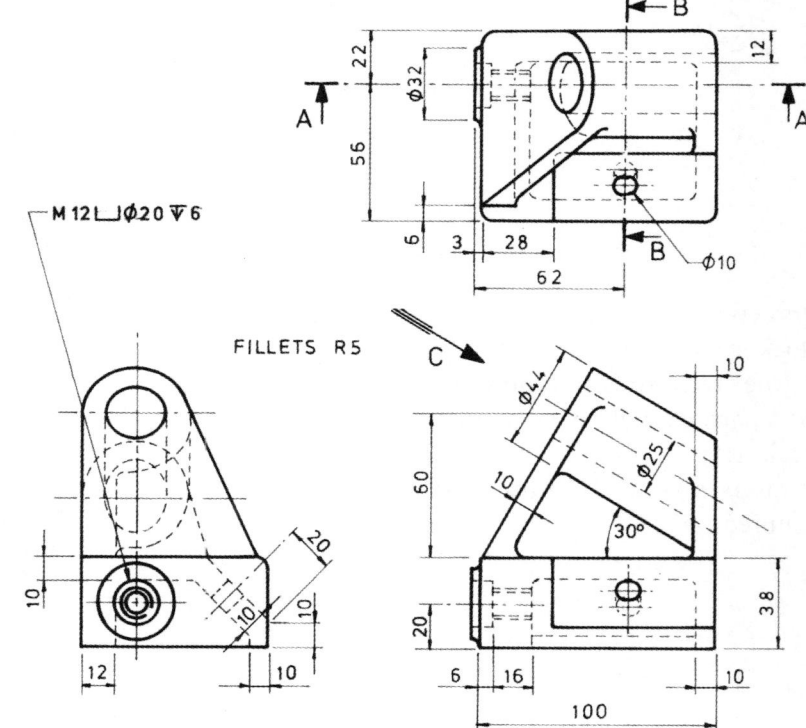

## 4.21 OBLIQUE SUPPORT BLOCK

Draw the following views of the oblique support block in third-angle projection:

(a) the front view from A
(b) the side view from B
(c) complete primary and secondary auxiliary views in order to show the true shape of the face abc (scale 1:1)

Omit dimensions and hidden detail.

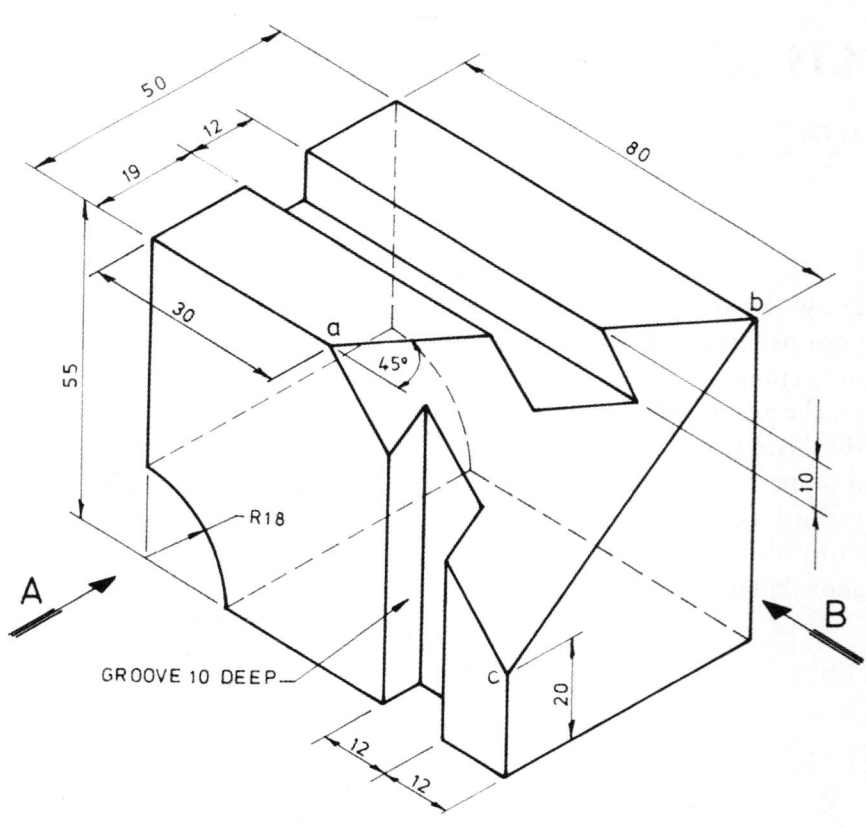

## 4.22 LOCATING SLIDE

Draw the following views of the locating slide in third-angle projection:

(a) the top view

(b) complete primary auxiliary view in line with the inclined face

(c) a partial secondary auxiliary view showing the true shape of the inclined face

(scale 1:1)

Omit dimensions and hidden detail.

**Note:** Draw partial top view, project primary view, and then project back to complete the top view. Refer to Figure 4.15.

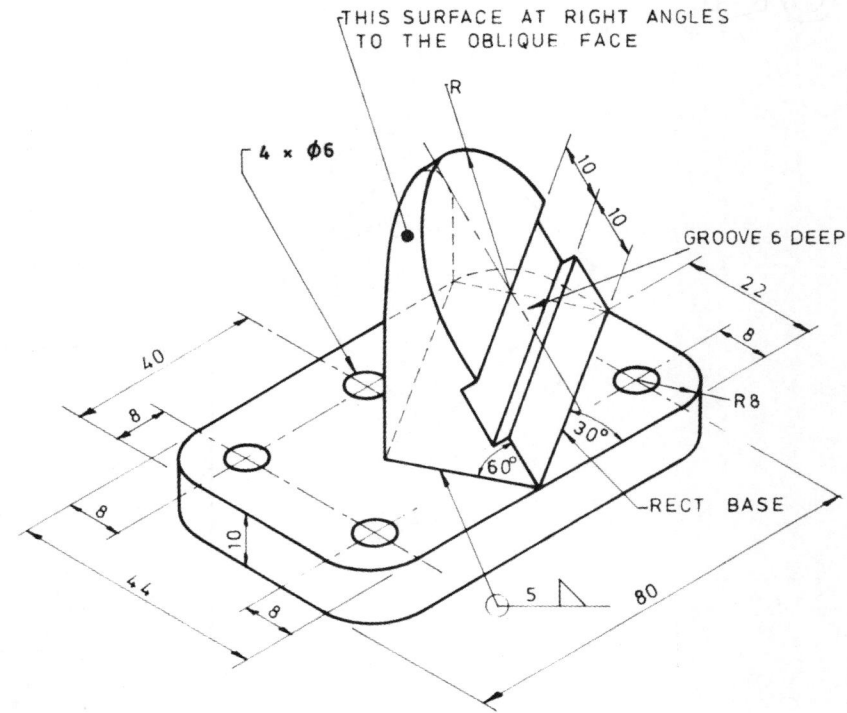

## 4.23 SPAR CLIP

Draw the following views of the spar clip in third-angle projection:

(a) a view showing the true shape of the slotted face

(b) primary and secondary auxiliary views necessary to show the true shape of the other face and to complete view (a)

(scale 1:2)

Fully dimension the views, but omit hidden detail.

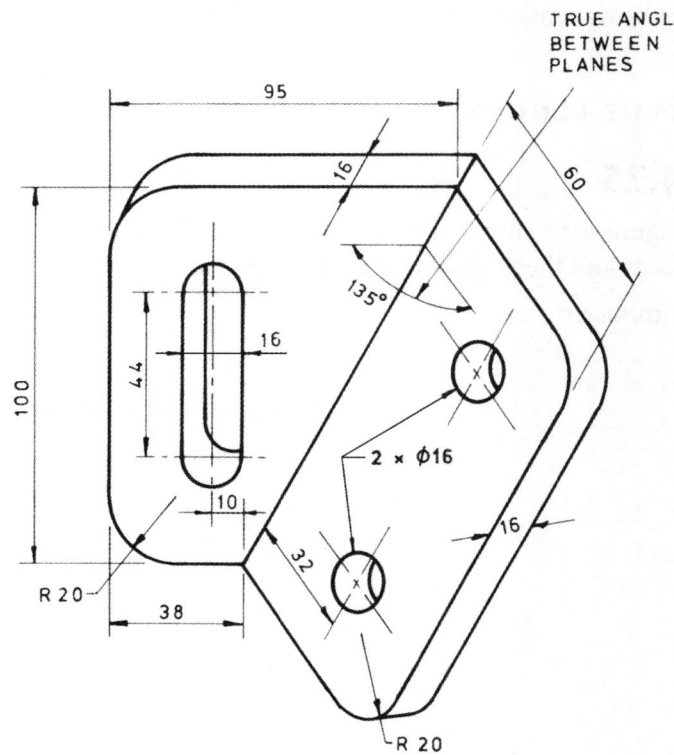

## 4.24 GAUGE BLOCK

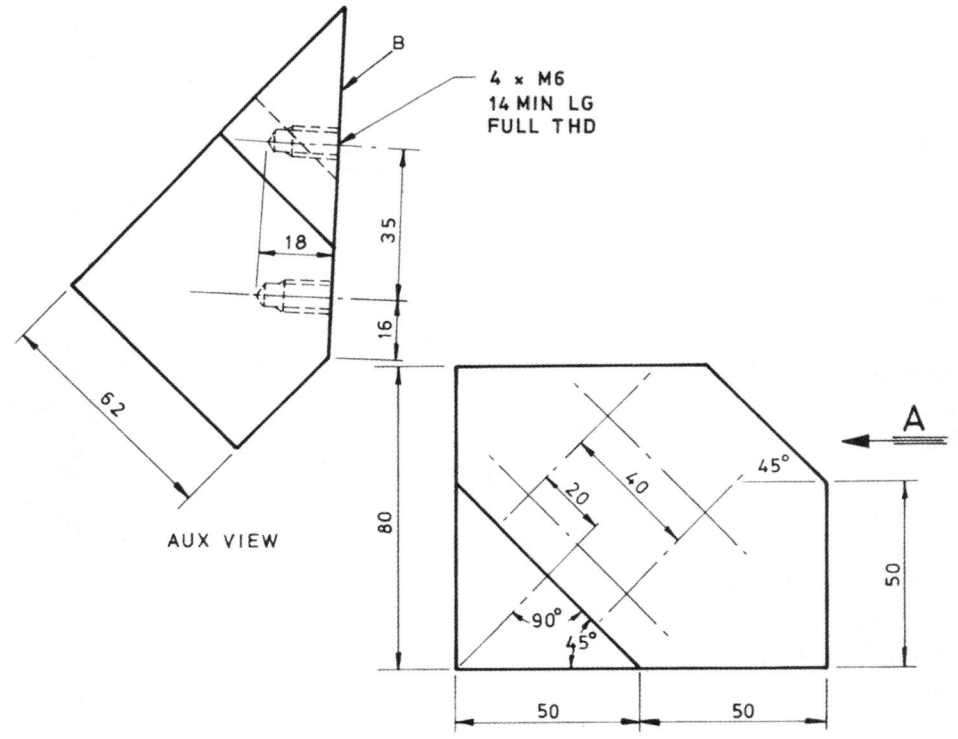

B

4 × M6
14 MIN LG
FULL THD

18    35    16

62

AUX VIEW

80    50    50

50    50

45°

20    40

90°    45°

A

Draw the following views of the gauge block in third-angle projection:
(a) the given front and auxiliary views
(b) the top view
(c) the side view from A
(d) the true shape of the inclined face showing the position of the screwed holes
(scale 1:1)
Fully dimension the views, but omit hidden detail.

## TRUE LENGTH

## 4.25

Figure 4.17 shows the front and side views of an underground mine shaft. Determine the length of the mine shaft and the angle it makes with ground level.

**FIGURE 4.17**

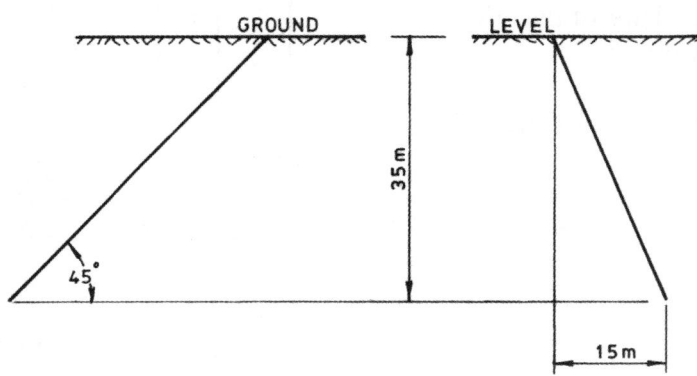

GROUND    LEVEL

35 m

45°

15 m

## 4.26

Figure 4.18 shows two views of a flag pole mounted on a 15° slope. Guy wires support the pole at a height of 13 m up the mast and are fixed to the ground at a radius of 6.5 m from the pole and spaced as shown in the top view. Determine the true length of each guy wire and its acute angle to the ground plane.

**FIGURE 4.18**

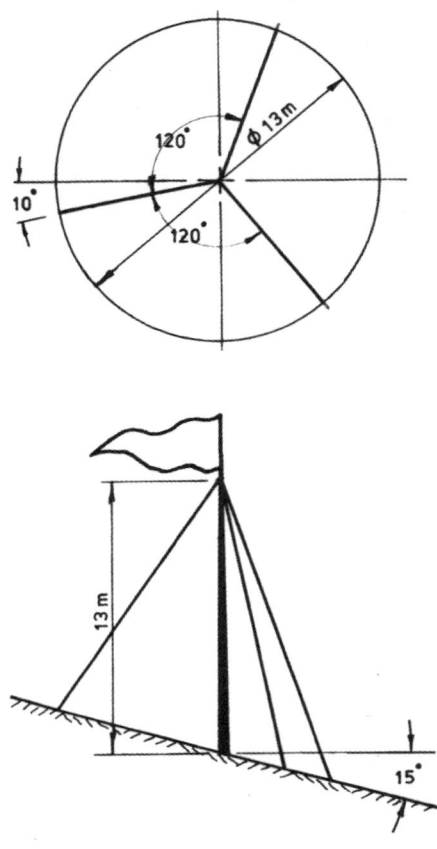

## 4.27

Two orthogonal views of an aeroplane wheel strut system are given in Figure 4.19. A, B and C are the location points of each strut on the fuselage. The neutral axes of the struts meet at D on the axis of the wheel axle. Determine the axial length of each strut and its angle of inclination to the horizontal plane.

**FIGURE 4.19**

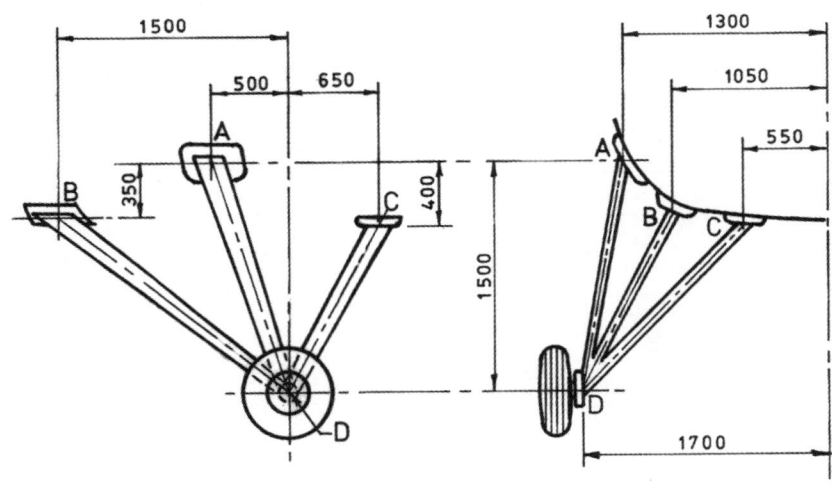

# 4.7 CAD corner

**4.28** Using the dimensions from Problem 4.19 on page 191 complete the following:

(a) A full auxiliary view in the direction of A.

(b) An offset section through A-A as shown below.

(c) A front view.

- Include centre lines and hidden detail in the front and auxiliary views.

- Do not dimension the views.

Hints:

- Features or parts of features should be drawn in the views that show true geometry and then projected to the correct positions in the other views.

- Use different colours to distinguish drawing entities from construction geometry.

**4.29** This question is suitable for people using a CAD system with 3D modelling capabilities.

(a) Using the dimensions shown in Problem 4.21 on page 192 model the oblique support block shown below.

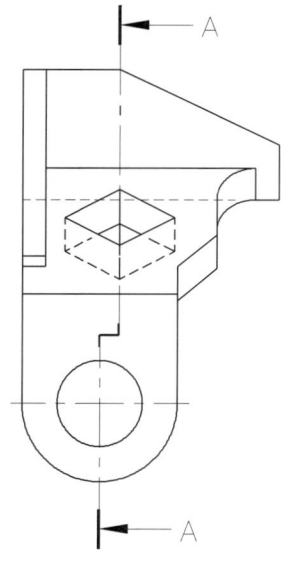

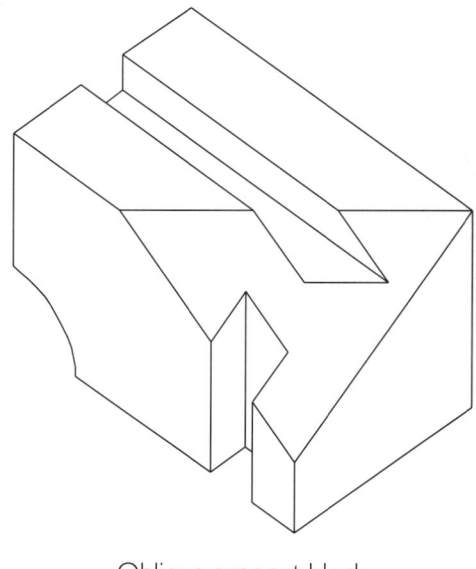

Oblique support block

(b) Set up the CAD coordinate system (X,Y) coplanar with the oblique face and view your model normal to the oblique face.

Trace the perimeter of the oblique face and then move the perimeter to the right of the model as shown below.

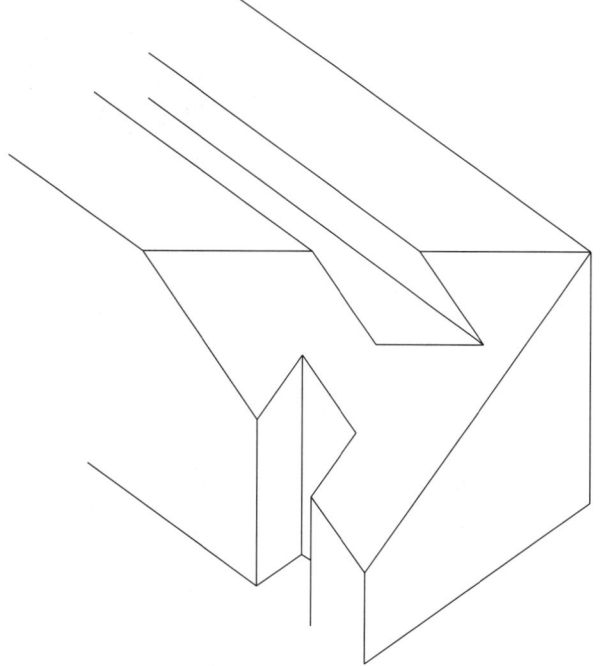

Shift the coordinate system (0,0) point to the top left corner of the perimeter and dimension, in X and Y, the position of each corner relative to the top left corner.

Note: One of the corners is shown dimensioned similar to that required.

**Challenge:** Use your CAD program to determine the length and area of the profile perimeter.

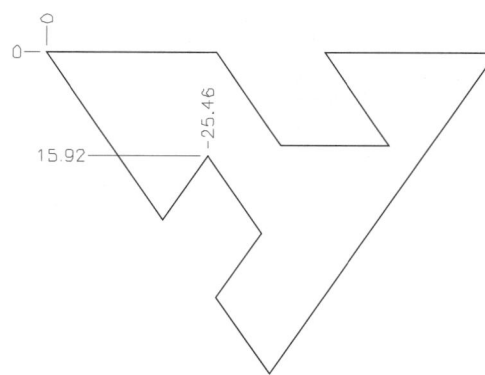

# Pictorial drawing:
## Isometric and oblique parallel projection

**5**

After studying this chapter and working through the problems you should be able to do the following:

- produce isometric and oblique pictorial drawings from orthogonal views
- select the best viewing direction when making a pictorial drawing
- understand the relationship between two- and three-dimensional drawing

# 5.1 Introduction

Pictorial views are not intended to transmit dimensions; hence they are not normally dimensioned. Sometimes, however, an engineer may wish to give a drafter a pictorial sketch of a design in mind, and will quite often add the dimensions which are applicable or considered necessary.

There are three general classifications of pictorial drawings:

1. axonometric projection
2. oblique projection
3. perspective projection

Perspective views are more complicated to produce than the first two, but are more realistic and are used mainly by architects. Engineers prefer either axonometric or oblique views.

# 5.2 Axonometric projection

This involves turning the object so that any three principal faces can be seen from the one viewing position. There are an infinite number of views possible, and they all result in shortening of the edges by varying degrees, depending on the angles involved.

Accordingly, certain positions have been classified as *isometric*, *dimetric* and *trimetric*, and one of these is used when an axonometric projection is required.

The most commonly used view of these three is the isometric; it will be described in detail. The other two are generally described in AS 1100 Part 101.

# 5.3 Isometric projection

The word 'isometric' means 'equal measure', and to produce an isometric projection it is necessary to view an object so that its principal edges are equally inclined to the viewer and hence are foreshortened equally. This is best illustrated by considering orthogonal views of an inclined cube. Figure 5.1 shows the front, top and side views of a cube resting on one of its faces with its horizontal edges inclined at 45° to the vertical plane. The front view shows that all horizontal edges are equally foreshortened, but the vertical edges are not altered.

**FIGURE 5.1** ▶ Foreshortening of cube edges

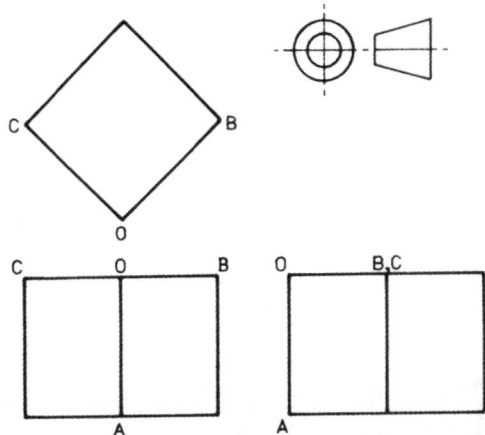

Now consider the cube to be pivoted on corner A and tilted forward until all the edges are equally inclined to the vertical plane. This is shown in Figure 5.2. (The angle between the edges and the vertical plane is approximately 35°15'.) In this position, edges OA, OB and OC make 120° with each other in the front view, and are called the *isometric axes*. The axis OA is vertical, OB is inclined at 30° to the right, and OC at 30° to the left. Any line parallel to one of these axes is called an *isometric line*, and can be drawn with a 30° set square. All other lines are called *non-isometric lines*, and must be plotted.

**FIGURE 5.2** ▶ Concept of isometric axes

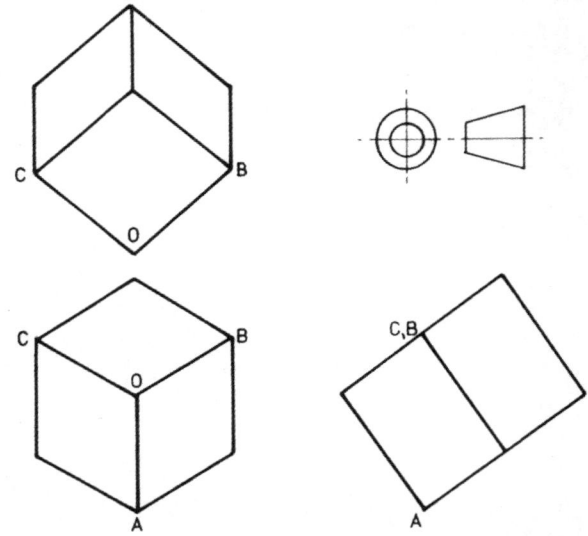

## ISOMETRIC SCALE

For correct isometric projection, a scale is used which allows for the foreshortening of isometric lines. The construction of such a scale is shown in Figure 5.3. When an isometric view is drawn using an isometric scale, it is termed an *isometric projection*.

FIGURE 5.3 ▶ Making an isometric scale

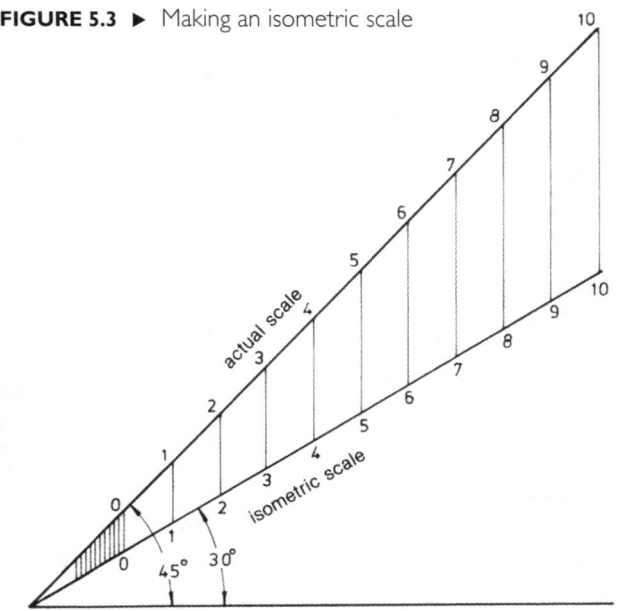

## ISOMETRIC DRAWING

An isometric drawing differs from an isometric projection in that it is prepared without shortening measurements. An isometric drawing gives about a 22.5 per cent larger view than the isometric projection, but the pictorial value of each view is the same. Hence for most purposes the isometric drawing is used.

## SELECTION OF ISOMETRIC AXES

The main purpose of an isometric view is to provide a pictorial view which reveals as much detail as possible, and this fact should be remembered when selecting the principal edges as the isometric axes.

Figures 5.4(a)–(h) show eight isometric views of the same block with the isometric axes intersecting at the circled point in each view. View (a) is preferred as it reveals more detail than the others.

The isometric axes can be rotated to make one axis horizontal, as shown in Figures 5.4(i) and (j). This is sometimes preferred for long, narrow objects, where the long axis can be placed horizontally for best effect.

## ISOMETRIC CIRCLES—ORDINATE METHOD

Circles may be drawn whole or in part in isometric view by the use of ordinates constructed on an orthogonal view and then transferred to the isometric view, as shown in Figure 5.5 (page 200). A smooth curve is drawn either freehand or with a French curve through the ends of the ordinates to give the isometric circle or curve.

FIGURE 5.4 ▶ Selection of isometric axes

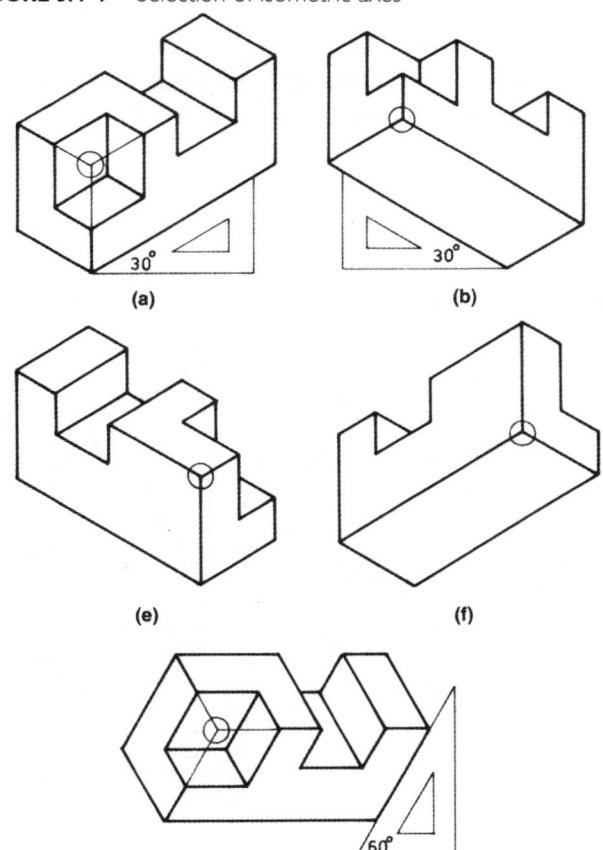

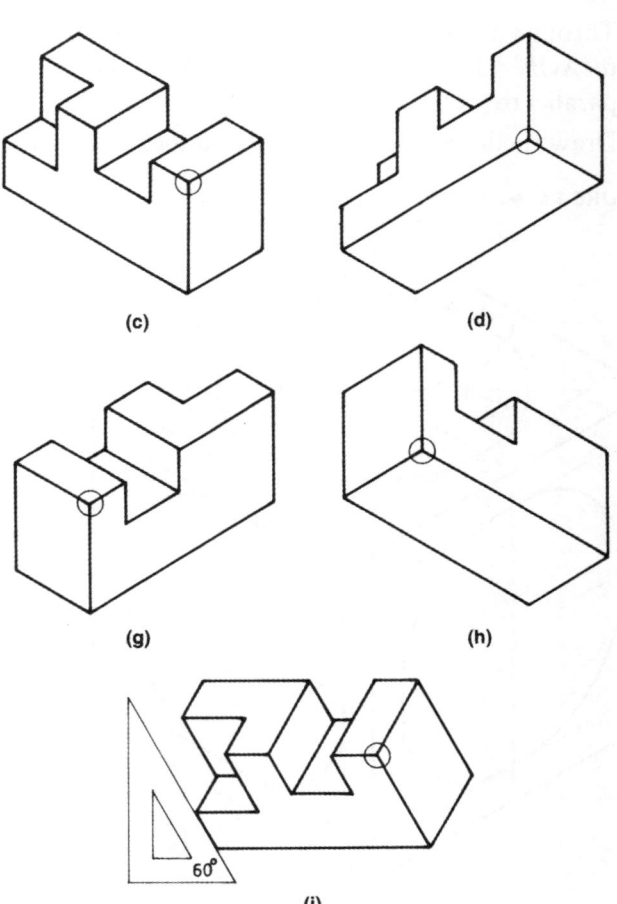

FIGURE 5.5 ▶ Isometric circles—ordinate method

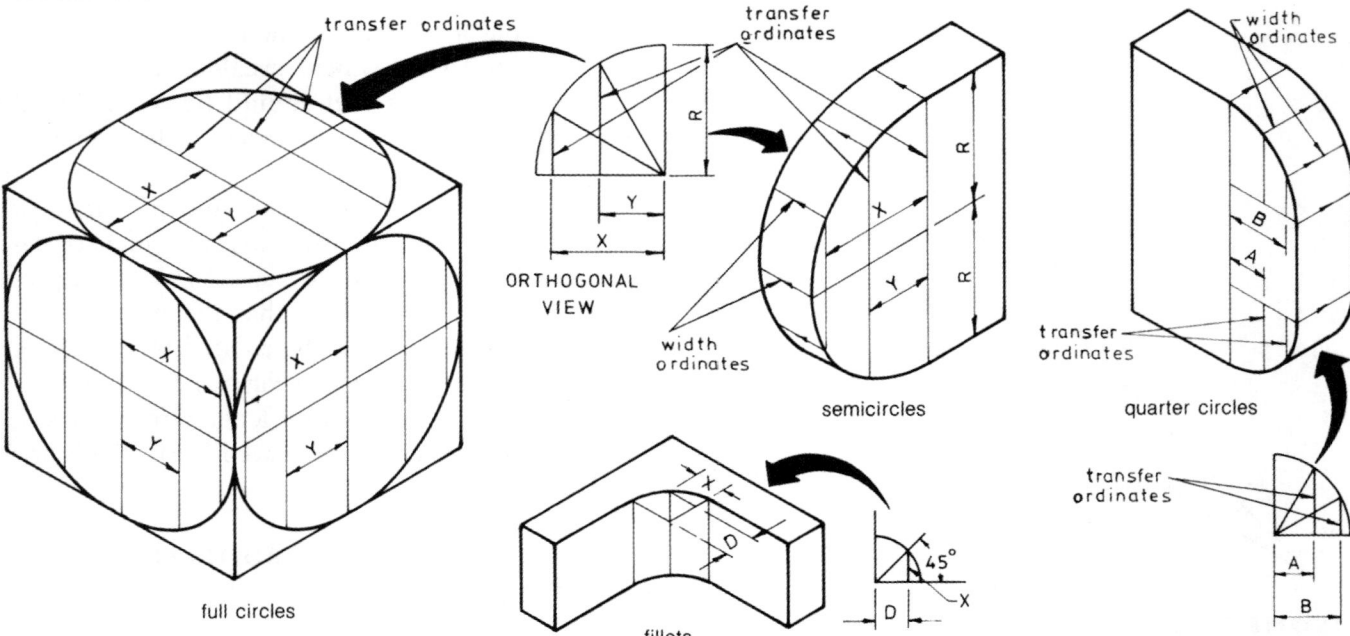

full circles

ORTHOGONAL VIEW

semicircles

quarter circles

fillets

## ISOMETRIC CIRCLES—FOUR-CENTRE METHOD

Referring to Figure 5.6(a) for full circles:

1. Draw the centre lines AOB and COD through O, the centre of the circle, so that AO = OB = CO = OD = the radius of the circle.

2. Through C and D draw FCG and EDH parallel to AOB. Through A and B draw FAE and GBH parallel to COD.

3. Draw the long diagonal FOH, and locate points J and K on it such that FJ = HK = the radius of the circle.

4. With centre G and radius $R_1$ = GA, draw an arc between GJ produced at L and GK produced at M. Similarly with centre E.

5. With centres J and K and radius $R_2$ = JL = KM, complete the figure.

Half and quarter circles may also be drawn by this method as shown in Figures 5.6(b) and (c) respectively, using part of the construction outlined above.

FIGURE 5.6 ▶ Isometric circles—four-centre method

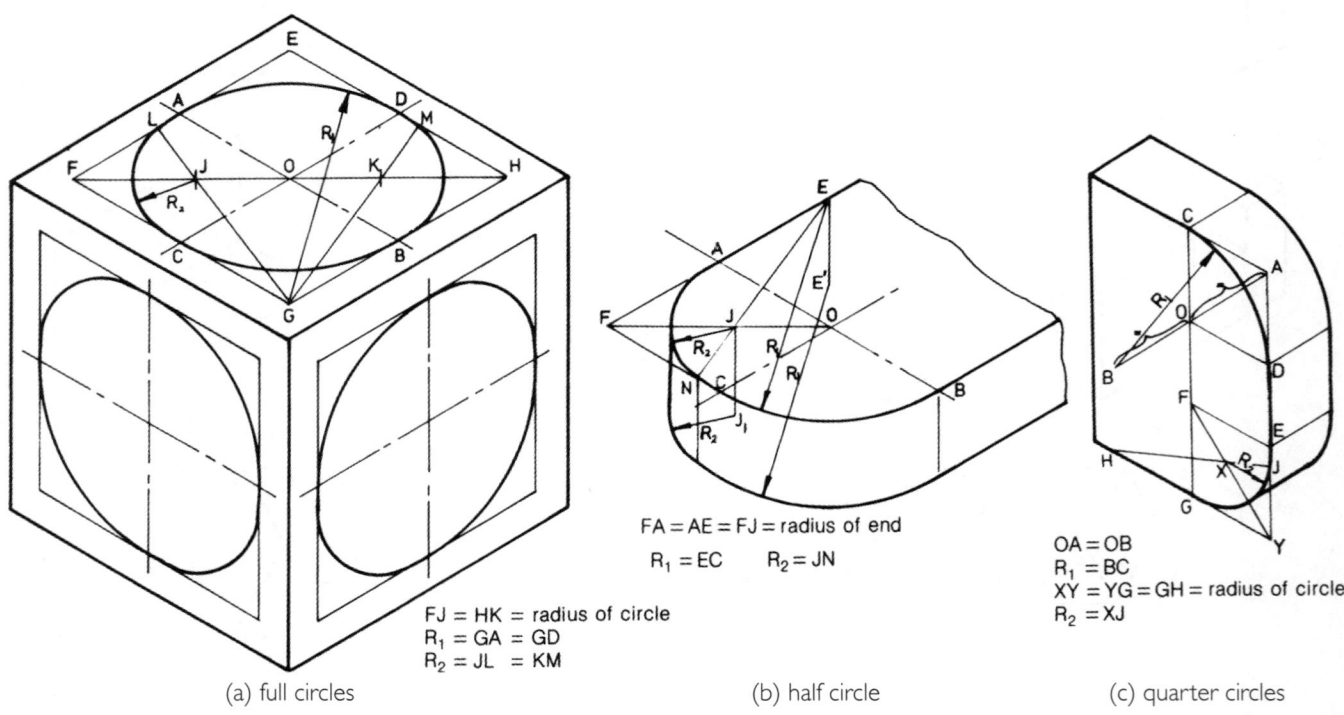

(a) full circles

FJ = HK = radius of circle
$R_1$ = GA = GD
$R_2$ = JL = KM

(b) half circle

FA = AE = FJ = radius of end
$R_1$ = EC    $R_2$ = JN

(c) quarter circles

OA = OB
$R_1$ = BC
XY = YG = GH = radius of circle
$R_2$ = XJ

## ISOMETRIC CURVES

Points on these curves are plotted by the method of ordinates taken from an orthogonal view, as shown in Figure 5.7.

A smooth curve is drawn through the plotted points, which are obtained by transferring lengths from the orthogonal view to the other by means of dividers.

**FIGURE 5.7 ▶** Isometric curves

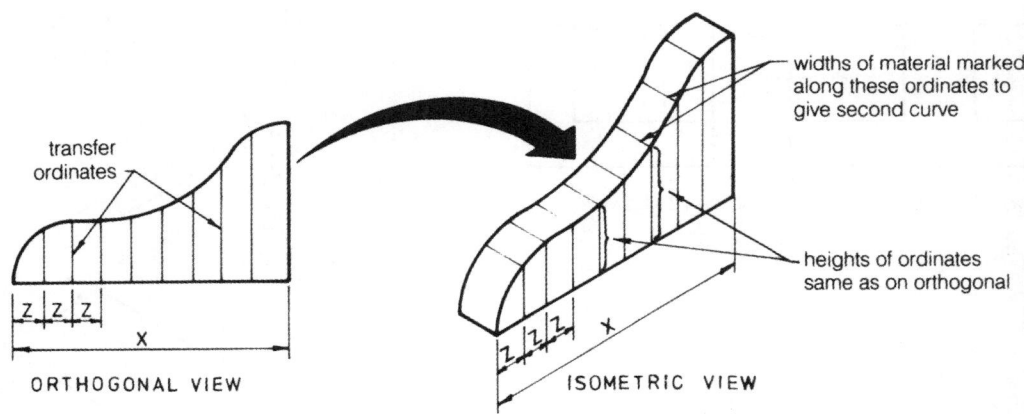

## ISOMETRIC ANGLES AND NON-ISOMETRIC LINES

These have to be plotted by the use of horizontal and vertical measurements as shown in Figure 5.8.

**FIGURE 5.8 ▶** Isometric angles and non-isometric lines

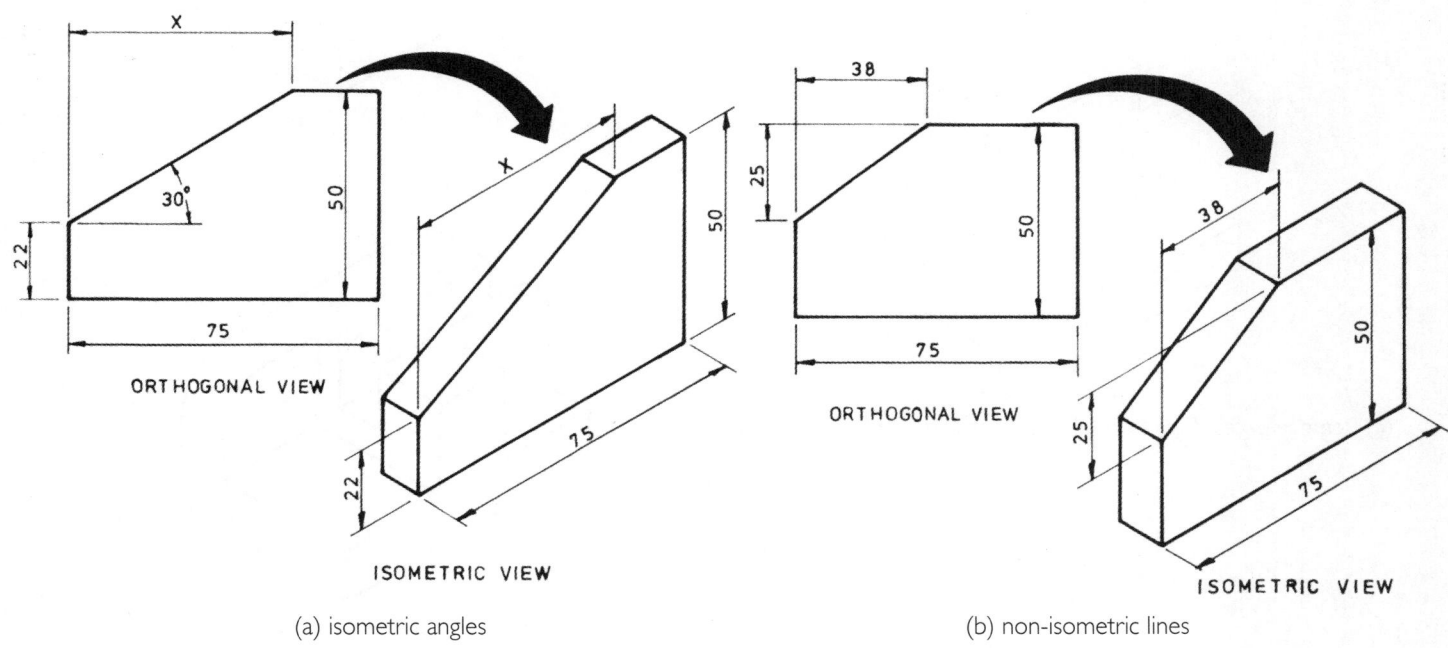

(a) isometric angles          (b) non-isometric lines

**FIGURE 5.9** ▶ Making an isometric drawing

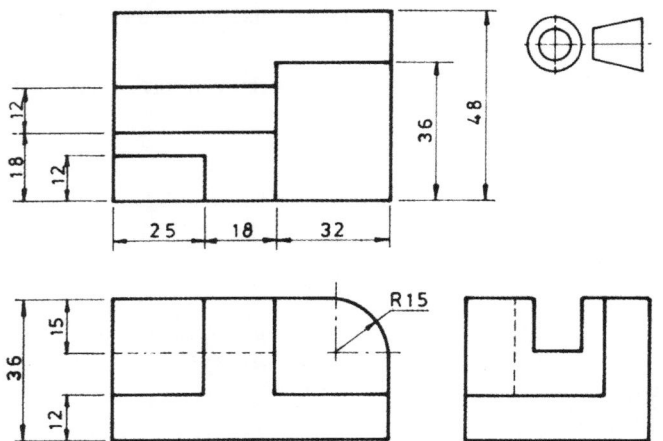

# 5.4 Making an isometric drawing

The series of five views in Figures 5.9(b)–(f) shows step-by-step production of a simple isometric drawing, the orthogonal views of which are shown in Figure 5.9(a).

The isometric axes meet at the circled point in Figure 5.9(b). This point is carefully chosen so that the view will reveal as much detail as possible.

(a) Orthogonal views for making the isometric drawing

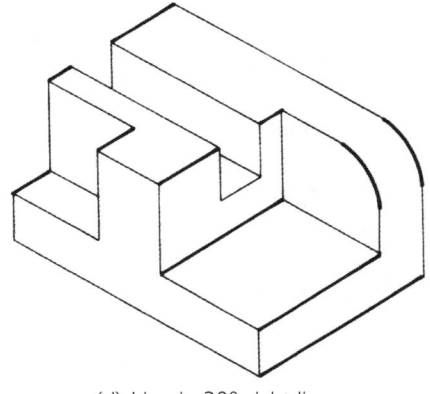

(d) Line in 30° right lines

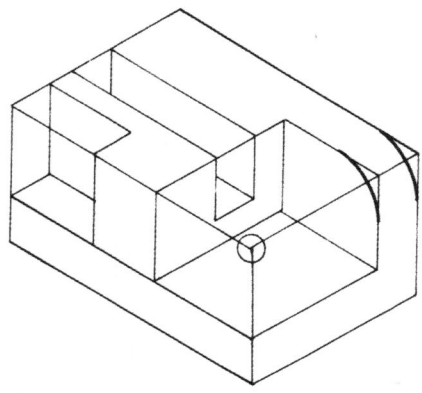

(b) Draw in light construction lines (circles and curves full thickness)

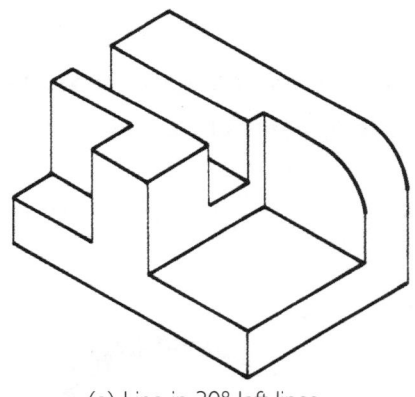

(e) Line in 30° left lines

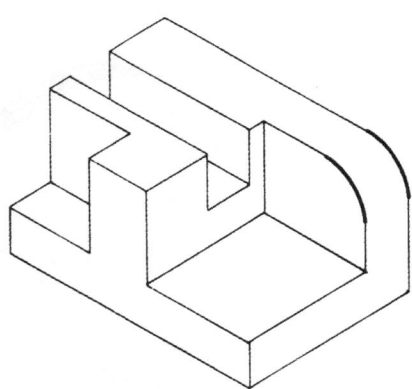

(c) Remove excess lines (simplified if construction lines lightly drawn)

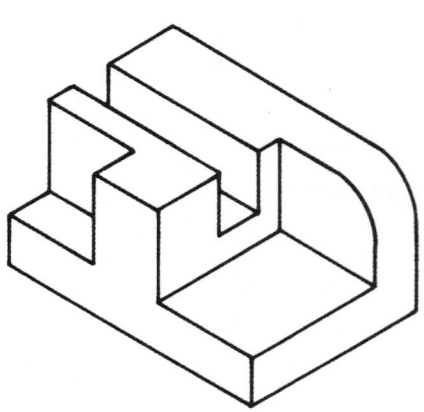

(f) Line in vertical lines to complete view

# 5.5 Representation of details common to pictorial drawings

## FILLETS AND ROUNDS

Filleted corners and rounded edges may be represented by either straight or curved lines as shown in Figure 5.10 using a type B (thin) line.

## THREADS

Threads may be represented by a series of ellipses or circles (depending on the type of drawing) evenly spaced along the centre line of the threaded section using a type B (thin) line (Fig. 5.11).

## SECTIONING

Pictorial drawings should be sectioned along centre lines, the cutting plane cutting parallel to one of the principal viewing planes of the object (Fig. 5.12(a)).

Hatching on half-sections should be drawn in the opposite direction on the adjacent cut faces coinciding at the axis (Fig. 5.12(b)).

## DIMENSIONING

Dimensioning on pictorial views may sometimes be required, and should follow the same general rules as for orthogonal views; that is, the dimension line, projection lines and the dimension itself should lie in the same plane.

One of the following two methods should be used:

1. unidirectional—where all dimensions are read from the bottom of the drawing (Fig. 5.13(a))
2. principal plane dimensioning—where dimensions lie in one or more of the three pictorial planes (Fig. 5.13(b)).

**FIGURE 5.11** ▶ Pictorial representation of threads

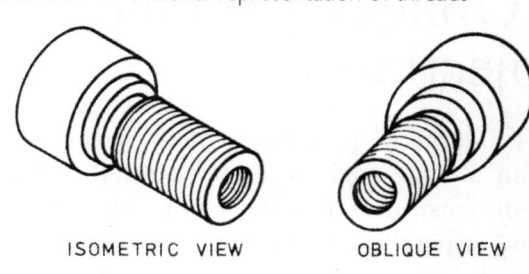

ISOMETRIC VIEW          OBLIQUE VIEW

**FIGURE 5.12** ▶ Pictorial representation of sections

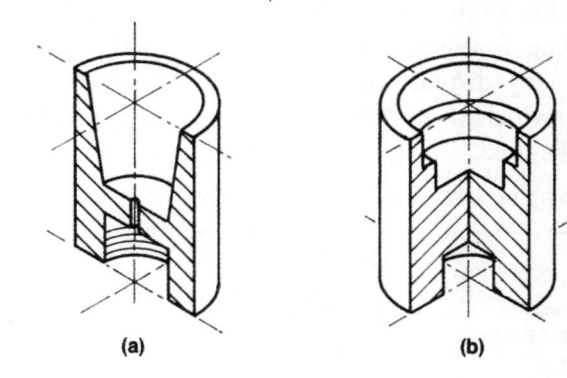

(a)                (b)

**FIGURE 5.13** ▶ Dimensioning pictorial views

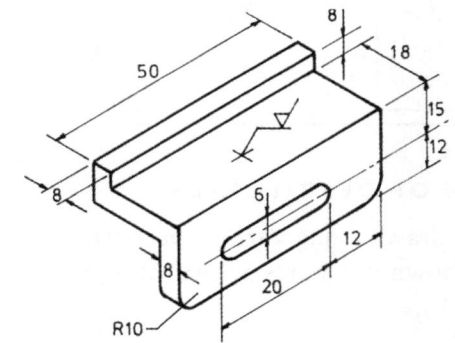

(a) unidirectional

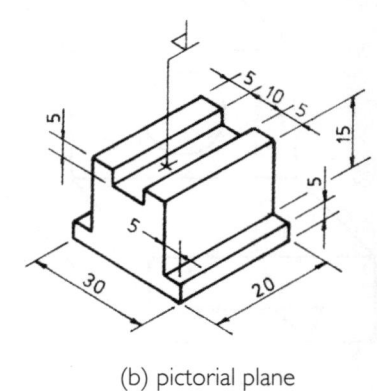

(b) pictorial plane

**FIGURE 5.10** ▶ Pictorial representation of fillets and rounds

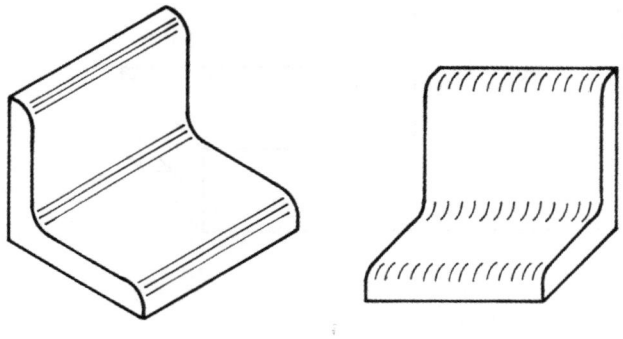

# 5.6 Oblique parallel projection

With this type of projection the object is viewed from an oblique angle so that the resulting view is three-dimensional. The view is produced by the drawing of parallel projectors from the object to the picture plane as shown in Figure 5.14. As the object is placed so that its front face is parallel to the picture plane, the oblique projectors will produce this face on the picture plane. Depth lines will also be reproduced, and their lengths will vary with the viewing angle. Depth lines are usually taken as receding at angles of 45°, 30° or 60° as these angles are easily drawn with set squares. However, any angle which shows the detail to the best advantage may be used.

FIGURE 5.14 ▶ Oblique parallel projection

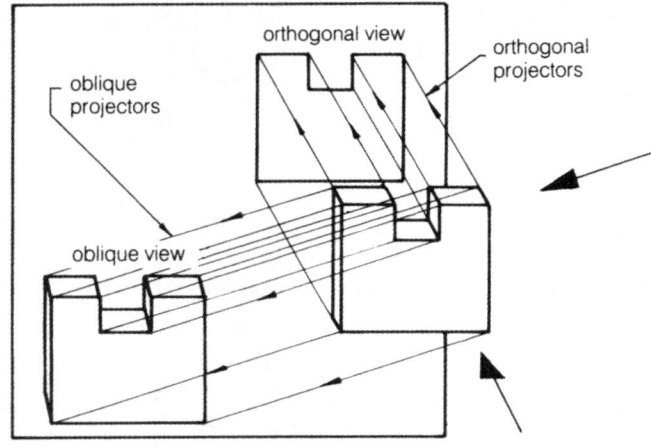

## LENGTH OF DEPTH LINES

A cube is drawn using various proportions of depth lines, as shown in Figures 5.15(a) and (b).

FIGURE 5.16 ▶ Irregular face parallel to picture plane

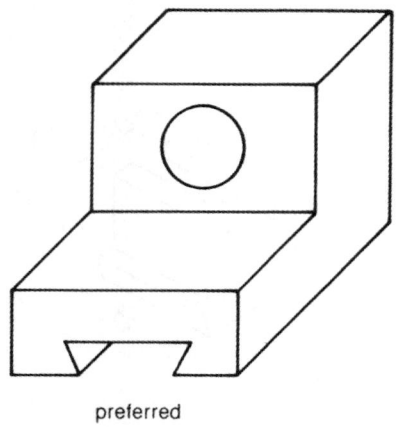

preferred

In Figure 5.15(a) the depth lines are not reduced, and it is noticed that the appearance is unnatural with the depth lines seeming too long and appearing to diverge. This type of drawing is known as cavalier projection. Another type of drawing, which eliminates some of the faults of cavalier projection, is cabinet projection. Here depth lines are shortened to half their length, as shown in Figure 5.15(b). This projection is used in most drawings.

Three rules are worth remembering when making an oblique drawing.

FIGURE 5.15 ▶ Length of depth lines in oblique parallel projection

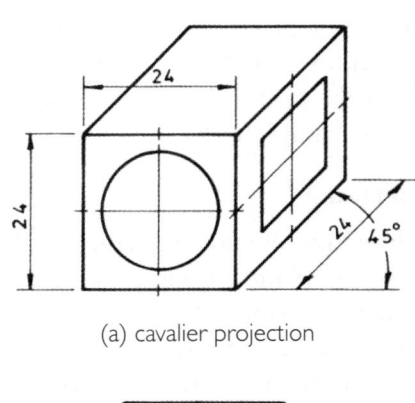

(a) cavalier projection

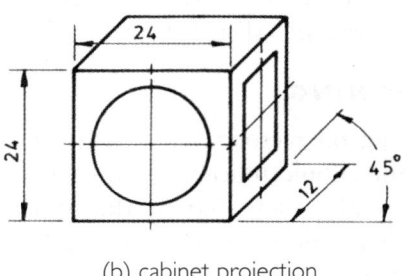

(b) cabinet projection

### Rule 1

Place the object so that the view with the most detail is parallel to the picture plane, especially if the view consists of arcs and circles. This is illustrated in Figure 5.16.

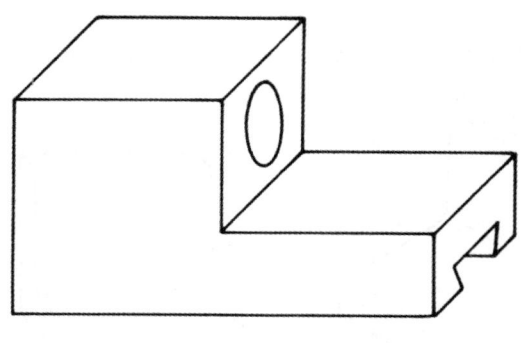

not preferred

## Rule 2

Place the object so that the longest dimension runs horizontally across the sheet, as shown in Figure 5.17.

**FIGURE 5.17** ▶ Longest dimension parallel to picture plane

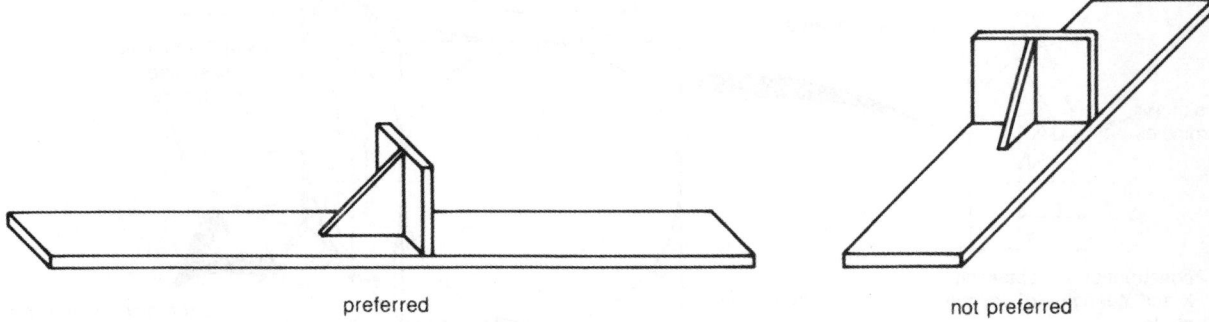

preferred

not preferred

## Rule 3

In some cases the above two rules conflict, and when this is so, Rule 1 has preference as the advantage gained by having the irregular face without distortion is greater than that gained by observing Rule 2. This rule is illustrated in Figure 5.18.

**FIGURE 5.18** ▶ Preferred pictorial view

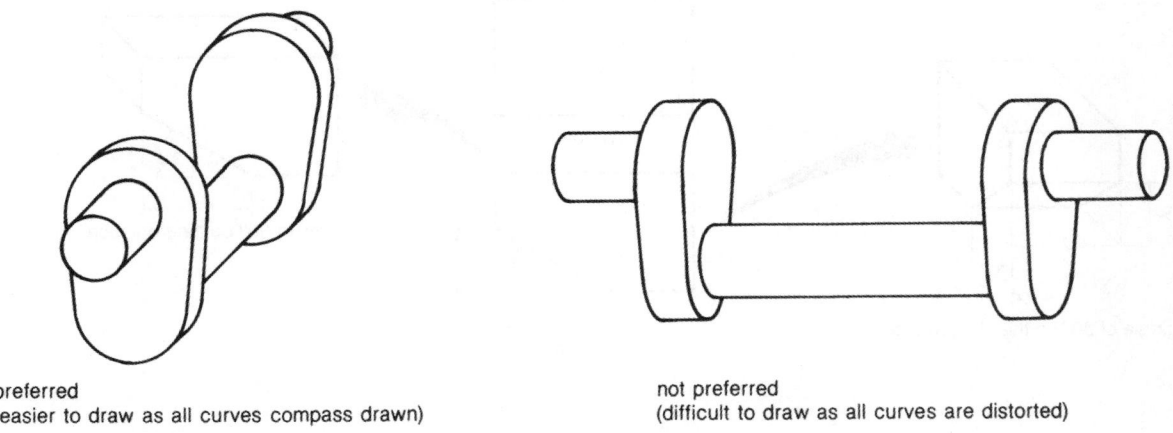

preferred
(easier to draw as all curves compass drawn)

not preferred
(difficult to draw as all curves are distorted)

## Rule 4

Decisions about viewing an object in oblique projection should aim to show the object so that its shape is most clearly presented and is conducive to showing its dimensions.

### CIRCLES ON THE OBLIQUE FACE

These circles are plotted using a plotting view, which consists of a true size quadrant of the circle, together with a half size quadrant on the same view (Fig. 5.19). The circles are plotted in a similar manner to isometric circles, except that measurements along the 45° axis are taken from the half size quadrant.

Alternatively, oblique circles may be plotted using true shape semicircles located on the edges of the oblique face and projecting points on the oblique circles as shown in Figure 5.19 (page 206).

### ANGLES ON OBLIQUE DRAWINGS

These are drawn as shown in Figure 5.20 (page 206).

### SELECTION OF THE RECEDING AXIS

A number of views which can be obtained by varying the angle of the receding axis are shown in Figures 5.21(a)–(d). Each view is chosen because it reveals the maximum amount of detail for that particular orientation of the object, taking into account Rules 1, 2, 3 and 4 mentioned previously.

The reference corner is circled and outlined above on each view.

**FIGURE 5.19** ▶ Oblique circles

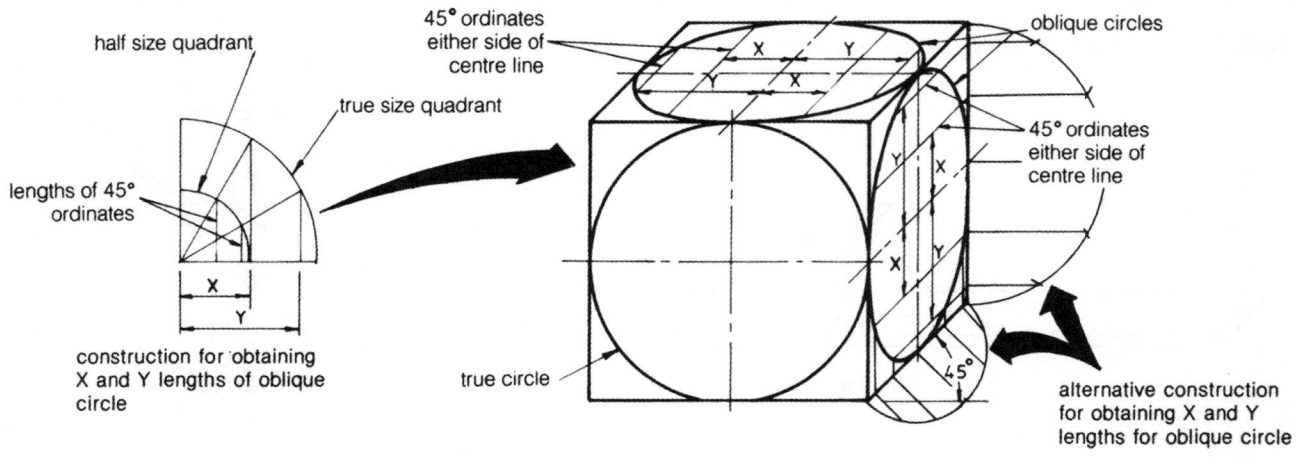

half size quadrant

true size quadrant

45° ordinates
either side of
centre line

oblique circles

lengths of 45°
ordinates

45° ordinates
either side of
centre line

X
Y

construction for obtaining
X and Y lengths of oblique
circle

true circle

45°

alternative construction
for obtaining X and Y
lengths for oblique circle

**FIGURE 5.20** ▶ Oblique angles

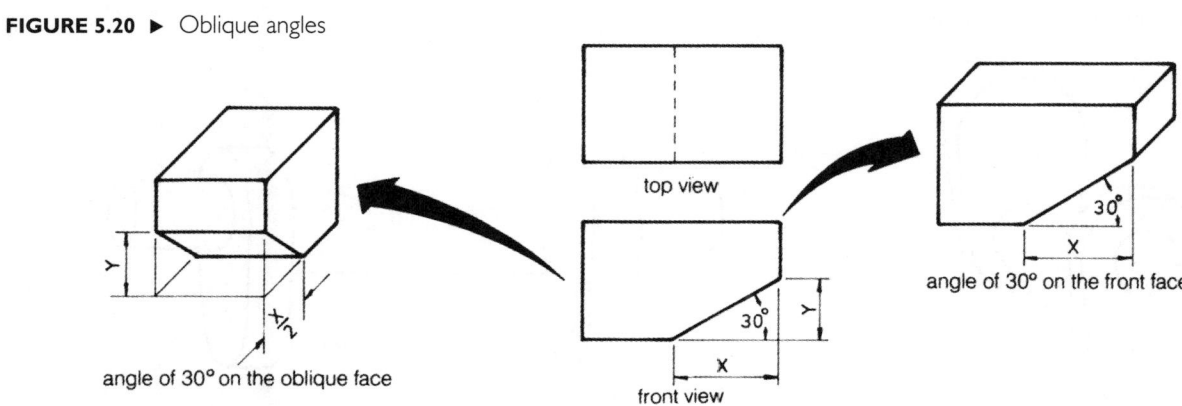

top view

front view

angle of 30° on the oblique face

angle of 30° on the front face

**FIGURE 5.21** ▶ Selection of oblique axes

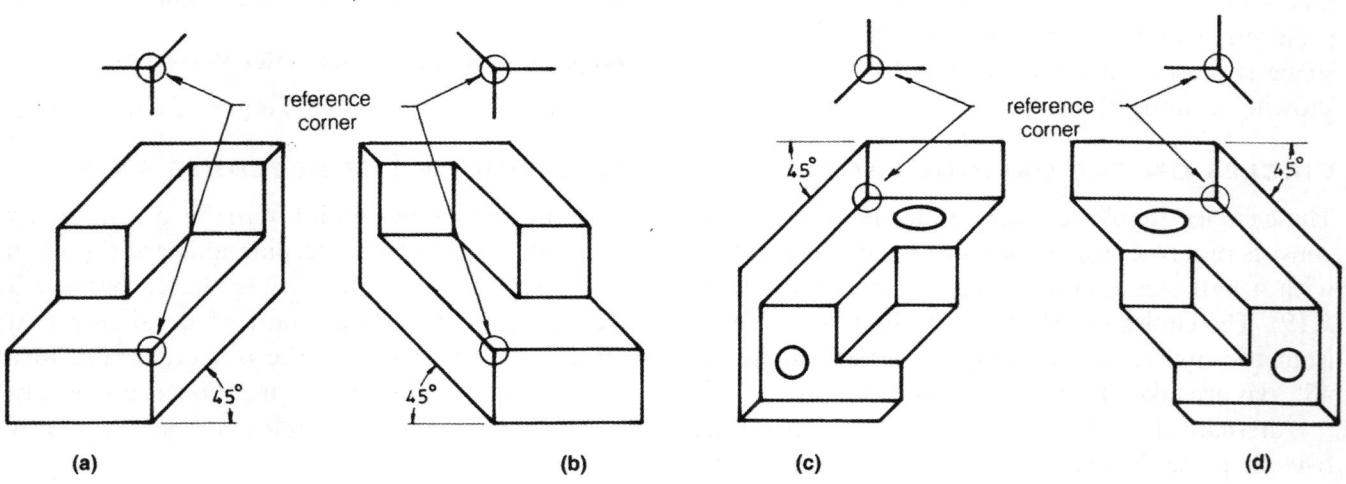

reference
corner

reference
corner

45°

45°

45°

45°

(a)

(b)

(c)

(d)

# 5.7 Problems

The following exercises may be drawn in either or both isometric or oblique parallel projection.

## 5.1 GUIDE PLATE

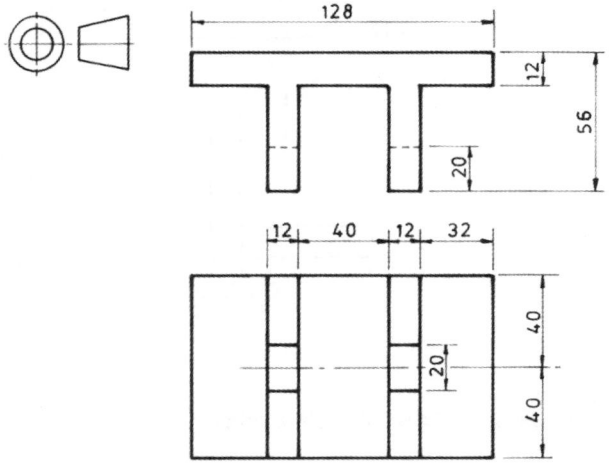

## 5.4 MS SLOT SUPPORT

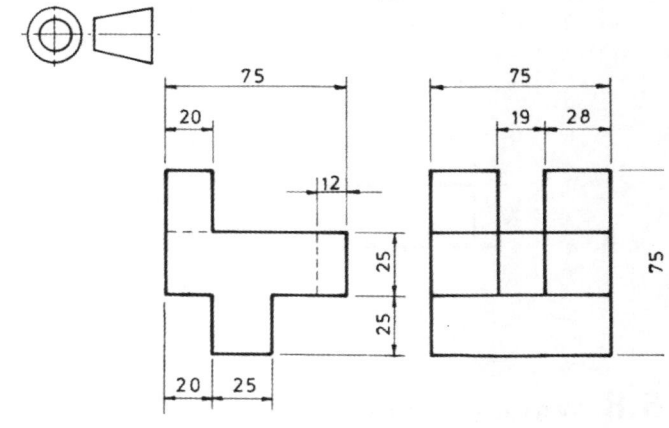

## 5.2 MS CONNECTING PIECE

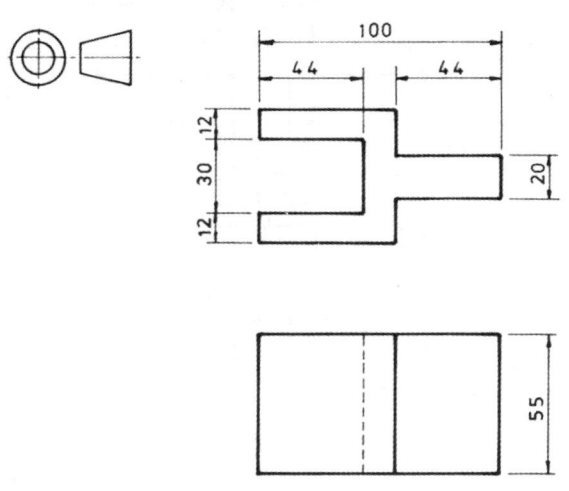

## 5.5 DOVETAIL SLIDE

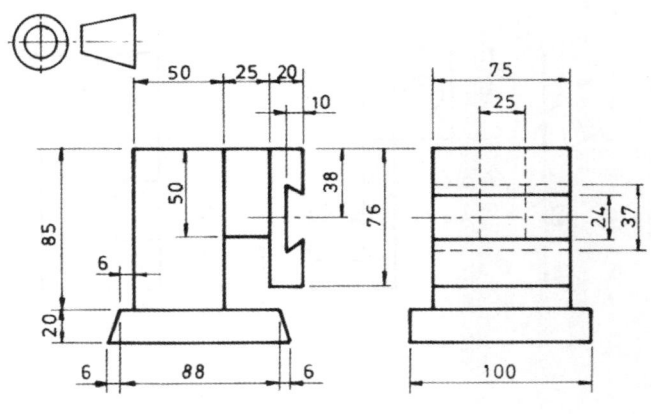

## 5.3 SUPPORT BRACKET

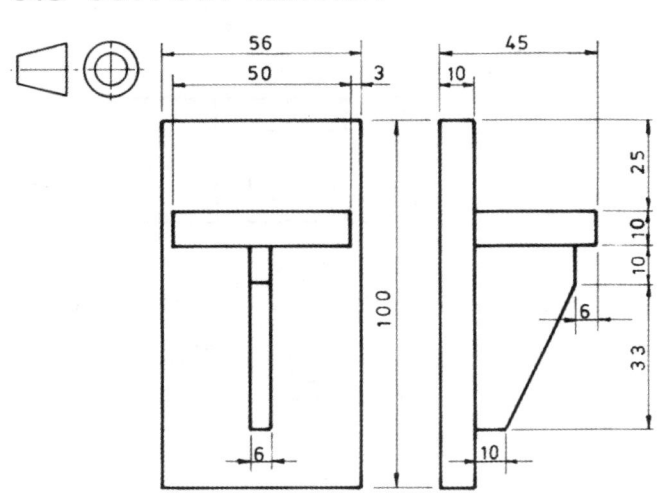

## 5.6 MACHINED BLOCK

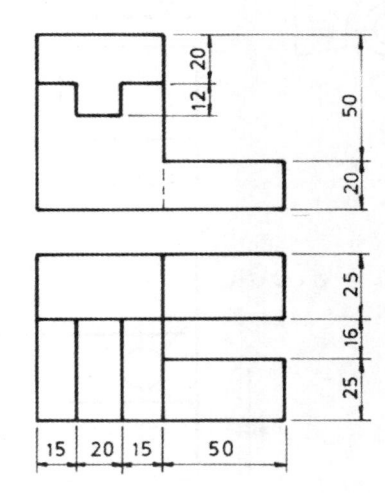

## 5.7 MS BAR BRACKET

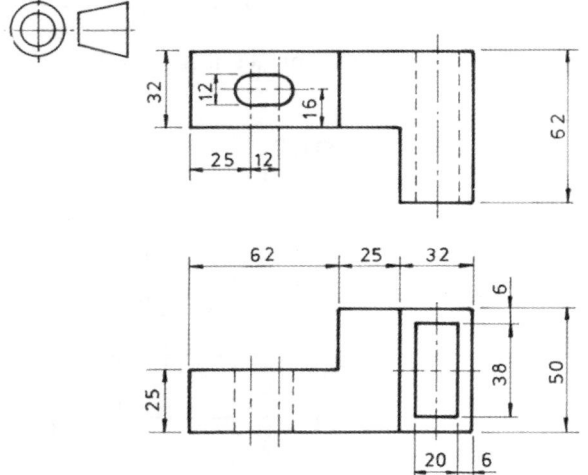

## 5.10 HOOK SUPPORT

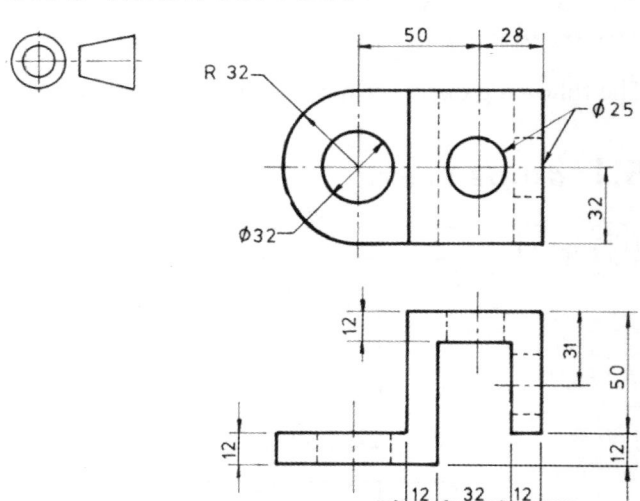

## 5.8 WALL BRACKET

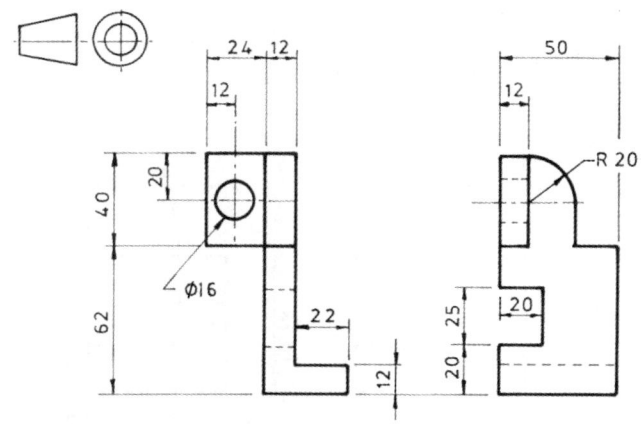

## 5.11 MILL TABLE FITTING

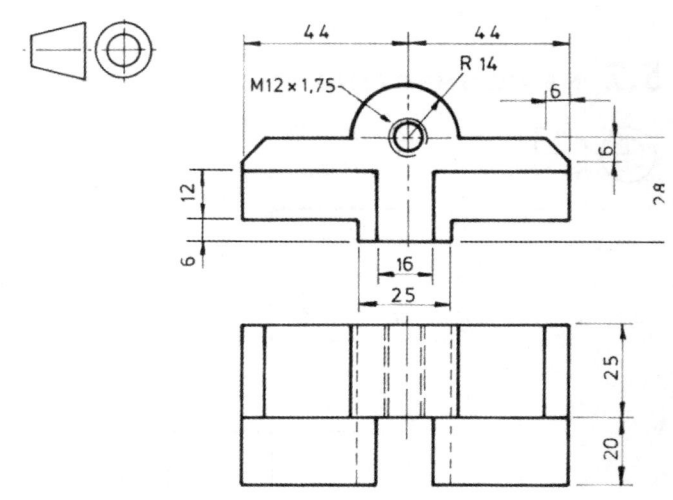

## 5.9 SLIDE BLOCK

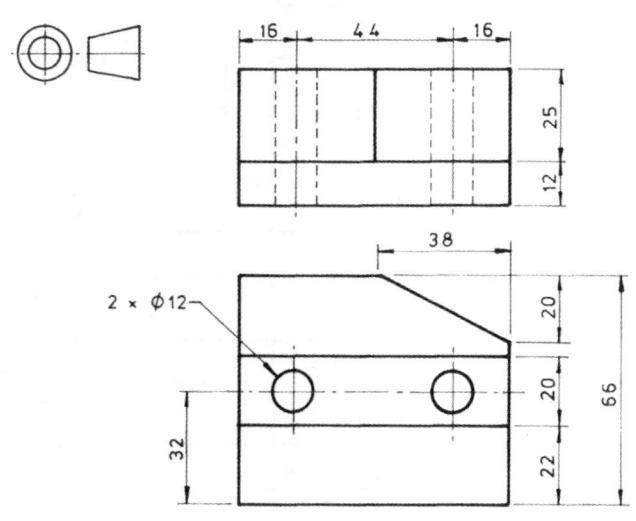

## 5.12 SQUARE SHAFT LINK

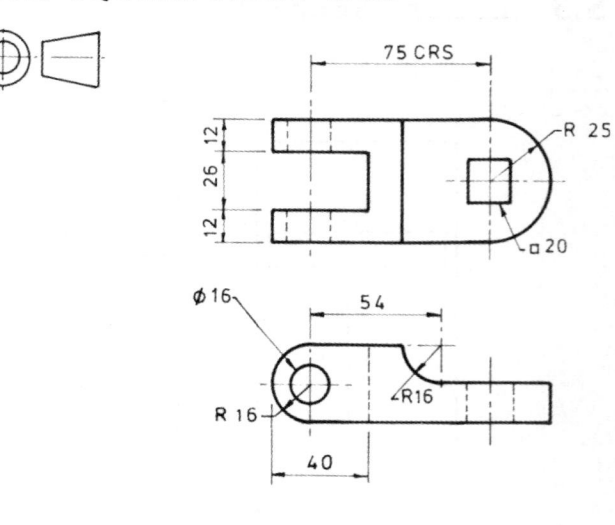

## 5.13 TABLE DOG

Draw an isometric or an oblique parallel view of the table dog.
Do not use an isometric scale.
(scale 1:1)

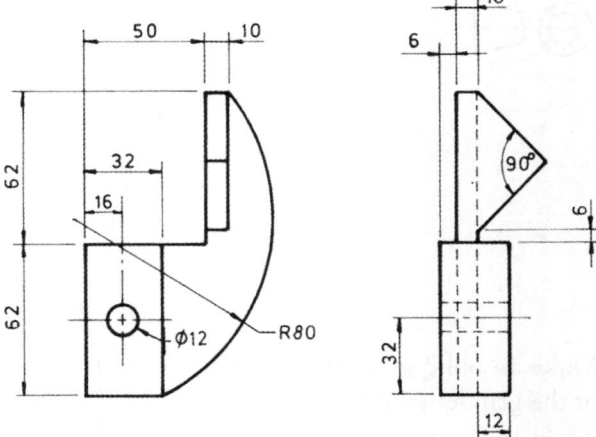

## 5.14 FABRICATED BRACKET

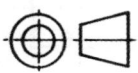

Draw the given side view of the fabricated bracket, and using this view as an aid make an oblique drawing of the bracket.
(scale 1:2)

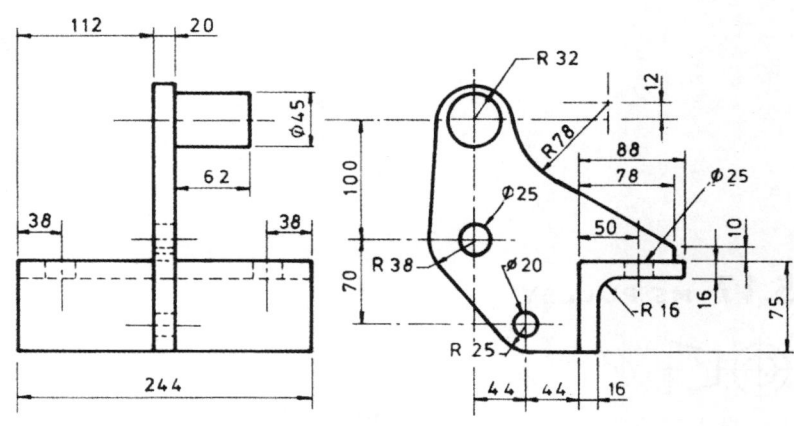

## 5.15 MOUNTING BRACKET

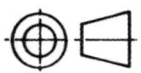

Draw, full size, an oblique parallel view of the mounting bracket.

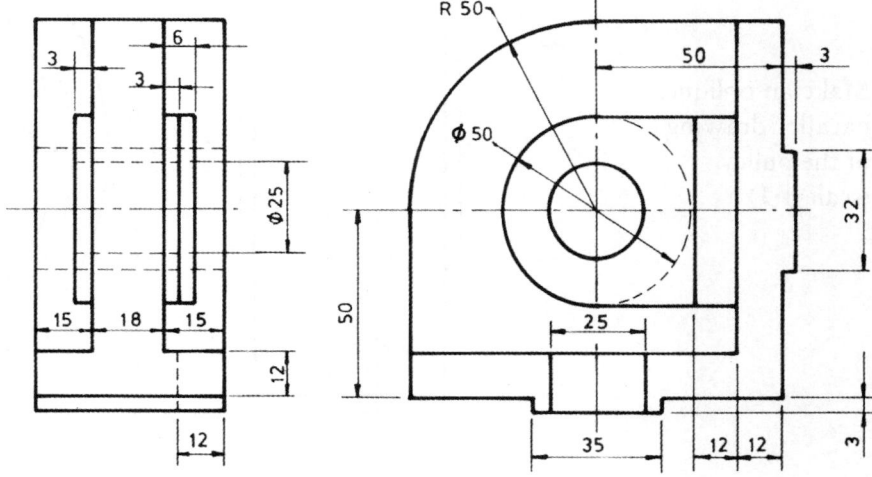

## 5.16 GRINDER GUIDE

Make an oblique parallel drawing of the grinder guide.
(scale 1:2)

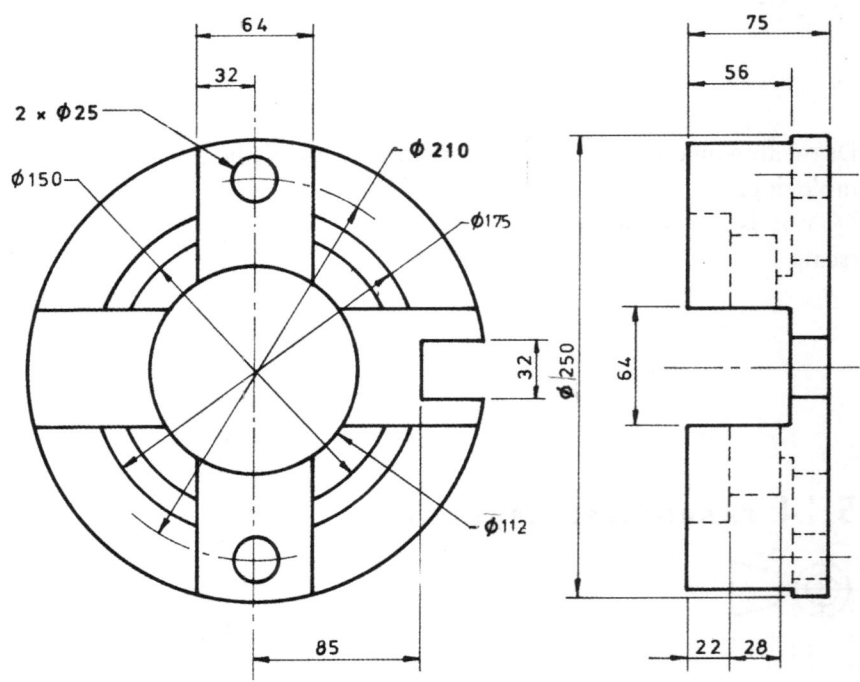

## 5.17 MS PULLEY

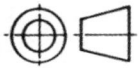

Make an oblique parallel drawing of the pulley.
(scale 1:1)

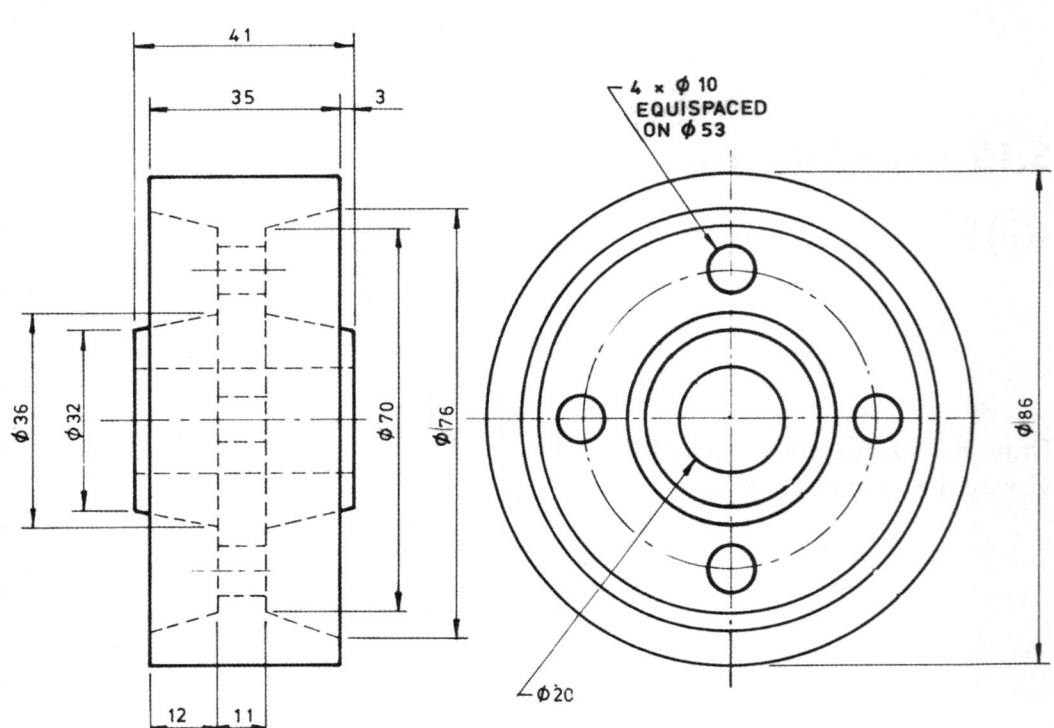

## 5.18 RAM ADAPTOR

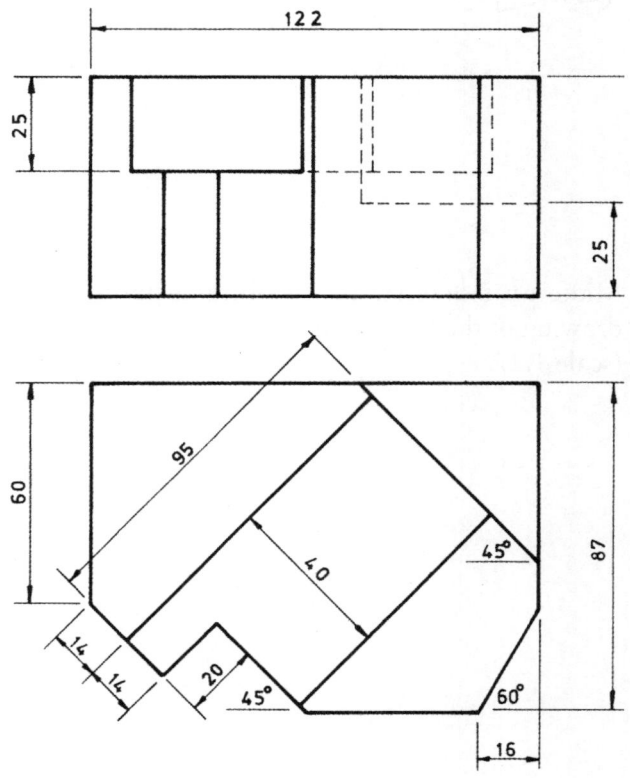

Make an isometric drawing of the ram adaptor.
(scale 1:1)
Show hidden detail where required for a complete
shape description.

## 5.19 TAPERING LINK

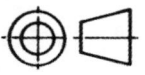

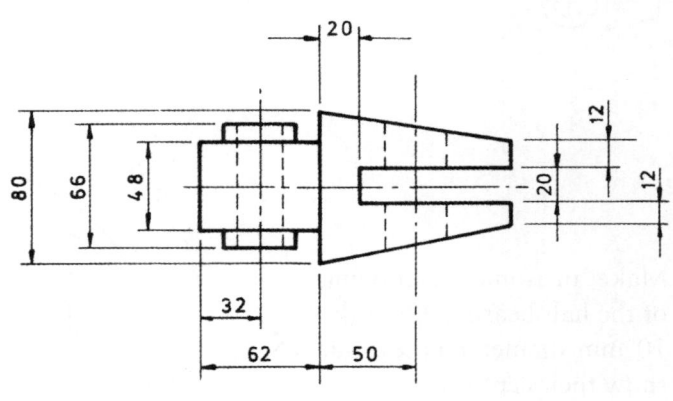

Make an oblique parallel drawing of the
tapering link.
(scale 1:1)

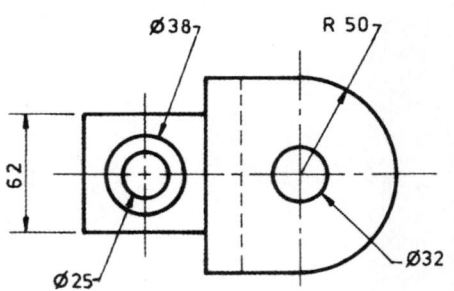

## 5.20 FORKED END

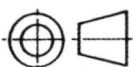

Make an isometric or an oblique parallel drawing of the forked end.
(scale 1:2)

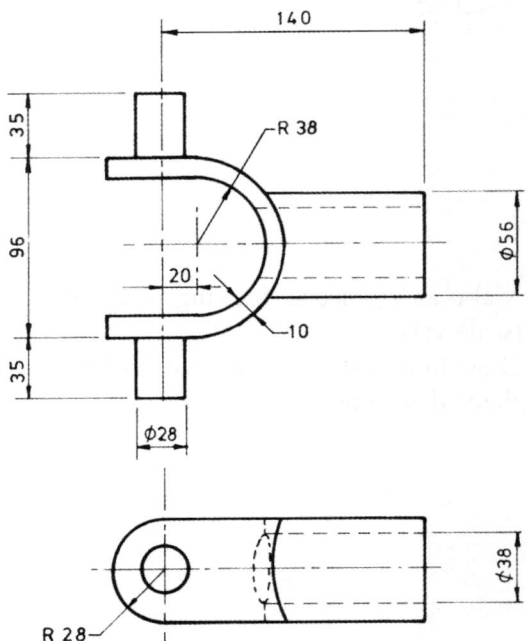

## 5.21 UNIVERSAL JOINT HALF BEARING

Make an isometric drawing of the half bearing. Omit the 10 mm diameter holes, but show their centres.
(scale 1:1)

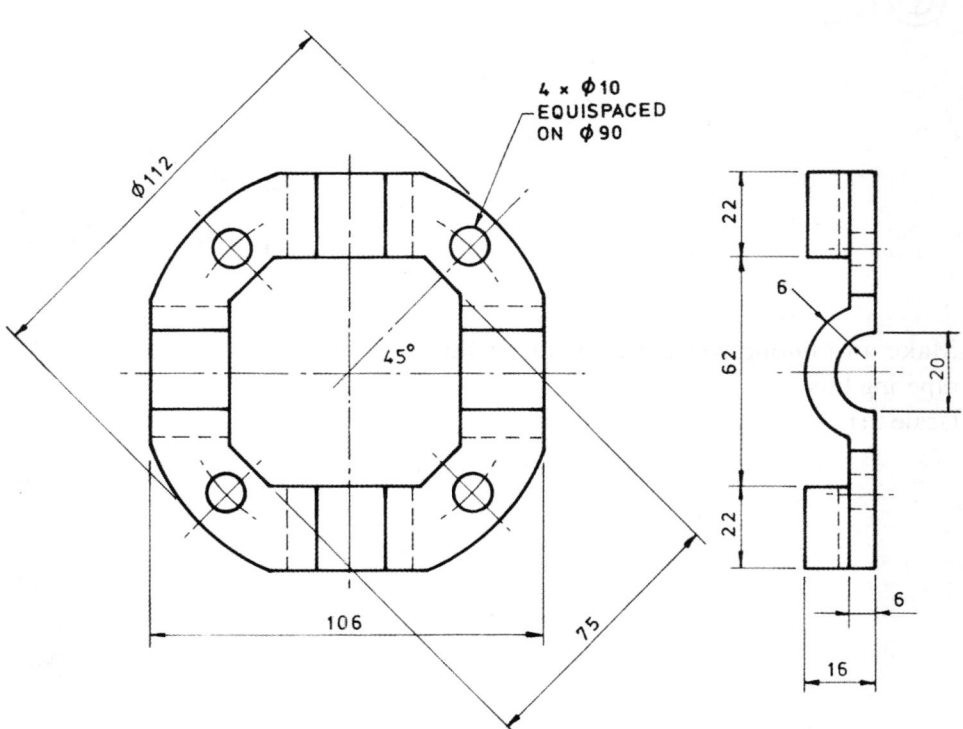

## 5.22 SUPPORT BRACKET

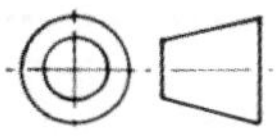

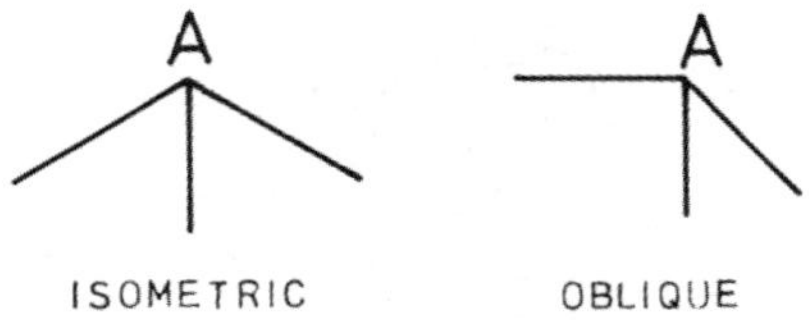

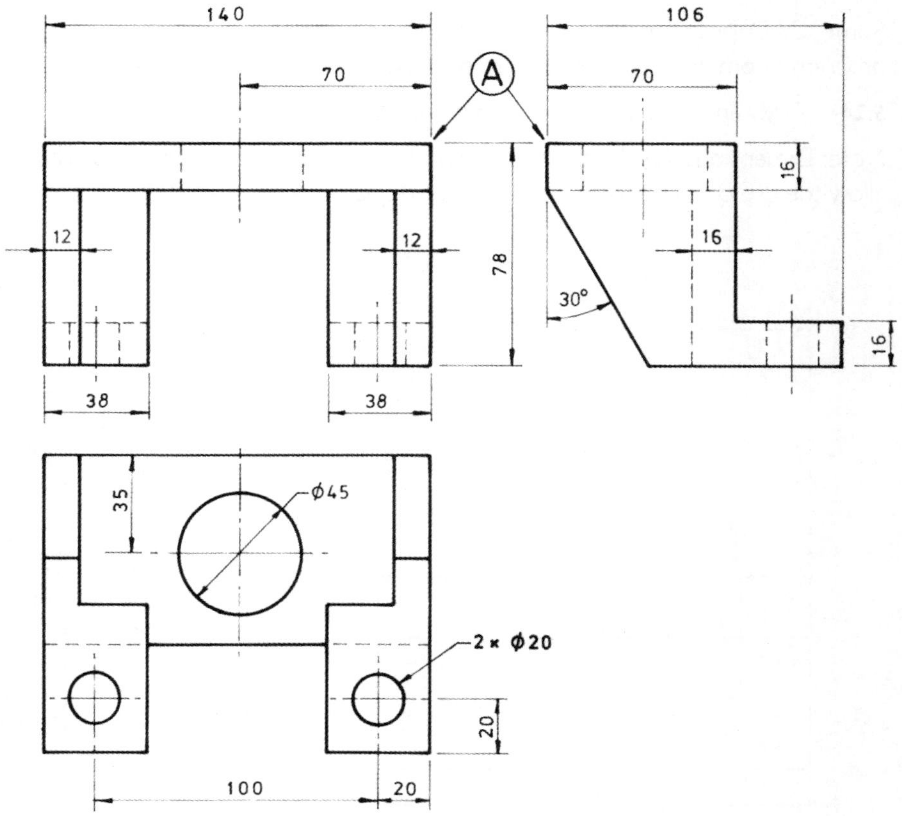

ISOMETRIC  OBLIQUE

Make either an isometric or an oblique parallel drawing of the support bracket using point A as the starting point, with the isometric or oblique axes disposed around it as shown above. (scale 1:1)

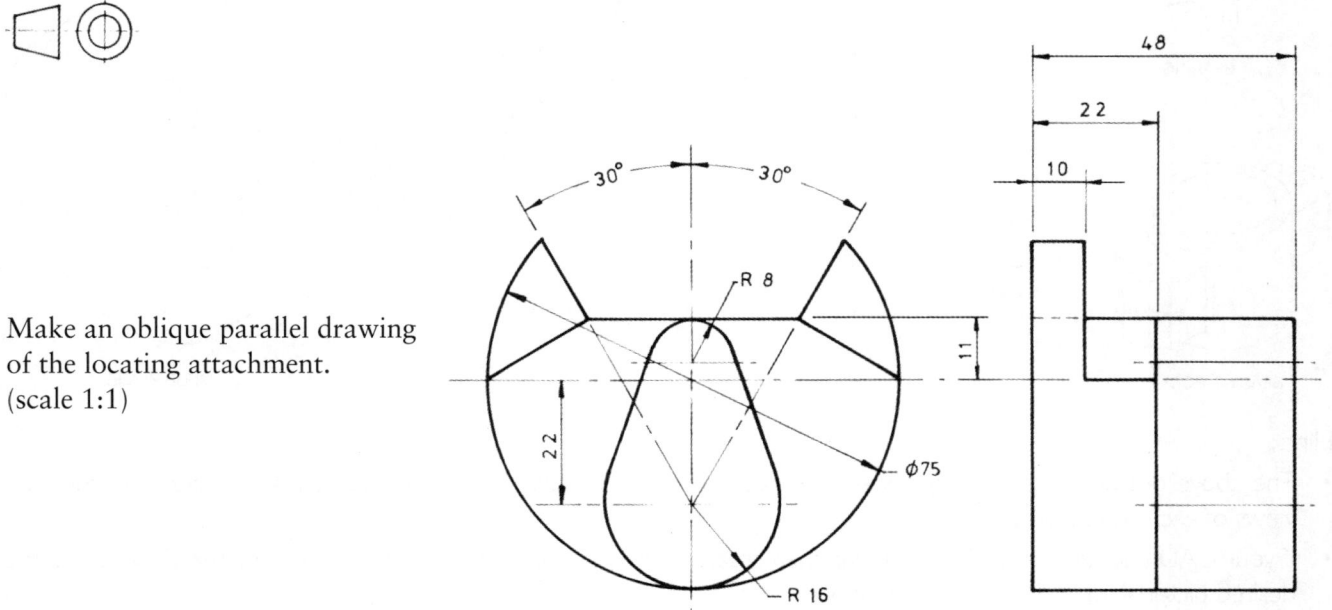

## 5.23 LOCATING ATTACHMENT

Make an oblique parallel drawing of the locating attachment. (scale 1:1)

# 5.8 CAD corner

Some CAD programs have excellent aids for isometric drawing. Before attempting this question, investigate your program's isometric potential and utilise these attributes to the fullest.

**5.24**  Draw an isometric view of the fabricated bracket shown.

Note: Dimensions, for the bracket, shown below, are in increments of 5. This may allow the use of snapping to points and/or displaying a grid pattern.

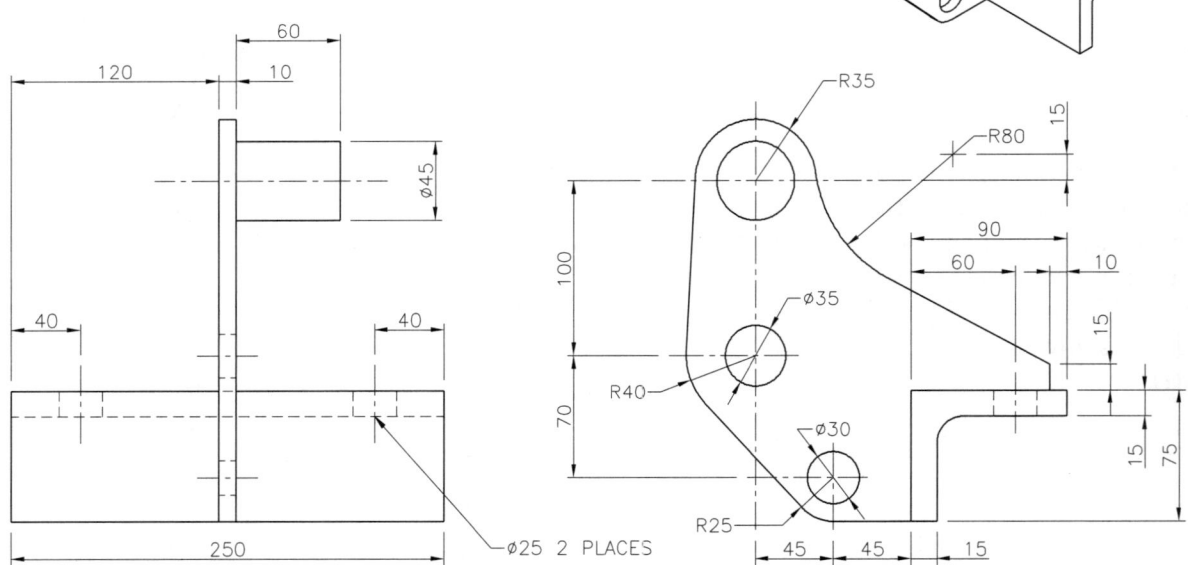

Modified fabricated bracket

**5.25**  Draw pictorial views of the components from Exercise 3.61. Illustrations of each component are shown below.

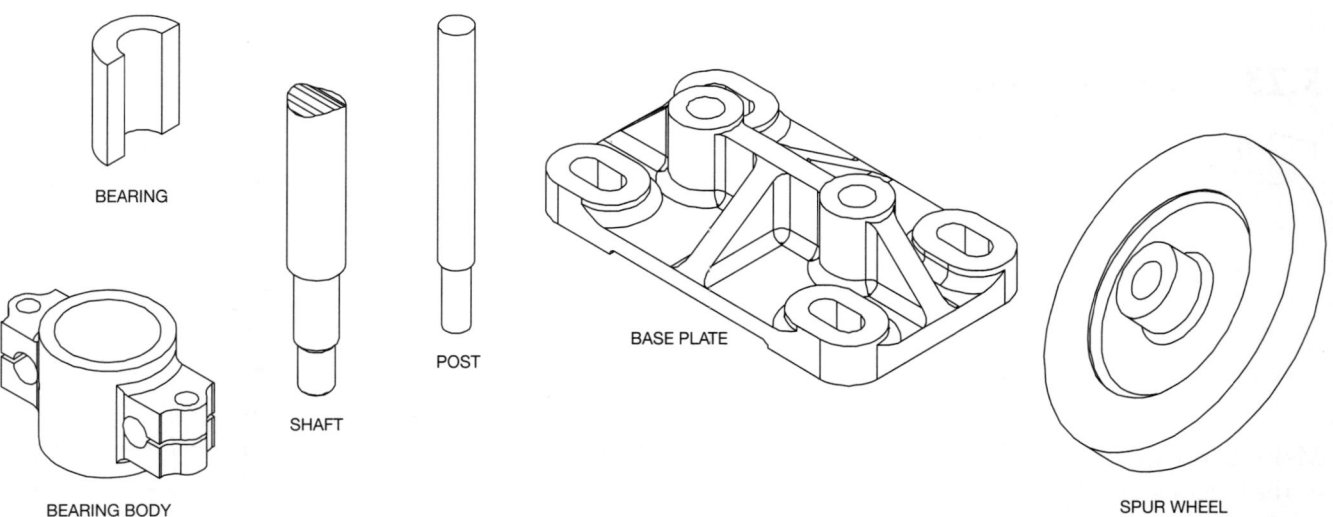

Hints:

- The above drawings are 'Isometric' views of solid models. The challenge in this question is to draw true isometric views of each component.
- If your CAD package allows 3D modelling then creating solids of each component and adjusting the view to give the desired pictorial image is the easier solution.

# Working drawings: Detail and assembly drawing

**6**

After studying this chapter and working through the problems you should be able to do the following:

- make a detail drawing following standard procedures
- make an assembly drawing following standard procedures
- prepare a set of working drawings

# 6.1 Detail drawings

The detail drawing is used as the main reference in the manufacture of individual components. It should contain sufficient information to manufacture the part as well as suitable, fully dimensioned orthogonal views of each part, together with other information that may be required in the manufacturing process. A complete detail drawing should contain at least the following information (not necessarily in order of importance):

1. sufficient orthogonal views of the part concerned
2. dimensions and instructional notes
3. scale used
4. projection used, for example first or third angle
5. drafting standard reference, for example AS 1100 Part 101
6. name or title of drawing
7. drawing numbers
8. dimensional units which apply
9. tolerances where necessary
10. surface texture requirements
11. special treatments needed (heat, metallic coatings, paint, etc.)
12. reference to a particular assembly if applicable
13. type of material used
14. names of drafter, checker, approver, etc.
15. relevant dates of action by those concerned
16. zone reference system when necessary
17. revisions or modifications
18. drawing sheet size
19. name of company or department as applicable
20. drawing sheet reference, for example sheet 1 of 2

It is preferable to draw only one item on a single drawing sheet, the sheet size depending on the item size and number of views required. However there are instances when multidetail drawings are used. Many of the problems in this section are multidetail drawings, where individual parts are simple and it is more convenient to group them on one sheet.

It is common practice for firms to print their own drawing sheets with a drawing frame and title block in order to standardise the general information provided and to ensure that such information is included on all drawings. Figure 6.1 illustrates the layout of three separate detail drawings of parts of a machine screw jack.

While the title block is shown in the bottom right-hand corner (the preferable location) in Figure 6.1,

AS 1100 Part 101 also recommends that the title block may be located in the top right-hand corner with the revisions table in the top left-hand corner when convenient for drawing layout.

In all cases the drawing number may be repeated in other corners or along the sides of the sheet to ensure that it is visible when the drawing is filed or folded.

A check of the drawings in Figure 6.1 against the list above will illustrate the points raised. Note that each of these drawings is referenced to the assembly drawing of the jack, shown in Figure 6.2 (page 219).

It will be seen that each drawing in Figure 6.1 was originally issued on 9.4.00, but that a revision was carried out to the thread on the 'Spindle' and 'Jack body' by changing it from a Whitworth to a metric thread form. These revisions have been inserted on the drawing, and a record of them tabulated in the top right-hand corner on 1.6.01. Minor revisions to components are an everyday occurrence in a drawing office, and when such a revision does not affect the interchangeability of a part, the revision may be carried out on the old drawing. Where interchangeability is affected, a new drawing number should be raised.

# 6.2 Assembly drawings

Assembly drawings are primarily used to show how a number of components are fitted together to make a complete product unit. The term subassembly is commonly applied to a product unit which combines with other subassemblies to make an assembly. For example, an assembly drawing of a motorcar engine would show a number of complete units included on the drawing, such as the distributor, generator, carburettor, etc. Each of these units is referred to as a subassembly of the engine assembly.

Assembly drawings may be divided into two categories depending on the proposed use:

1. *general assembly*—where the main purpose is to identify the individual components and show their working relationship
2. *working or detailed assembly*—a combined detail and general assembly drawing which fulfils the function of both types

Figure 6.2 is a general assembly of a machine screw jack. Detailed drawings of the individual parts of the jack are shown in Figure 6.1.

Features of a *general assembly drawing* are:

1. Views are selected which show how the parts fit together and indicate how the unit may function.

**FIGURE 6.1** ▶ Detail drawings

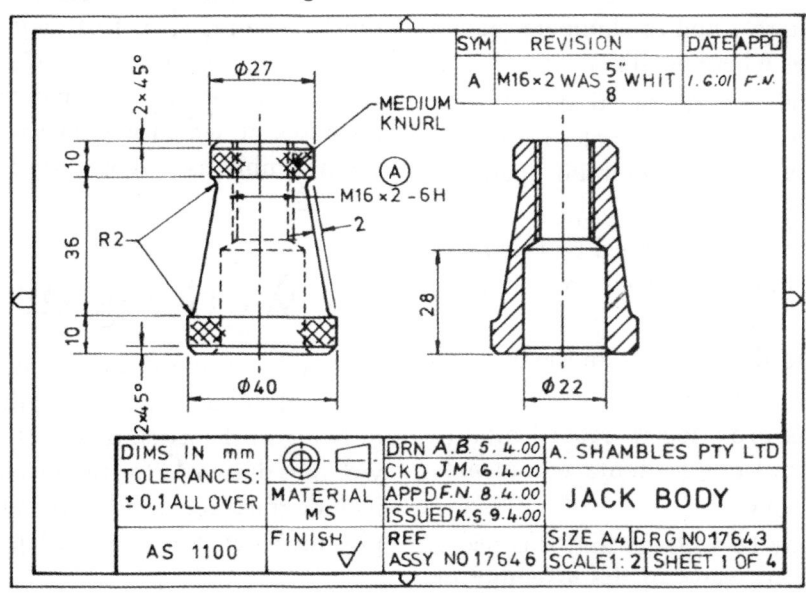

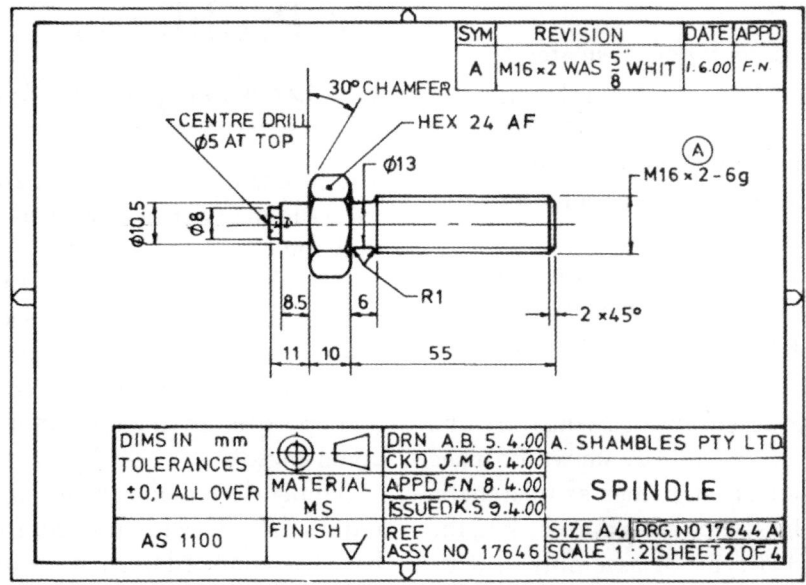

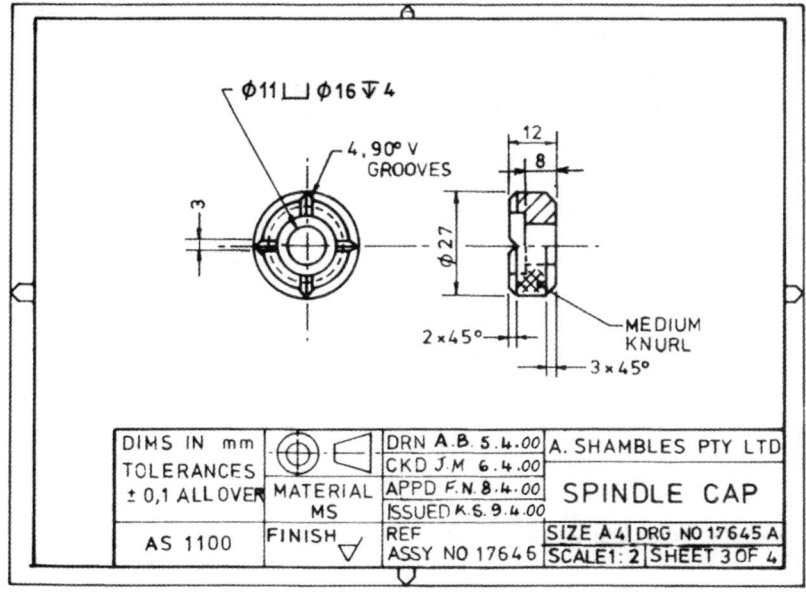

2. Sectional views are used extensively in this type of drawing to facilitate 1 and to eliminate the use of hidden detail lines where possible.

3. Dimensions which relate to the function of the unit as a whole are indicated; for example, Figure 6.2 indicates the maximum and minimum operating heights of the jack.

4. Individual components are identified by the use of numbers contained in circles which are connected by leaders to the related parts.

5. A parts list relates to the numbers on the drawing and identifies the component.

6. A revisions table is provided to record modifications to individual components which may occur from time to time.

7. Some assemblies may be so large that it is necessary to draw different views of the assembly on separate sheets.

Features of a *working assembly drawing* are:

1. Only simple assemblies are drawn in this manner, as views have to be chosen which show the assembly relationship as well as sufficient dimensional details of individual components to enable their manufacture.

2. It is ideally suited to furniture construction drawings where the assembly views are not complex and details of joints may be enlarged and shown as partial views.

Examples of working assembly drawings are the fan, and pulley and shaft assemblies shown on page 228. These are not complete drawings as tolerances, title blocks, material lists, etc. are omitted, but the general principle of this type of assembly will be appreciated.

The information provided on a general assembly drawing is somewhat different from that required on a detail drawing. Information on the manufacture of individual parts is not required, for example surface finish, tolerances, or treatments. However, assembly instructions (see note zone B2, Fig. 6.2) are required, as are dimensions which may be used for installation purposes or which are relevant to the operation of the assembly as a working unit (see note zone B4, Fig. 6.2).

# 6.3 Working drawings

A set of working drawings includes detail drawings of the individual parts together with an assembly drawing of the assembled unit. For example, a set of working drawings for the machine screw jack would include the three detail drawings shown in Figure 6.1 plus the assembly drawing, Figure 6.2.

# 6.4 Problems (*working drawings*)

The problems in this section are intended to provide the student with practice in detail and assembly drawings.

Standard size drawing frames should be used along with standard title blocks and parts lists. The layout of views within the frame area is an important consideration; it should be planned by the student and approved by the instructor before the drawing is commenced.

Dimensioning may be unidirectional or aligned as required. Surface texture requirements and tolerances have been omitted from the examples for convenience; where they are required they may be assessed in consultation with the instructor. The sheets of details show quantities of parts required for one assembly. Such information is normally provided in a parts list.

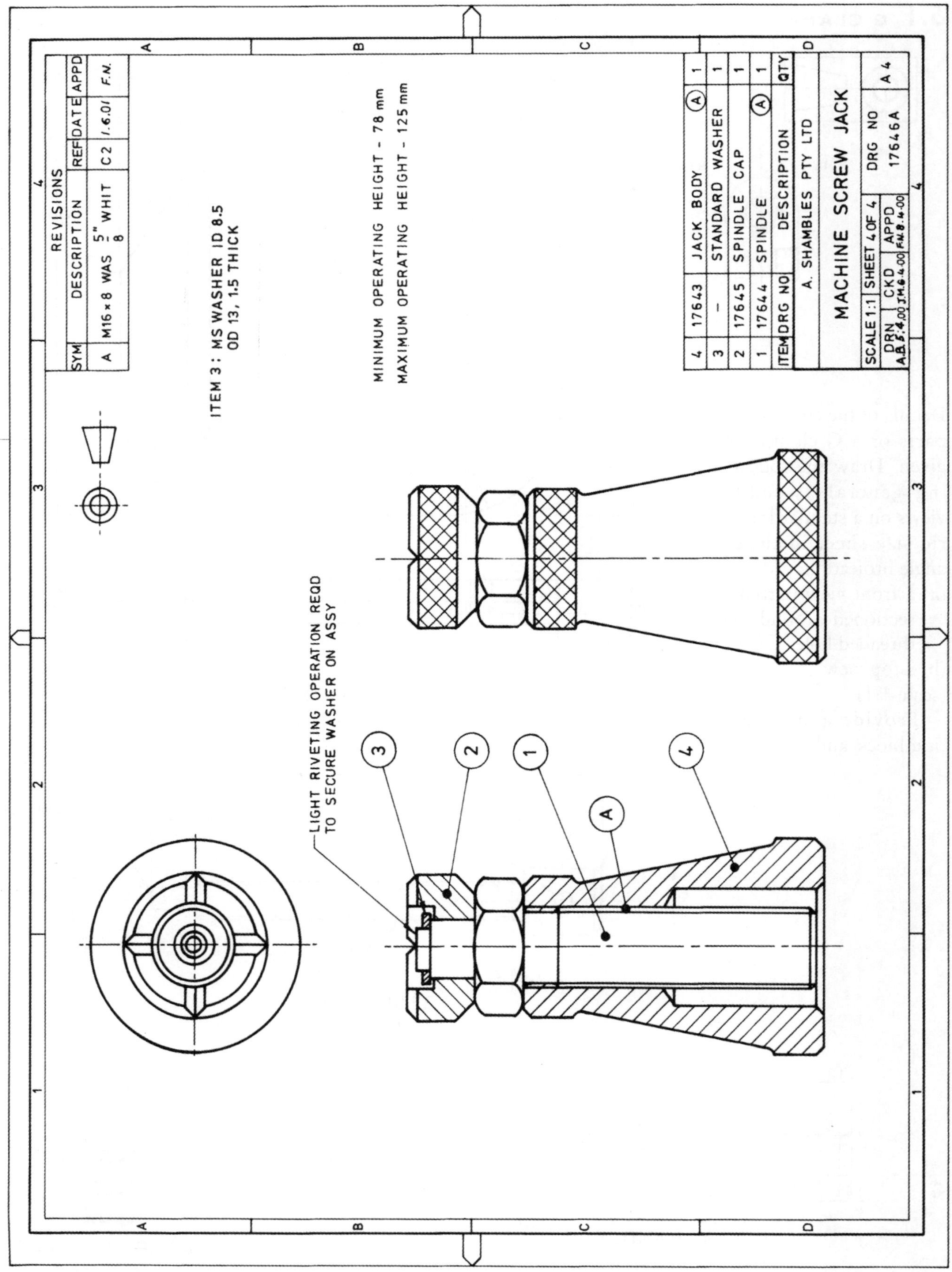

**FIGURE 6.2** ► Assembly drawing

## 6.1 G CLAMP

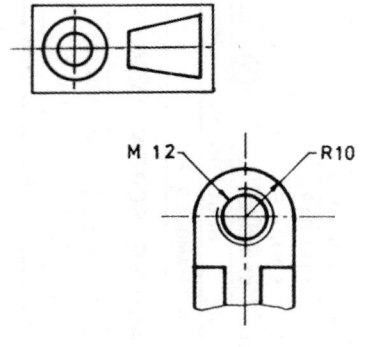

Details of the component parts of a G clamp are given. Draw the following general assembly views on a standard metric size sheet in third-angle projection:

(a) a front view partially sectioned around the threaded hole

(b) a top view

(scale 1:1)

Provide a standard title block and parts list.

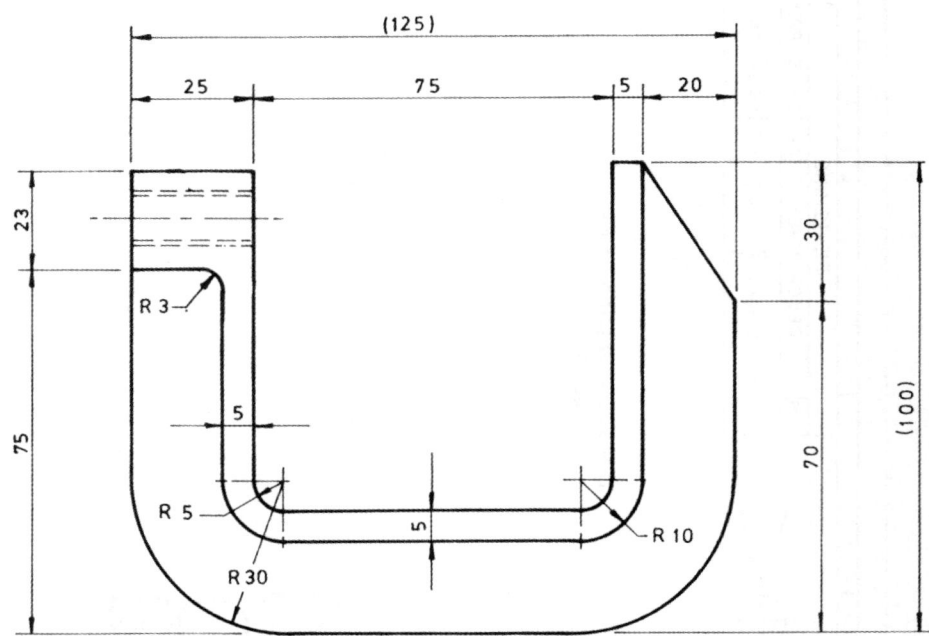

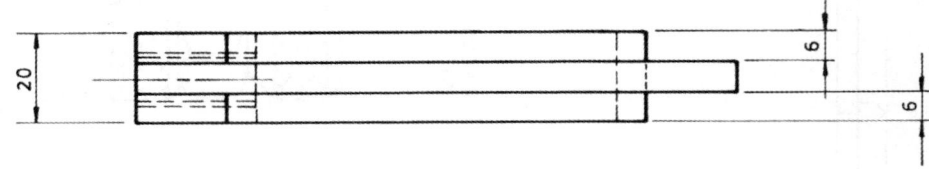

①CLAMP JAW
(MS — 1 REQD)

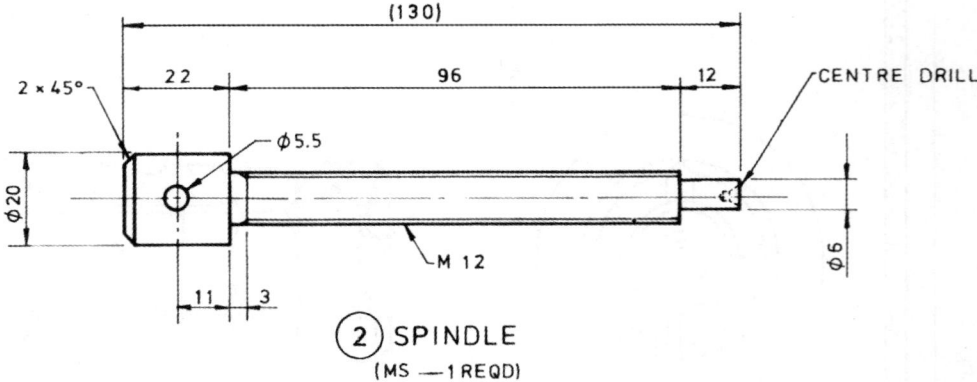

② SPINDLE
(MS — 1 REQD)

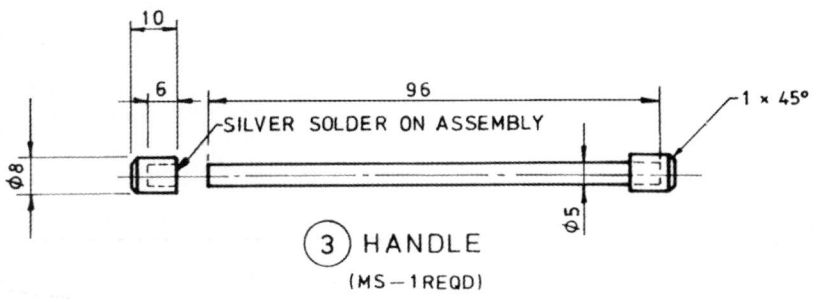

③ HANDLE
(MS — 1 REQD)

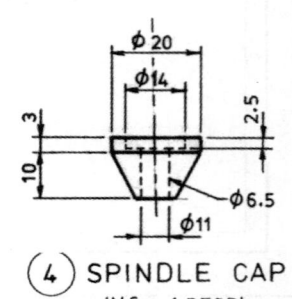

④ SPINDLE CAP
(MS — 1 REQD)

## 6.2 PLUMB-BOB

Details of a plumb-bob are given, together with manufacturing instructions. Sketch on squared paper, or draw on cartridge paper, the following assembled views:

(a) a sectional front view (showing the lead filling) positioned with the axis horizontal

(b) a top view

(scale 1:1)

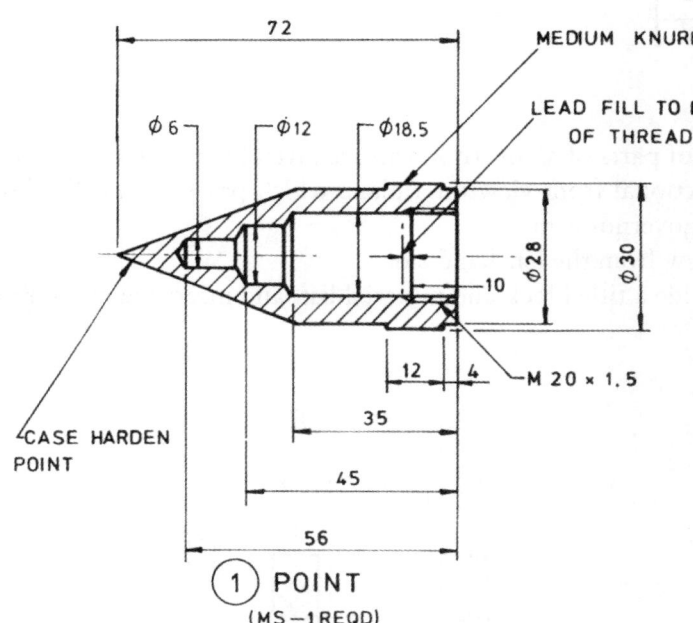

MEDIUM KNURL

LEAD FILL TO END OF THREAD

Ø 6    Ø 12    Ø 18.5

Ø 28    Ø 30

10

72

M 20 × 1.5

12    4

35

45

56

CASE HARDEN POINT

**① POINT**
(MS – 1 REQD)

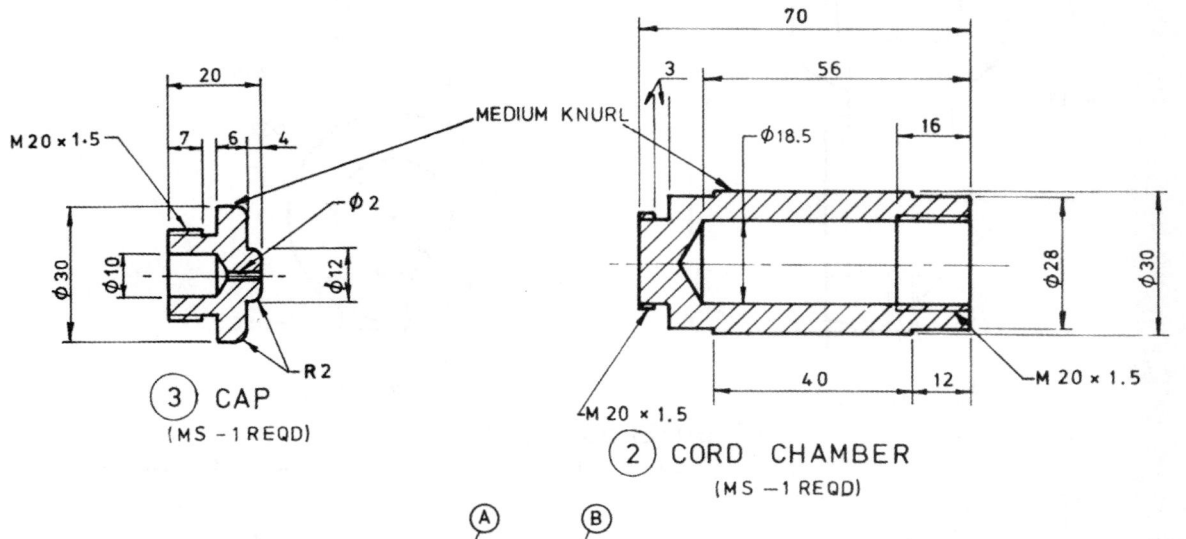

M 20 × 1.5

20

7    6    4

Ø 2

Ø 30    Ø 10    Ø 12

R 2

**③ CAP**
(MS – 1 REQD)

MEDIUM KNURL

70

3    56

16

Ø 18.5

Ø 28    Ø 30

M 20 × 1.5

40    12

M 20 × 1.5

**② CORD CHAMBER**
(MS – 1 REQD)

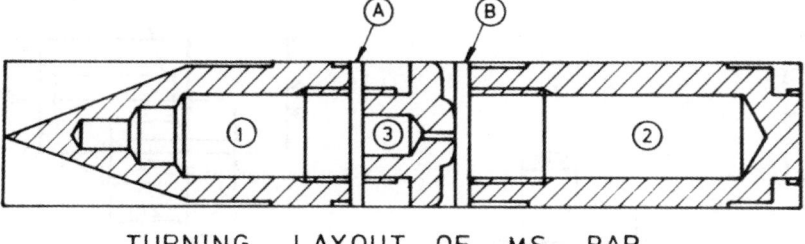

Ⓐ    Ⓑ

①    ③    ②

TURNING    LAYOUT    OF    MS    BAR

MATERIAL REQUIRED:

1 PIECE BRIGHT MILD STEEL 172 LONG × 30 DIA

PROCEDURE:

1. TURN END AND BODY OF ②

2. PART OFF AT Ⓐ

3. DRILL AND TAP ① AND SCREW ② INTO

4. TURN END OF ③ AND PART OFF AT Ⓑ

5. COMPLETE ②, SCREW ③ INTO AND FINISH

6. REVERSE JOB, TURN POINT AND KNURL

## 6.3 GOVERNOR ARM

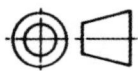

Details of parts of a governor arm are given below. Draw, full size, the following general assembly views:
(a) a sectional front view on a plane which passes vertically through the axis of the 32 mm diameter hole of the governor arm
(b) a view from the left-hand side
    Provide a title block and material list, and insert the main dimensions.

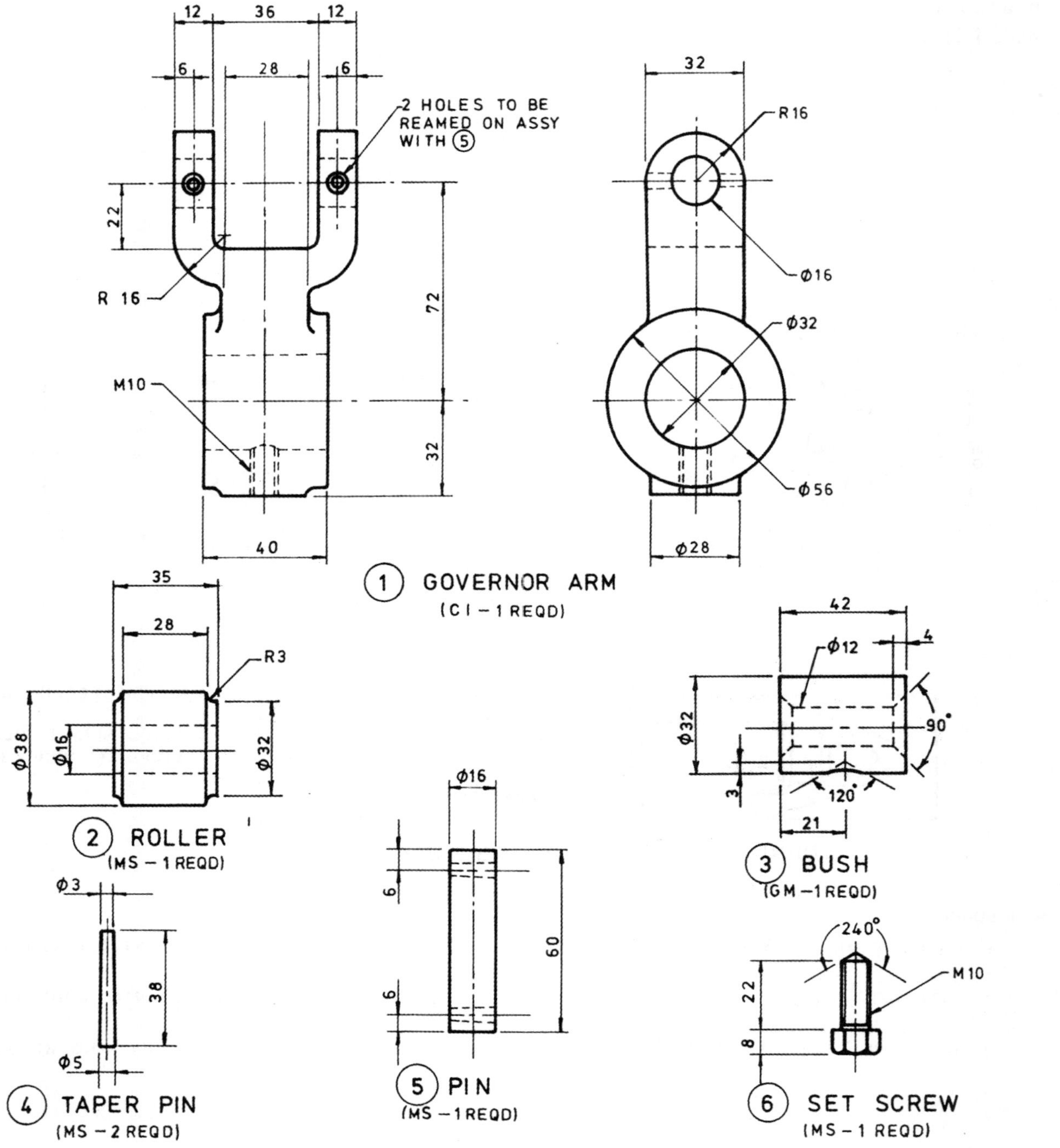

12  36  12

6  28  6

2 HOLES TO BE
REAMED ON ASSY
WITH ⑤

22

R 16

M10

72

32

40

32

R 16

Ø16

Ø32

Ø56

Ø28

① GOVERNOR ARM
(CI – 1 REQD)

35

28

R3

Ø38  Ø16  Ø32

② ROLLER
(MS – 1 REQD)

Ø16

6

60

6

⑤ PIN
(MS – 1 REQD)

42

Ø12  4

Ø32

90°

3  120°

21

③ BUSH
(GM – 1 REQD)

Ø3

38

Ø5

④ TAPER PIN
(MS – 2 REQD)

240°

22

8

M10

⑥ SET SCREW
(MS – 1 REQD)

## 6.4 MACHINIST'S JACK

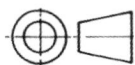

Details of components of a machinist's jack are given. Draw the necessary views to show how these components are assembled together. One sectioned view is sufficient.
(scale 1:1)

Use a standard size sheet, and provide a title block and material list.

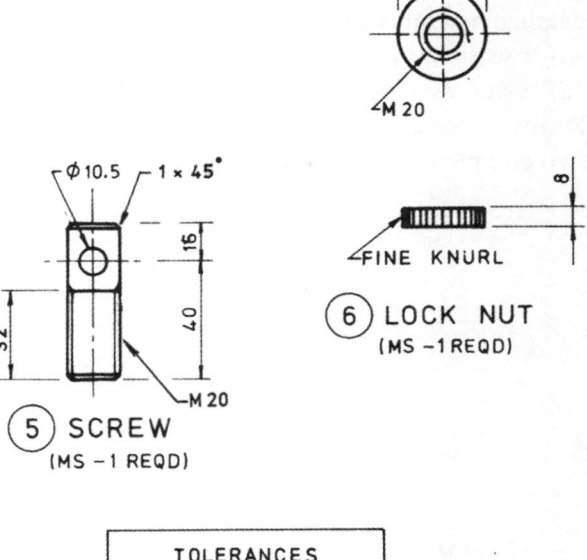

⑤ SCREW
(MS – 1 REQD)

⑥ LOCK NUT
(MS –1 REQD)

FINE KNURL

```
TOLERANCES
±0.5 EXCEPT
AS STATED
```

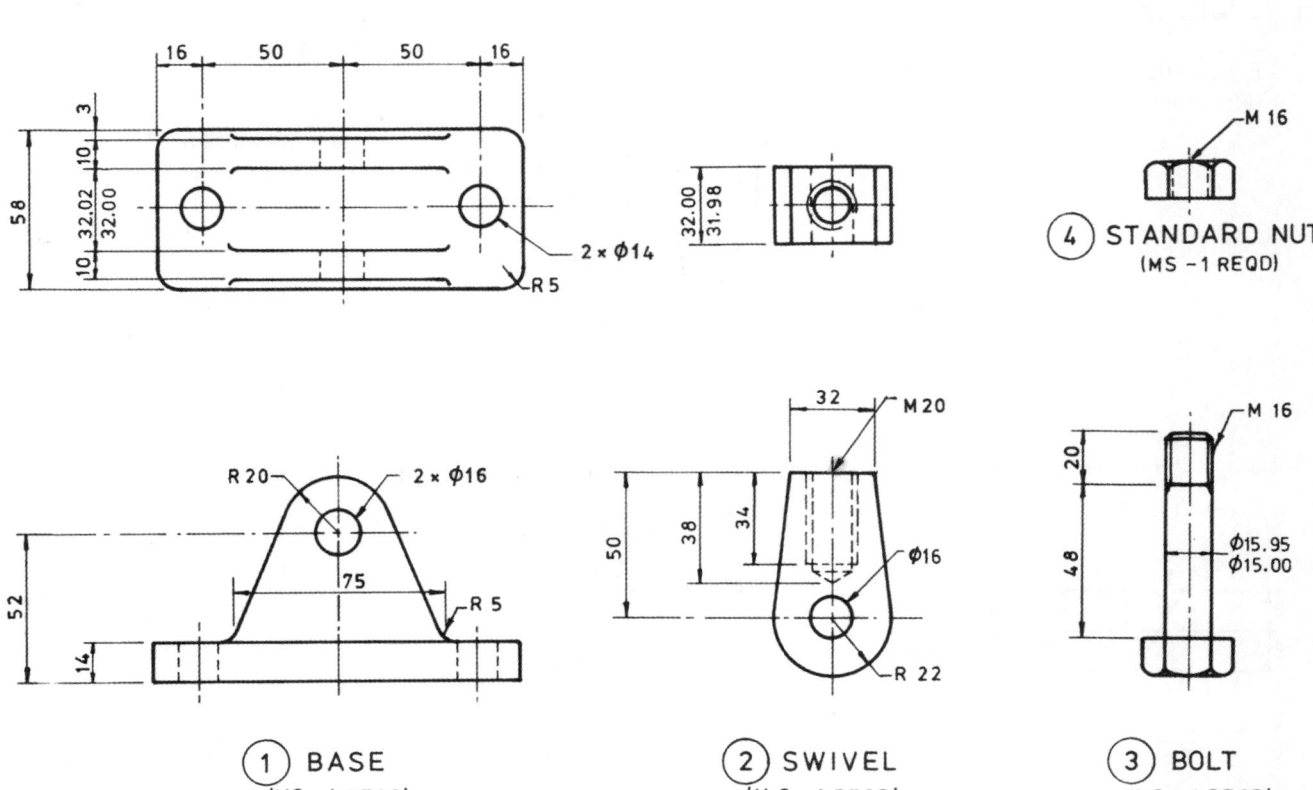

2 × Ø14

R 5

④ STANDARD NUT
(MS – 1 REQD)

M 16

① BASE
(MS –1 REQD)

② SWIVEL
(M S –1 REQD)

③ BOLT
(MS –1 REQD)

## 6.5 SPRING-LOADED SAFETY VALVE

Details of a spring-loaded safety valve are given.
(a) Draw, full size, in third-angle projection a
    detailed assembly drawing comprising:
    (i) a sectional front view
    (ii) a side view
(b) Draw, twice full size, a true projection of
    two complete turns of the spring.

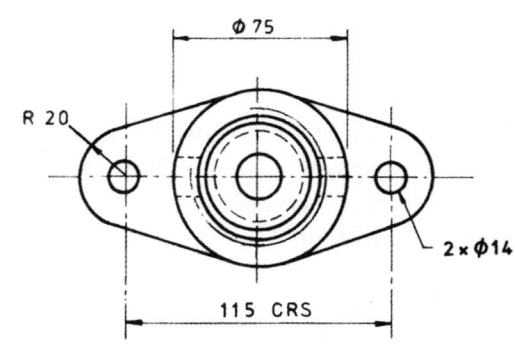

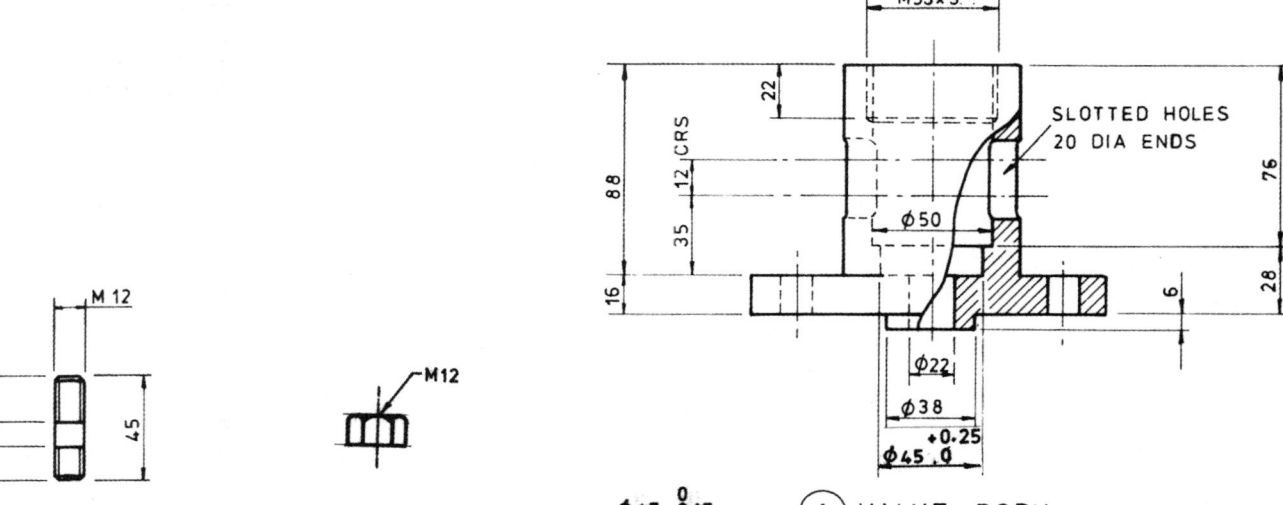

(1) VALVE BODY
(CI – 1 REQD)

```
TOLERANCES
±0.5 EXCEPT
AS STATED
```

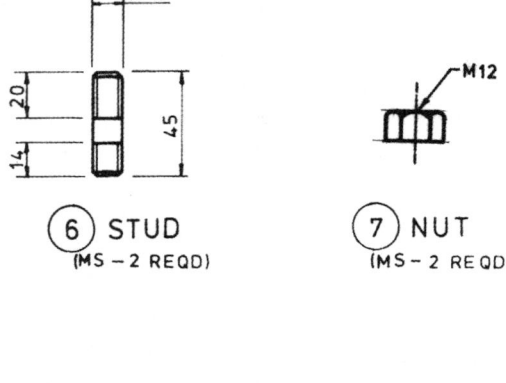

(6) STUD
(MS – 2 REQD)

(7) NUT
(MS – 2 REQD)

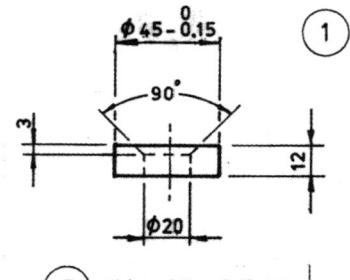

(5) VALVE SEAT
(BRS – 1 REQD)

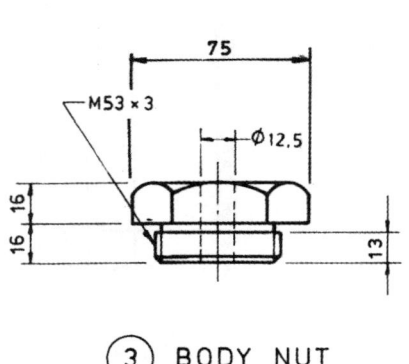

(3) BODY NUT
(BRS – 1 REQD)

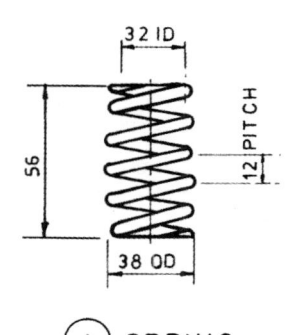

(4) SPRING
(SPR STL – 1 REQD)

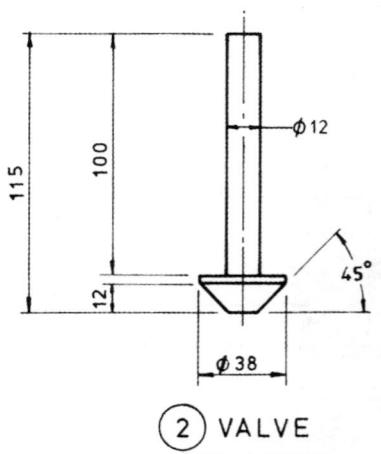

(2) VALVE
(MS – 1 REQD)

## 6.6 ROLLER BRACKET

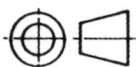

Details of a roller bracket are included below. Draw, full size, in third-angle projection, a general assembly drawing showing a sectional front view on the centre line of the base.

Provide also a standard title block and material list.

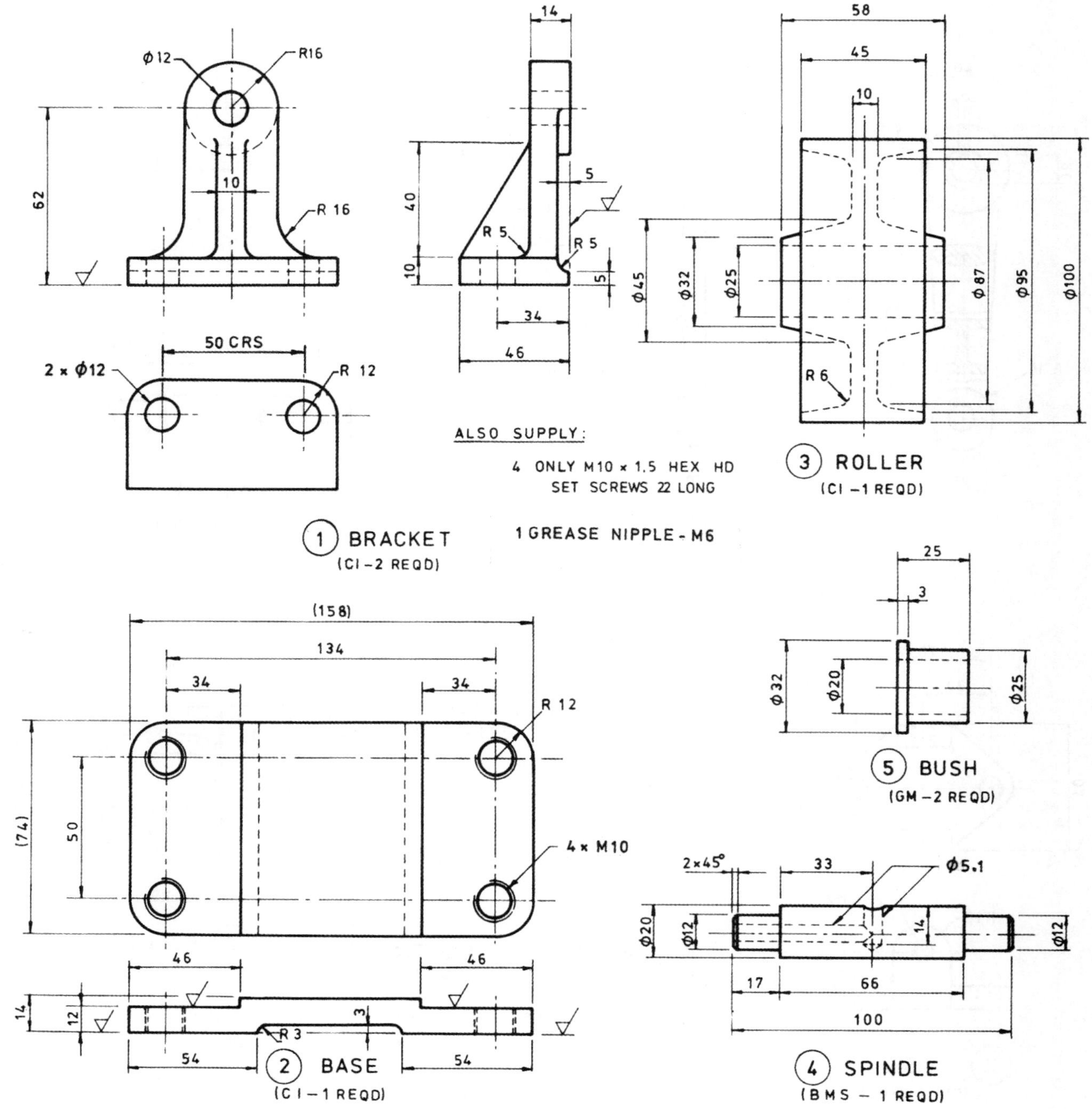

ϕ12
R16
62
10
R 16
50 CRS
2 × ϕ12
R 12
14
40
10
5
R 5
R 5
5
34
46
ALSO SUPPLY:
4 ONLY M10 × 1.5 HEX HD
SET SCREWS 22 LONG
1 GREASE NIPPLE - M6
1 BRACKET
(CI - 2 REQD)
58
45
10
ϕ45
ϕ32
ϕ25
ϕ87
ϕ95
ϕ100
R 6
3 ROLLER
(CI - 1 REQD)
25
3
ϕ32
ϕ20
ϕ25
5 BUSH
(GM - 2 REQD)
(158)
134
34
34
R 12
(74)
50
4 × M10
46
46
14
12
3
R 3
54
54
2 BASE
(CI - 1 REQD)
2×45°
33
ϕ5.1
ϕ20
ϕ12
14
ϕ12
17
66
100
4 SPINDLE
(BMS - 1 REQD)

## 6.7 BELT TENSIONER

(a) Draw the following general assembly view of the belt tensioner, component details of which are given below (scale 1:1):
    (i) a sectional front view taken on the centre line of the frame
       Use a standard drawing frame size.
       Complete the layout with a title block and material list, and record a revision assuming that the length of the M12 thread on the pin, item 6, was 10 mm
(b) Prepare a detail drawing of the 'shaft', item 4, allocating tolerances to dimensions where required (scale 1:1).
    Provide a title block, and record a revision assuming that the M16 thread was previously $^5/_8$ inch Whitworth.

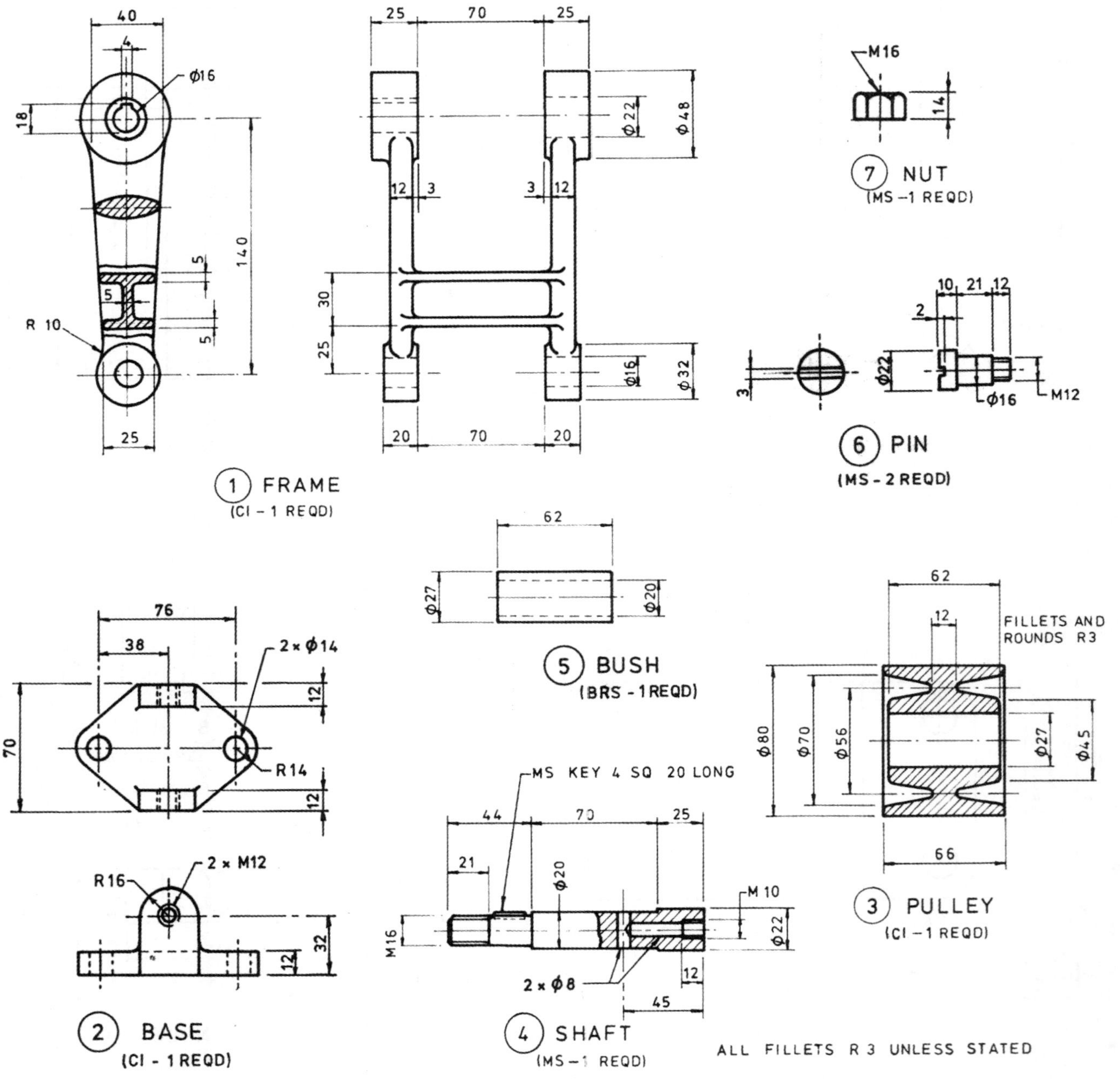

① FRAME
(CI – 1 REQD)

⑦ NUT
(MS – 1 REQD)

⑥ PIN
(MS – 2 REQD)

② BASE
(CI – 1 REQD)

⑤ BUSH
(BRS – 1 REQD)

③ PULLEY
(CI – 1 REQD)

④ SHAFT
(MS – 1 REQD)

FILLETS AND ROUNDS R3

ALL FILLETS R 3 UNLESS STATED

# 6.8 ADJUSTABLE BEARING

Details of an adjustable bearing are given below. Make a general assembly drawing of the unit comprising a half sectional front view and a side view.

Bolts, nuts, washers and screws may be omitted. A title block and material list are to be included, and the layout contained within a standard drawing frame.
(scale: full size)

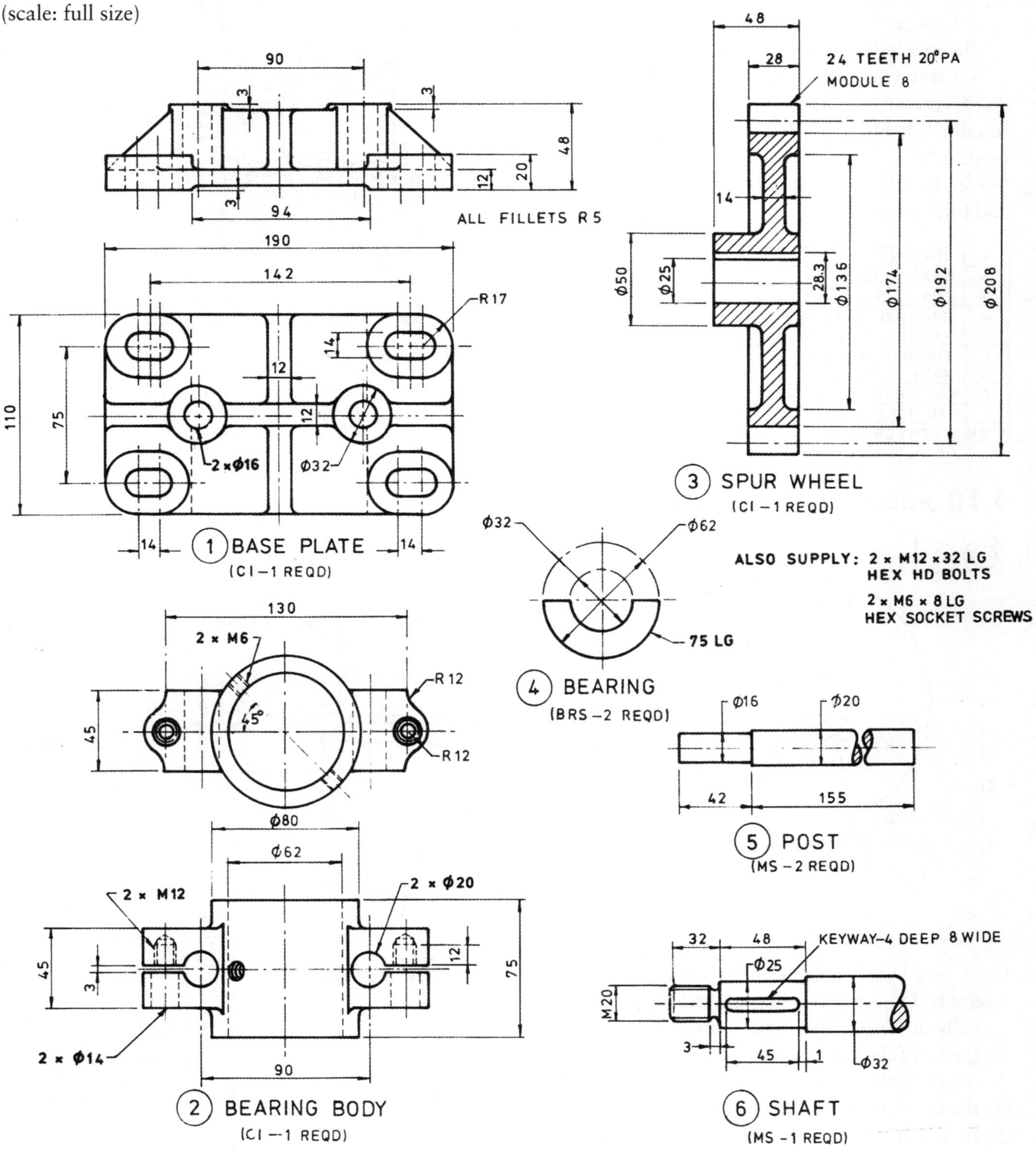

90

3          3

94

48

12      20

ALL FILLETS R 5

190

142

14

110      75

12

12

2 × Ø16      Ø32

R 17

14      1  BASE PLATE
14      (CI —1 REQD)

130

2 × M6

45      45°

R 12

R 12

Ø80

Ø62

2 × M12      2 × Ø20

45

3

12

75

2 × Ø14

90

2  BEARING BODY
(CI —1 REQD)

48

28      24 TEETH 20°PA
MODULE 8

14

Ø50  Ø25      28.3  Ø136  Ø174  Ø192  Ø208

3  SPUR WHEEL
(CI —1 REQD)

Ø32      Ø62

75 LG

4  BEARING
(BRS —2 REQD)

ALSO SUPPLY: 2 × M12 × 32 LG
HEX HD BOLTS

2 × M6 × 8 LG
HEX SOCKET SCREWS

Ø16      Ø20

42      155

5  POST
(MS —2 REQD)

32      48      KEYWAY-4 DEEP 8 WIDE

Ø25

M20

3      45      1      Ø32

6  SHAFT
(MS —1 REQD)

## 6.9 FAN ASSEMBLY

Make detailed drawings of the following components of the fan assembly:

(a) the fan pulley showing two views fully dimensioned with suitable tolerances for the bearing housings

(b) the fan shaft showing suitable tolerances for the steps on which the two ball bearings are housed

Set out your views on a standard size sheet, and provide a suitable title block.

(scale 1:1)

(Refer to Ch. 1 for standard fits and tolerances.)

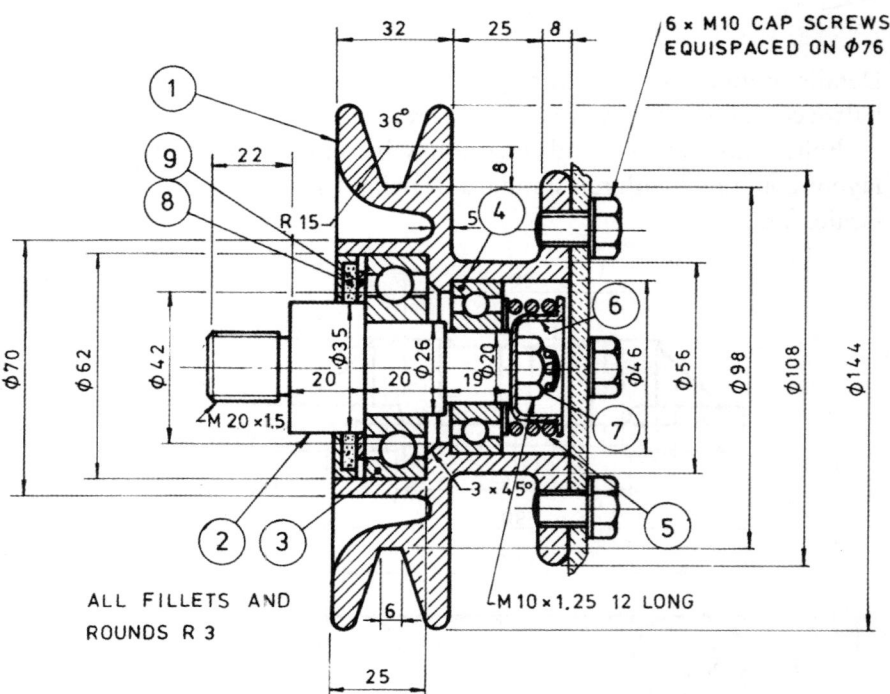

ALL FILLETS AND ROUNDS R 3

| 9 | SPACER |
| 8 | SHAFT SEAL |
| 7 | CUP LOCKING NUT |
| 6 | FRICTION DRIVE CUP |
| 5 | FRICTION DRIVE SPRING |
| 4 | SUPPORT BEARING |
| 3 | MAIN BEARING |
| 2 | FAN SHAFT |
| 1 | FAN PULLEY |
| ITEM | DESCRIPTION |

## 6.10 PULLEY AND SHAFT ASSEMBLY

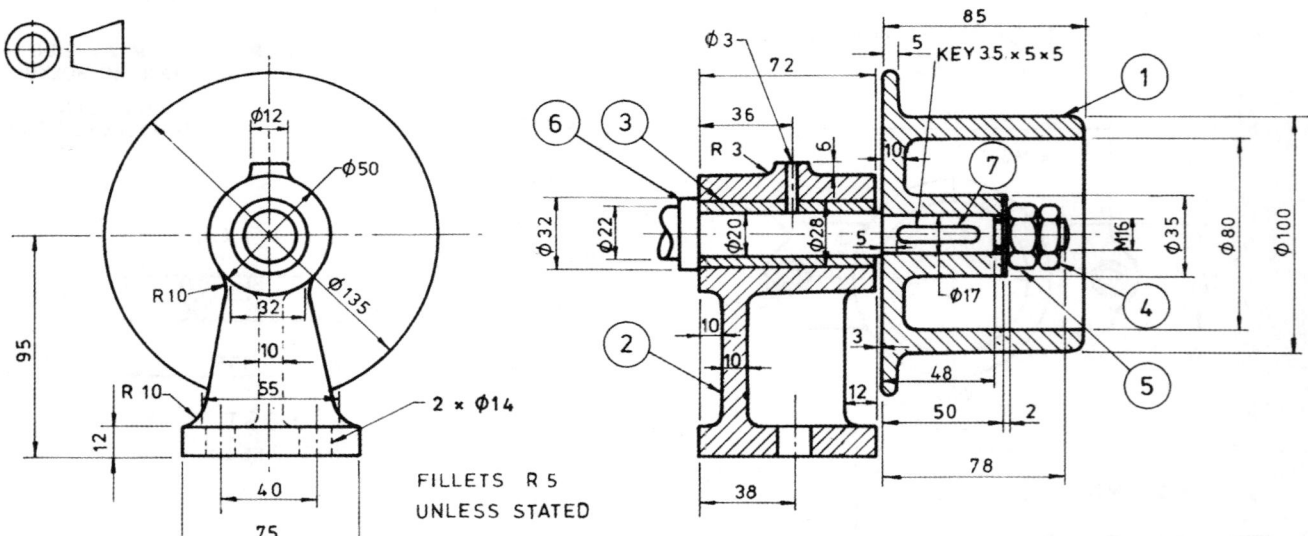

FILLETS R 5
UNLESS STATED

Draw full-sized detailed drawings of the components of the pulley and shaft assembly. Omit the key, nuts and washer.

Fully dimension all parts, and show suitable tolerances for:

(a) the fit of the bush in the housing

(b) the fit of the shaft in the bush

(c) the fit of the shaft in the pulley

(Refer to Ch. 1 for standard fits and tolerances.)

| 7 | KEY |
| 6 | SHAFT |
| 5 | LOCKNUT |
| 4 | NUT |
| 3 | BUSH |
| 2 | BRACKET |
| 1 | PULLEY |
| ITEM | DESCRIPTION |

# 6.11 BELT DRIVE ASSEMBLY

Make detailed drawings of the following components of the belt drive assembly: housing, shaft, pulley and gear.
(scale 1:2)

There are four classes of fit which need to be allowed for when detailing the above components. Enter these in the table shown below and insert the tolerances on the appropriate dimensions on your drawings.

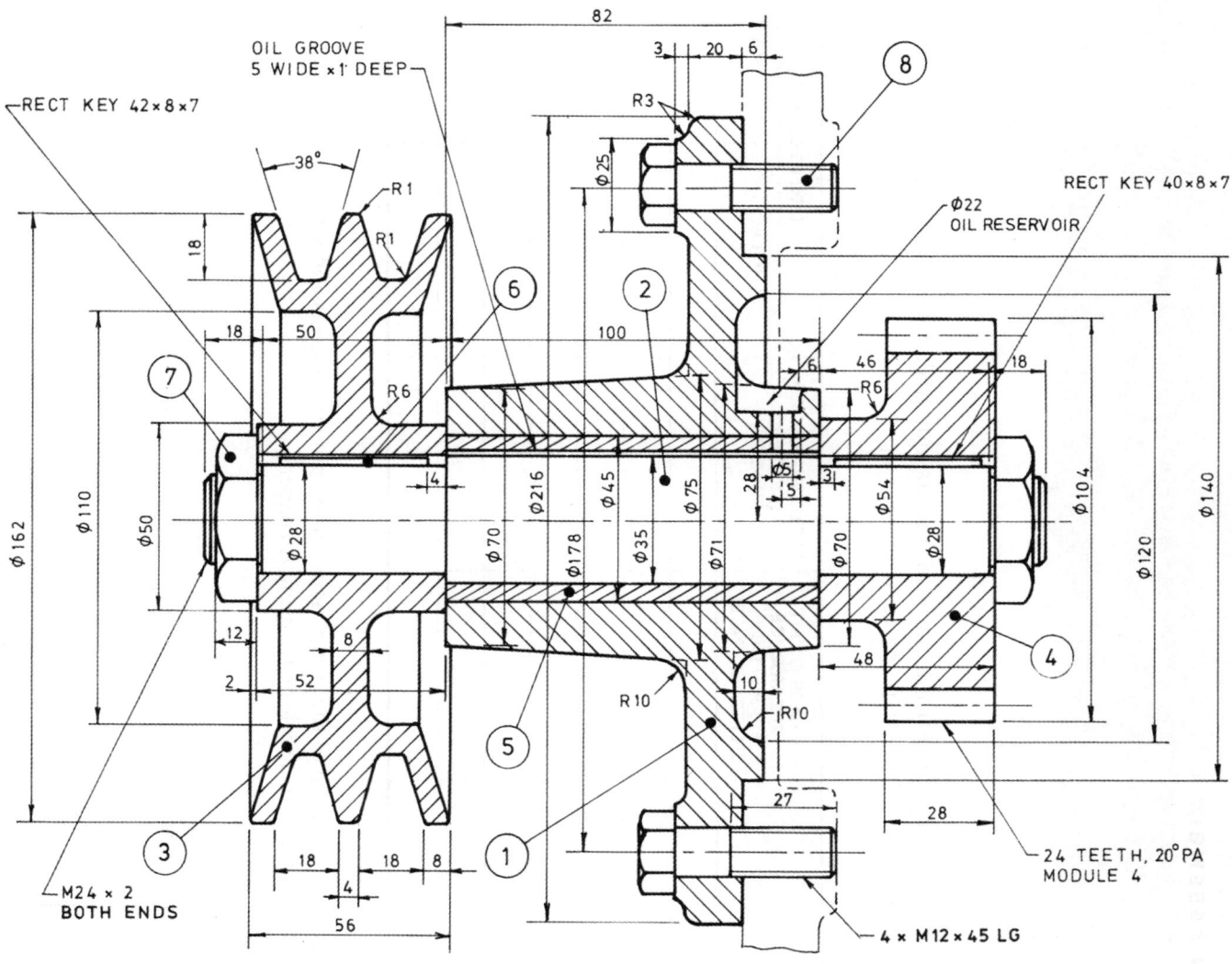

| FIT BETWEEN | CLASS OF FIT | DESIGNATION | LIMITS OF SIZE | |
|---|---|---|---|---|
| | | | HOLE | SHAFT |
| HOUSING — BUSH | | | | |
| BUSH — SHAFT | | | | |
| PULLEY & GEAR — SHAFT | | | | |
| SHAFT LENGTH — HOUSING | | | | |

| | |
|---|---|
| 8 | BOLT |
| 7 | NUT |
| 6 | KEY |
| 5 | BEARING BUSH |
| 4 | GEAR |
| 3 | PULLEY |
| 2 | SHAFT |
| 1 | HOUSING |
| ITEM | DESCRIPTION |

# 6.12 POWER TRANSMISSION ASSEMBLY

(a) Complete the following table of linear tolerances for the fits indicated. Transfer these toleranced dimensions to the appropriate component drawing prepared in answer to part (b).

| FIT BETWEEN | CLASS OF FIT (CLEARANCE, TRANSITION, INTERFERENCE) | FIT DESIGNATION (LETTER GRADES) | LIMITS OF SIZE | |
|---|---|---|---|---|
| | | | HOLE | SHAFT |
| BRACKET—BUSH Ø30 | | | | |
| BUSH—SHAFT Ø24 | | | | |
| BRACKET—SHAFT Ø24 | | | | |
| PULLEY AND GEAR—SHAFT Ø18 | | | | |
| SHAFT LENGTH—BRACKET 140 | | | | |
| GLAND—SHAFT Ø24 | | | | |
| GLAND—BRACKET Ø36 | | | | |

(b) Make detailed drawings of each of the components of the power transmission assembly shown below to a scale of 1:1 using A2 paper or to a scale of 1:2 using A3 paper.

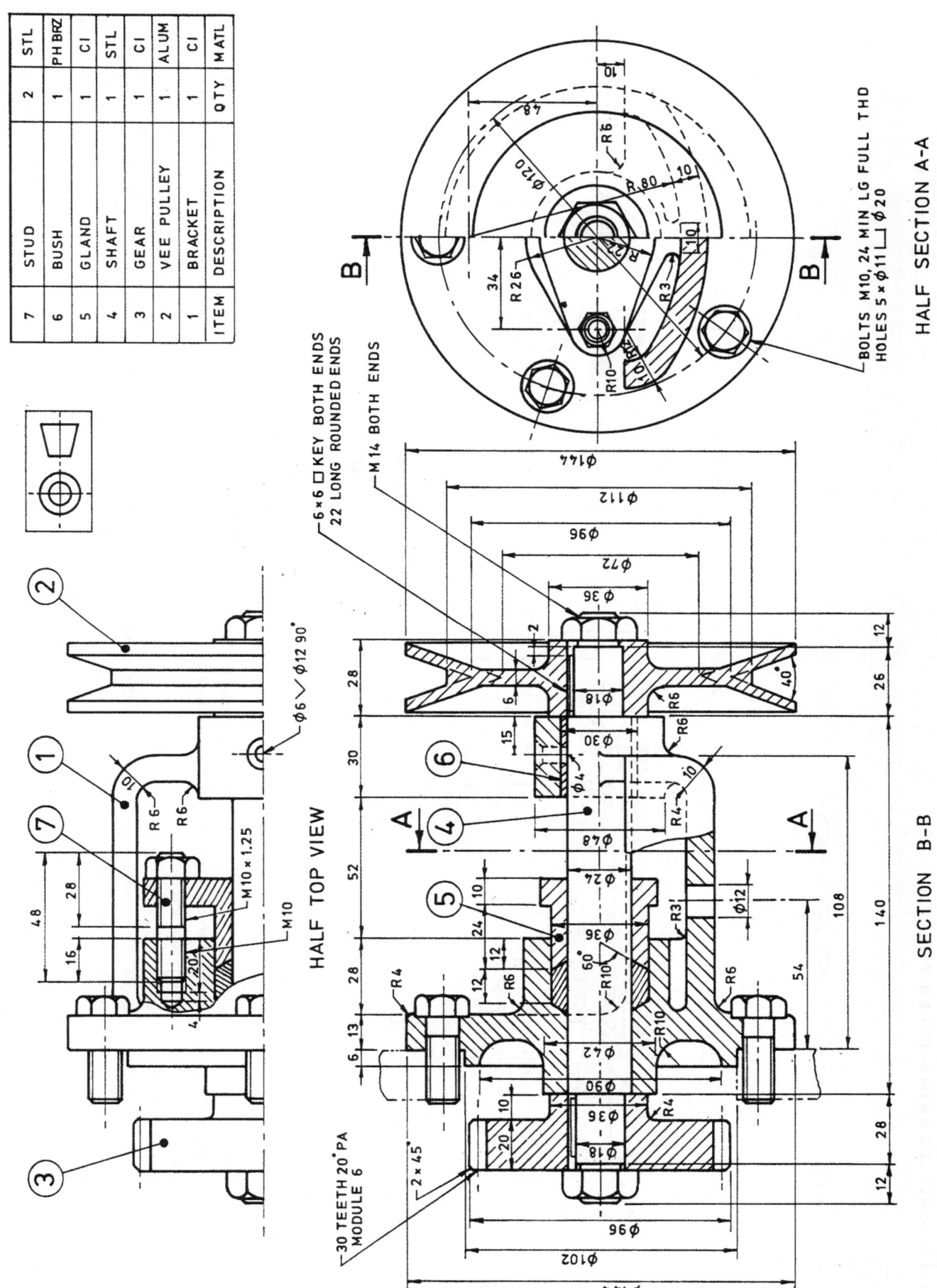

| 7 | STUD | 2 | STL |
|---|---|---|---|
| 6 | BUSH | 1 | PH BRZ |
| 5 | GLAND | 1 | CI |
| 4 | SHAFT | 1 | STL |
| 3 | GEAR | 1 | CI |
| 2 | VEE PULLEY | 1 | ALUM |
| 1 | BRACKET | 1 | CI |
| ITEM | DESCRIPTION | QTY | MATL |

HALF SECTION A-A

BOLTS M10, 24 MIN LG FULL THD
HOLES 5 × ⌀11 ⌴ ⌀20

6 × 6 □ KEY BOTH ENDS
22 LONG ROUNDED ENDS

M14 BOTH ENDS

HALF TOP VIEW

SECTION B-B

30 TEETH 20° PA
MODULE 6

## 6.13 BALL-BEARING IDLER PULLEY

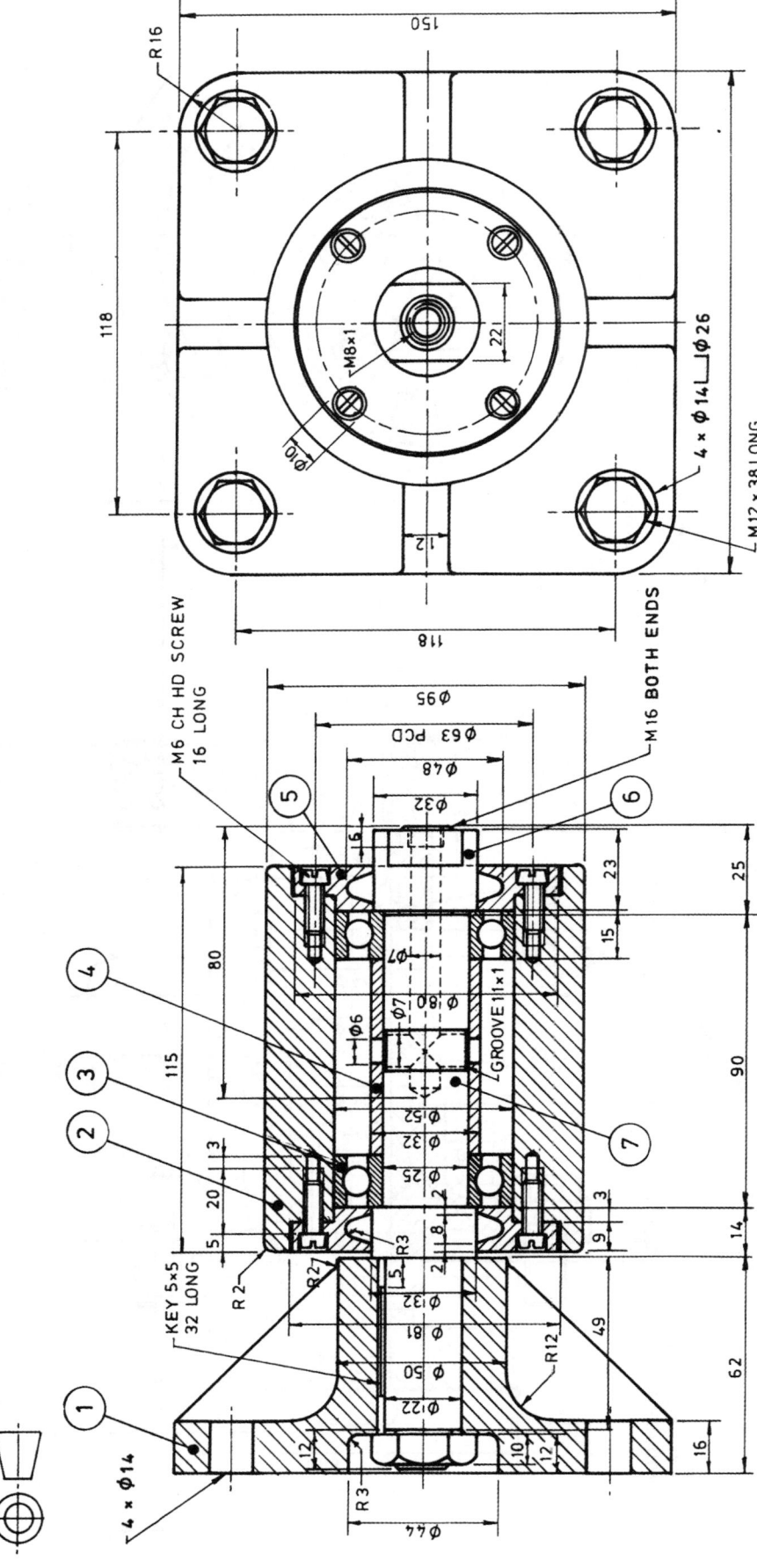

| 7 | SHAFT | 1 |
| 6 | SPECIAL NUT | 1 |
| 5 | FLANGE | 2 |
| 4 | SLEEVE | 1 |
| 3 | BALL BEARING | 2 |
| 2 | PULLEY | 1 |
| 1 | BRACKET | 1 |

Make detailed drawings of the various components of the pulley assembly, ensuring that the appropriate dimensions are toleranced to provide a suitable type of fit for the following:

(a) the bearings and sleeve on the shaft diameter

(b) the bearings in the pulley housing

(c) the accumulation of tolerances of the two bearings, sleeve and the inside shoulders of the flanges, all measured along the shaft, to ensure that allowance is made to enable the two flanges to seat in their respective housings

(d) the shaft in the bracket for both the diameter and lengthwise fit

(e) the shaft in the pulley housing to ensure that the special nut bears against the inner race of the ball bearing and not the shoulder of the shaft (scale: full or half size as appropriate for the sheet size selected)

# 6.14 REFRIGERATOR VALVE

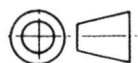

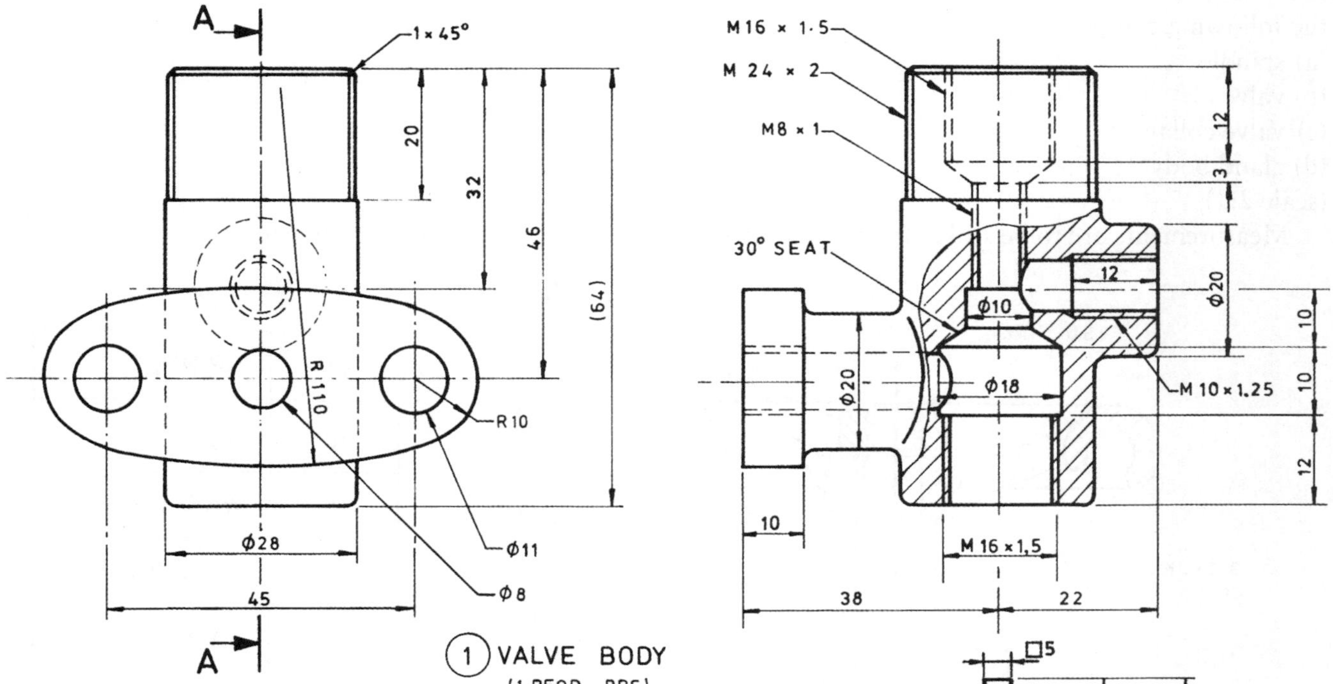

A

1 × 45°

20

32

46

(64)

R 110

R 10

Ø28

45

Ø11

Ø8

A

**1 VALVE BODY**
(1 REQD - BRS)

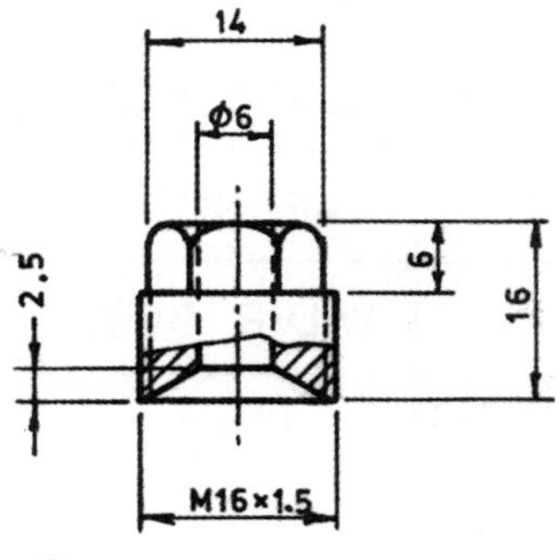

M 16 × 1·5

M 24 × 2

M8 × 1

30° SEAT

Ø10

Ø20

Ø18

Ø20

12

3

Ø20

10

10

12

M 10 × 1.25

10

M 16 × 1.5

38

22

14

Ø6

2.5

6

16

M16×1.5

**4 GLAND NUT**
(1 REQD - BRS)

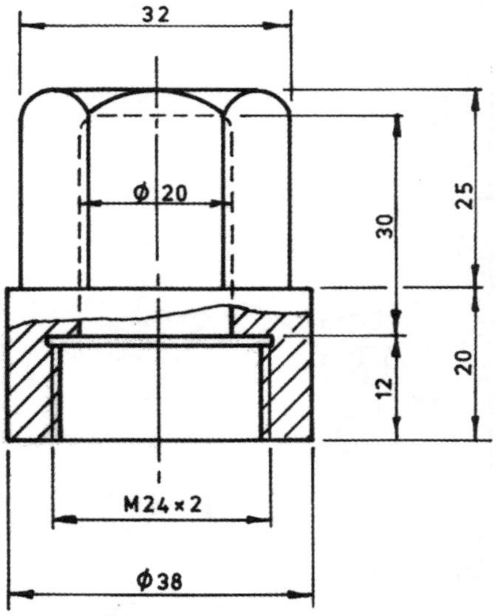

32

Ø 20

30

25

20

12

M24×2

Ø38

**5 CAP**
(1 REQD - BRS)

□5

6

Ø6

38

74

M8×1

12

Ø6

10

Ø12

30°

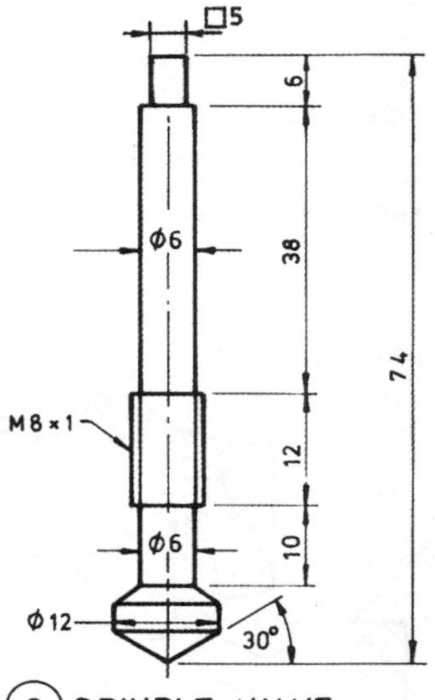

**2 SPINDLE VALVE**
(1 REQD - SS )

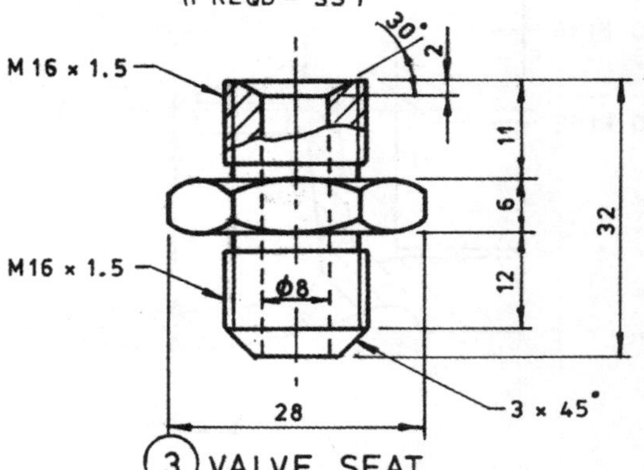

M 16 × 1.5

30°

2

11

6

32

M 16 × 1.5

Ø8

12

28

3 × 45°

**3 VALVE SEAT**
(1 REQD - BRS)

Details of components of a refrigerator valve are given. Draw the following assembled views in third-angle projection:

(a) a sectional front view on A-A showing the valve closed onto the valve seat, item 3
(b) a side view looking at the flange
(scale 2:1)

Provide a title block and a parts list.

# 6.15 GLOBE VALVE

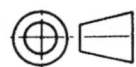

A sectional assembly drawing of a globe valve is given below. Prepare fully dimensioned detailed drawings of the following components:
(a) spindle
(b) valve
(c) valve collar
(d) gland body
(scale 2:1)

    Measurements are to be scaled from the drawing using the scale at the foot of the page.

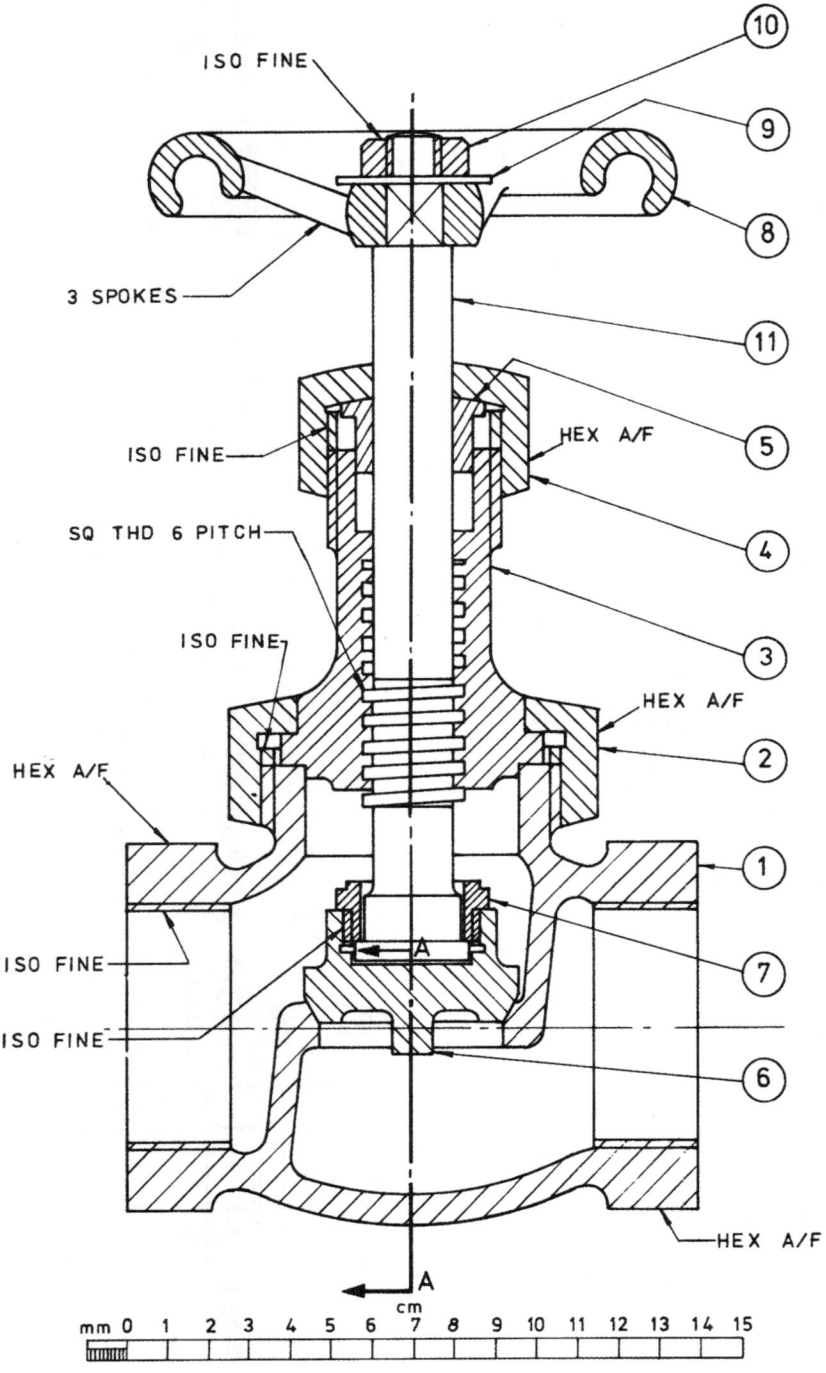

| ITEM NO | ITEM |
|---------|------|
| 11 | SPINDLE |
| 10 | NUT |
| 9 | WASHER |
| 8 | HANDWHEEL |
| 7 | VALVE COLLAR |
| 6 | VALVE (circ.) |
| 5 | GLAND BUSH |
| 4 | GLAND NUT |
| 3 | GLAND BODY |
| 2 | CAP NUT |
| 1 | VALVE BODY |

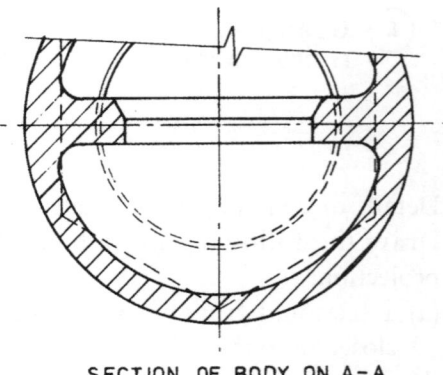

SECTION OF BODY ON A-A

## 6.16 SWIVEL TOOL HOLDER

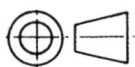

Draw full-size detailed drawings in third-angle projection of the four components of the swivel tool holder assembly. Set out each drawing within a standard frame size, and provide a title block.

The fit of the various parts is designated by the hole-basis system (refer to Table 1.24(a), pp. 56–57). When detailing the tolerances for the fit of the bearing pin in the holder and the base, convert to a shaft-basis system of tolerancing (refer to Table 1.24(b), pp. 58–59) to achieve a single diameter size of the pin.

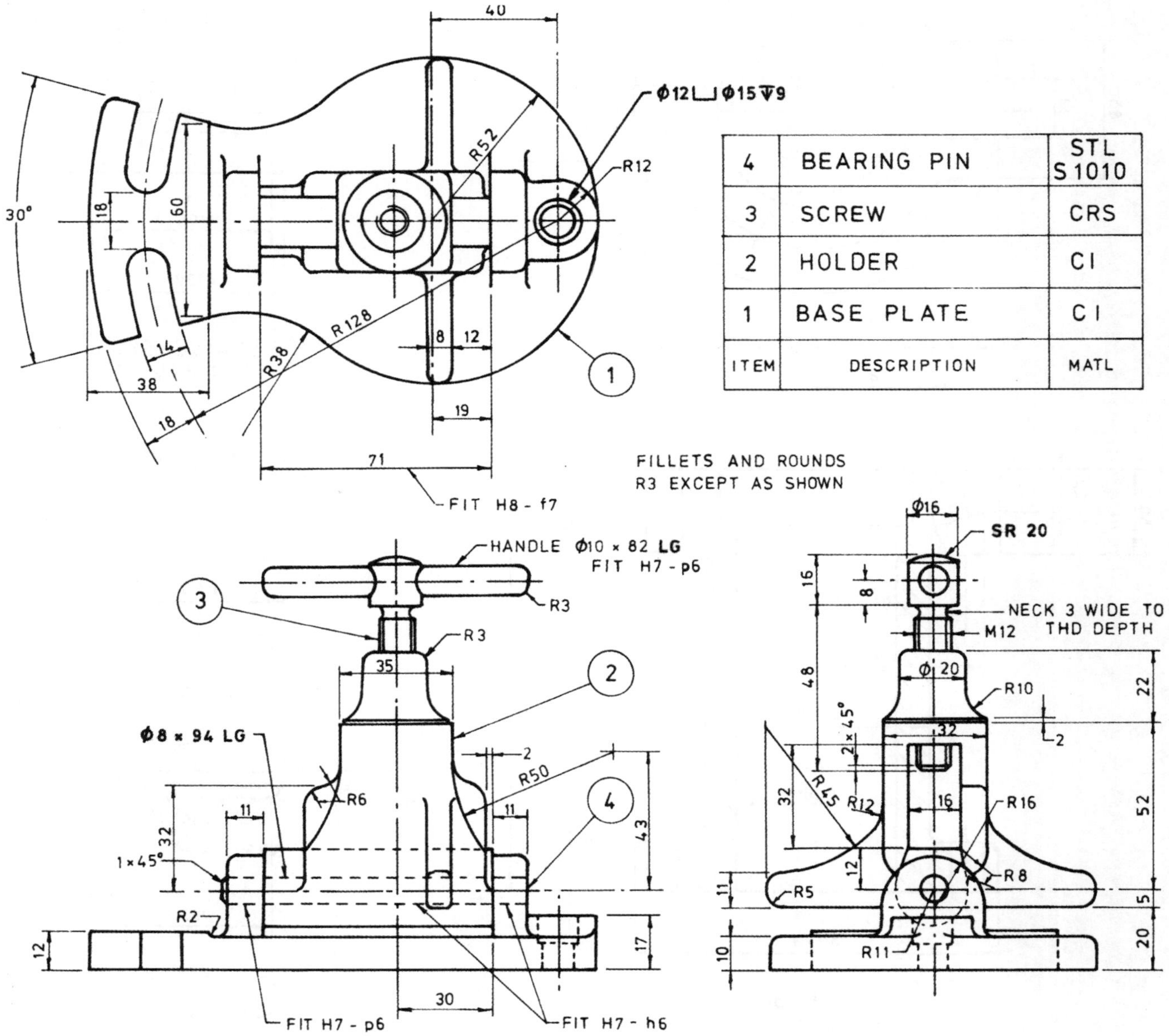

Ø12 ⌴ Ø15 ↧9
R12
R52
R128
R38
30°
18
60
14
38
18
8
12
19
40
71
FIT H8 - f7
FILLETS AND ROUNDS
R3 EXCEPT AS SHOWN
HANDLE Ø10 × 82 LG
FIT H7 - p6
R3
35
Ø8 × 94 LG
R6
R50
2
11
1×45°
32
R2
12
43
17
30
FIT H7 - p6
FIT H7 - h6
Ø16
SR 20
16
8
NECK 3 WIDE TO
THD DEPTH
M12
48
Ø 20
R10
22
2×45°
32
2
R45
R12
16
R16
52
R8
R5
5
11
10
R11
20


| ITEM | DESCRIPTION | MATL |
|---|---|---|
| 4 | BEARING PIN | STL S 1010 |
| 3 | SCREW | CRS |
| 2 | HOLDER | CI |
| 1 | BASE PLATE | CI |

# 6.17 LEVER TYPE SAFETY VALVE

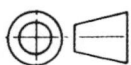

The following are details of a lever-type safety valve. Make a general assembly drawing using a standard size sheet comprising a front and top view. The front view should be sectionalised to illustrate detail more clearly. Scrap sections may be used. Show four main dimensions. Calculate the 'blow-off' pressure for the position of the CI weight shown on the lever.
(scale 1:2)

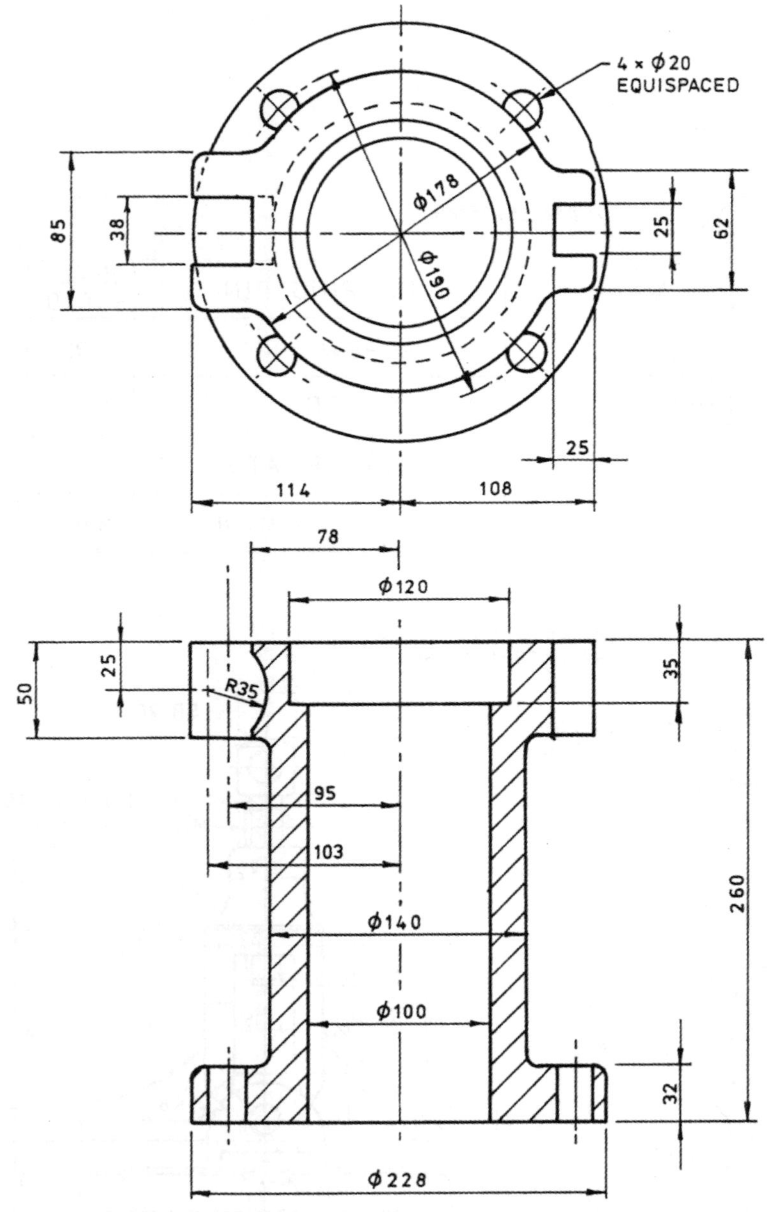

4 × Ø20
EQUISPACED

Ø178
Ø190

85
38
25
62
114
108
25

① VALVE BODY
(1 REQD-CI)

78
Ø120
50
25
R35
35
95
103
Ø140
Ø100
260
32
Ø228

Ø100
45°
5
12
38
Ø105
Ø120

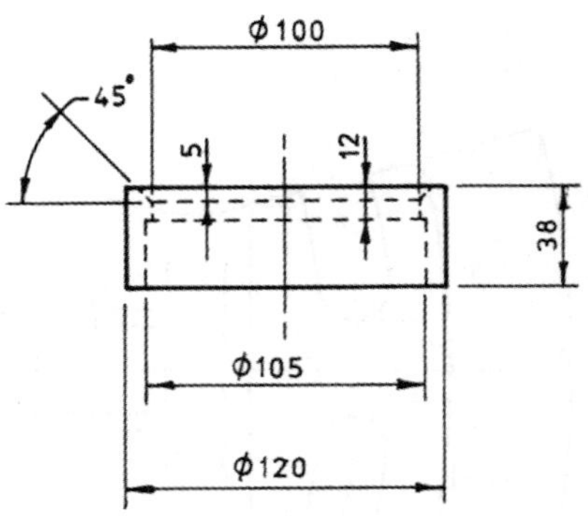

⑥ VALVE SEAT
(1 REQD- BRS)

FILLETS AND ROUNDS
R10 UNLESS STATED

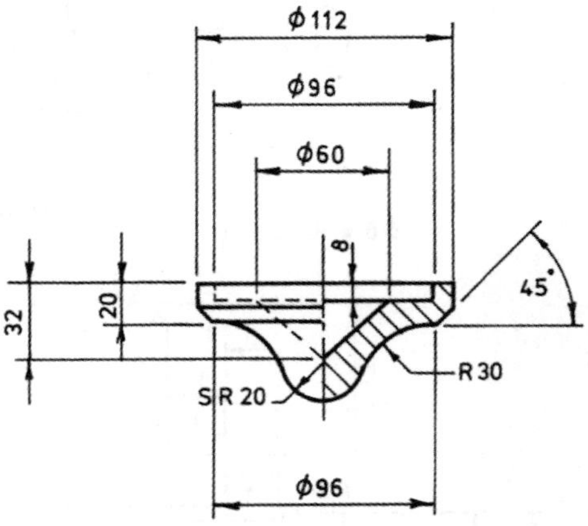

Ø112
Ø96
Ø60
32
20
8
45°
S R 20
R30
Ø96

② VALVE
(1 REQD-BRS)

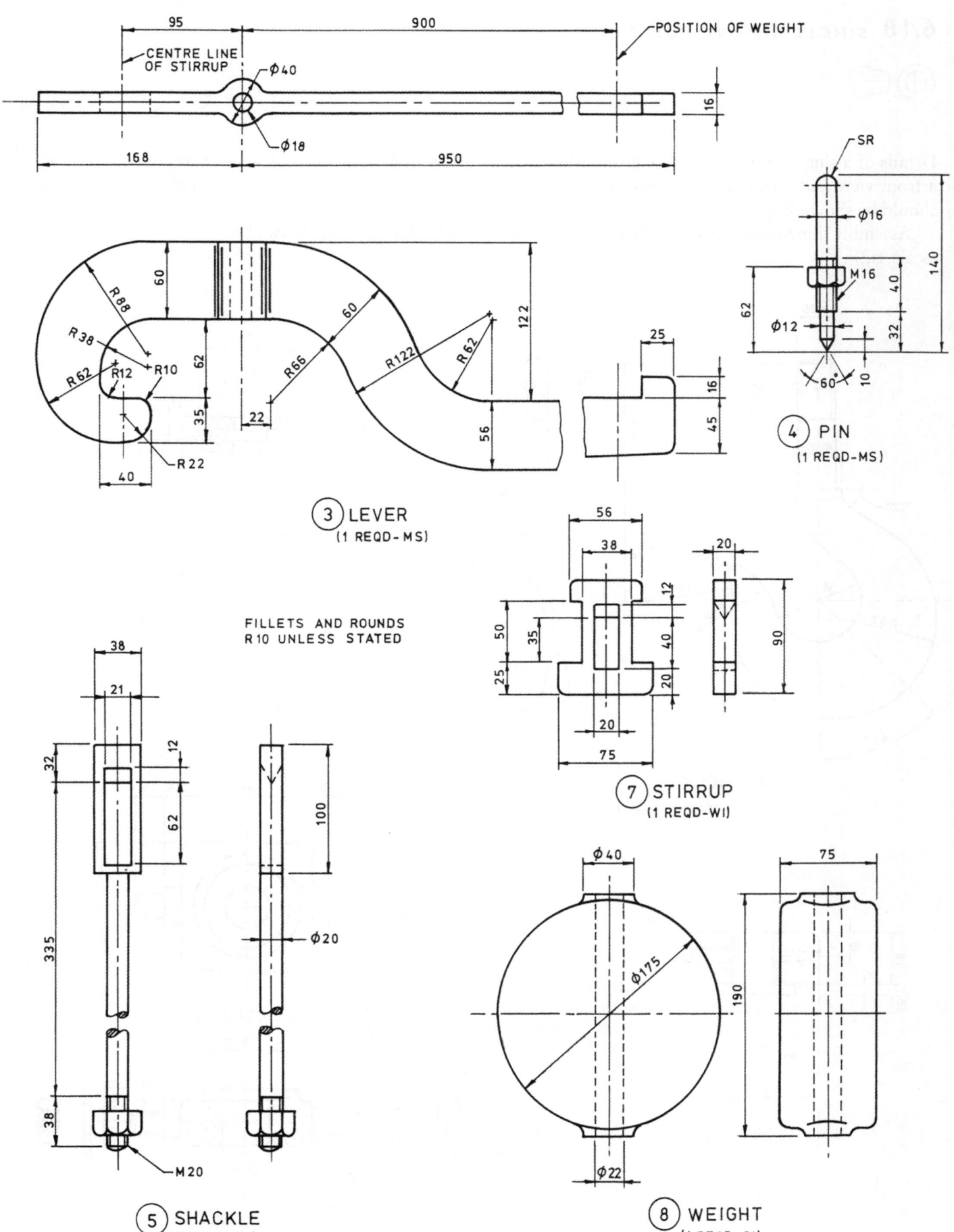

95
900
POSITION OF WEIGHT
CENTRE LINE OF STIRRUP
Ø40
16
168
Ø18
950

R88
60
R38
R62
R10
R62
R12
62
35
R22
40
60
R66
22
R122
R62
122
56
25
16
45

③ LEVER
(1 REQD- MS)

SR
Ø16
140
M16
62
Ø12
40
32
60°
10

④ PIN
(1 REQD-MS)

FILLETS AND ROUNDS
R10 UNLESS STATED

56
38
20
50
35
12
40
20
25
20
90
75

⑦ STIRRUP
(1 REQD-WI)

38
21
32
12
62
100
335
Ø20
38
M20

⑤ SHACKLE
(1REQD-WI)

Ø40
75
Ø175
190
Ø22

⑧ WEIGHT
(1 REQD-CI)

# 6.18 SINGLE SHEAVE PULLEY

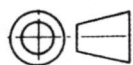

Details of a single sheave pulley are given. Make a general assembly drawing on standard size paper, showing a front view and a sectional side view. Parts not detailed but noted, for example grease nipple and taper pin, should be shown in position.

Assembly dimensions are to be shown, together with a title block and a material list.
(scale 1:2)

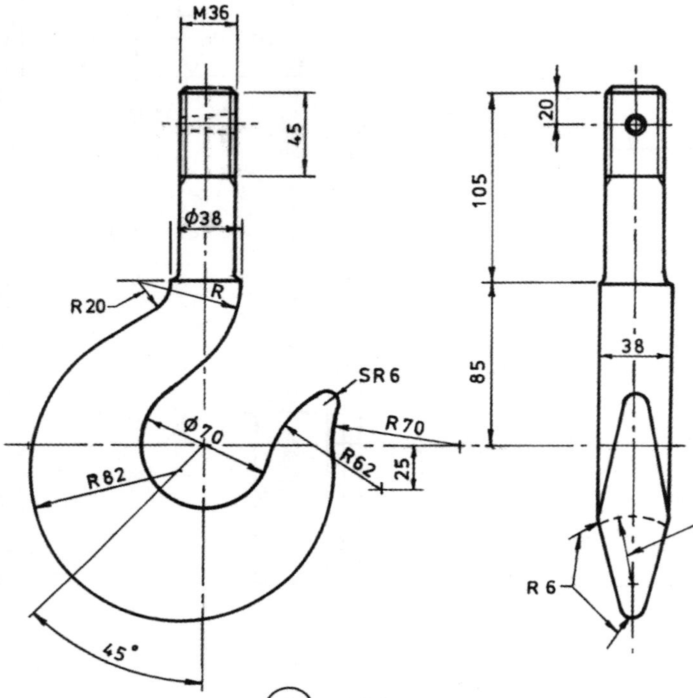

**1 HOOK**
(1 REQD - FORGED STEEL)

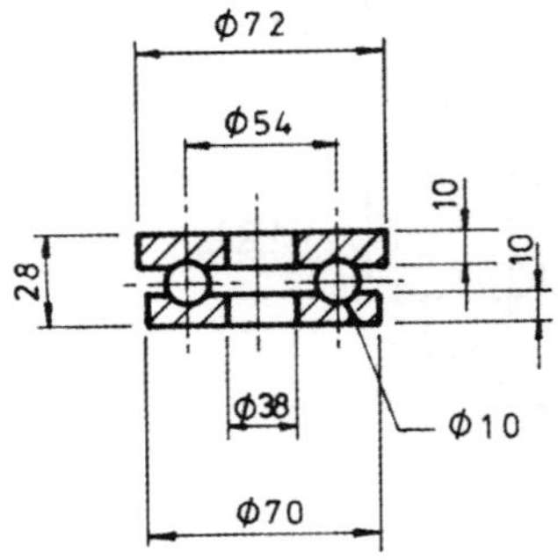

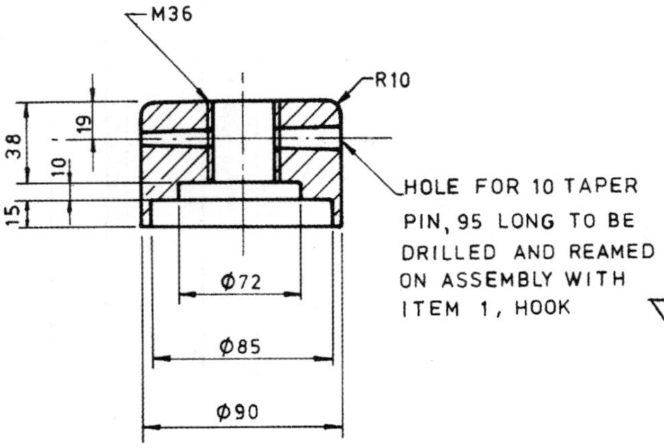

**2 NUT**
(1 REQD - MS)

HOLE FOR 10 TAPER PIN, 95 LONG TO BE DRILLED AND REAMED ON ASSEMBLY WITH ITEM 1, HOOK

**3 THRUST BEARING**
(1 REQD)

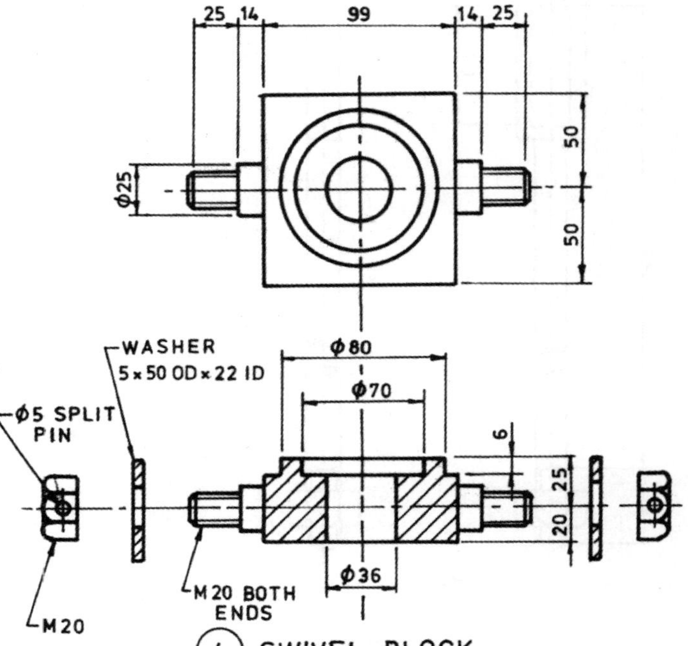

**4 SWIVEL BLOCK**
(1 REQD - MS)

WASHER 5 × 50 OD × 22 ID

Ø5 SPLIT PIN

M20

M20 BOTH ENDS

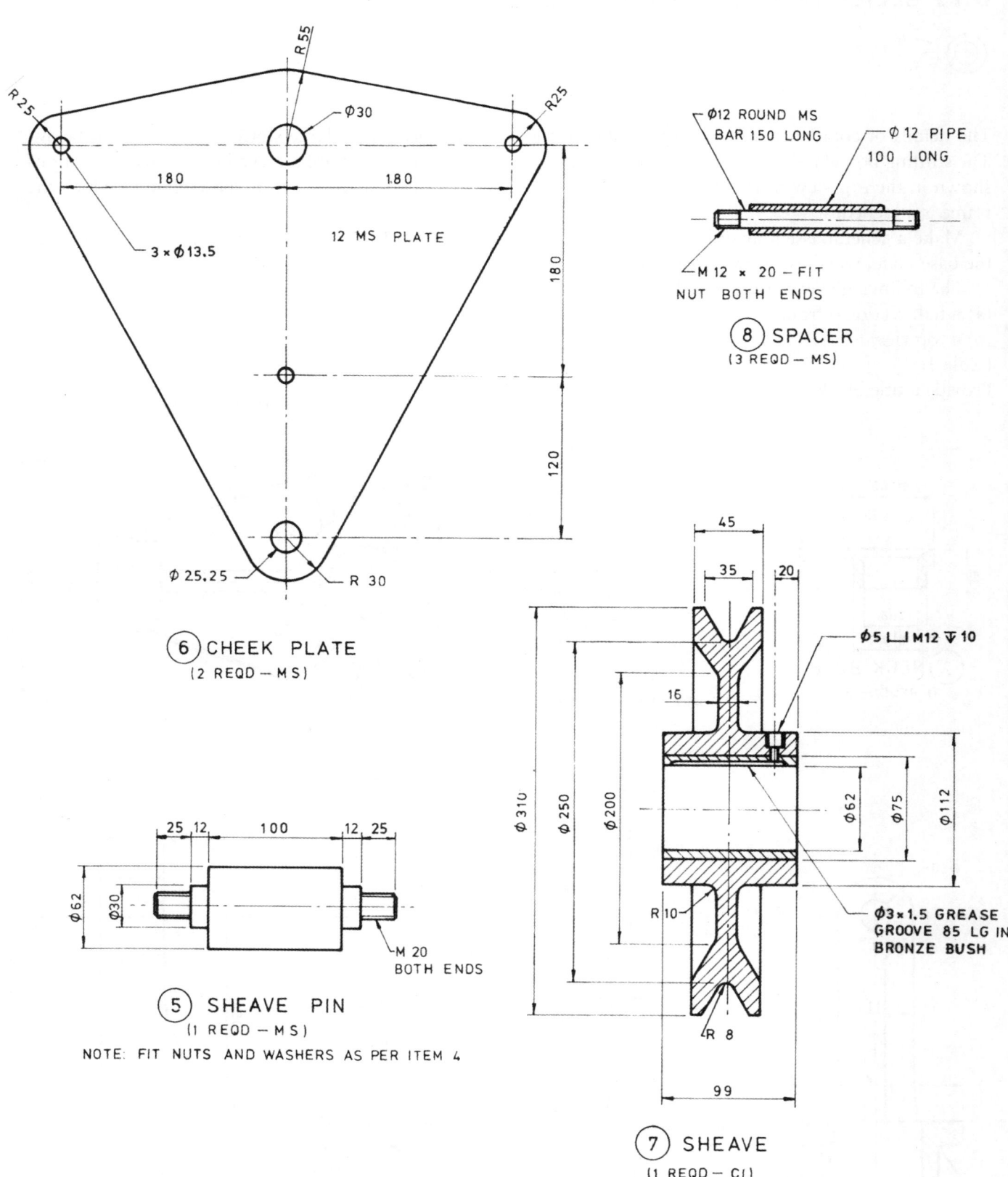

R 55

R 25

Ø30

R 25

180        180

12 MS PLATE

180

3 × Ø13.5

120

Ø 25.25        R 30

⑥ CHEEK PLATE
(2 REQD — MS)

Ø12 ROUND MS
BAR 150 LONG

Ø 12 PIPE
100 LONG

M 12 × 20 — FIT
NUT BOTH ENDS

⑧ SPACER
(3 REQD — MS)

45

35        20

16

Ø5 ⌴ M12 ⊽ 10

25  12        100        12  25

Ø 62

Ø 30

M 20
BOTH ENDS

⑤ SHEAVE PIN
(1 REQD — MS)

NOTE: FIT NUTS AND WASHERS AS PER ITEM 4

Ø 310

Ø 250

Ø 200

Ø 62

Ø 75

Ø 112

Ø3 × 1.5 GREASE
GROOVE 85 LG IN
BRONZE BUSH

R 10

R 8

99

⑦ SHEAVE
(1 REQD — CI)

# 6.19 SLUICE-VALVE OPERATING GEAR

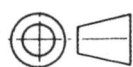

The details of components required to make up the operating gear for a large sluice valve are given below. The stuffing box gland is held in position by means of two eyebolts which are attached to the two projections shown in the top view of the frame. The collar on the spindle sits in a recess on the base plate to which the frame is attached.

Make a general assembly drawing on a standard drawing sheet of the operating gear, showing it bolted to the base plate, which is 75 mm thick.

The following views are required:
(a) a half sectional front view
(b) a top view
(scale 1:5)
Provide a title block and material list.

Ø140

30°

56

38

Ø88

Ø108

② NECK BUSH
(1 REQD – GM)

4 – 20 WEBS AT 30°
TO VERT CL

25

114

356

6 × Ø11 EQUISPACED ON Ø280

Ø 35

(800)

95          610          95

Ø50

FILLETS AND
ROUNDS R20

114

254

35

44

25

R 208

Ø198

Ø190

Ø140

305

R 205

R 225

480

38

Ø108

① FRAME
(1 REQD – C I )

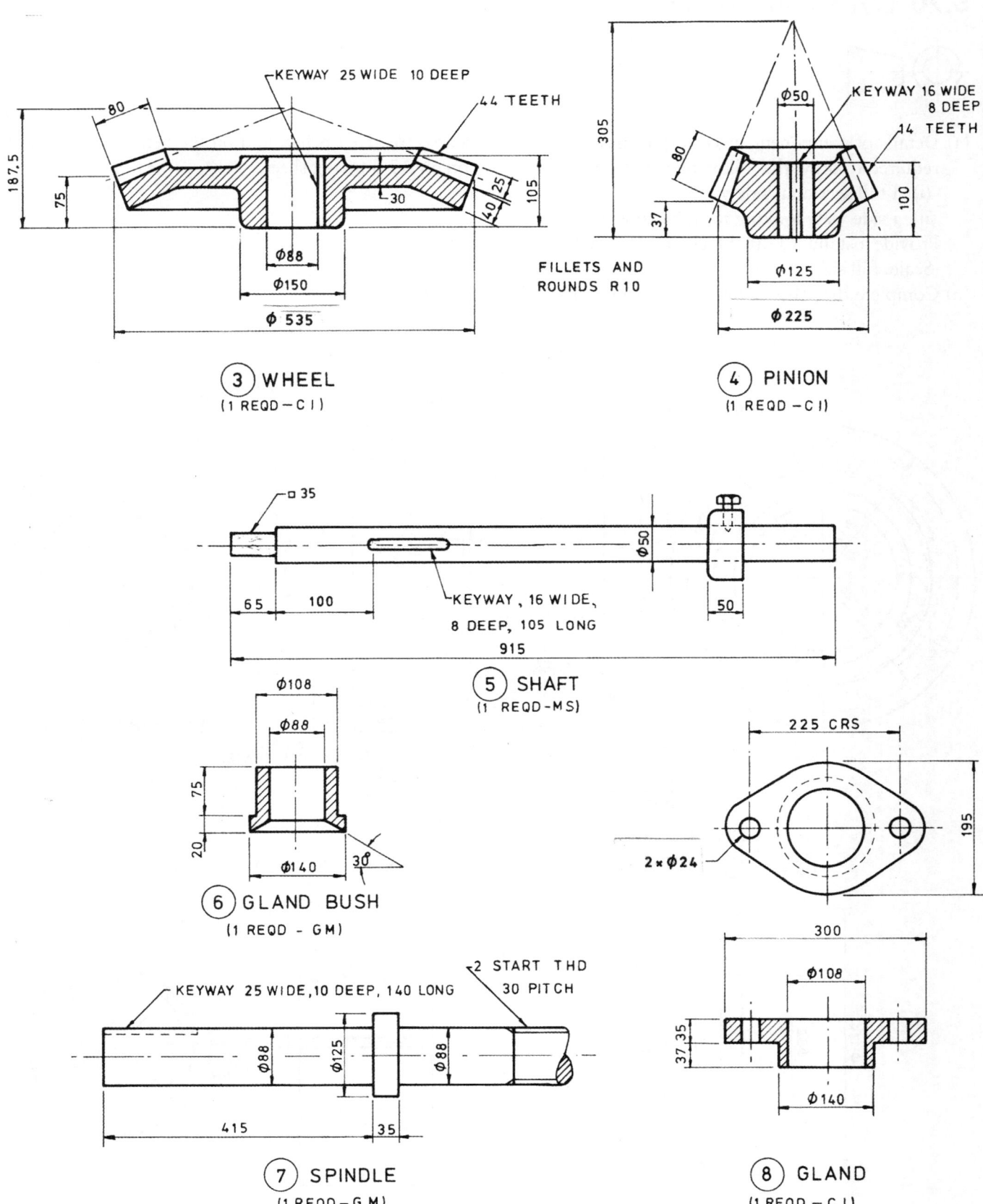

KEYWAY 25 WIDE 10 DEEP

80

187.5

75

44 TEETH

25

30

105

40

Ø88

Ø150

Ø 535

③ WHEEL
(1 REQD — C I)

305

Ø50

KEYWAY 16 WIDE
8 DEEP

14 TEETH

80

37

100

Ø125

Ø 225

FILLETS AND
ROUNDS R 10

④ PINION
(1 REQD — C I)

□ 35

Ø50

65

100

KEYWAY, 16 WIDE,
8 DEEP, 105 LONG

50

915

⑤ SHAFT
(1 REQD-MS)

Ø108

Ø88

75

20

Ø140

30°

⑥ GLAND BUSH
(1 REQD - GM)

225 CRS

195

2 × Ø24

KEYWAY 25 WIDE,10 DEEP, 140 LONG

2 START THD
30 PITCH

Ø88

Ø125

Ø88

415

35

⑦ SPINDLE
(1 REQD — G M)

300

Ø108

37 35

Ø140

⑧ GLAND
(1 REQD — C I)

## 6.20 LIVE LATHE CENTRE

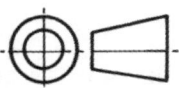

(a) Details of components necessary to make a live lathe centre are given below and on the next page. You are required to draw the following general assembly views on an A2 drawing sheet:
  (i)  a full sectional side view
 (ii)  a side view looking into the cone centre
Provide a standard title block and material list.
(Scale: full size)

(b) Complete the table of fits relating to the live lathe centre assembly.

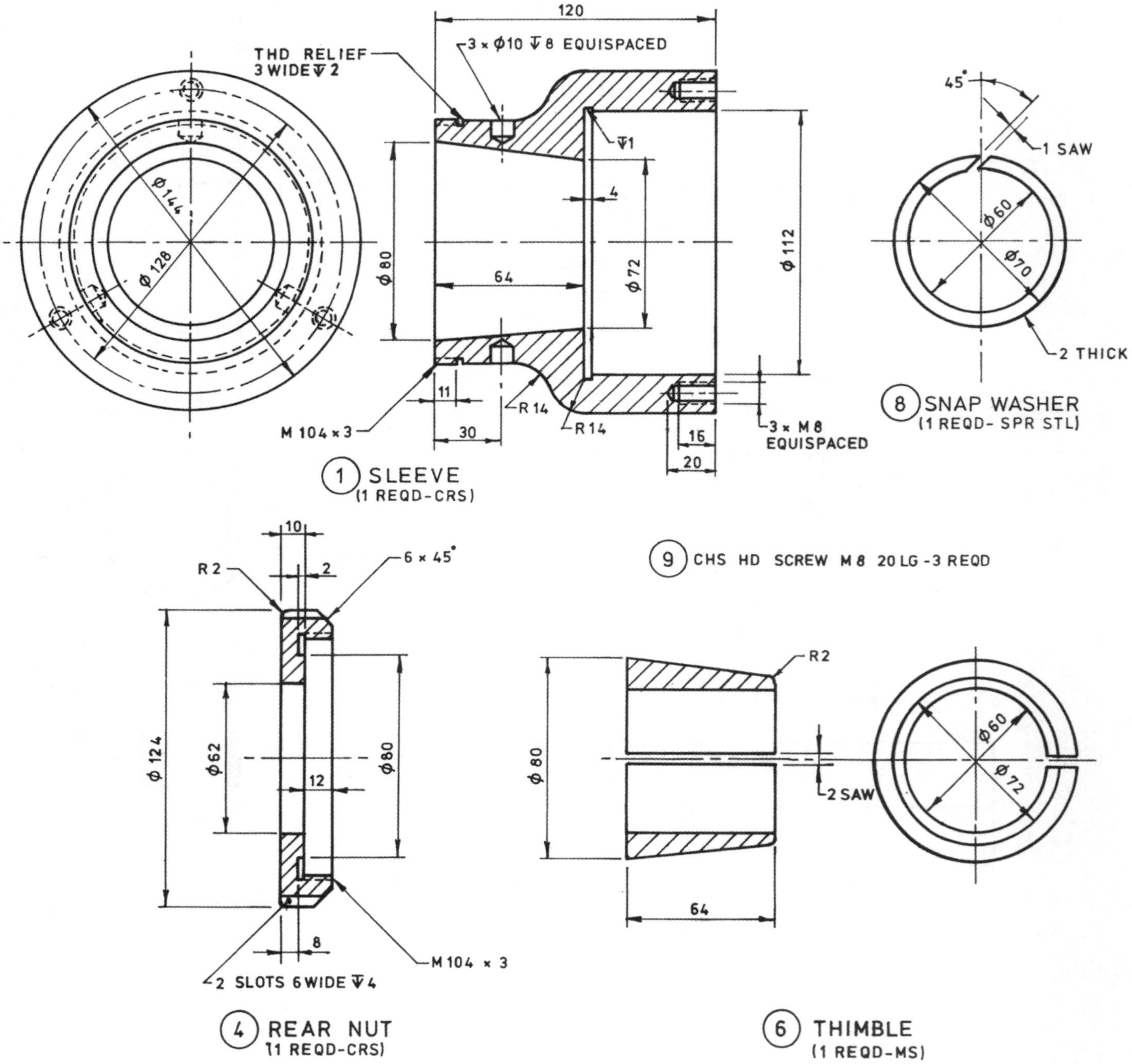

THD RELIEF
3 WIDE ▽ 2

⌀144
⌀128

M 104 × 3

120

⌐3 × ⌀10 ▽ 8 EQUISPACED

▽ 1

⌀80
⌀72
⌀112

4
64
11
30
R 14
R 14
16
20

⌐3 × M 8
EQUISPACED

**(1) SLEEVE**
(1 REQD-CRS)

45°
⌐1 SAW
⌀60
⌀70
⌐2 THICK

**(8) SNAP WASHER**
(1 REQD- SPR STL)

**(9) CHS HD SCREW M 8 20 LG -3 REQD**

10
R 2
2
6 × 45°

⌀124
⌀62
⌀80
12
8

M 104 × 3

⌐2 SLOTS 6 WIDE ▽ 4

**(4) REAR NUT**
(1 REQD-CRS)

R 2
⌀80
⌀60
⌀72
⌐2 SAW
64

**(6) THIMBLE**
(1 REQD-MS)

| FIT BETWEEN | CLASS OF FIT | FIT DESIGNATION | LIMITS OF SIZE | |
|---|---|---|---|---|
| | | | HOLE | SHAFT |
| CONE CENTRE—BEARING | LIGHT PUSH | | | |
| BEARING—SLEEVE | LIGHT PUSH | | | |
| WASHER—SLEEVE | NORMAL RUNNING | | | |
| FRONT PLATE—SLEEVE | LOCATION | | | |

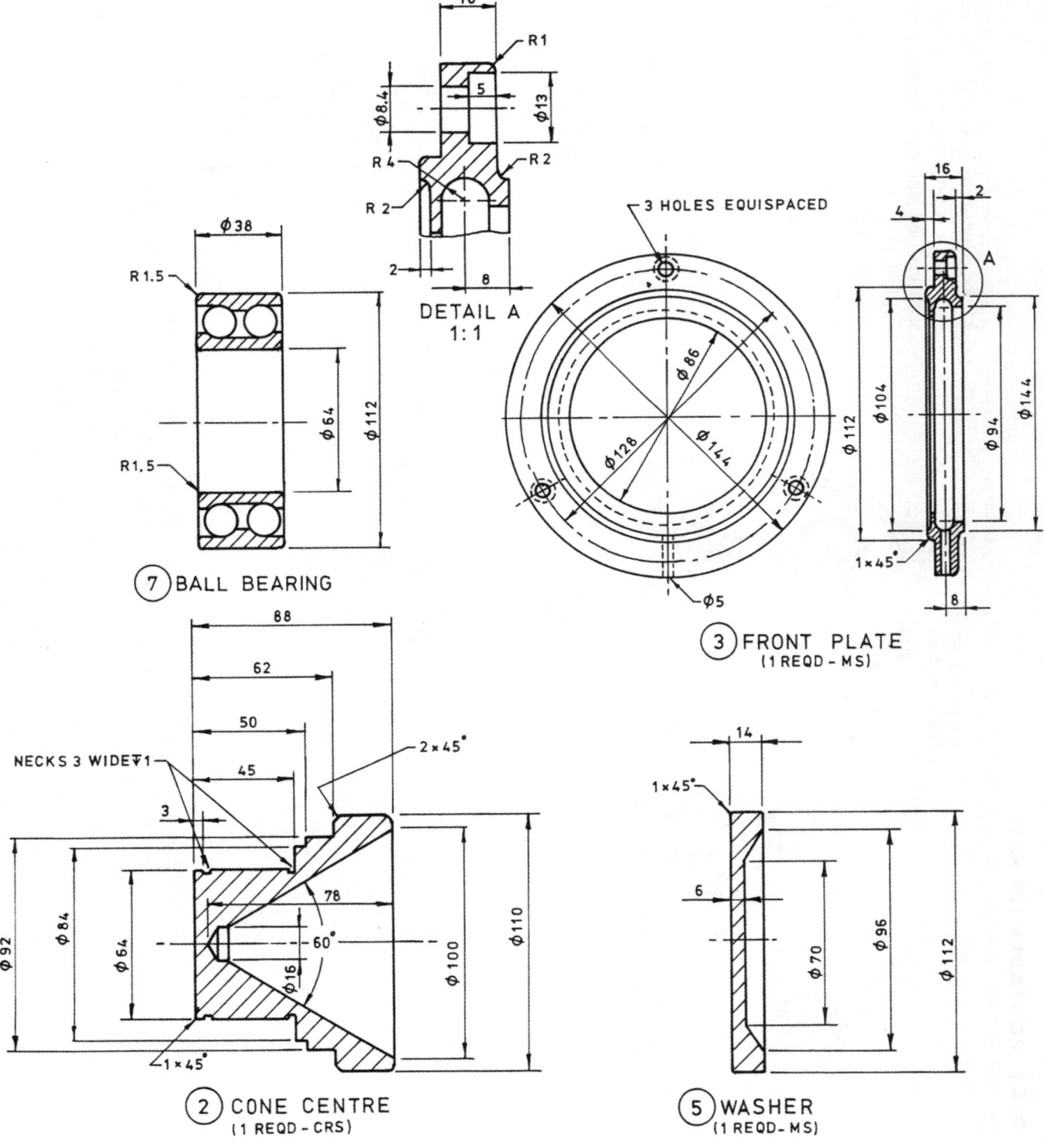

DETAIL A
1:1

⑦ BALL BEARING

3 HOLES EQUISPACED

③ FRONT PLATE
(1 REQD - MS)

② CONE CENTRE
(1 REQD - CRS)

⑤ WASHER
(1 REQD - MS)

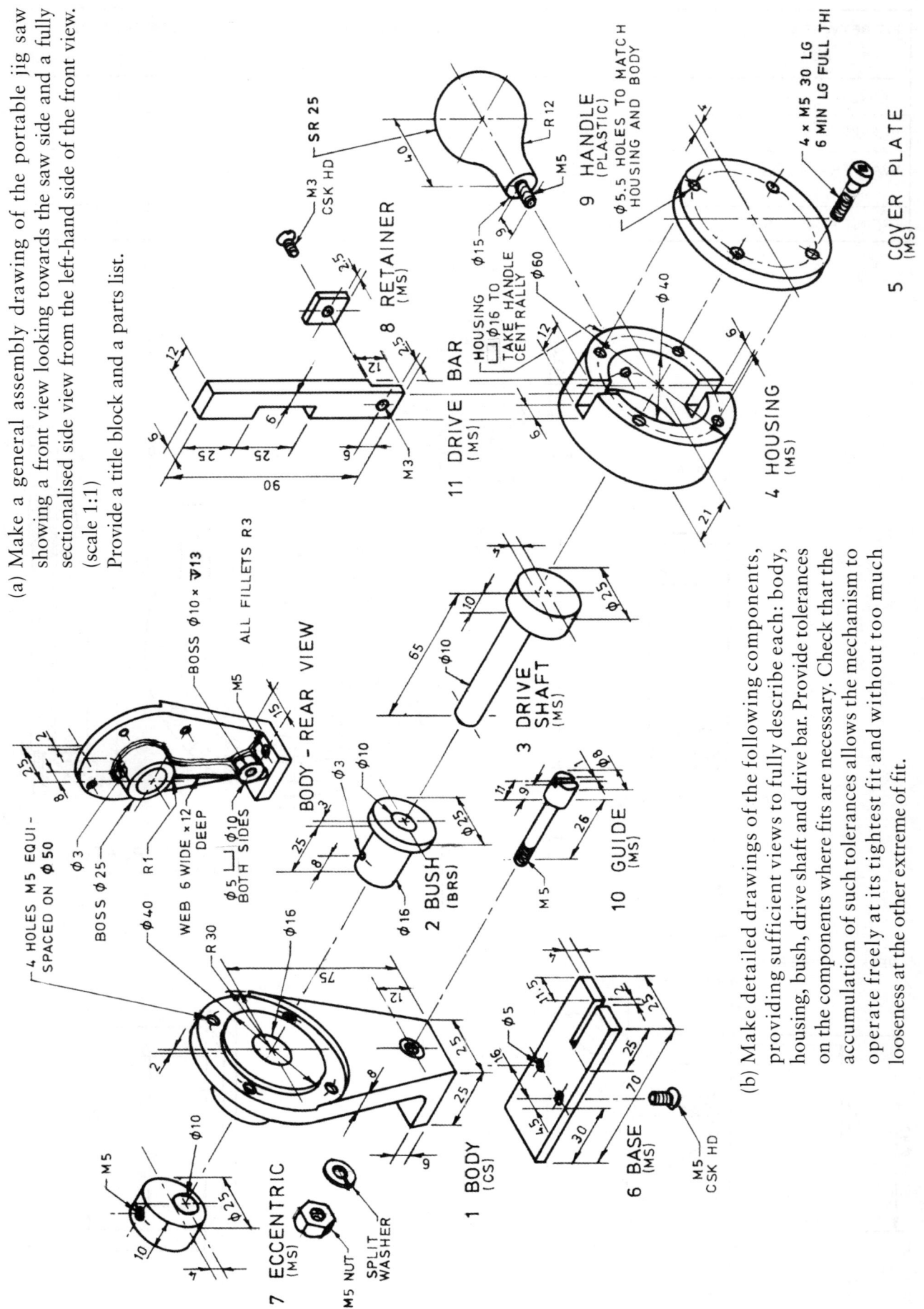

(a) Make a general assembly drawing of the portable jig saw showing a front view looking towards the saw side and a fully sectionalised side view from the left-hand side of the front view. (scale 1:1)

Provide a title block and a parts list.

**BODY – REAR VIEW**

4 HOLES M5 EQUI-
SPACED ON Ø 50
BOSS Ø10 × ▽13
ALL FILLETS R3
BOSS Ø25
M5
15
Ø3
R1
WEB 6 WIDE × 12 DEEP
Ø5 ⌴ Ø10 BOTH SIDES
Ø40
R 30
Ø16
75

**1 BODY** (CS)

**7 ECCENTRIC** (MS)
M5
Ø10
25
10
M5 NUT
SPLIT WASHER

**2 BUSH** (BRS)
Ø3
Ø10
25
Ø25
Ø16

**3 DRIVE SHAFT** (MS)
10
65
Ø10
Ø25

**10 GUIDE** (MS)
Ø8
11
26
M5

**6 BASE** (MS)
11.5
Ø5
16
25
25
70
30
M5 CSK HD

**8 RETAINER** (MS)
M3 CSK HD
SR 25
2.5
2.5

**9 HANDLE** (PLASTIC)
R12
M5
SR 25
Ø15

**11 DRIVE BAR** (MS)
12
6
6
25
25
90
M3

**4 HOUSING** (MS)
HOUSING ⌴ Ø16 TO TAKE HANDLE CENTRALLY
Ø5.5 HOLES TO MATCH HOUSING AND BODY
Ø60
Ø40
12
6
21

**5 COVER PLATE** (MS)
4 × M5 30 LG
6 MIN LG FULL THI

(b) Make detailed drawings of the following components, providing sufficient views to fully describe each: body, housing, bush, drive shaft and drive bar. Provide tolerances on the components where fits are necessary. Check that the accumulation of such tolerances allows the mechanism to operate freely at its tightest fit and without too much looseness at the other extreme of fit.

# 6.5 CAD corner

**6.22** Using your A2 drawing sheet from the first chapter and the adjustable bearing components from Exercise 6.8, create three views of the assembled components. The views are to have the following parameters:

(a) General

- The views are to be 3rd angle projection.
- Finished plot scale 1:2.
- Numbered call out balloon for each component (the numbers must match the numbers used in the detail drawing).
- Fasteners where required from the symbol library stored in your A2 drawing sheet.
- Component list incorporated in the title block similar to that shown below.
- Completed title block.

| 10 | | HEX SOCKET SCREW M6x8 LG | 2 | STEEL | |
|----|----|----|----|----|----|
| 9 | | KEY   8x8x45 LONG | 1 | KEYSTEEL | |
| 8 | | HEX HD BOLTS M12x32 LG | 2 | M.S. | |
| 7 | | HEX NUT M20 + ø20 FLAT WASHER | 1 | M.S. | |
| 6 | | SHAFT | 1 | M.S. | CHP–03–01–01 |
| 5 | | POST | 2 | C.R.S. | CHP–03–01–01 |
| 4 | | BEARING | 2 | BRASS | CHP–03–01–01 |
| 3 | | SPUR WHEEL | 1 | C.I. | CHP–03–01–01 |
| 2 | | BEARING BODY | 1 | C.I. | CHP–03–01–01 |
| 1 | | BASE PLATE | 1 | C.I. | CHP–03–01–01 |
| ITEM | PART | PART NAME | QTY | MATERIAL | REFERENCE |

### A.W. BOUNDY

ADJUSTABLE BEARING ASSEMBLY

(b) Top view

- Fully sectioned top view with the cutting plane through the centre of the bearing body.

(c) Front view

- Fully sectioned front view with the cutting plane through the centres of the posts.

(d) Side view

**Challenge:** Assemble the pictorial views of the components from Exercise 5.25 (page 214) and Copy/Paste a pictorial view of the complete assembly in the top right corner of the A2 sheet.

**6.23** (a) Investigate your CAD program to determine the following:

(i) The ability to apply tolerance values to a dimension.

(ii) Methods of displaying toleranced dimensions.

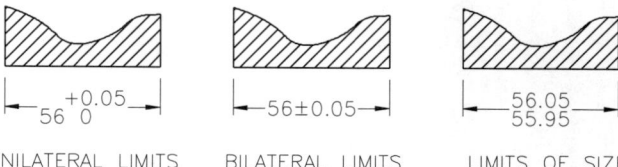

UNILATERAL LIMITS    BILATERAL LIMITS    LIMITS OF SIZE

(iii) Ability to display basic dimensions and datum features.

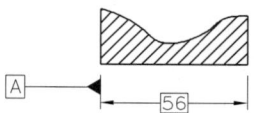

BASIC DIMENSIONS AND DATUMS

(iv) Methods of creating and displaying geometry tolerance frames.

GEOMETRY TOLERANCE FRAME

(v) Methods of applying machining symbols.

MACHINING SYMBOL

Note: If any of the above is not a function within your CAD program you may be able to add appropriate symbols to the symbol library created in the A2 drawing sheet from Exercise 1.22.

(b) Using the A2 drawing sheet answer question 6.12 with the following additions:

- Show machining symbols with a surface roughness value of 0.8 on all surfaces requiring machining.
- Apply a positional geometry tolerance with a value of Dia. 0.05 to the 5 drilled holes in the CI bracket.

# Drawing analysis

**7**

After studying this chapter and working through the problems you should be able to do the following:

- understand the terminology of engineering elements commonly found on engineering drawings

- obtain a pictorial comprehension of a complex component by analysing and combining the information on multiple two-dimensional views

- analyse a drawing and 'take off' information relevant to its manufacture

# 7.1 Sample analysis

The following is a description of the detail labelled on Figure 7.1.

1. A *counterbored hole* is used to house a screw or bolt head so that it does not project from the surface. It also provides a surface, square to the hole axis, for bolt head seating.

2. A *bolt* is designated by the material, head shape, ISO metric thread diameter (mm) and the length (mm) of its shank.

3. A *spigot* is a piece of material (usually circular) which projects from the face of a member. It is used to locate a member precisely when assembling it with another member. It may also be used to carry any shear load which may be applied to bolts holding the two members together.

4. Note that as this is a sectional view, the cross-hatch lines pass over the *internal thread* section.

5. A *recess* allows a member to engage right to the bottom of a hole without interference from a rounded corner. A recess can also be used externally, for example when turning a thread up to a shoulder.

6. A *centre line* is a light, long-short dash line (type G) which is used to indicate axes of holes and the centres of part and full circles.

7. A *countersunk hole* in this case is used as an oil hole, but mostly it would be used to house the countersunk head of a screw.

8. Note that the hatch lines do *not* pass over the *assembled threads*, but where the thread stands alone, item 4 above applies.

9. A *stud* is a member, threaded both ends and screwed firmly into the main part. Studs are used to attach coverplates and housings as shown.

10. A *seal* is generally a plastic ring seal which, when compressed against the main housing, squeezes against the rotating shaft and prevents entry of dust and grit into the main bearing. It also prevents lubricant from leaking out.

11. A *chamfer* is generally 45°, its purpose being to eliminate the sharp edge.

12. A *shaft* is a rotating member used to transmit torque. Note the chamfer on the end and the method of showing a break in the shaft, that is, the shaft actually extends beyond the length shown in the drawing.

13.–14. A *washer* (13) is used for assembly with the *nut* (14) on to the stud. It prevents scoring of the plate when the nut is tightened up.

15. A *housing* is a general term used to describe the location of items such as seals, bearings, gears, etc. Shown here is a *seal housing*.

16. A *clearance hole* is a hole just a little larger than the diameter of the stud, so that assembly is made easy. Recommended diameters of clearance holes for various sizes of metric thread diameter are given in Table 1.10.

17. *Leaders* are used to indicate where dimensions or notes are intended to apply. They are thin full lines which terminate in arrowheads or dots. Arrowheads terminate on a line, dots should fall within the outline of the object, as shown by items 30, 28, 23 and 10.

18. An *external* or *male thread* is the representation of the outside view of a threaded member.

19. A *projection line* is a thin full line (type B) extending from the outline, but not touching it. These lines denote the extremities of a dimension and should extend a little beyond the dimension line.

20. A *dimension line* is a thin full line (type B) extending between projection lines. It has arrow-heads on either end to indicate the length of the dimension, which is placed above the dimension line and approximately in the centre.

21. A *runout* is used to indicate the intersection of two surfaces which do not meet at a sharp corner.

22. A *surface texture symbol* indicates the finish of the surface to which it is applied. See page 45 for more details.

23. A *spotface* is an area around a hole which is machined perpendicular to the hole axis. It provides a flat true seating for the head of a nut or bolt.

24. *Flange* is a term used to describe a section of a member that carries holes through which bolts or screws pass to fasten the member.

25. A *boss* is a raised or extra portion of metal machined on top to support the screw head. The term 'boss' can be applied to extra projections of metal which provide additional support as well as an extension of the function; for example, shaft bosses provide extra bearing length, and screw or bolt bosses provide for adequate thread length.

26. *Pitch circle diameter* (PCD) is a long-short dash circle which passes through the centres of a series of holes. The holes are generally pitched evenly around its circumference.

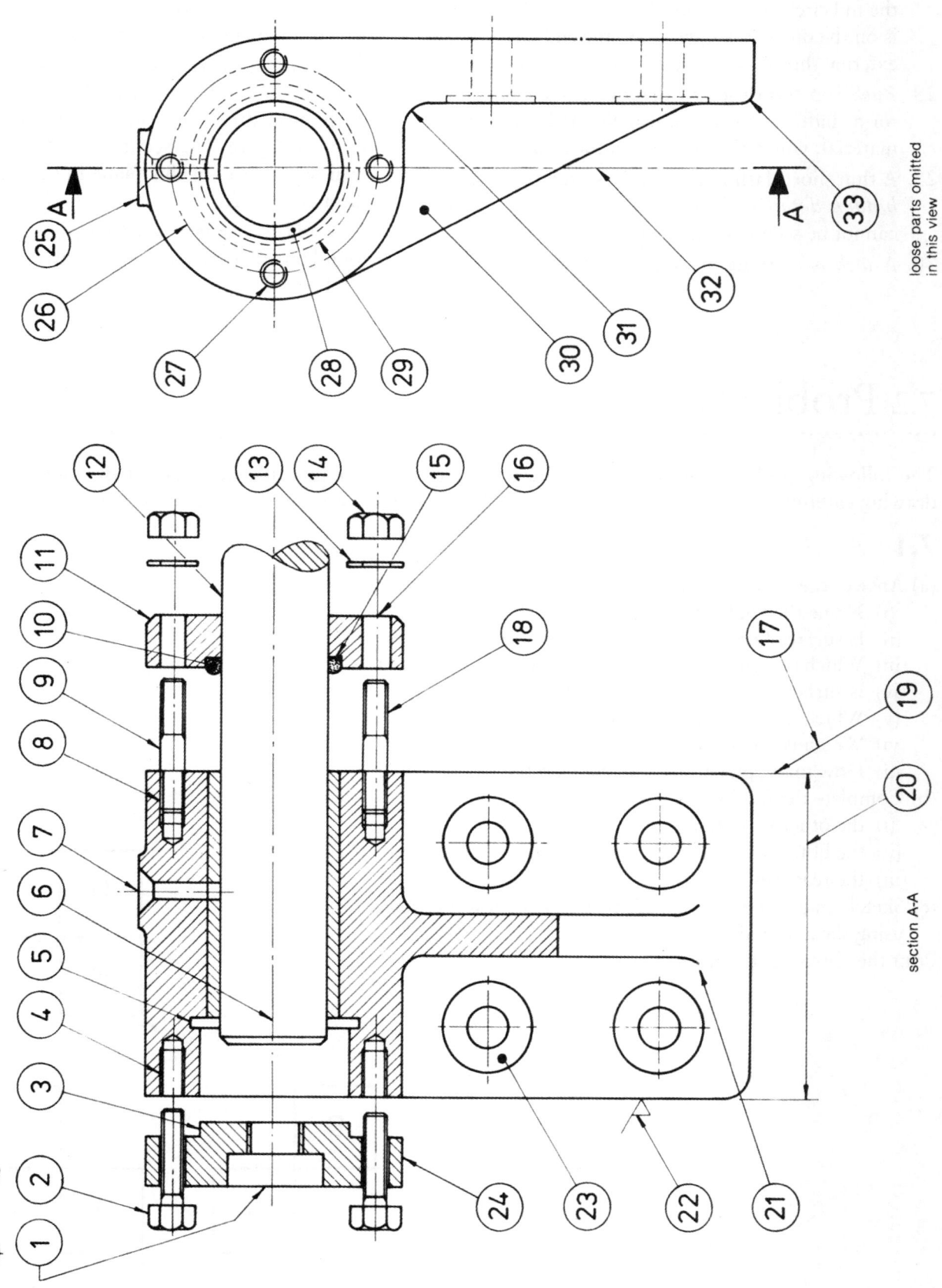

**FIGURE 7.1** ▶ Features found on engineering drawings

loose parts omitted
in this view

section A-A

27. Note in the end view of an *internal thread* that the full circle is on the inside and the broken circle is on the outside as opposed to the end view of an external thread. See page 28 for more details.

28. *Bush* is a term used to describe a plain bearing for a shaft. It is a sleeve, usually made of bronze material, which fits tightly into the housing.

29. A thin short-dash line (type E) is used to indicate *hidden detail* such as corners or edges which cannot be seen from the outside.

30. A *web* is a strengthening or stiffening member.

31. All castings have *fillets* on internal corners to prevent the formation of stress fatigue cracks which originate in sharp corners.

32. The *course of a cutting plane* is indicated by a chain line (type H), thick at the ends and where it changes direction, but thin elsewhere. The view in Figure 7.1 (section A-A) reveals detail seen at the level of this plane in the direction of the arrows A-A.

33. A *round* is similar to a fillet, but is normally found on external corners of a casting.

# 7.2 Problems

The following problems are designed to test a student's ability to read and interpret drawings. Study each drawing carefully, then answer the questions and either sketch or draw the views if they are required.

## 7.1

(a) Answer the following questions relating to the drawing:
   (i) Name the angle of projection.
   (ii) Is surface D above or below surface H?
   (iii) Which two surfaces are on the same horizontal plane?
   (iv) Is surface J in front of or behind surface G?
   (v) Which is the highest surface?
   (vi) Which is the nearest surface?
   (vii) How many plane surfaces make up the whole block?
(b) Complete the following views:
   (i) the other side view
   (ii) the bottom view
   (iii) the rear view
(c) Sketch an isometric view with corner Y nearest the viewer using the axes indicated.
Print the correct letter on each face.

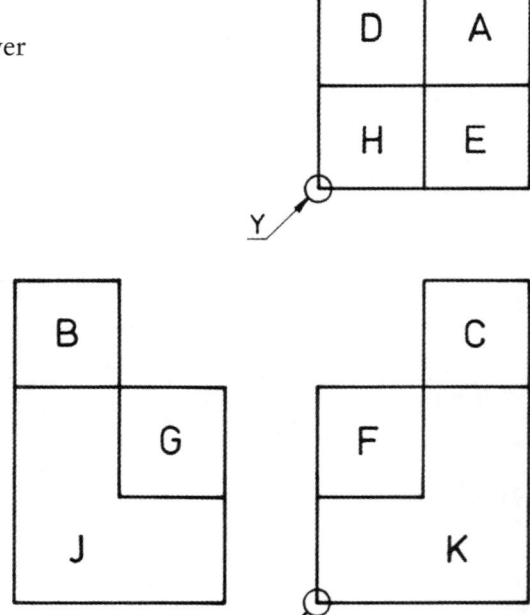

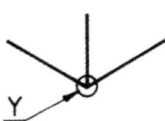

## 7.2

(a) Answer the following questions relating to the drawing:
   (i) What is the angle of projection?
   (ii) Is surface H in front of surface G?
   (iii) Which surface is nearer the viewer, K or J?
   (iv) Is surface E below surface F?
   (v) Which of the surfaces G, H, K or J are in the same plane?
   (vi) Which surfaces are the highest?
   (vii) What surface is level with surface D?
   (viii) What surfaces are shown in hidden detail on the side view?
   (ix) What surfaces are shown in hidden detail on the top view?
   (x) Which of the surfaces M, N, L, O and P are in the same plane?
(b) Make an isometric sketch of the block shown using the axes indicated. Place the correct letter on each surface.

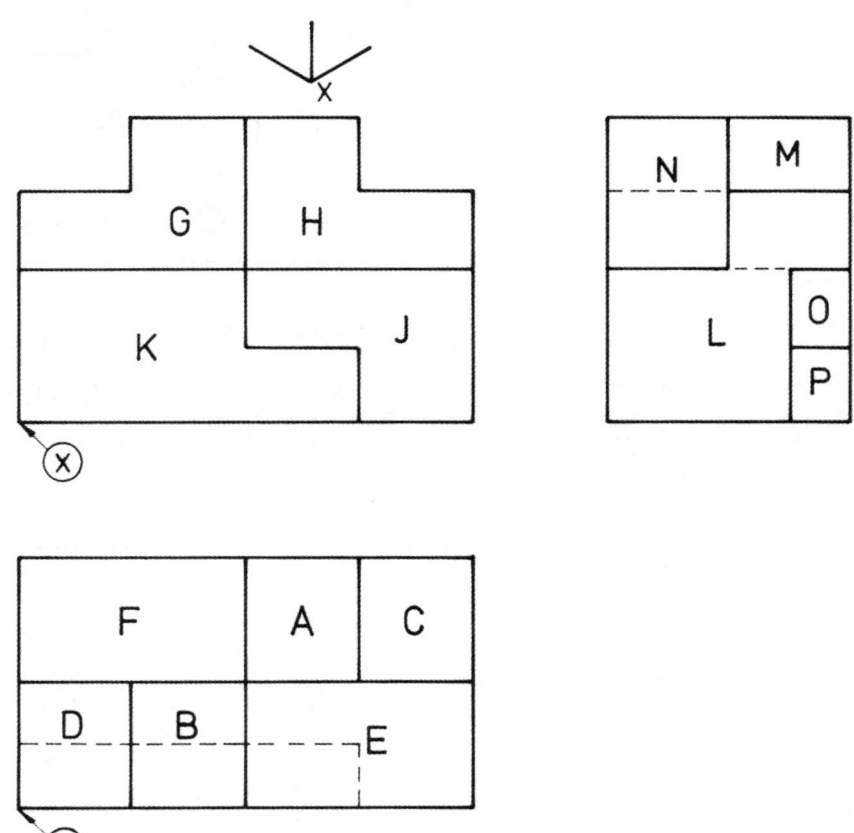

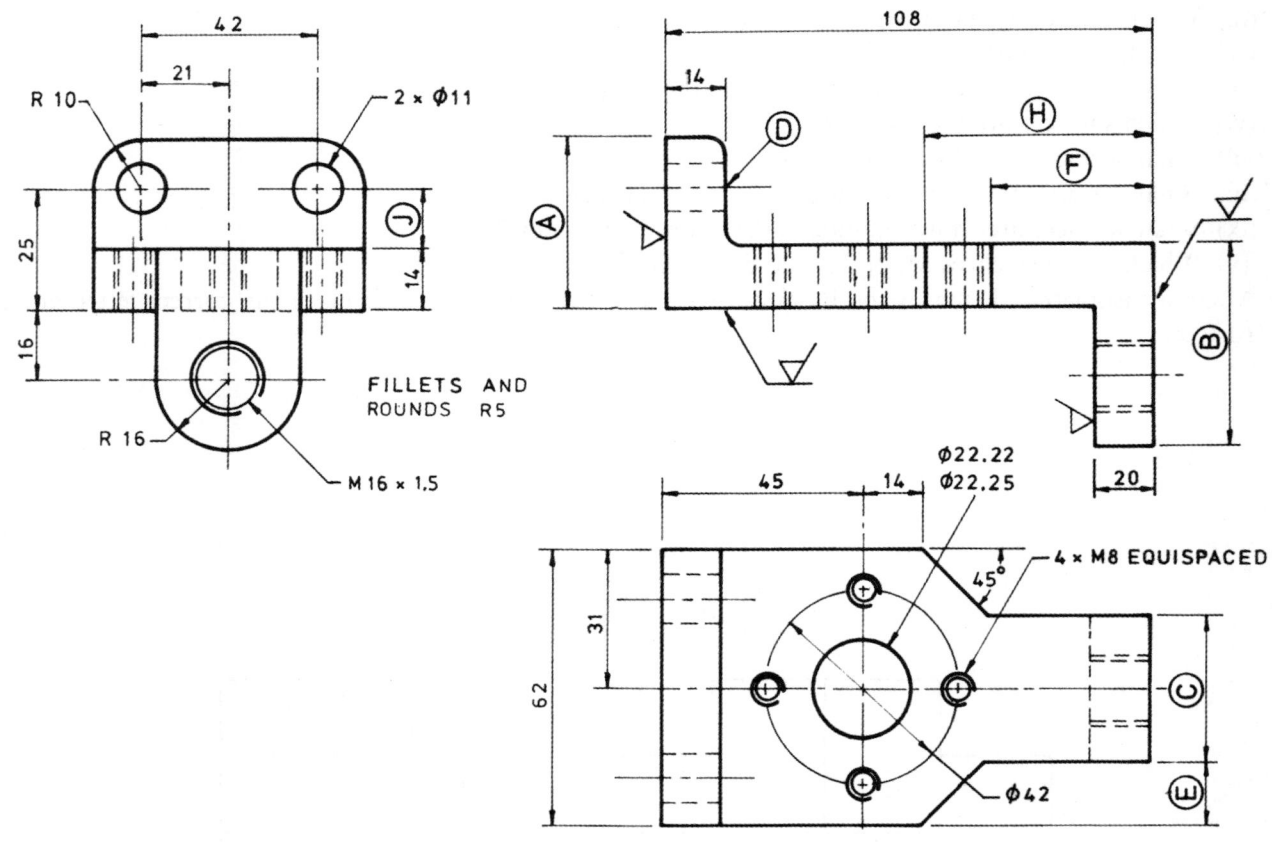

FILLETS AND
ROUNDS R5

| | | | | | | TOLERANCES ± 0.5 EXCEPT AS STATED | MATERIAL C I | DRN | A SHAMBLES PTY LTD | |
|---|---|---|---|---|---|---|---|---|---|---|
| | | | | | | | | CKD | **MOUNTING BRACKET** | |
| | | | | | | | FINISH PAINT - RED | APPD | | |
| 2 | THREAD | WAS M16 | aw8 | 7:01:01 | | | | | | |
| 1 | THREAD | WAS M8 × 1 | aw8 | 7:01:01 | | | HEAT TREAT | ISSUED | SCALE 1 : 2 | DRAWING NO A−161 | A1 |
| CHANGE NO | ITEM | CHANGE | CKD | OK | DATE | DATE 12.03.00 | | | | |

| | | | |
|---|---|---|---|
| (a) How many holes require drilling? | | (k) What is dimension F in mm? | |
| (b) What is the overall height of the part? | | (l) What is the tolerance of the large hole? | |
| (c) What is dimension A in mm? | | (m) What material is the bracket made of? | |
| (d) What is the tapping size of the M8 holes? | | (n) What surface finish is required? | |
| (e) What is dimension B in mm? | | (o) What is the radius of fillets and rounds? | |
| (f) What general tolerance is specified? | | (p) What is dimension H in mm? | |
| (g) What is dimension C in mm? | | (q) How many threaded holes are on the component? | |
| (h) What is the thread series for the 16 mm threaded hole? | | (r) What is dimension J in mm? | |
| (i) What is dimension E in mm? | | (s) How many surfaces are machine finished? | |
| (j) What is the specified diameter of hole D? | | (t) What dihedral angle is used for projection of views? | |

## 7.4

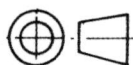

Answer the questions in the lower half of the page, and find the appropriate dimension for each letter down the left-hand side of the diagram.

Draw the following views in third-angle projection:
(a) a sectional front view on A-A
(b) a side view from B
(c) a top view
(scale: full size)
Fully dimension and subtitle your drawing.

| A | |
|---|---|
| B | |
| C | |
| D | |
| E | |
| F | |
| G | |
| H | |
| J | |
| K | |
| L | |

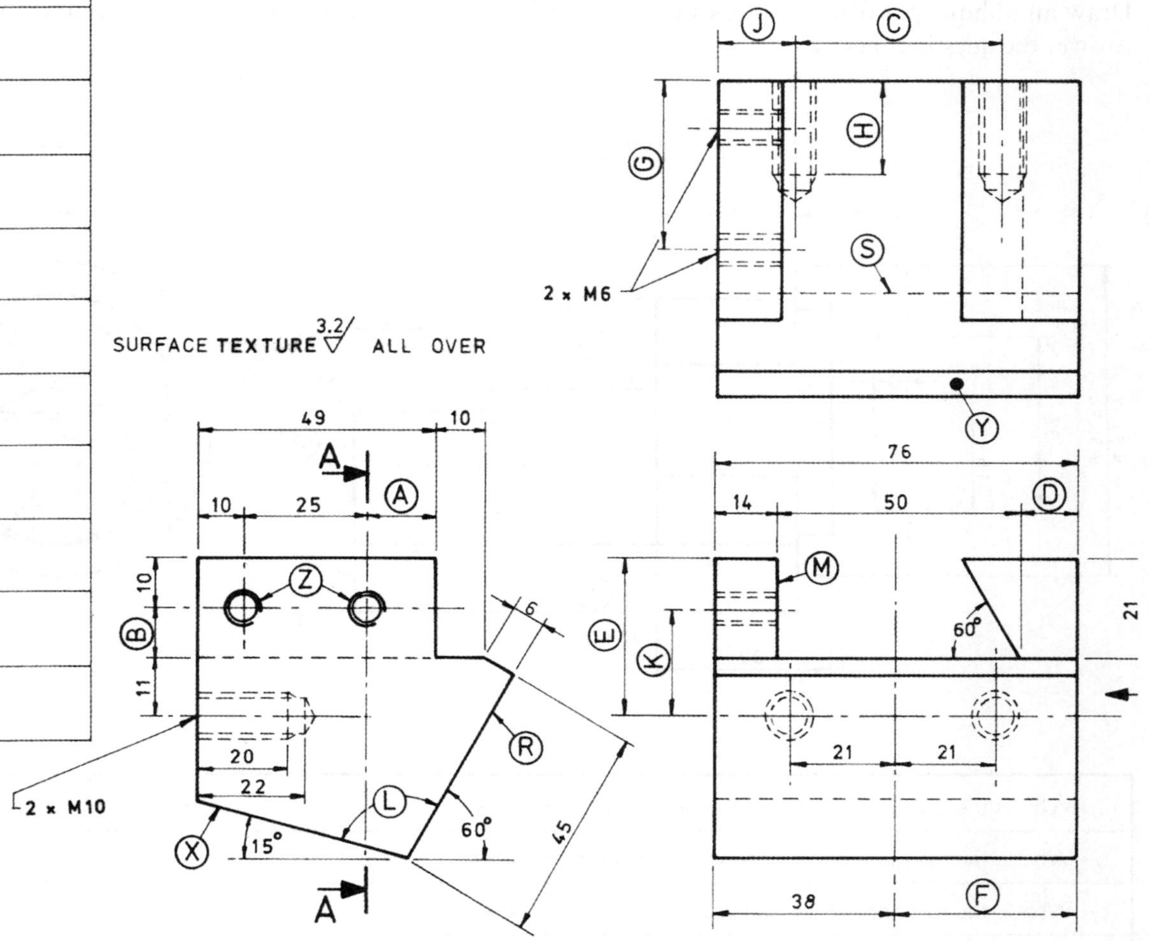

SURFACE **TEXTURE** $\frac{3.2}{\bigtriangledown}$ ALL OVER

2 × M6

49    10

A

10    25    (A)

(Z)

(B)    10

11

20

22

2 × M10

(X)    15°

(L)    60°

A

45

(J)    (C)

(G)    (H)

(S)

(Y)

76

14    50    (D)

(M)

60°

21

(E)

(K)

21    21

38    (F)

R

| (a) | What is the centre-to-centre distance of the 6 mm threaded holes? | | (g) | What are the dimensions of the face marked R? | |
|---|---|---|---|---|---|
| (b) | What does the dotted line S represent? | | (h) | What are the dimensions of surface Y? | |
| (c) | What tap drill is required for the holes marked Z? | | (i) | In what dihedral angle is the drawing made? | |
| (d) | How many surfaces are machine finished? | | (j) | What are the dimensions of face M? | |
| (e) | Is the true shape of face R shown on the drawing? | | (k) | Would the bare-faced dovetail slot be turned, shaped, milled or ground? | |
| (f) | Which view shows the true shape of surface X? | | (l) | In the surface texture symbol we have 3.2 units. What are they? | |

## 7.5 ROLLER STUD

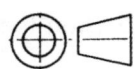

Make a detailed drawing of the roller stud within a standard drawing frame in third-angle projection, showing the following views:
(a) the given front view
(b) a side view from X
(c) a top view
(scale: 1:1)
Fully dimension the drawing, and provide a title block.
Draw an oblique parallel view looking at the small end of the roller stud. Omit the 8 mm diameter hole.
Answer the questions below.

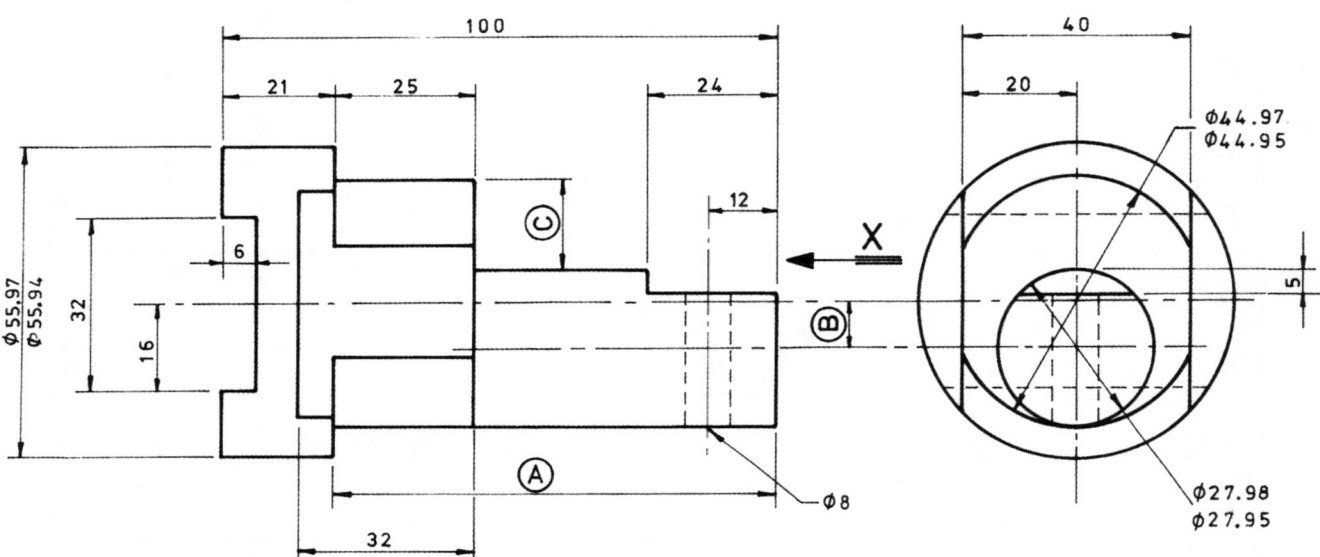

| | |
|---|---|
| (a)  What is the tolerance of the nominal 56 mm diameter roller? | |
| (b)  What is the tolerance of the nominal 45 mm diameter roller? | |
| (c)  What is the tolerance of the nominal 28 mm diameter roller? | |
| (d)  What is the maximum measurement of eccentricity B? | |
| (e)  What is the minimum measurement of eccentricity B? | |
| (f)  What are the minimum and maximum measurements of C? | |
| (g)  What type of fit is designated for the three roller diameters above? | |
| (h)  What is dimension A? | |
| (i)  Give suitable machine finishes for the following:<br>　(i)  all flat surfaces to be milled<br>　(ii)  all curved surfaces to be turned | |
| (j)  If the 28 mm diameter spigot end of the roller stud had to fit a hole dimensioned $\frac{28.03}{28.00}$ what would be the maximum and minimum clearances between stud and hole? | |

# 7.6 TOOL HOLDER

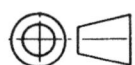

Make a detailed drawing of the tool holder within a standard drawing frame in third-angle projection, showing the following views:
(a) the front view as shown
(b) a right side view
(c) a top view
(scale: 1:1)
Fully dimension the drawing, and provide a title block. Insert on your drawing what you consider to be suitable machine surface finish symbols for a toolmaker's workshop.
Answer the questions below.

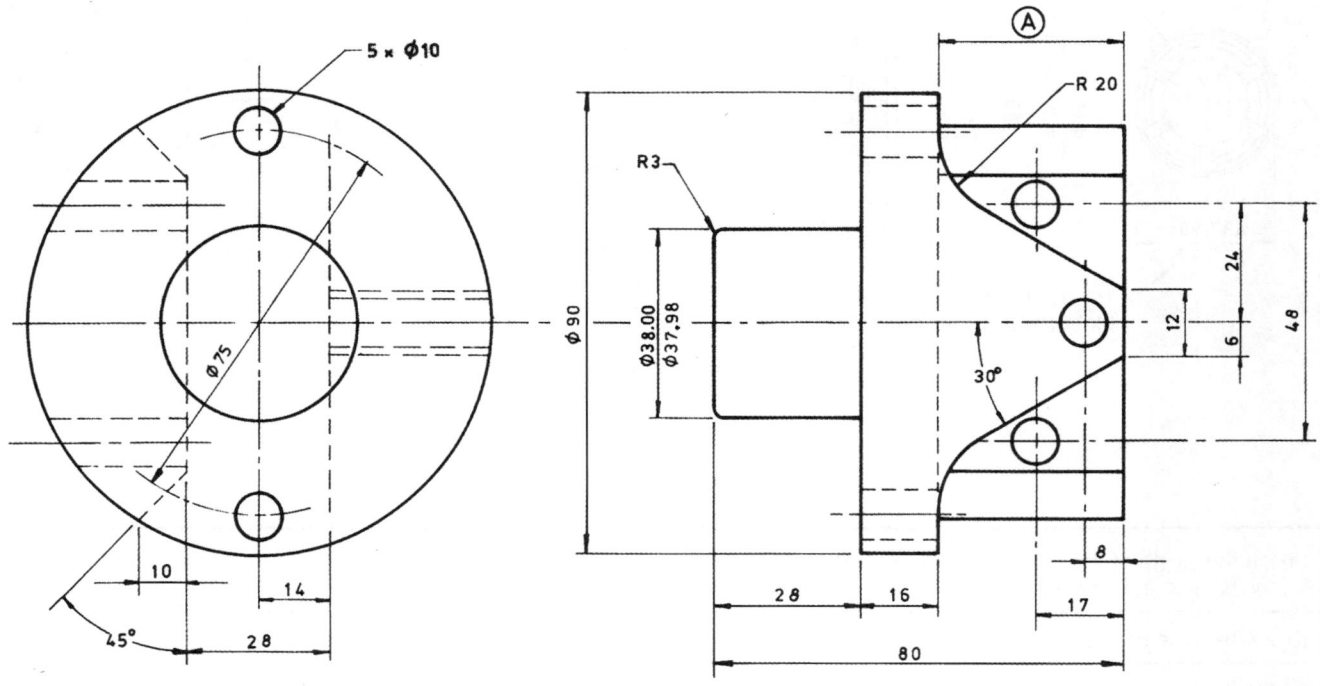

| (a) | What is the tolerance of the 38 mm diameter spigot? | |
|-----|------|---|
| (b) | What type of tolerance is it? | |
| (c) | What is dimension A? | |
| (d) | What material would you suggest for this holder? | |
| (e) | What heat treatment would you recommend, if any? | |
| (f) | Give a toleranced dimension for a hole to give a close locational fit on the spigot end of the holder. | |
| (g) | What extremes of fit between the spigot and the hole would you suggest? | |

# 7.7 EXPANDING GROOVE COLLET

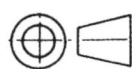

Two views of an expanding groove collet are shown below, together with a part view of a rod end, which is used to expand the collet. The rod end is drawn in line for assembly with the collet. Answer the following questions.

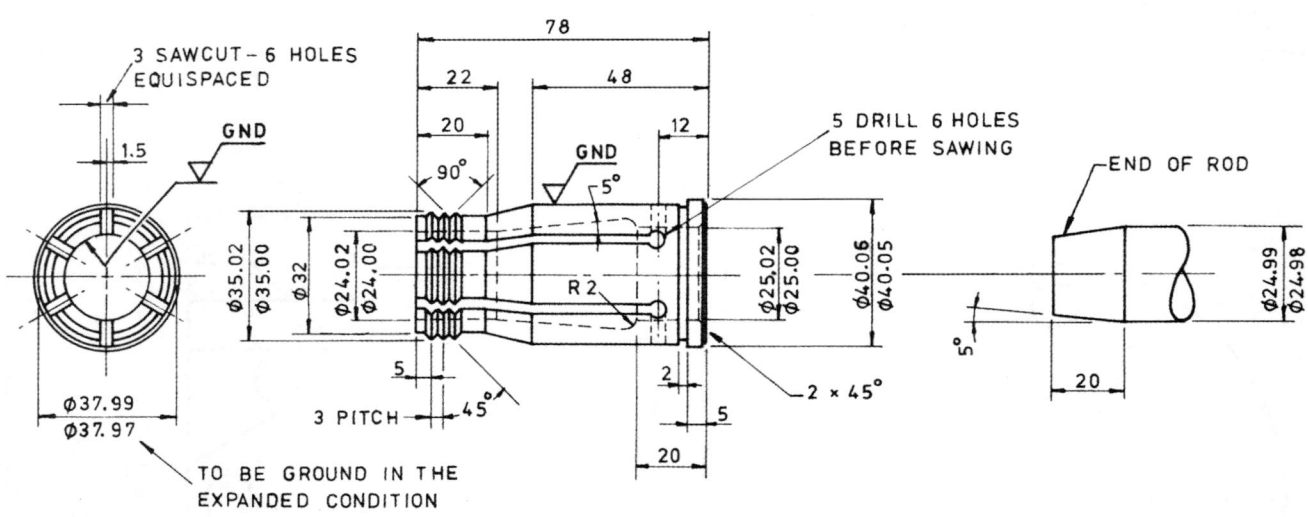

| | |
|---|---|
| (a) In the length of the collet there is a redundant dimension for part of the exterior surface. What is it and how long is it? | |
| (b) What type of tolerances are used throughout the drawing? | |
| (c) Why should the six holes be drilled before sawing? | |
| (d) Why specify six sawcuts? Why not five or seven? | |
| (e) The end of the rod which expands the collet is shown in line for assembly; when pushed into the collet it expands the grooves. What minimum expanded diameter of the crest of the grooves would it be possible to have, based on the tolerances? | |
| (f) What maximum expanded diameter of the crest of the grooves would it be possible to have, based on the tolerances? | |
| (g) Would the 90° grooves be cut before or after the sawcuts? Why? | |
| (h) What are the maximum and minimum clearances between the rod and the 25 mm nominal hole where the rod enters the collet? | |
| (i) Why have the 2 mm wide relief or, for that matter, any relief at al? | |
| (j) Why do you think it is necessary to grind the nominal 38 mm dimension while the collet is in the expanded condition? | |

# 7.3 CAD corner

**7.8**   The analysis of a hard copy of a drawing produced with CAD is no different to that of a manually drafted drawing, however drawing geometry can be analysed in CAD drawings, which cannot be achieved from a manual drawing without the aid of a calculator and the aid of various formulae.

In this exercise you will analyse drawing geometry to obtain information for the component manufacture.

(a) Draw the slide lock plate shown below complete with dimensions and centre lines.

(b) Use your CAD tools to find the following information about the slide lock plate geometry. All of the information can be displayed on the drawing in the form of notes or dimensions.

- The outer perimeter.
- The total length of cut including the two Dia. 15 (15 mm diameter) holes.
- The area of the plate (make sure you subtract the area of the two Dia. 15 holes).
- The volume of the plate given that the plate thickness is 6.0 mm.
- Given that the origin of X and Y (0,0) is located at the bottom left corner of the plate, as shown below, the (X,Y) coordinate locations of all holes and arc centres.
- The angle of the 10 mm wide slot in relation to the horizontal.

Note: The plate is going to be cut out using a CNC controlled laser cut profile machine.

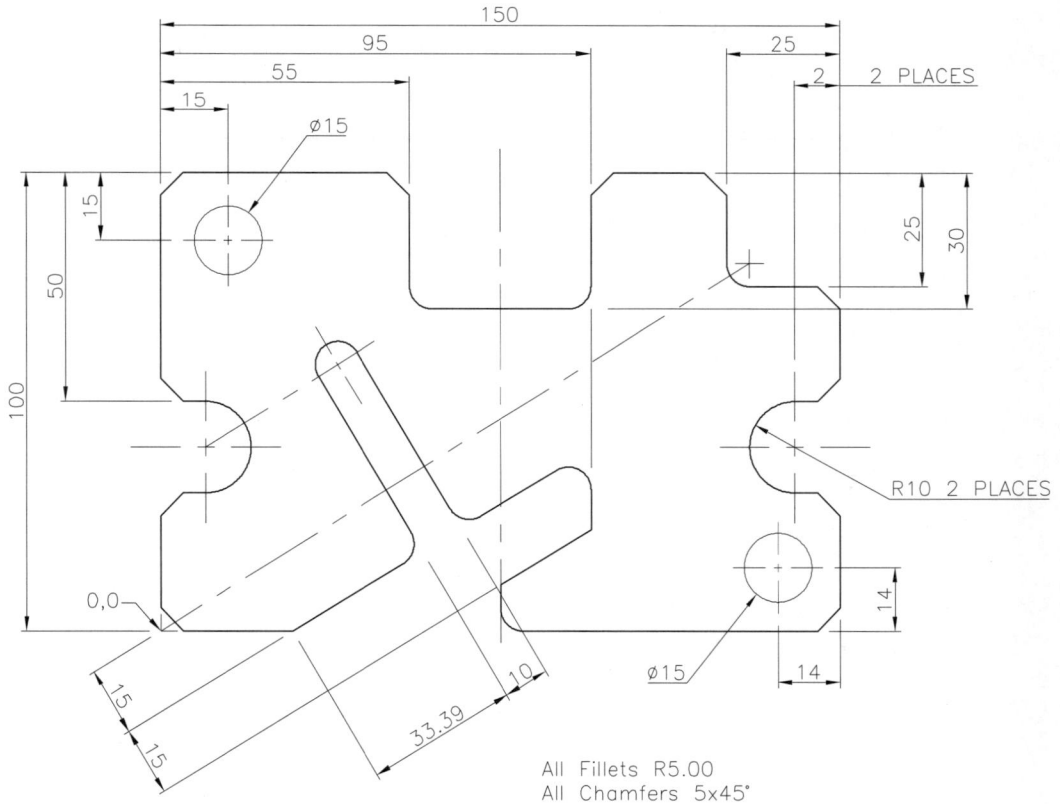

Slide lock plate

**7.9**    In this exercise you will calculate how many metres of gutter/kerbing can be obtained from a cubic metre of concrete.

(a) Draw the profile of the gutter/kerb shown below. Do not include centre lines dimensions or hatching.

(b) Analyse the profile to find its area and then calculate the volume of a 1000 mm length.

    Divide 1 cubic metre by the volume of 1 metre of guttering/kerbing (be careful with your units). This will result in the length in metres of gutter/kerbing from 1 cubic metre of concrete.

**Challenge:** If your CAD program has solid modelling capabilities you should be able to extrude the profile 1000 mm long and then analyse the solid to obtain its volume directly.

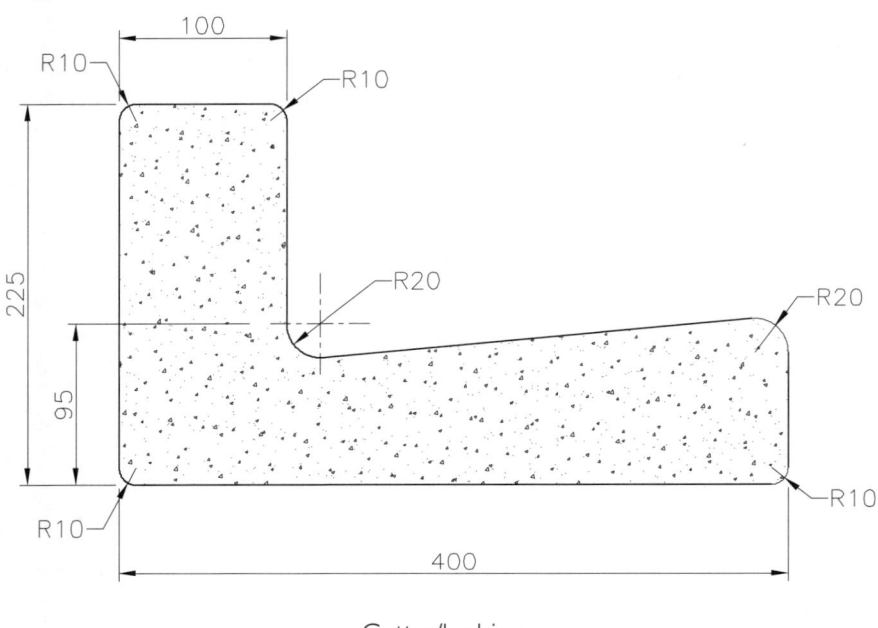

Gutter/kerbing

# Intersections
## & development of surfaces

**8**

After studying this chapter and working through the problems you should be able to do the following:

- construct parallel line developments of prisms and cylinders
- construct radial line developments of pyramids and cones
- draw the line of intersection between geometric surfaces and construct their development
- construct the development of transition pieces using triangulation

# 8.1 Development of prisms

## RECTANGULAR RIGHT PRISM

Figure 8.1(a) shows a pictorial view of a rectangular right prism with open ends. This prism consists of four rectangular sides which, when folded out on to a flat surface, form the area necessary to make the prism. This area is called the *development of the prism* or the *pattern for the prism*. Figure 8.1(b) is a view showing the prism unfolding on to a flat surface, while Figure 8.1(c) is the complete layout of the surface of the prism when it is unfolded. It can easily be seen that the development of the rectangular right prism is a rectangle whose dimensions are the perimeter of the end and the length of the prism.

## TRUNCATED RIGHT PRISM

Figure 8.2 illustrates the development of a truncated right prism shown on the left of the figure. To obtain the development, follow these steps:

1. Draw the orthogonal views of the truncated prism as an aid, showing the line of truncation and the joint XX, which is usually positioned midway along the shortest side.
   **Note:** Only one orthogonal view (e.g. the side view) is normally required. In this case the others are included for clarity.

2. Number the corners 1, 2, 3 and 4 on the orthogonal views.

3. Project horizontally to the right (or left) from the side view. These projectors define the heights of the development.

4. Commencing at joint XX, mark off the sides of the prism along the bottom projector, making sure to finish with joint X. These distances are best taken from the top view.
   **Note:** X1 and 4X are half of the side 14.

5. Draw vertical lines to intersect the other projectors at X, 1, 2, 3, 4 and X as shown.

6. Join the points X, 1, 2, 3, 4 and X to complete the development.
   **Note:** Lines 1-1, 2-2, 3-3 and 4-4 are called fold lines, that is, the flat development is 'folded' or 'bent' along these lines to form the required prism.

## RECTANGULAR PRISM PIPE ELBOW

A practical application of a truncated prism is shown in Figure 8.3, which illustrates an elbow in rectangular pipe. The development of one half of the elbow is shown on the right. Note that in an elbow of this nature, the junction of the two branches of the elbow is on a line which bisects the total angle of the elbow, in this case 120° as shown on the side view. This is necessary as the cross-sectional shape of each piece of the elbow has to be the same to match, and this is only the case at the bisection of the total angle.

**FIGURE 8.1** ▶ Rectangular right prism

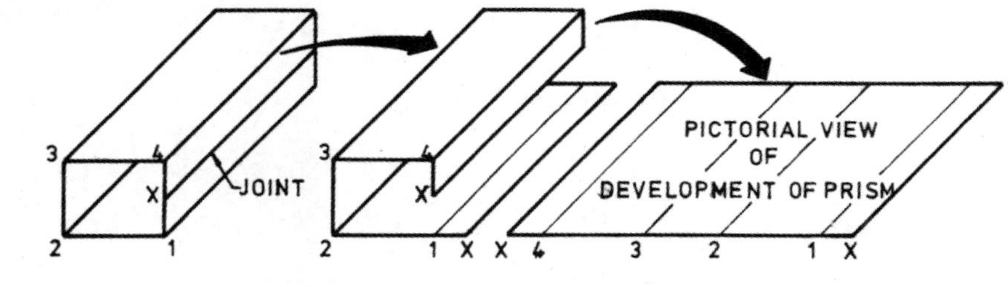

(a) rectangular prism 'folded up'  (b) prism unfolding  (c) prism unfolded

**FIGURE 8.3** ▶ Rectangular prism pipe elbow

PICTORIAL VIEW OF ELBOW

DEVELOPMENT OF (A)

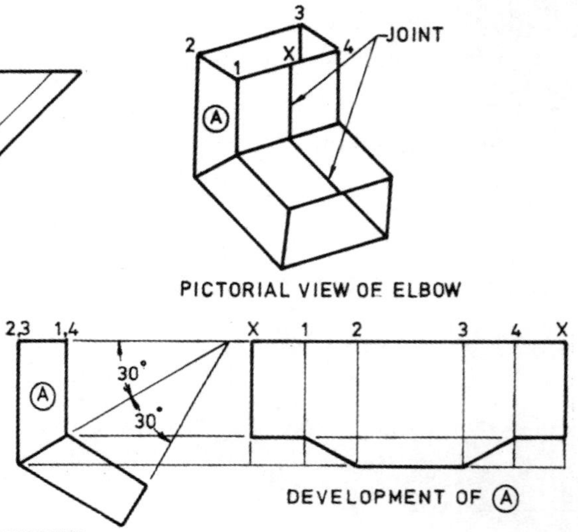

**FIGURE 8.2** ▶ Truncated right prism

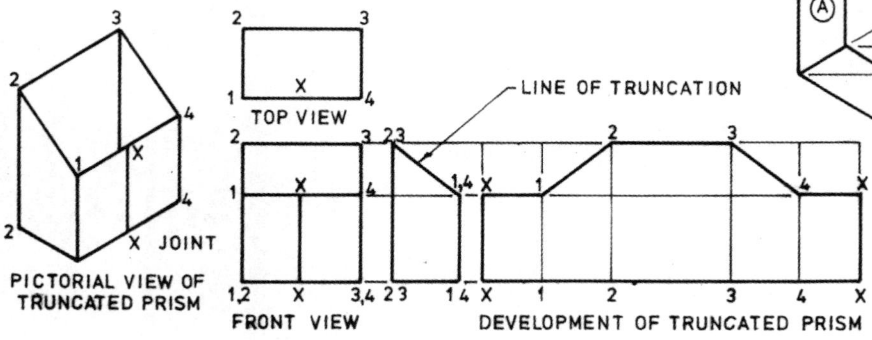

PICTORIAL VIEW OF TRUNCATED PRISM

FRONT VIEW    DEVELOPMENT OF TRUNCATED PRISM

## HEXAGONAL RIGHT PRISM

Figure 8.4(a) is a pictorial view of a hexagonal right prism with open ends. This prism consists of six rectangular sides. Figures 8.4(b) and (c) illustrate how the development of this prism is obtained. The area required for its development consists of a rectangle whose dimensions are the perimeter of the prism end and the length of the prism sides.

## TRUNCATED HEXAGONAL RIGHT PRISM

Figure 8.5 shows the development of a truncated hexagonal right prism. It is constructed in a manner similar to the development of the truncated right prism in Figure 8.2.

## TRUNCATED OBLIQUE HEXAGONAL PRISM

Figure 8.6 shows the development of a truncated oblique hexagonal prism. To obtain it, follow these steps:

1. Draw the front view as an aid, showing the true lengths of the prism sides. The top view is also shown.
2. Number the side corners at the top 1, 2, 3 and the joint line 0, as shown.
3. Project the side lengths at right angles to the sides of the front view.
4. Commencing at 0, mark off the base edge lengths

0-1, 1-2, 2-3, etc. on to the appropriate projectors, and join up the points to give the top of the development. The base edge lengths are taken from the top view.
5. Project the fold lines from the top to the bottom projectors to give the bottom end of the development.

## OTHER PRISMATIC SHAPES

Square, pentagonal and octagonal, right and oblique prisms are developed in a similar manner. Problems 8.11 and 8.12 (page 288) are two lobster-back bends made up of truncated square and hexagonal prisms respectively, called segments. In problem 8.11 there is a half segment at each end of the bend, while in problem 8.12 the bend consists of three whole segments.

It can be seen that at the centre of each full segment in problem 8.11, the cross-sectional area of the bend is the same as at the inlet and outlet. In problem 8.12, the inlet and outlet cross-sectional areas are the same as at the junction of the segments, while the cross-sectional area halfway along each segment is smaller.

Hence if it is not desirable to have a reduction in cross-sectional area of the bend, the segments must be designed and fitted according to problem 8.11. More segments may be inserted in the bend than shown, in order to make the change of direction smoother and to approximate a radial bend.

**FIGURE 8.4** ▶ Hexagonal prism

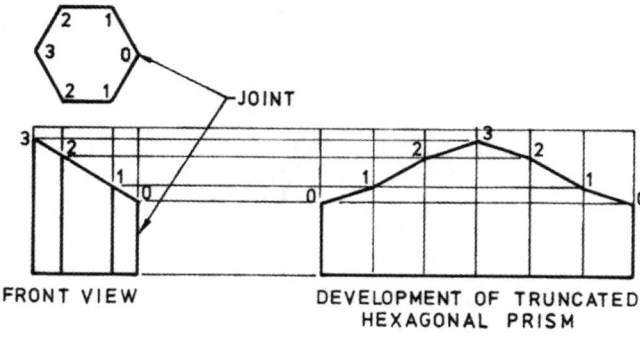

(a) hexagonal right prism 'folded up'     (b) prism unfolding

PICTORIAL VIEW OF DEVELOPMENT OF HEXAGONAL PRISM

(c) prism unfolded

**FIGURE 8.5** ▶ Truncated hexagonal right prism

FRONT VIEW     DEVELOPMENT OF TRUNCATED HEXAGONAL PRISM

**FIGURE 8.6** ▶ Oblique hexagonal prism

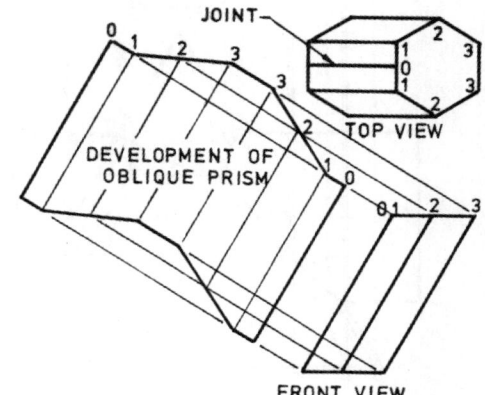

DEVELOPMENT OF OBLIQUE PRISM

JOINT

TOP VIEW

FRONT VIEW

# 8.2 Line of intersection— cylinders and cones

The line of intersection of two or more intersecting surfaces has to be determined in order to develop any of the surfaces. Methods used in drawing lines of intersection are as follows.

## I. ELEMENT METHOD

This involves the use of line elements drawn on the surfaces of the intersecting shapes, and passing through the area where the line of intersection occurs.

*Cone and cylinder intersection (Fig. 8.7)*

1. Draw the front, side and top views of the cone, showing the intersection of the cylinder on the side view, the intersection being a circle.
2. Draw two identical sets of three elemental lines on the side view from the apex to cut the base at a, b and c, and cutting the cylinder at 0, 1, 2, 3, 4, 5 and 6.
   **Note:** The lines to c are tangential to the circle at 3.
3. Project these elemental lines on to the top view and then down on to the front view as indicated by the arrows.

4. Project points 0, 1, . . ., 6 from the side view to intersect the elemental lines on the front view and then up to the top view to give corresponding points, 0, 1, . . ., 6 on the line of intersection of the cylinder and cone on each view.
5. Draw a smooth curve through the points to give the line of intersection on each view.

## 2. CUTTING PLANE METHOD

This involves drawing a series of horizontal cutting planes, each of which cuts through both the intersecting surfaces, for example a cone (to give a circle) and a cylinder (to give a rectangle).

*Cone and cylinder intersection (Fig. 8.8)*

1. Draw the front, side and top views of the cone, showing the intersection of the cylinder on the side view, the intersection being a circle.
2. Divide the end view of the cylinder into twelve equal parts numbered 0, 1, 2, 3, 4, 5 and 6.
3. Project these points across to the front view to represent a series of horizontal cutting planes through the cylinder and cone.
4. Project the cutting planes from the front view on to the top view, where they are represented by circles.

**FIGURE 8.7** ▶ Line of intersection of cylinder and cone—element method

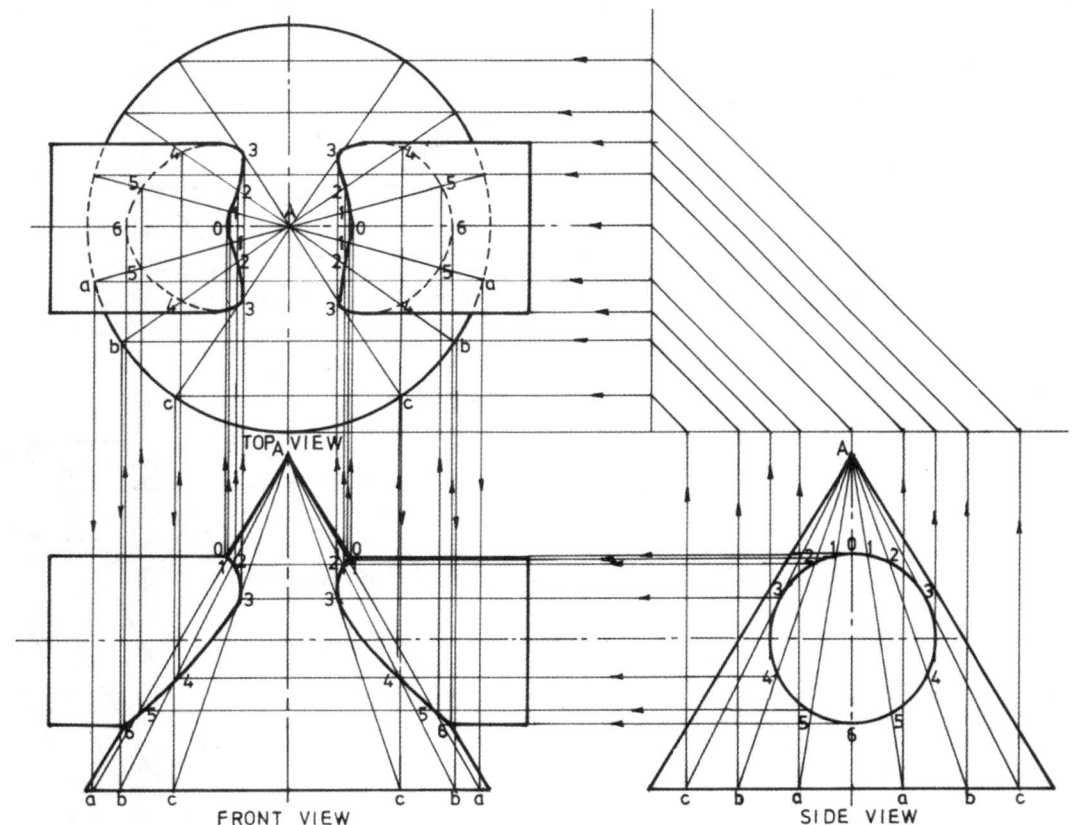

TOP VIEW

FRONT VIEW

SIDE VIEW

5. Project the points 0, 1, . . ., 6 from the side view up to the top view and along the cylinder. The distances between lines of similar numbers, for example 1-1 on the top view, is the width of the cylinder section at that level.

6. The intersections of lines 0, 1, . . ., 6 drawn along the cylinder on the top view with the circles drawn by projecting the cross-sections of the cone at these levels represent points on the line of intersection. Join these points with a smooth curve to give the line of intersection.

7. Project the points 0, 1, . . ., 6 from the line of intersection on the top view down to the front view to intersect the corresponding line on that view to give points on the line of intersection. Join them with a smooth curve.

### 3. COMMON SPHERE METHOD

When intersecting cylinders and cones envelop a common sphere, the line(s) of intersection are straight when viewed from the side.

#### *Cone and cylinder intersection (Fig. 8.9(a))*

1. Draw the front and side views of the cylinder and cone, showing the two views of the common sphere touching both surfaces in each view.

**Note:** The common sphere in the side view is also the end view of the cylinder.

2. The point of tangency indicated on the side view is horizontal to where the lines of intersection meet on the front view.

**Note:** It is not really necessary to draw the side view in order to find the line of intersection on the front view. The two straight lines which form the intersection may simply be drawn from one side to the other as shown, and they will cross at the point of tangency.

#### *Cone and two cylinders intersection (Fig. 8.9(b))*

1. Draw the side view showing the three surfaces A, B and C in their correct positions, each touching the common sphere. Note the axes all meet at a common point O, which is the centre of the common sphere.

2. Project each surface on until it intersects both of the other two surfaces. These intersections are shown as points a, b, c, d, e and f.

3. The two cylinders B and C alone would have intersected along ad, while the cylinder B and cone A alone would have intersected along be, and cylinder C and cone A alone along fc.

**FIGURE 8.8** ▶ Line of intersection of cylinder and cone—cutting plane method

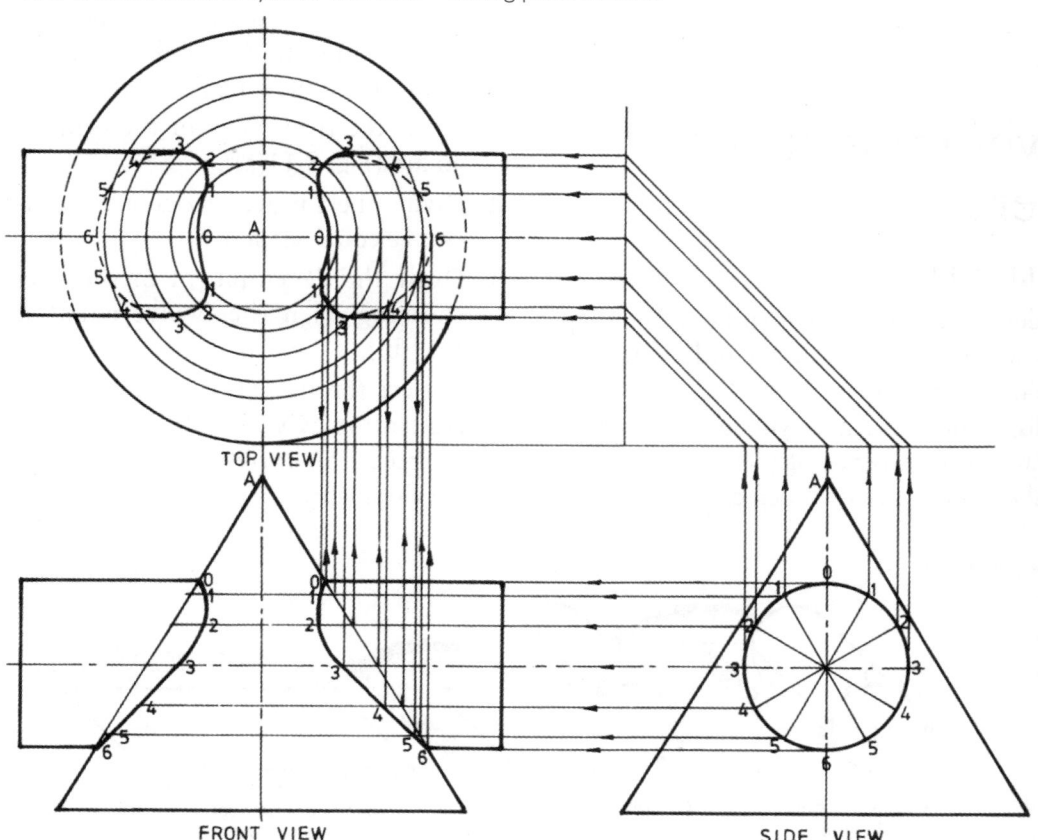

TOP VIEW

FRONT VIEW          SIDE VIEW

4. These lines of intersection cross at a common point labelled X.
5. The portion of these lines which form the line of intersection of A, B and C combined is made up of three parts (aX, cX and eX) shown outlined in Figure 8.9(b).

**FIGURE 8.9** ▶ Line of intersection of cylinder and cone—common sphere method

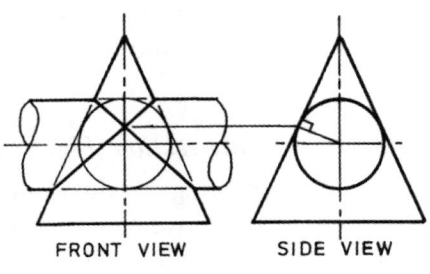

FRONT VIEW    SIDE VIEW

**(a)**

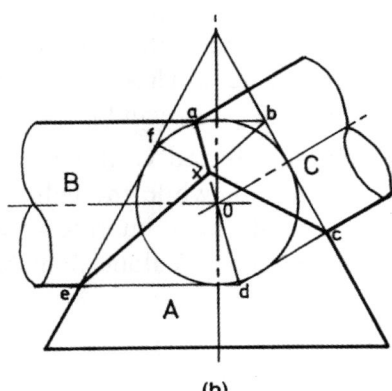

**(b)**

# 8.3 Development of cylinders

## RIGHT CYLINDER

A right cylinder is a closed circular surface. The shape of the cross-section of all right cylinders at right angles to the axis is circular.

The development of a right cylinder is illustrated in Figure 8.10, where a series of pictorial views (a), (b) and (c) show how the cylinder can be unrolled from the formed position at (a) to the flat rectangular surface at (c), the dimensions of which are equal to the circumference and length of the cylinder. For practical development purposes, the circumference is usually found by dividing the circle representing the cross-section of the cylinder into twelve equal parts and transferring one of these on to a straight line twelve times.

However, this method of determining the length of the cylinder stretchout circumference is always short because the chordal length, rather than the exact length, of one-twelfth of the circumference is used. An accurate method is to determine the circumference mathematically $(\pi D)$. If truncations and/or intersections must be plotted on the development, this length should be divided geometrically into twelve equal parts to create the surface element lines. The method described at 2.15 is ideal for dividing the circumference into twelve equal parts.

## TRUNCATED RIGHT CYLINDER

A truncated right cylinder is one which is cut at an angle to the axis. A practical example of its use is in the construction of the cylindrical elbow shown in Figure 8.11(a). This is made up of two truncated right cylinders joined together along the axis of truncation to form the included angle of the bend, in this case 120°. The development is more complicated than for the plain right cylinder because the line of truncation on the development is curved and has to be plotted. It is obtained by following these steps:

1. Draw the front view of the cylindrical elbow as an aid (Fig. 8.11(a)).
2. Divide the surface of one arm of the elbow into twelve equal sections.
3. Draw the development of the overall cylinder, showing on it the twelve equal sections (Fig. 8.11(b)).
4. Project the points of intersection of the line of truncation XX and the section lines across to intersect the corresponding section lines on the development.

**FIGURE 8.10** ▶ Right cylinder

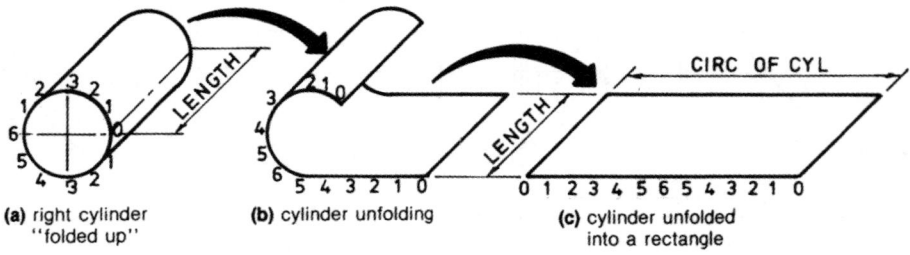

**(a)** right cylinder "folded up"    **(b)** cylinder unfolding    **(c)** cylinder unfolded into a rectangle

5. Join these points with a smooth curve to complete the development.

Note: The joint line is normally located along the shortest section of the surface.

## OBLIQUE CYLINDER

An oblique cylinder can be defined as a closed curved surface in which the shape of the cross-section at right angles to the axis is elliptical. One particular cross-section at an angle to the axis is circular, and it is at this cross-section that the joining of right cylindrical pipes takes place. See Figure 8.12(a).

To obtain the development, follow these steps (refer to Fig. 8.12(b)):

1. Draw the front view of the oblique cylinder as an aid.

2. Divide the surface into twelve equal sections.

3. Project the joint line 6-6 to the side of the front view parallel to the axis of the oblique cylinder.

4. From one end of 6-6, say the top, strike an arc equal to one-twelfth the circumference of the base, and project point 5 across from the front view to intersect the arc to give point 5 on the development.

5. Continue striking arcs equal to one-twelfth the base circumference and projecting across the next point around the circumference until the top curve of the development is plotted.

6. Draw a smooth curve through these points.

7. To plot the bottom curve, project the base points across from the front view to intersect lines drawn from corresponding points on the top curve.

**FIGURE 8.11** ▶ Truncated right cylinder

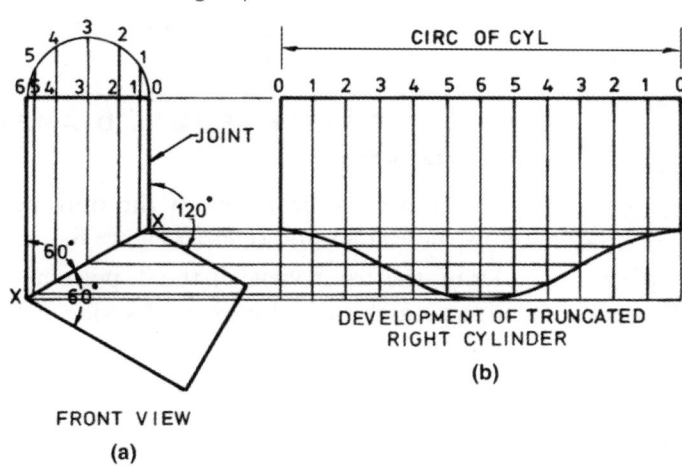
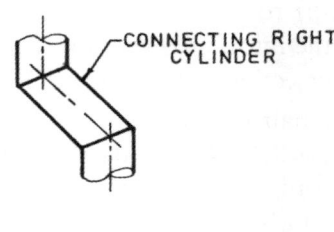

**FIGURE 8.12** ▶ Oblique cylinder

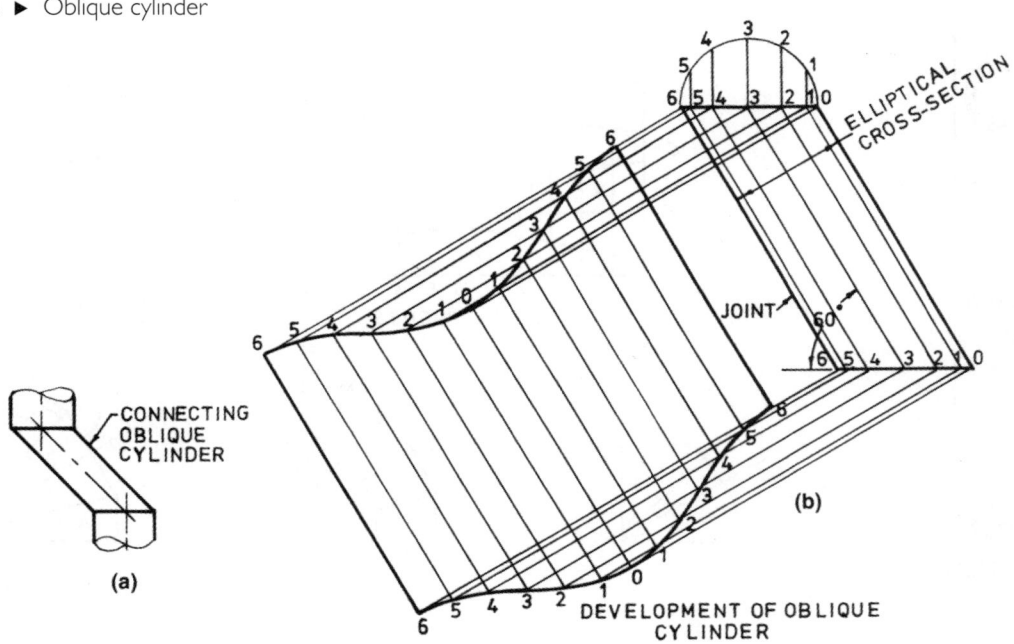

## ELBOWS

Elbows are used to change the direction of pipes in round, square, hexagonal and other cross-sections and are normally made of two, three, four or more pieces. The number of pieces depends on the cross-sectional shape and area.

To develop the four-piece round elbow shown in Figure 8.13, follow these steps:

1. Draw the front view (Fig. 8.13 (a)) showing the heel and throat radii.

2. Divide the bend angle into the correct number of pieces. The middle pieces are twice as large as the end pieces. The angles in view (b) are determined by the following formula.

    Number of angular spaces
    $$= (\text{number of pieces} \times 2) - 2$$
    $$= (4 \times 2) - 2 = 6$$
    each angle $= 90° \div 6 = 15°$

3. Draw tangents to the heel and throat radii (Fig. 8.13 (c)) to intersect on radials 1', 3' and 5'. Draw the mitre lines on radii 1', 3' and 5' joining the intersection of the tangents.

4. Draw a half view of the end of the pipe as shown in Figure 8.13(d). Divide the view into six equal parts and project surface elements from the points of division parallel to the surfaces of pieces C and D. There is no need to repeat on B and A as these are the same as C and D.

5. Set the stretchout line of the development equal to the circumference of the pipe ($\pi D$) and divide it into twelve equal parts.

6. Draw surface element lines perpendicular to the stretchout line and mark off the appropriate lengths taken from Figure 8.13(d) to plot the developments of A, B, C and D. Note the four pieces have alternate seams on the throat and heel (Fig. 8.13(d)), which allows the curves on the development to be common for two pieces. Each piece is a truncated right cylinder.

# 8.4 Development of T pieces

The development of cylindrical T pieces involves finding the line of intersection of the two cylinders, and then drawing the development as shown in Figure 8.11 (page 265).

### OBLIQUE T PIECE—EQUAL DIAMETER CYLINDERS

To develop both branches of the oblique T piece, follow these steps and refer to Figure 8.14:

1. Draw the front view of the T piece as an aid to drawing the development. (A side view is also shown.)

**FIGURE 8.13** ▶ Development of a four-piece elbow

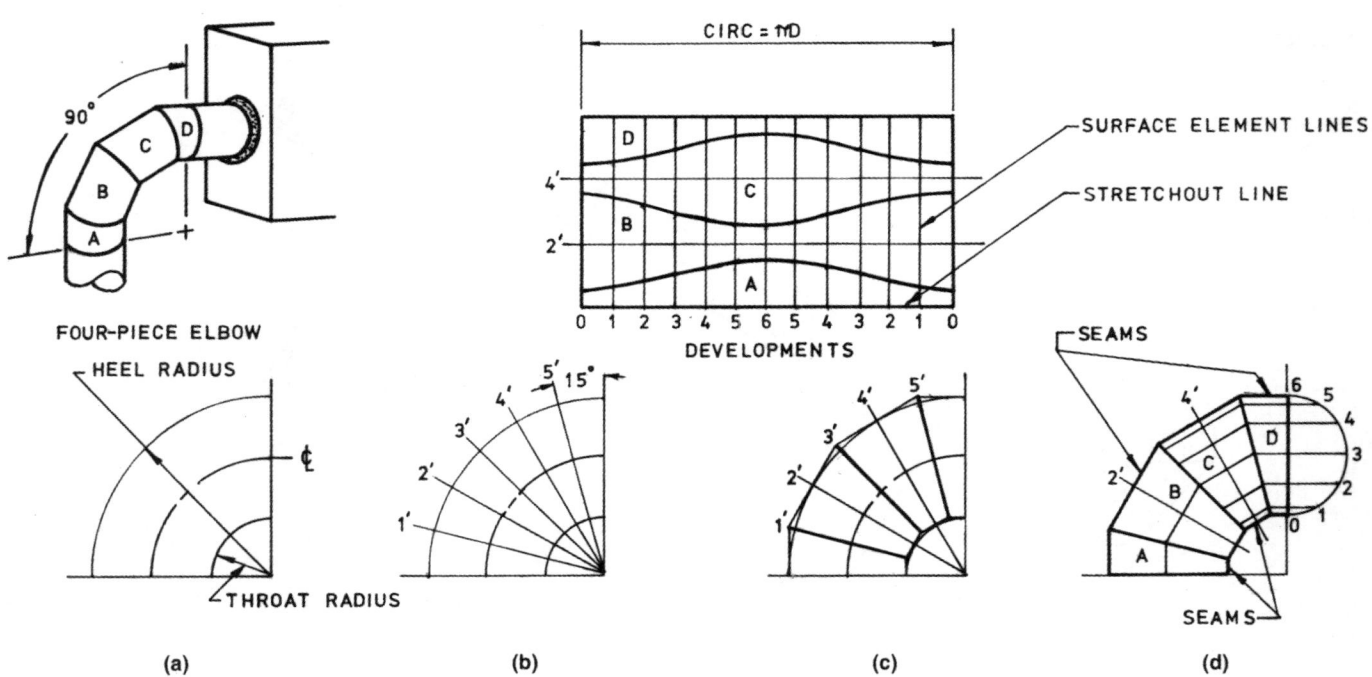

(a)  (b)  (c)  (d)

**FIGURE 8.14** ▶ Oblique T piece—equal diameter cylinders

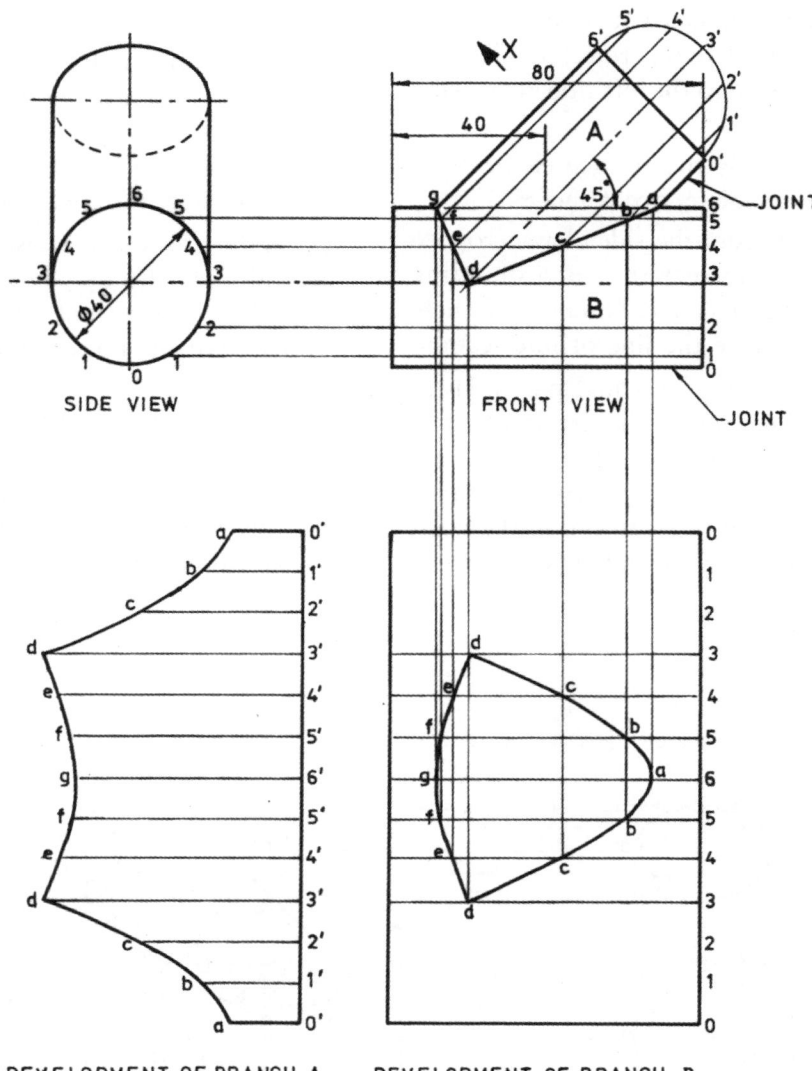

DEVELOPMENT OF BRANCH A    DEVELOPMENT OF BRANCH B

2. Draw the line of intersection of the two branches, in this case two straight lines meeting on the centre line at d (common sphere method).

3. Divide both branches A and B into twelve equal parts, and draw surface element lines which will meet the line of intersection at b, c, d, e and f.

4. The development of B (a rectangle) is projected below the front view, and the surface element lines which meet the line of intersection, namely 3, 4, 5 and 6, are drawn.

5. The shape of the line of intersection on the development of branch B is found by projecting points a, b, c, d, e, f and g from the front view onto the relevant lines of the development; a curve is drawn through the points so found.

6. The development of A would normally be projected above the front view in the direction of

arrow X at right angles to the centre line, but has been placed at the bottom of the page for the sake of layout.

7. The development of branch A (see Fig. 8.11) is found by developing the overall rectangle for the cylinder, drawing on it the surface element lines, and projecting points a, b, c, d, e, f and g on to the relevant element lines to give the required points for the intersecting end of branch A.

## OFFSET OBLIQUE T PIECE—UNEQUAL DIAMETER CYLINDERS

To develop both branches of the offset oblique T piece, refer to Figure 8.15 (page 268) and follow these steps:

1. Draw both the front and side views of the T piece as an aid in drawing the developments.

2. The line of intersection of A and B is first determined. Divide A on the front view into twelve equal surface elements 0' to 6'. Project the elliptical view of the end of A on to the left side view using transfer ordinates.

3. On the side view, draw the surface elements to intersect branch B at a, b, c, d, e, f and g.

4. Project these points from the side view across to the front view to intersect the corresponding surface elements at a, b, c, d, e, f and g. Join these points as shown to give the line of intersection.

5. Divide branch B into twelve equal parts. These are the long-dash lines. Draw the development of branch B below the front view, showing the long-dash dividing lines. An extra line is drawn tangential to the line of intersection on the front view at g and intersecting the end at X between 4

and 5. Plot this line on the development to aid in drawing the looping curve which just touches it.

6. Project the points from the front view where the long-dash lines cut the line of intersection down to the corresponding long-dash lines on the development to give points on the curved hole. Draw a smooth curve through these points.

7. As in the previous exercise, the development of A would normally be projected at right angles to branch A on the front view, but in Figure 8.15 it is placed at the bottom for better layout. Project the rectangle for the development of A, marking on it the surface element lines 0' to 6' as shown. Project the points a, b, c, d, e, f and g on to the relevant element lines to give the required points for the intersecting end of branch A. Draw a curve through these points.

**FIGURE 8.15** ▶ Offset oblique T piece—unequal diameter cylinders

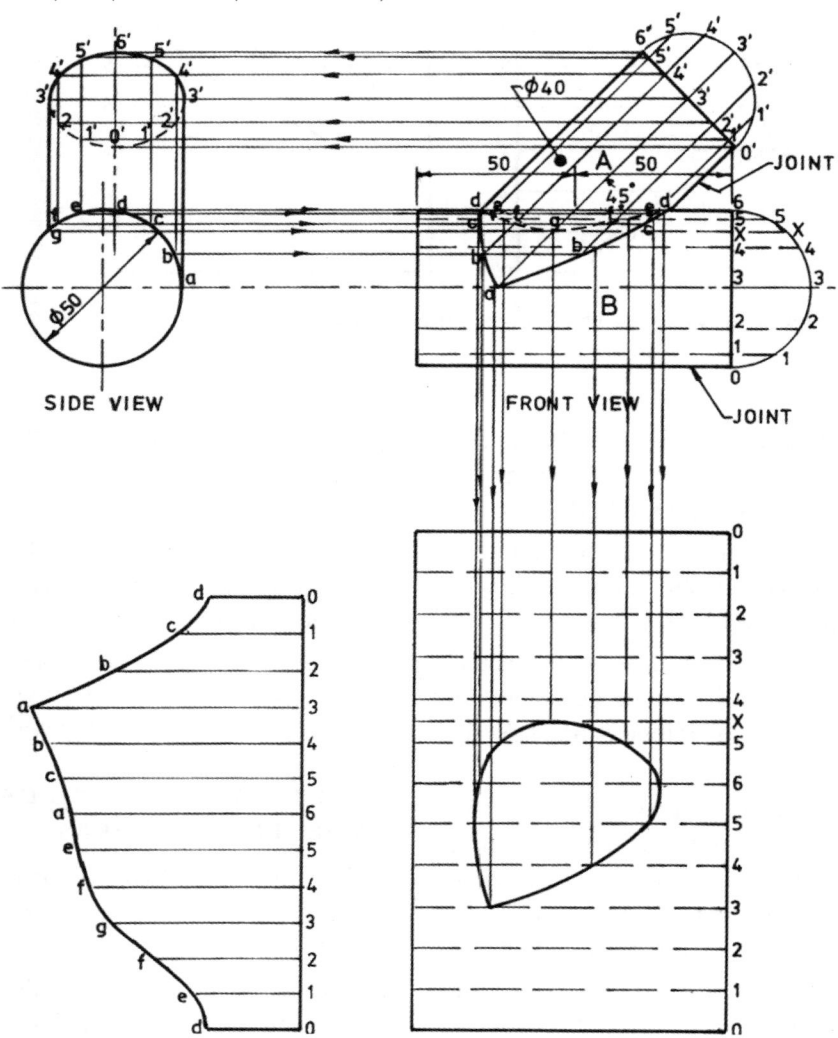

## OBLIQUE CYLINDRICAL CONNECTING PIPE

Figure 8.16 (page 270) illustrates the development of an oblique cylindrical connecting pipe with a cylindrical pipe insert. The development of the connecting pipe without the hole for the insert is described on page 265. The line of intersection between the insert and the connecting pipe for this problem must be determined before the development can be completed.

1. Draw the front and top views as shown.

2. To determine the line of intersection, use the cutting plane method described on pages 262–263. Consider a horizontal section B-B cutting both pipes. Project this section on to the top view where it is represented by two circles, centres f and g, intersecting at b. Project b down on to B-B to give b', a point on the required line of intersection.

3. Draw another section C-C, project it on to the top view to give c, and project c back to the section C-C to give c', another point on the line of intersection. In the same way indicate d' and as many more points as are required to plot a satisfactory curve.

4. Draw the development of the oblique cylinder as described on page 265.

5. Points a' and e' are projected directly across to the centre line, 6, on the development.

6. Draw surface element lines on the front view parallel to the axis and passing through b', c' and d' to meet the end of the pipe at X and Y. Note b' and d' are on the one line.

7. Project X and Y on to the end of the development (two positions each), and draw surface element lines from these points across the development.

8. Project points b' and d' from the front view on to the two surface element lines from X. Project c' onto the two lines from Y. Join up the points plotted on the development to give the egg-shaped hole shown.

**Note:** As this is a symmetrical development, each point found on the front view represents two points on the development. If the cylindrical insert were offset (e.g. if its axis were not in the same vertical plane as the axis of the connecting pipe), two curves would be required on the front view to give a non-symmetrical hole on the development similar to the problem for the offset oblique T piece development of Figure 8.15.

The development of the right cylindrical insert is shown on the right-hand side of Figure 8.16.

Figure 8.17 (page 271) illustrates a second method of obtaining the line of intersection between the insert and the connecting pipe.

1. Draw the front and half top views.

2. Instead of using horizontal sections as in Figure 8.16, a series of vertical sections are taken through the insert and the connecting pipe. The small diagram illustrates how one such section A-A through points 4 and 2 on the front side of the insert pipe is used to determine points c and e (circled) on the line of intersection. (The shaded area represents the side view of the vertical section taken along plane A-A.) Points c and e can also be found by using points 4 and 2 on the back side, as is done on the large diagram when only a half plane is used.

3. Other points on the line of intersection are determined in a manner similar to c and e.

# 8.5 Development of pyramids

## RIGHT PYRAMID

A right pyramid may be defined as a surface with a number of identical triangular sides which have a common apex situated vertically above the centre of the base. An important fact to remember about all right pyramids is that the sloping edges may be totally contained within the surface of an enveloping cone. This is illustrated in Figure 8.18(a); Figure 8.18(b) is a pictorial view of a hexagon-based right pyramid—its development is described as follows on page 271.

1. Draw the front and half top views of the pyramid, either looking across the points (Fig. 8.18(c)) or across the flats (Fig. 8.18(d)). Figure 8.18(c) gives the true length of the edges (called the slant height) directly. Figure 8.18(d) requires the top view of the edge to be rotated into the base line, and this point joined to the apex A to give the slant height.

2. To develop the pyramid, an arc of radius equal to the slant height is described, and one of the base edges (taken from the half top view) is marked around the arc six times (Fig. 8.18(e)). These points are joined to the centre of the arc A, and represent the fold lines along which the development is bent when 'forming up' the pyramid (Fig. 8.18(b)).

**FIGURE 8.16** ▶ Oblique cylindrical connecting pipe

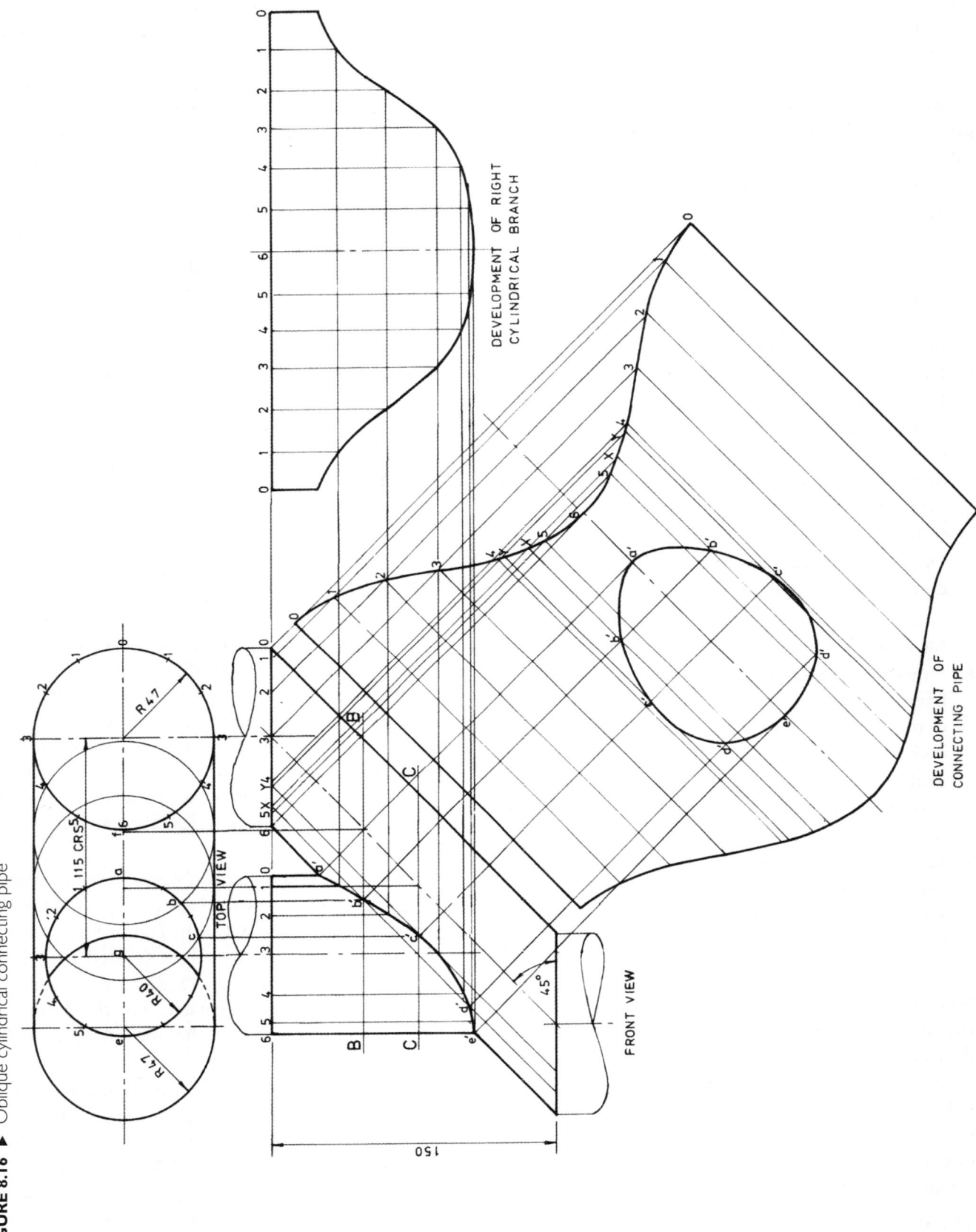

**FIGURE 8.17** ▶ Alternative method of determining the line of intersection

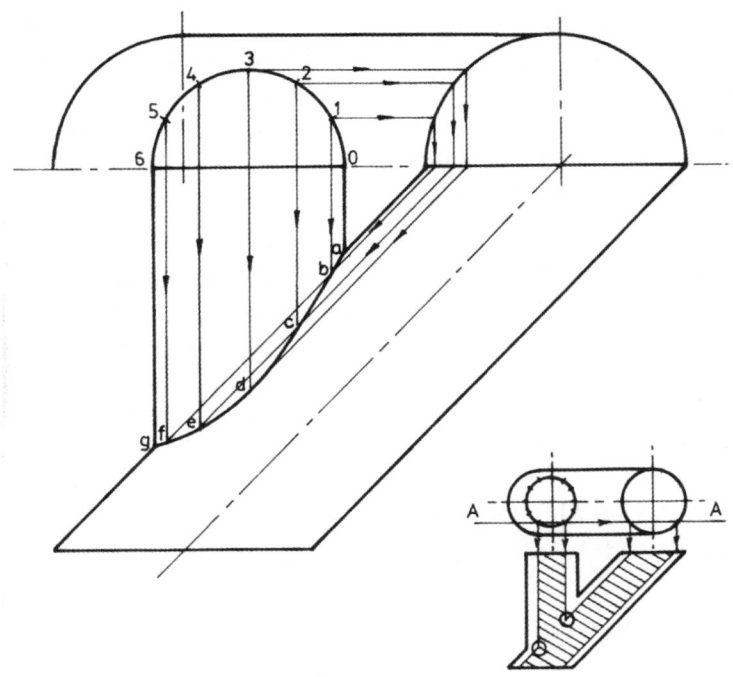

**FIGURE 8.18** ▶ Right pyramid

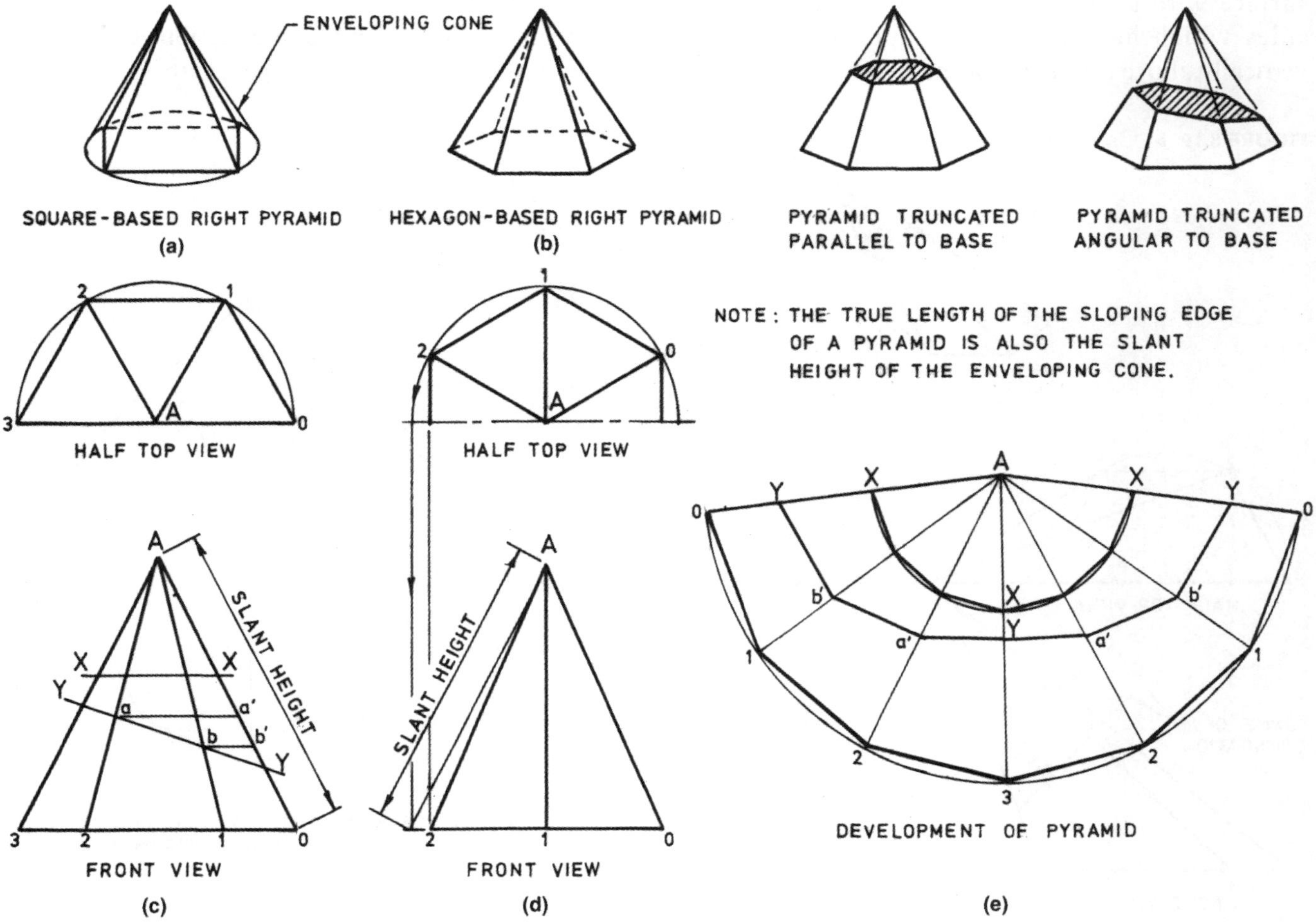

ENVELOPING CONE

SQUARE-BASED RIGHT PYRAMID
(a)

HEXAGON-BASED RIGHT PYRAMID
(b)

PYRAMID TRUNCATED
PARALLEL TO BASE

PYRAMID TRUNCATED
ANGULAR TO BASE

HALF TOP VIEW

HALF TOP VIEW

NOTE: THE TRUE LENGTH OF THE SLOPING EDGE
OF A PYRAMID IS ALSO THE SLANT
HEIGHT OF THE ENVELOPING CONE.

SLANT HEIGHT

FRONT VIEW
(c)

SLANT HEIGHT

FRONT VIEW
(d)

DEVELOPMENT OF PYRAMID

(e)

3. A pyramid may be truncated by a plane either parallel to the base (X-X) or at an angle to the base (Y-Y) (Fig. 8.18(c)). In the first case, the portion of the slant height (AX) from the apex to the truncating plane is used to describe an arc on the development, cutting the edge lines at points which are joined to give the line of truncation (XXX).

4. The angular truncation Y-Y which intersects the sloping edges at various distances from the apex needs to be plotted on the development. Project the points a and b (where Y-Y cuts the sloping edges) horizontally on to the slant height, points a' and b' (Fig. 8.18(c)). Then Aa' and Ab' are the true lengths of that part of the edges A2 and A1 cut off by the plane Y-Y.

5. Transfer these lengths (Aa' and Ab') from the front view onto the lines A2 and A1 respectively on the development.

6. The two lengths of AY on the front view are transferred to the edges A0 and A3 on the development to complete the points which when joined give the line of truncation (YYY).

## OBLIQUE PYRAMID

The oblique pyramid (Fig. 8.19) may be defined as a surface with a number of flat unequal triangular sides which have a common apex not situated vertically above the centre of the base.

Refer to Figure 8.19 as you read through these steps:

1. Draw the front and half top views of the pyramid.

2. Construct a true length diagram at the side of the front view in order to obtain the true length of the sloping edges. The diagram is based on the true length triangle described in Figure 4.5 and is constructed as follows.

3. Draw AP, the vertical height of the triangle, also equal to the vertical difference (VD) of the sides. From P plot P1 and P2 equal to A1 and A2 respectively on the half top view. Join A1 and A2 on the true length diagram, and these are the true lengths of the sloping edges A1 and A2. The true lengths of A0 and A3 may be taken directly from the front view.

4. Set down a length A0 on either the right or left side of the development.

5. Point 1 is found by the intersection of two arcs of radius A1 (taken from the true length diagram) and 1-2 which is equal to an edge of the base taken from the half top view.

6. Similarly, point 2 is found by the intersection of A2 from the true length diagram and the base edge length 1-2.

7. Point 3 is found by the intersection A3 from the front view and the base edge length 1-2.

**FIGURE 8.19** ▶ Oblique pyramid—whole and truncated

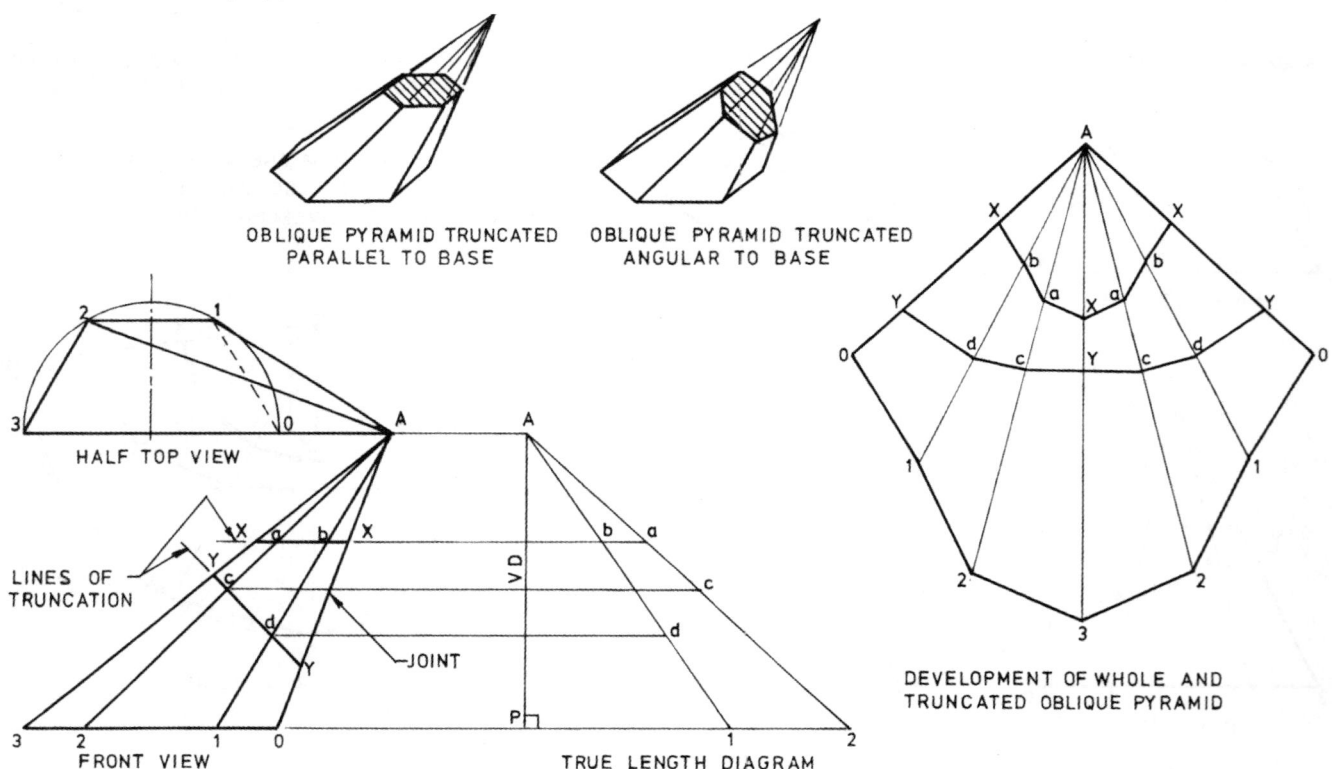

8. The second half of the development is symmetrical to the half already plotted, and may be constructed by projection or by intersecting arcs similar to the first half commencing at A3 and finishing at A0.

9. A truncation parallel to the base such as XX is projected across to the true length diagram to determine the true lengths of Aa and Ab. These lengths are then transferred to the development along A2 and A1 respectively. The two lengths of AX taken from the front view are also plotted along A0 and A3 to complete the line of truncation (XXX).

10. A truncation angular to the base, such as YY, is projected across to the true length diagram to determine the true lengths of Ac and Ad. These lengths are then transferred to the development

along A2 and A1 respectively. The two lengths of AY taken from the front view are also plotted along A0 and A3 to complete the line of truncation (YYY).

# 8.6 Development of cones

## RIGHT CONE

A right cone can be defined as a surface which has a circular base and a curved sloping side which radiates from a point situated vertically above the centre of the base. This point is called the apex of the cone. The length of any straight line drawn down the sloping side from the apex to the base is constant and is called the slant height of the cone. Figure 8.20(a) shows a pictorial view of the right cone.

**FIGURE 8.20** ▶ Right cone

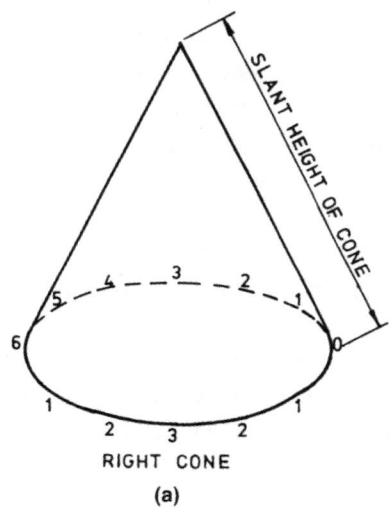

RIGHT CONE
(a)

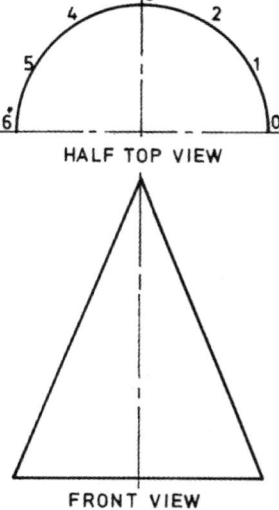

HALF TOP VIEW

FRONT VIEW

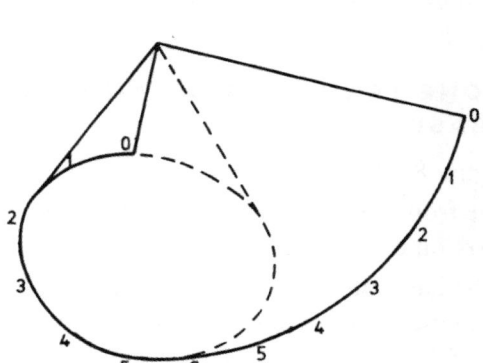

PICTORIAL VIEW OF CONE UNFOLDING
(b)

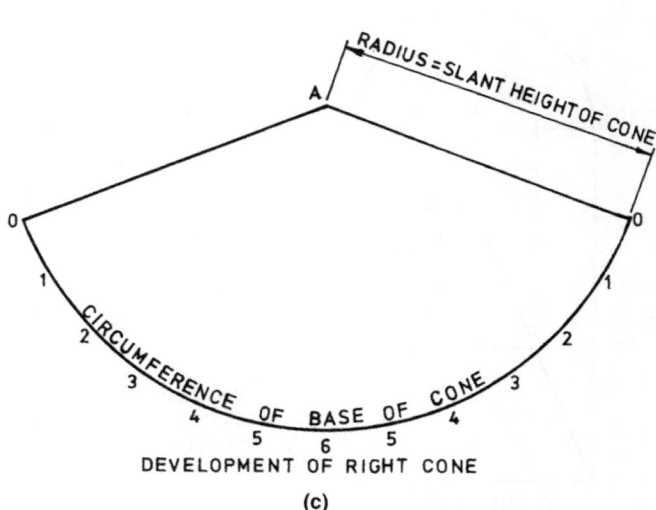

DEVELOPMENT OF RIGHT CONE
(c)

The shape of a plane surface required for the development of a right cone is shown in Figure 8.20(b), and is part of a circle, the radius of which is equal to the slant height of the cone.

Refer to Figure 8.20(c) for the development which is obtained as follows:

1. Draw the front and half top views of the cone, dividing the base into twelfths of its circumference.
2. With radius A0, the slant height of the cone, describe an arc and mark off one-twelfth of the base circumference around this arc twelve times. Join the ends of the arc 0 to A to complete the development.

## RIGHT CONE TRUNCATED PARALLEL TO THE BASE

Refer to Figure 8.21.

1. First draw the development of the whole cone as described in the previous exercise.
2. Describe a second arc on the development (R2) equal in length to the slant distance of the line of truncation from the apex A. The portion of the development between R1 and R2 is the truncated portion.

## RIGHT CONE TRUNCATED AT AN ANGLE TO THE BASE

Refer to Figure 8.22.

1. Draw the front and top views, showing the line of truncation on each view.
2. Divide the base into twelve equal parts, and draw surface element lines connecting these points to the apex, A1, A2, etc.
3. The true lengths of the surface element lines

between the apex A and the line of truncation 0'6' are found by projecting horizontally on to the slant height. That is A5', A4', A3', etc. are the true lengths of these elements.

4. With centre A on the development and radius the slant height of the cone, describe an arc. Mark one-twelfth the base circumference around the arc twelve times, and join these points to A. These are the surface element lines on the development corresponding to those on the front view.
5. Taking the true lengths A0', A1', A2' . . . A6' from the slant height, mark them off successively along the corresponding surface element line on the development. Join the points with a smooth curve to complete the line of truncation.

## RIGHT CONE–VERTICAL CYLINDER INTERSECTION

Refer to Figure 8.23.

1. Draw the front and half top views.
2. The line of intersection is determined by the method of horizontal sections. Draw X-X, a horizontal section cutting both surfaces.
3. Project this section onto the half top view, where it is represented by the two semicircles intersecting at x. Project x back to section X-X to point x', a point on the required line of intersection.
4. Continue taking a series of horizontal sections, all of which cut both surfaces, until sufficient points on the line of intersection have been found to draw a satisfactory curve (say four points). The line of intersection could also have been determined by the surface element method (Fig. 8.7).

**FIGURE 8.21** ▶ Right cone truncated parallel to the base

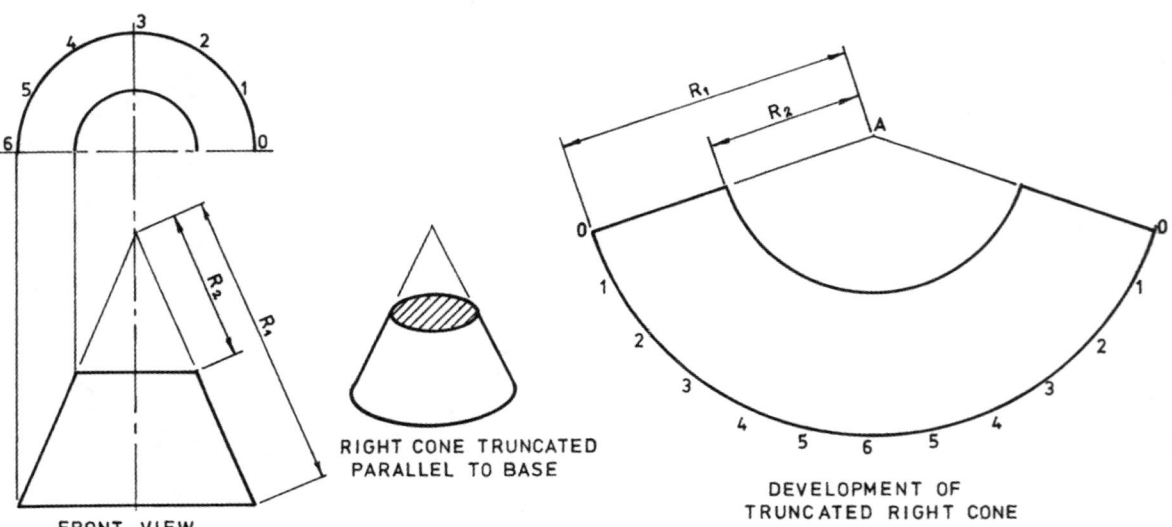

FRONT VIEW

RIGHT CONE TRUNCATED
PARALLEL TO BASE

DEVELOPMENT OF
TRUNCATED RIGHT CONE

**FIGURE 8.22** ▶ Right cone truncated at an angle to the base

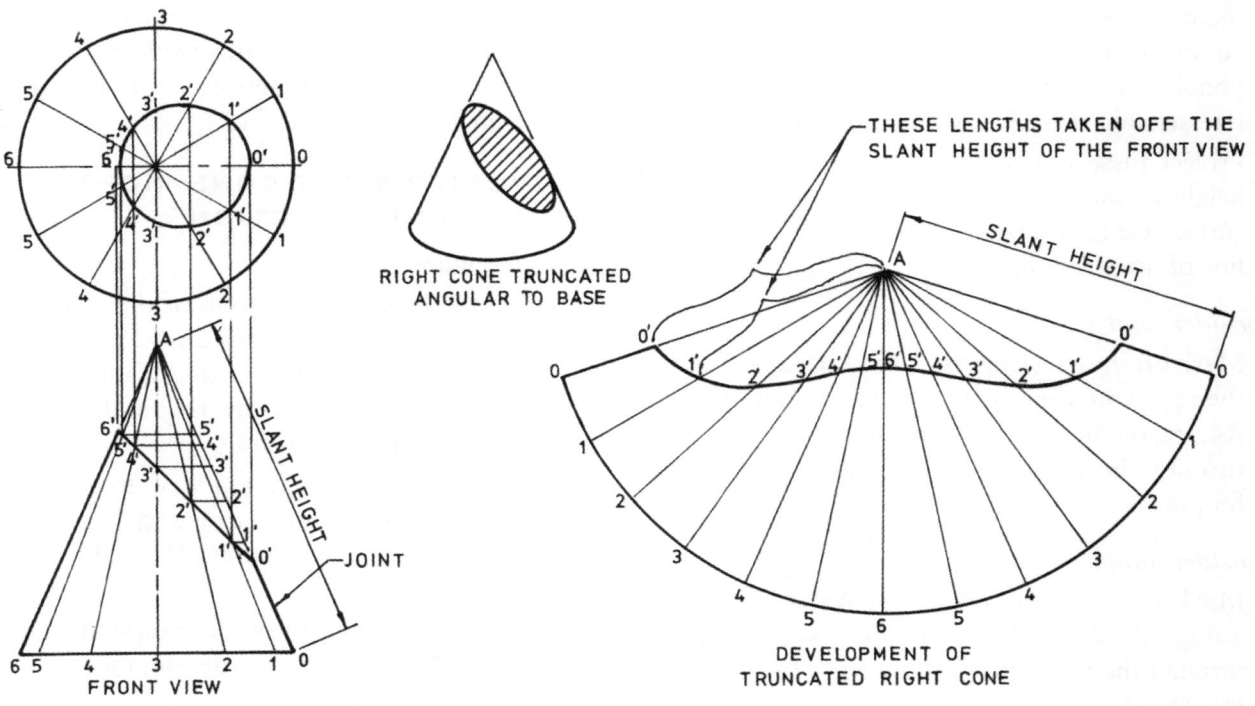

RIGHT CONE TRUNCATED
ANGULAR TO BASE

THESE LENGTHS TAKEN OFF THE
SLANT HEIGHT OF THE FRONT VIEW

SLANT HEIGHT

SLANT HEIGHT

JOINT

FRONT VIEW

DEVELOPMENT OF
TRUNCATED RIGHT CONE

**FIGURE 8.23** ▶ Right cone–vertical cylinder intersection

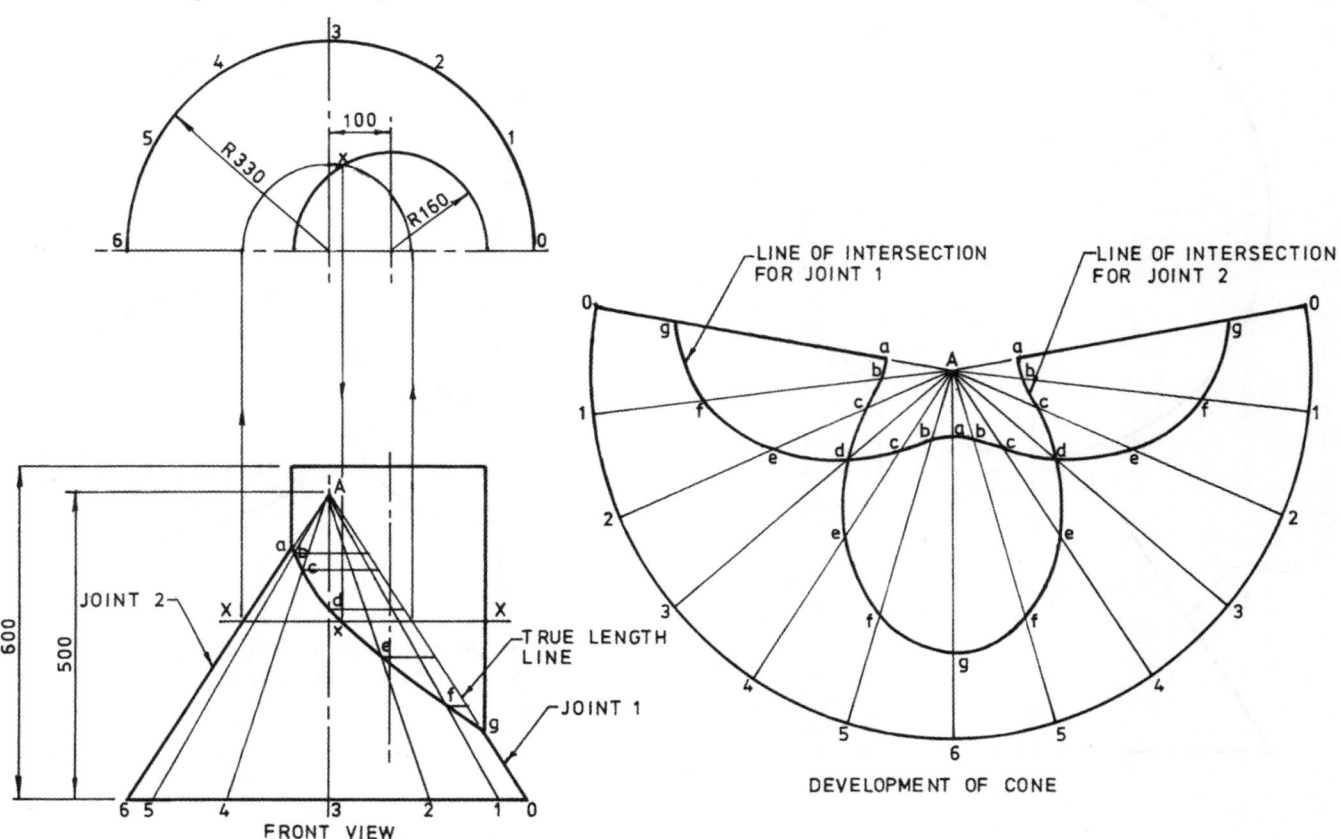

JOINT 2

JOINT 1

TRUE LENGTH
LINE

FRONT VIEW

LINE OF INTERSECTION
FOR JOINT 1

LINE OF INTERSECTION
FOR JOINT 2

DEVELOPMENT OF CONE

5. The development of the right cone is drawn, including surface element lines radiating from equal positions around the cone base. Corresponding surface element lines intersect the line of intersection at b, c, d, e and f on the front view.

6. Project these points horizontally on to the slant height of the cone to find the true lengths of the surface element lines between the apex A and the line of intersection.

### Consider joint 1

7. Mark off the true lengths of Aa, Ab, Ac, etc. along the corresponding surface element lines, A6, A5, A4, etc. on the development. Draw a smooth curve through the points to give the line of intersection for joint 1.

### Consider joint 2

8. Mark off the true lengths of Aa, Ab, Ac, etc. along A0, A1, A2, etc. Draw a smooth curve through the points to give the line of intersection for joint 2.

The effect on the development of using one or the other of the two joints is shown. For joint 1 the line of intersection narrows the development at the ends and widens it at the centre. For joint 2 the reverse applies, and the development is wider at the ends and narrow in the centre. Joint 1 is more desirable as the joint is shorter and the development easier to handle with narrow ends and a wide centre.

## TRUNCATED RIGHT CONE–RIGHT CYLINDER INTERSECTION

Refer to Figure 8.24.

1. Draw the line of intersection of the cylinder and cone as described on pages 262–263 and shown in Figure 8.8, using the cutting plane method. If the developments only are required, a half top view is all that is necessary.

2. Join the apex A on the top view to the cylinder intersection points b, c, d, e and f, and extend onto the base circle at points 1', 2', 4', 5' and 3' respectively.

3. Draw the development of the truncated cone with A0 as the centre line of the development.

4. Step the distances 0'1', 0'2', 0'3', etc. taken from the base on the top view to either side of 0' on the development. Join these points to A to form surface elements.

**FIGURE 8.24** ▶ Truncated right cone–right cylinder intersection

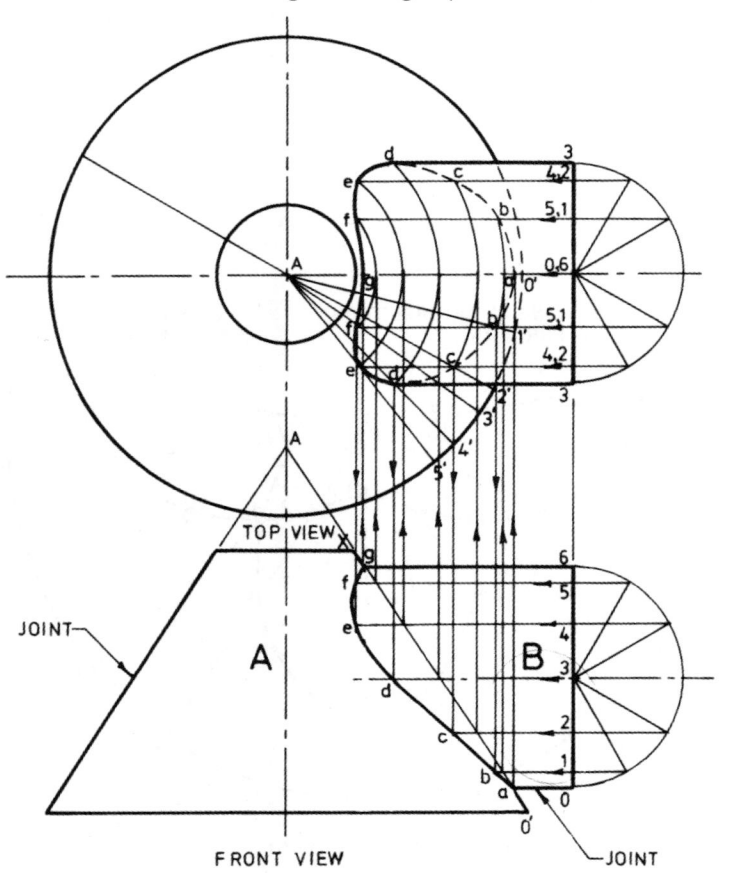

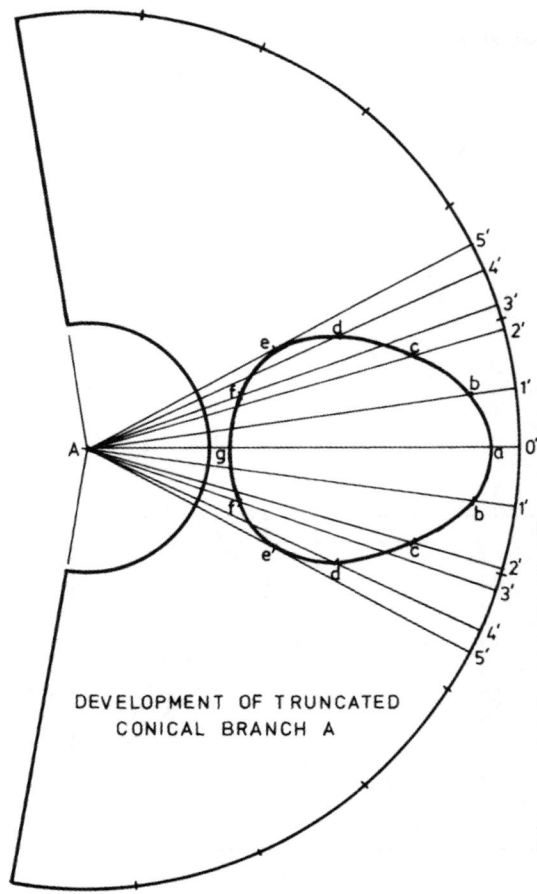

DEVELOPMENT OF TRUNCATED
CONICAL BRANCH A

5. Step along these elements from A the corresponding true lengths obtained from the slant height on the front view. Join the points determined to give the developed shape of the cylinder intersection.

## RIGHT CONE–RIGHT CYLINDER, OBLIQUE INTERSECTION

Refer to Figure 8.25.

1. It is necessary to draw an auxiliary view of the cone and cylinder showing the true shape of the cross-section of the intersecting cylinder in order to plot the line of intersection on the front view.

2. On the auxiliary view, draw surface element lines which pass through one half of the view of the cylinder to intersect the base at 0, a, b and c. There is no need to draw lines through the other half as the line of intersection is symmetrical about A0.

3. Project a, b and c from the auxiliary view across to the base of the front view and join to the apex.

4. Project the intersections of the cylinder and the surface element lines on the auxiliary view across to the corresponding surface element lines on the front view to give points on the line of intersection. Join with a smooth curve.

5. Draw the overall development of the cone and plot the points, a, b and c on it. This is achieved by projecting a, b and c from the front view up to the top view and transferring these positions on to the development.

6. Determine points on the line of intersection on the top view by projecting points on the line of intersection on the front view to the corresponding lines A0, Aa, Ab and Ac on the top view. Join these points on the top view with a smooth curve.

7. Project points from the line of intersection on the front view across to the slant height to give the true lengths of these elements which are in turn transferred to the corresponding elements on the development to give points on the line of intersection. Draw a smooth curve through the points.

**FIGURE 8.25** ▶ Right cone–right cylinder, oblique intersection

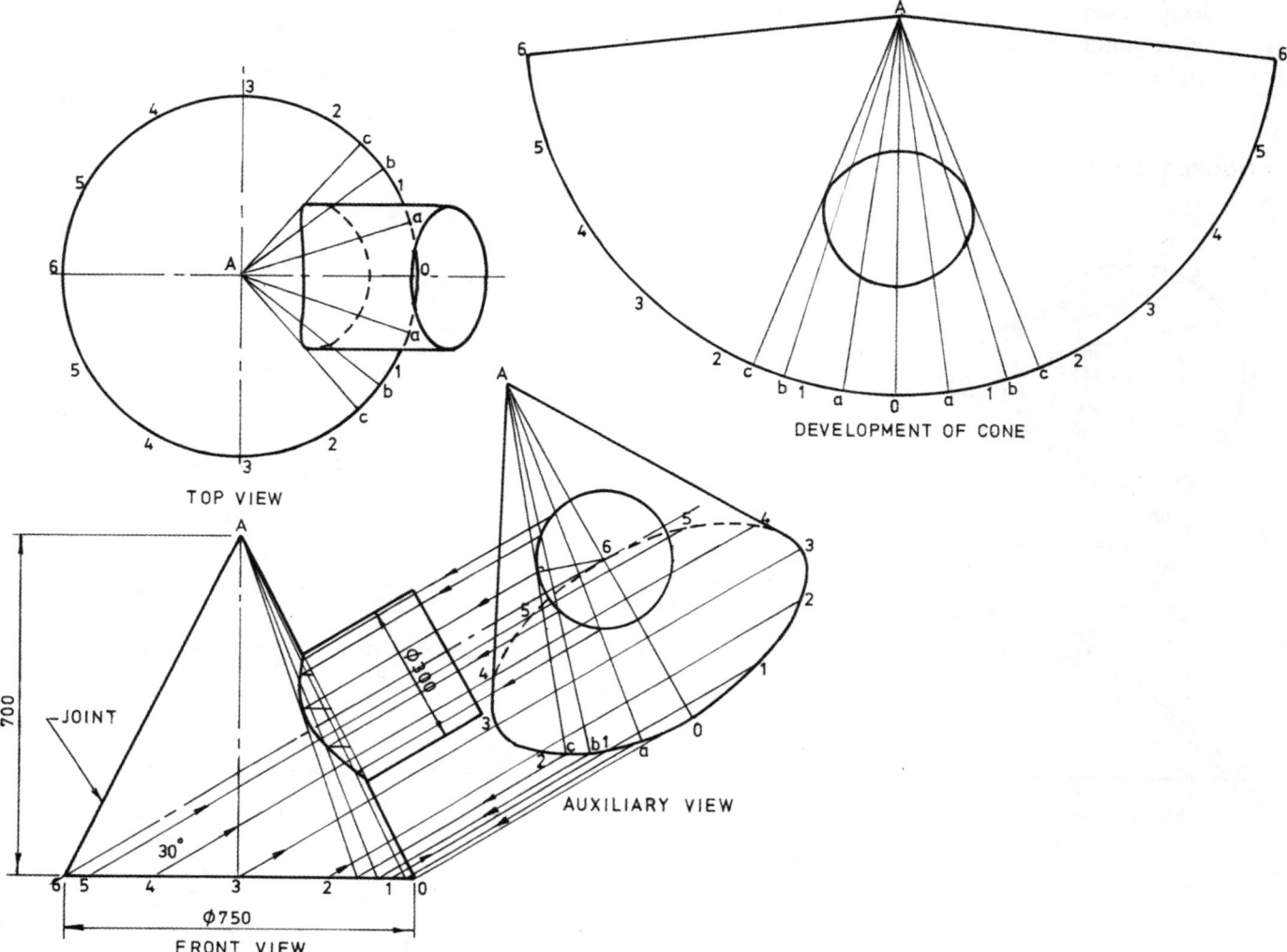

## OBLIQUE CONE

An oblique cone can be defined as a surface which has a circular base and a curved sloping side which radiates from a point not situated vertically above the centre of the base. The length of any straight line drawn down the sloping side from the apex to the base is not constant; hence, the oblique cone does not have a constant slant height, and its development is somewhat more complicated than that of the right cone. Refer to Figure 8.26.

1. Draw the front and top views of the oblique cone, showing surface element lines connecting the apex to the twelve divisions around the base.

2. Construct the true length diagram to the side of the front view based on the true length triangle (Fig. 4.5). Draw AP, the vertical difference (VD) in the heights of all surface element lines. From P, mark off P1, P2, P3, P4 and P5 equal in length to A1, A2, A3, A4 and A5 respectively on the top view.

3. Join points 1, 2, 3, 4 and 5 to A on the true length diagram. These are the true lengths of the corresponding surface element lines on the front and top views.

4. The development of the whole cone is now drawn. Set down A0, one side of the development taken from the right-hand side of the front view.

5. Each point is now successively located by describing two arcs to intersect; for example, point 1 is determined by describing arc A1 taken from the true length diagram to intersect with arc 01 equal to one-twelfth of the base circumference taken from the top view. Points 2, 3, 4, 5 and 6 are plotted similarly, although A6 is taken from the left-hand side of the front view.

6. The second half of the development, which is symmetrical to the first, is determined by projection or is plotted in a manner similar to the first half commencing at A6 and finishing at A0.

7. A truncation parallel to the base, such as X-X, is projected across to the true length diagram to determine the true lengths of Aa, Ab, Ac, Ad and Ae, which are those portions of the surface element lines between the line of truncation and the apex.

8. These true lengths are transferred to the development along A1, A2, A3, A4 and A5 respectively. The two lengths of AX taken from the front view are also plotted along A0 and A6 to complete the line of truncation (XXX).

**FIGURE 8.26** ▶ Oblique cone—whole and truncated

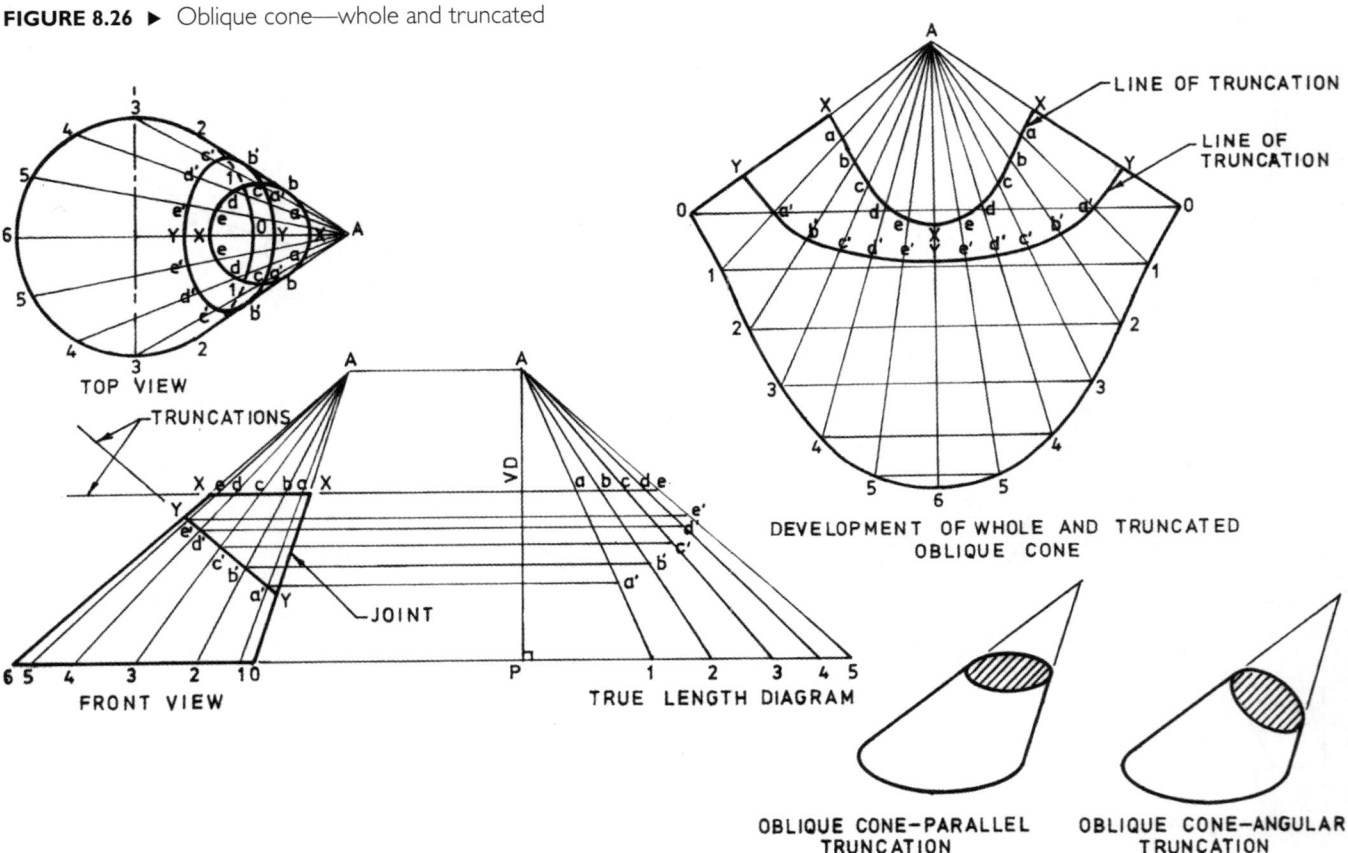

TOP VIEW

TRUNCATIONS

JOINT

FRONT VIEW

VD

TRUE LENGTH DIAGRAM

LINE OF TRUNCATION

LINE OF TRUNCATION

DEVELOPMENT OF WHOLE AND TRUNCATED OBLIQUE CONE

OBLIQUE CONE—PARALLEL TRUNCATION

OBLIQUE CONE—ANGULAR TRUNCATION

9. A truncation angular to the base, such as Y-Y, is projected across to the true length diagram to determine the true lengths of Aa', Ab', Ac', Ad' and Ae', which are those portions of the surface element lines between the line of truncation and the apex.

10. These true lengths are transferred to the development along A1, A2, A3, A4 and A5 respectively. The two lengths of AY taken from the front view are also plotted along A0 and A6 to complete the line of truncation (YYY).

## OBLIQUE CONE–OBLIQUE CYLINDER INTERSECTION

Refer to Figure 8.27.

1. Draw the front and top views. These are required to draw the line of intersection between the two surfaces and hence plot it on the development.

2. The line of intersection is determined by the method of sections. Draw any horizontal section A-A cutting both surfaces of the cone and cylinder. Where the section cuts the axis of the cone, construct a semicircle on a diameter equal to the cross-section of the cone at this level.

3. Similarly construct another semicircle on a diameter equal to the cross-section of the cylinder at this level to cut the first semicircle at point a.

4. Project point a downwards on to the section line at point a' to give a point on the line of intersection.

5. Project point a upwards to the centre line of the top view, and mark off a distance on either side of this line equal to aa' taken from the front view. This locates two points on the line of intersection on the top view.

6. A sufficient number of horizontal sections are taken to provide enough points to enable smooth curves to be drawn on the front and top views to give the lines of intersection of the cone and cylinder. Usually four or five sections are required.

**FIGURE 8.27** ▶ Oblique cone–oblique cylinder intersection

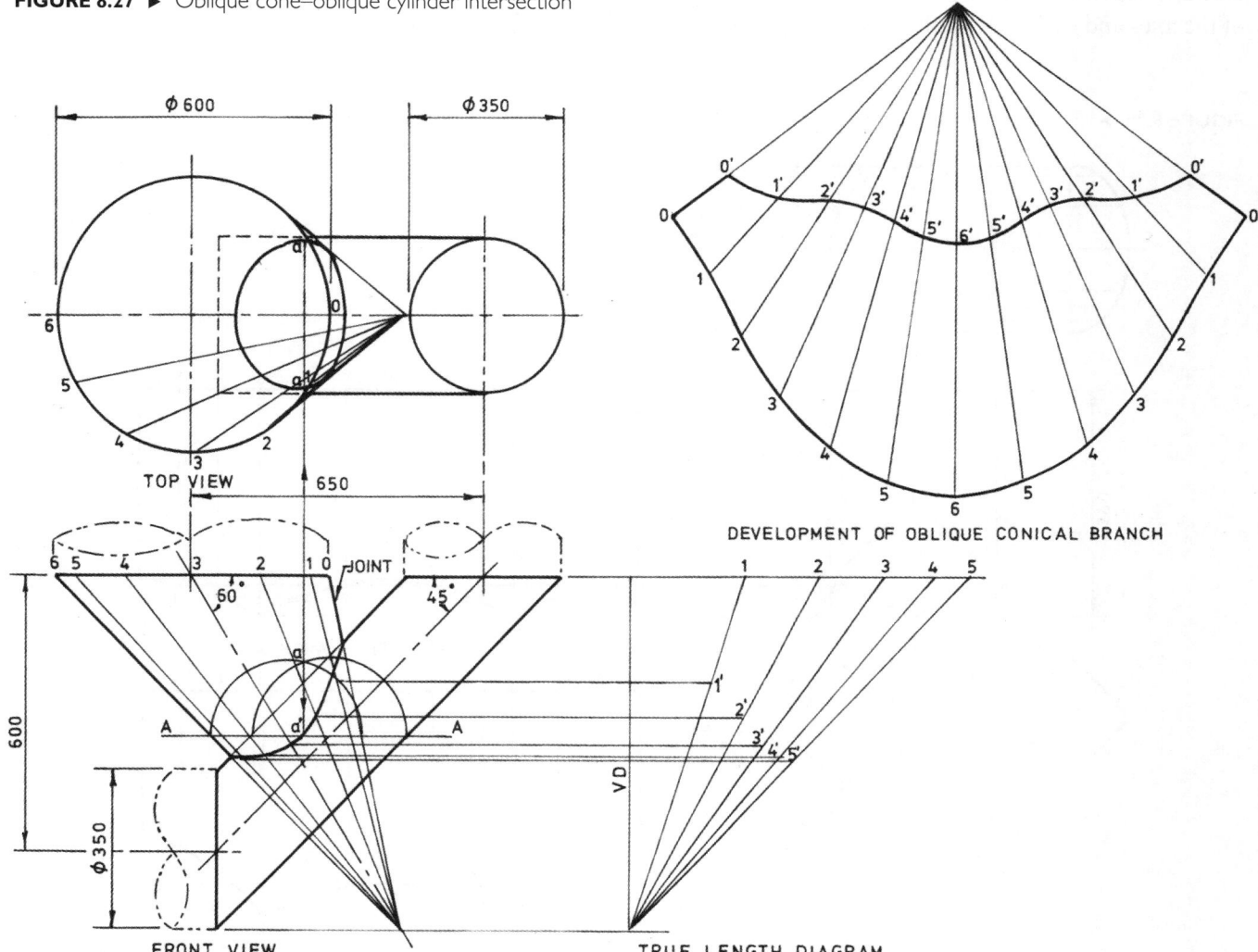

DEVELOPMENT OF OBLIQUE CONICAL BRANCH

TOP VIEW

FRONT VIEW

TRUE LENGTH DIAGRAM

7. The development of the oblique cone can now proceed. First develop the whole cone, then plot the line of truncation on it as described on page 278. Note that the cone is inverted in this case, as is the true length diagram.

# 8.7 Development of breeches or Y pieces

The breeches piece is a three-way junction between cylindrical pipes or between cylindrical pipes and conical sections. The angles between the various branches can be equal or unequal. The main requirement when drawing the front view and determining the line of intersection of the branches is that each branch should envelop a common sphere represented on the front view by a circle. This is shown on each of the three exercises of Figures 8.28, 8.29 and 8.30. On breeches pieces involving equal cylinders (Figs 8.28 and 8.29), the common point of intersection of the three cylinders on the front view is also coincident with the intersection of the axes and the centre of the common sphere.

Once the front view has been drawn and the line of truncation determined, the development of the branches is merely that of truncated cylinders and cones.

### BREECHES PIECE—EQUAL ANGLE, EQUAL DIAMETERS; UNEQUAL ANGLE, EQUAL DIAMETERS

This method applies to both Figures 8.28 and 8.29.

1. Draw the front view of the breeches piece, determining the line of intersection of the three branches A, B and C by the common sphere method outlined on page 263 and illustrated in Figure 8.9(b).

2. The development of the branches is determined by the surface element method described on page 264 and illustrated in Figure 8.11.

**Note:** The development of branches B and C would have been more conveniently projected at right angles to the side of B or C, as was the development of A. However, the positions have been chosen for the sake of page layout. The top view has been included for clarity.

**FIGURE 8.28** ▶ Breeches piece—equal angle, equal diameters

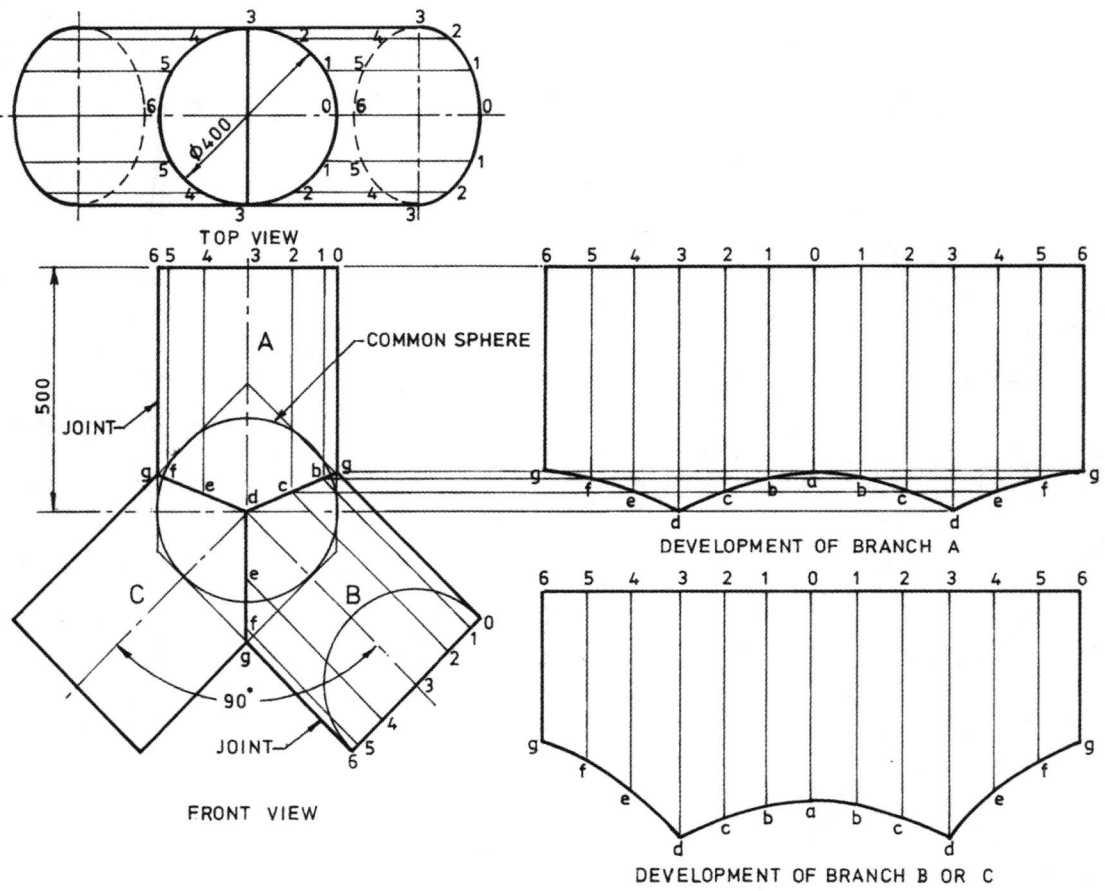

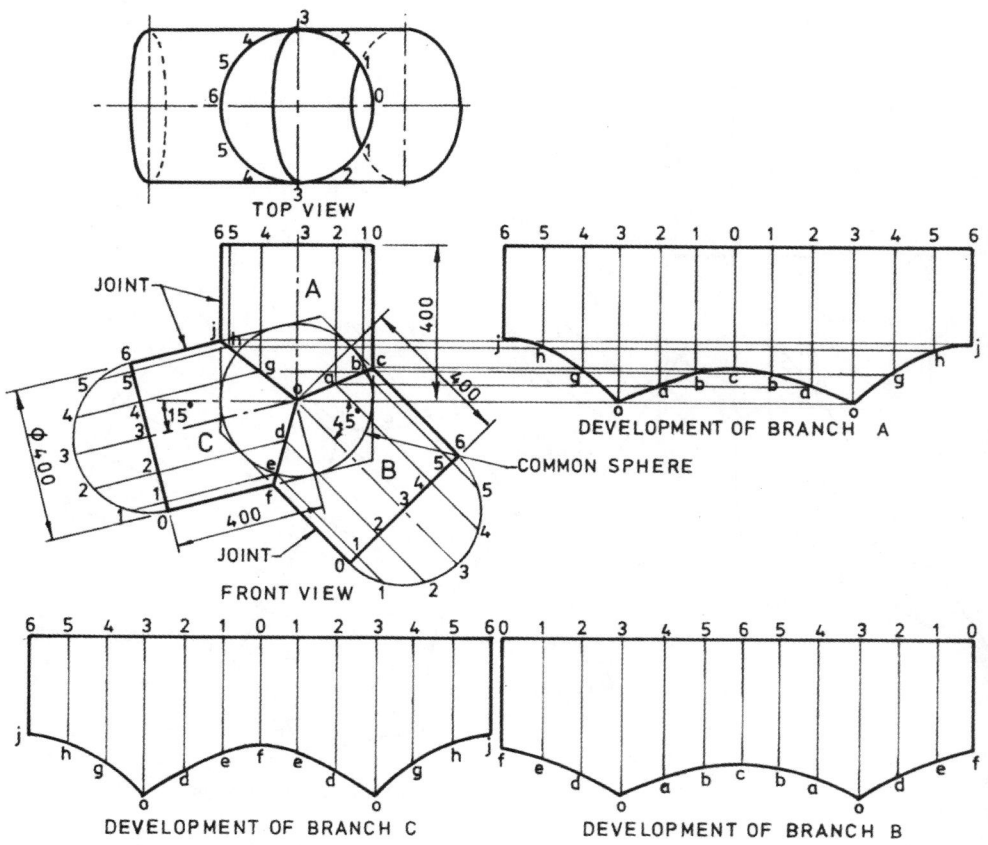

**FIGURE 8.29** ▶ Breeches piece—unequal angle, equal diameters

TOP VIEW

JOINT

FRONT VIEW

JOINT

COMMON SPHERE

DEVELOPMENT OF BRANCH A

DEVELOPMENT OF BRANCH C

DEVELOPMENT OF BRANCH B

## BREECHES PIECE—CYLINDER AND TWO CONES, EQUAL ANGLE

Refer to Figure 8.30.

1. Draw the front view of the breeches piece determining the line of intersection of the three branches A, B and C by the common sphere method outlined on pages 263–264 and illustrated in Figure 8.9(b).

2. The developments of the conical branches B and C are identical. The method used is the surface element method described on pages 264–265 and illustrated in Figure 8.11.

**Note:** The junction of the lines of intersection, point d on the front view, does not fall on a surface element line of the cone, and an extra line is drawn through d to intersect the base at D. Point D is then plotted on the development between points 2 and 3, and the surface element line AD is drawn in.

# 8.8 Development of transition pieces

Often in industry it is necessary to connect tubes and ducts of different cross-sectional shapes and areas, especially in air conditioning, ventilation and fume extraction applications. The required change in shape and/or area is achieved by developing a transition piece with an inlet of a certain shape and cross-sectional area, and an outlet of a different shape and/or area; for example square-to-round. The transition is achieved by a technique called triangulation which involves dividing the transition surface into a suitable number of triangular segments, finding the true shape of each and then laying these down side by side to form the true surface development.

## SQUARE-TO-RECTANGLE TRANSITION PIECE

Figure 8.31 shows the front and top views of a transition hopper required for transferring bird seed from a rectangular to a square cross-section suitable for bagging. Draw the development of the transition piece.

1. Draw the front and top views of the transition piece. Label the top corners 1, 2, 3 and 4 and the bottom corners a, b, c and d. Join the piece along xy.

2. Triangulate the quadrilaterals forming the sides by drawing one diagonal on each: x2, b3, c4, d1 and x1. Note that the joint line xy effectively makes two quadrilaterals out of the front face 12ba.

FIGURE 8.30 ▶ Breeches piece—cylinder and two cones, equal angle

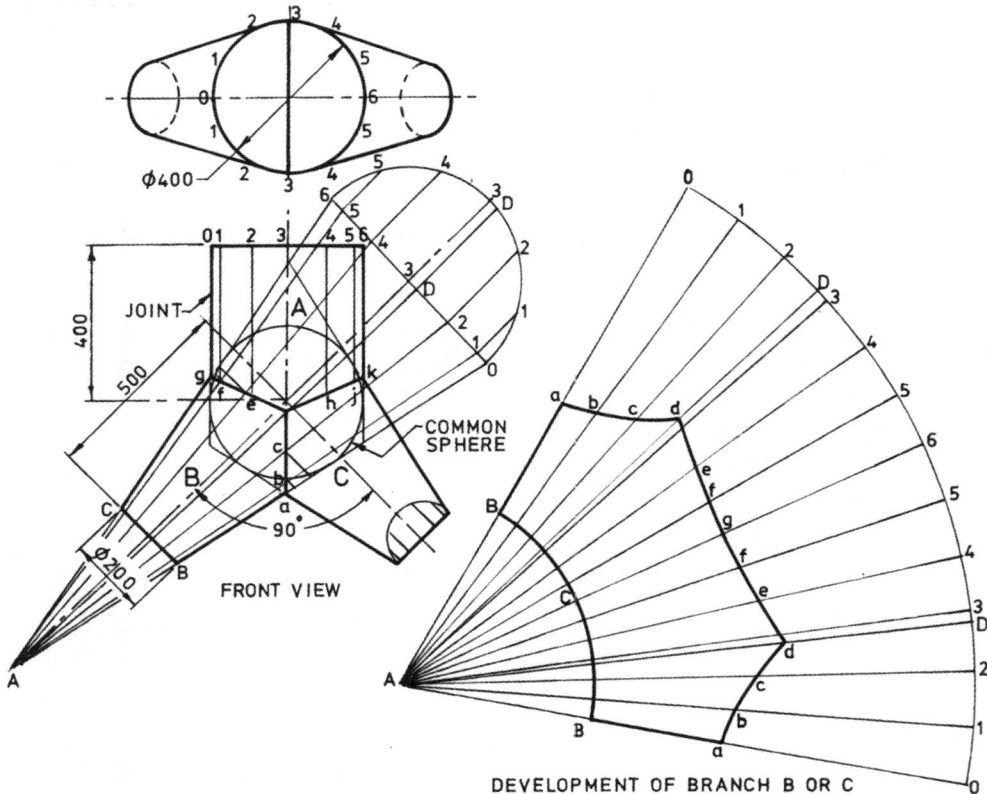

DEVELOPMENT OF BRANCH B OR C

3. Construct a true length diagram to the right of the front view by transferring the top view length of each of the sloping sides along the base line and joining to the vertical difference (VD) line to give the true lengths. Identify each line as it is constructed.

FIGURE 8.31 ▶ Square-to-rectangle transition piece

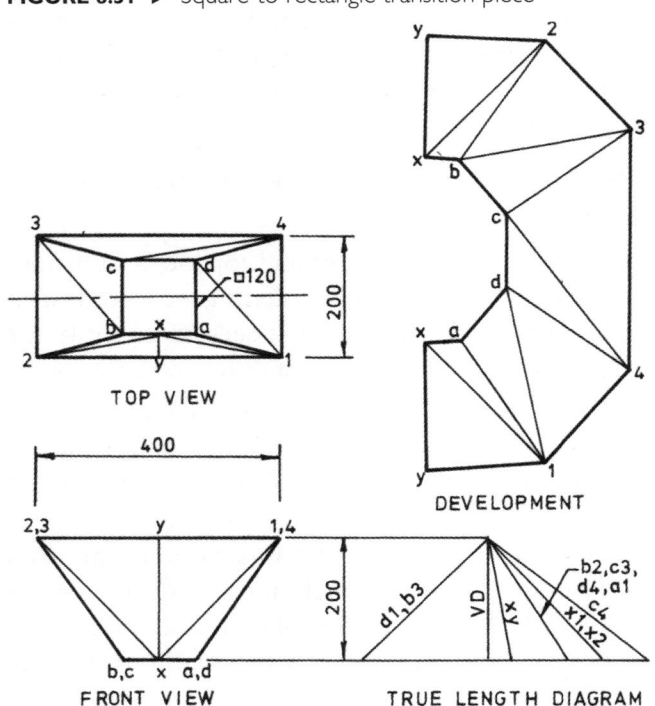

TOP VIEW

DEVELOPMENT

FRONT VIEW

TRUE LENGTH DIAGRAM

4. Proceed with the development by constructing the true shape of the triangles formed and laying them side by side in correct sequence. Begin with cd vertical, for example, then construct triangle cd4 followed by c43. Complete the bottom of the development to the joint xy, and the top to the joint xy. Remember that the ends 1234 and abcd are true shapes and hence their sides are true lengths.

### ROUND-TO-ROUND TRANSITION PIECE

A pictorial view of the transition piece which connects two circular sections not in parallel planes is shown in Figure 8.32(a). Its development is obtained as follows (refer to Fig. 8.32):

1. Draw the front and top views of the transition piece to be used as an aid to triangulate the curved surface, that is, to consider the surface as consisting of a number of flat triangles lying side by side and having their bases at one end or the other.

2. Divide the two openings into twelve equal parts 0, 1, 2, 3, etc. and a, b, c, d, etc. with 0 and a located on the joint line.

3. These divisions are now joined with surface element lines in a zig-zag pattern which has the effect of dividing the surface into the triangular

pattern. It is convenient to distinguish the surface element lines by making them alternately dash and full lines in order to avoid confusion on the true length diagram.

4. Commencing at 0 on both the front and top views, draw a full line up to b, dash line down to 1 and so on until the final line is a full line up to g. A line is considered to connect g6 although it is not shown.

5. Construct the true length diagram to the side of the front view. A common vertical difference (VD) line is used, and the heights of the points b, c, d, e, f and g are projected across to it.

6. The top view lengths of the full lines are now taken off the top view, set off to the right of the VD line, and numbered 0, 1, 2, 3, 4 and 5.

7. These numbers are now joined to the corresponding top points b, c, d, e, f and g to give the true length of the full element lines. For example, 0 joins b, 1 joins c (notice these cross over), 2 joins d, etc.

8. Similarly, a true length diagram is constructed for the dash lines to the left of the VD line.

9. The development can now be drawn. Set down a0 taken from the right-hand side of the front view (it is a true length). Draw the true shape of the first triangle a0b as follows: from a describe an arc ab equal to one-twelfth of the top opening

circumference; from 0 describe another arc equal to 0b taken from the full line side of the true length diagram to cut the first arc at b; join 0b with a full line.

10. Now draw the true shape of the second triangle 0b1 as follows: from 0 describe an arc 01 equal to one-twelfth of the bottom opening circumference; from b describe another arc equal to b1 taken from the dash line side of the true length diagram to cut the first arc at 1; join b1 with a dash line.

11. Continue constructing all the triangles until line g6 is set down. Its true length is taken from the left-hand side of the front view. Line g6 represents the dividing line of the development.

12. The second half of the development is best obtained by projecting points across at right angles to g6 and using transfer ordinates to plot the second half.

## SQUARE-TO-ROUND TRANSITION PIECE

The development is carried out in much the same way as the previous exercise. The triangulation of the surface is somewhat different because, as can be seen from Figure 8.33, there are four flat triangles whose bases correspond to the sides of the square end. Only the curved surfaces are triangulated.

1. Draw the front and top views to be used as an aid to triangulate the curved surfaces.

**FIGURE 8.32** ▶ Round-to-round transition piece

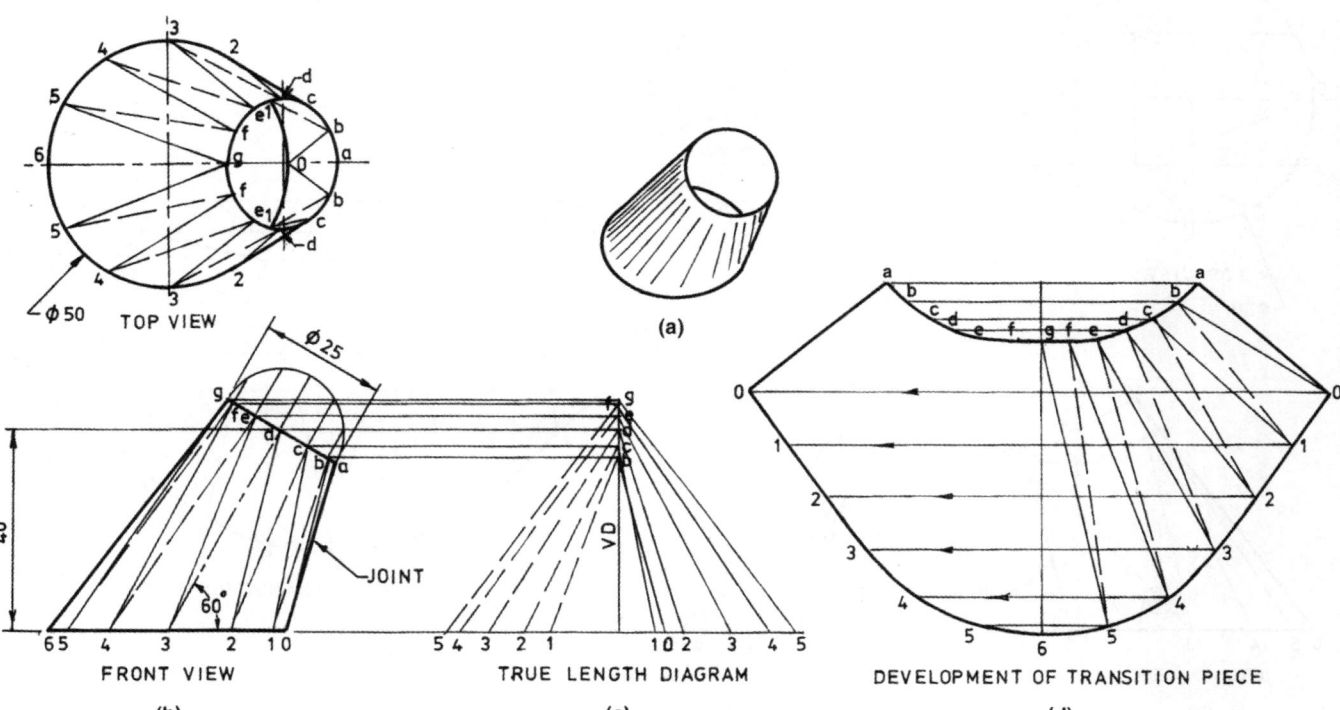

TOP VIEW    φ 50

(a)

FRONT VIEW

(b)

TRUE LENGTH DIAGRAM

(c)

DEVELOPMENT OF TRANSITION PIECE

(d)

2. Divide the circular end into twelve equal divisions such that 0, the bottom of the joint, is at the apex of the shortest triangular side and is also one of the circular end divisions.

3. Draw two sets of three triangles on the front view, each set having a common apex at c and b and bases coinciding with a division of the circular end as shown. The pictorial view in Figure 8.33(a) identifies the triangles more clearly.

4. Draw these triangles on the top view as well.

5. Construct the true length diagram to the side of the front view. Care must be taken to ensure that the top view lengths set out to the right of the VD line are joined to the correct height point. If difficulty is experienced, make one set of triangles full lines and the other set dash lines to distinguish between them more easily. Alternatively put one set on the left and one set on the right of the VD line.

6. The development is now set out, commenc-ing at line a0, whose length is taken from the right-hand side of the front view. The true shape of triangle ab0 is found as follows: from a des-cribe an arc equal to ab taken from the top view; from 0 describe an arc equal to 0b taken from the true length diagram to intersect the first arc at b.

7. The three triangles b01, b12 and b23 are constructed in a similar manner.

8. Triangle b3c is found by describing an arc from b equal to bc taken from the top of the front view (not the top view) and intersecting it with another arc from 3 equal to 3c taken from the true length diagram.

9. The second set of triangles is now constructed as before, then finally triangle cd6, which is an isosceles triangle with cd equal to twice ab taken from the top view. The remainder of the development is completed by projection and the use of transfer ordinates.

## OBLIQUE HOOD

Refer to Figure 8.34.

1. Draw the front and top views, showing the lines of triangulation joining equal divisions of the top and base. Note that the true shape of the base is elliptical and the top view is circular.

2. Construct the true length diagram using a common base along the top. Top view lengths of the element lines taken from the top view are set off along the common base line on either side of the VD line to give points a, b, c, d, e and f. The dash lines are set off to the left of the VD line and full lines to the right to lessen confusion.

**FIGURE 8.33** ▶ Square-to-round transition piece

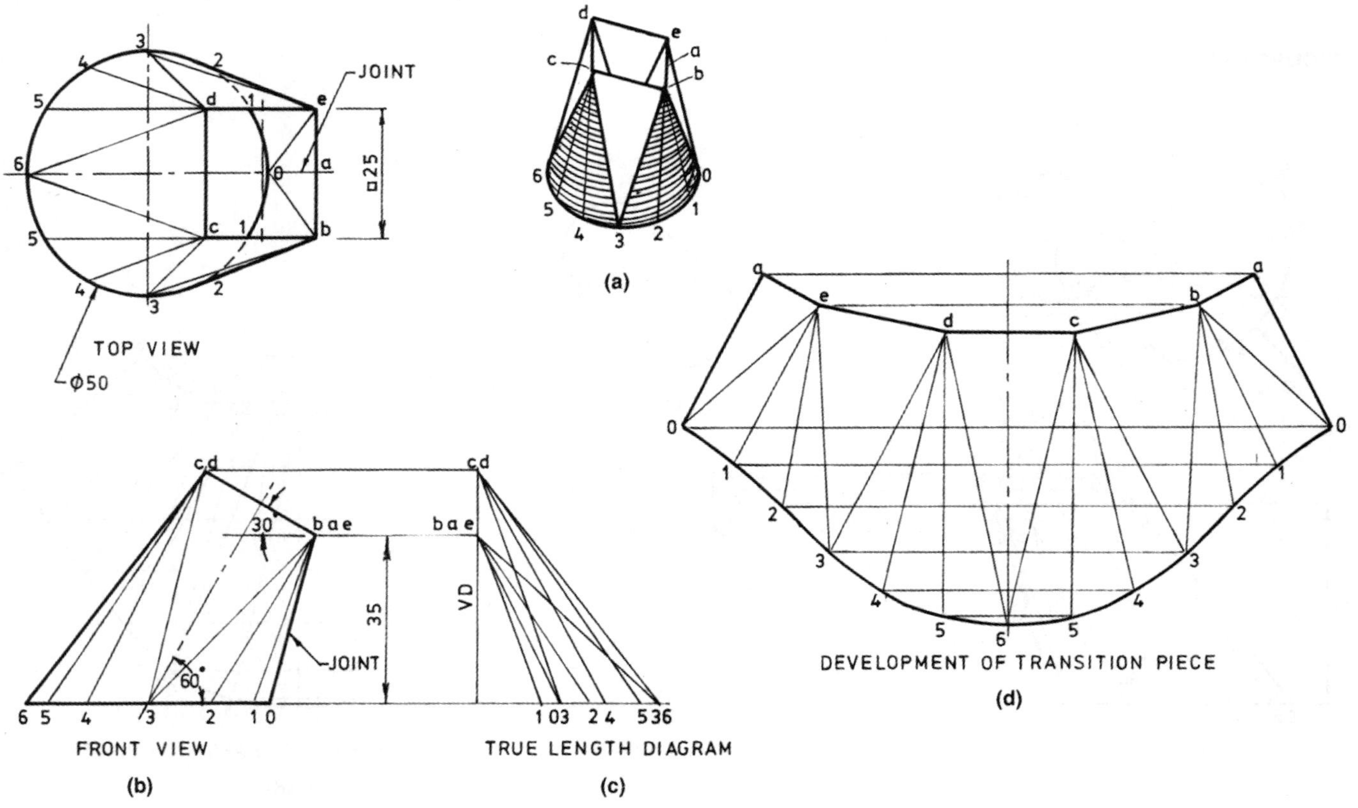

TOP VIEW
φ50

(a)

FRONT VIEW
(b)

TRUE LENGTH DIAGRAM
(c)

DEVELOPMENT OF TRANSITION PIECE
(d)

FIGURE 8.34 ▶ Oblique hood

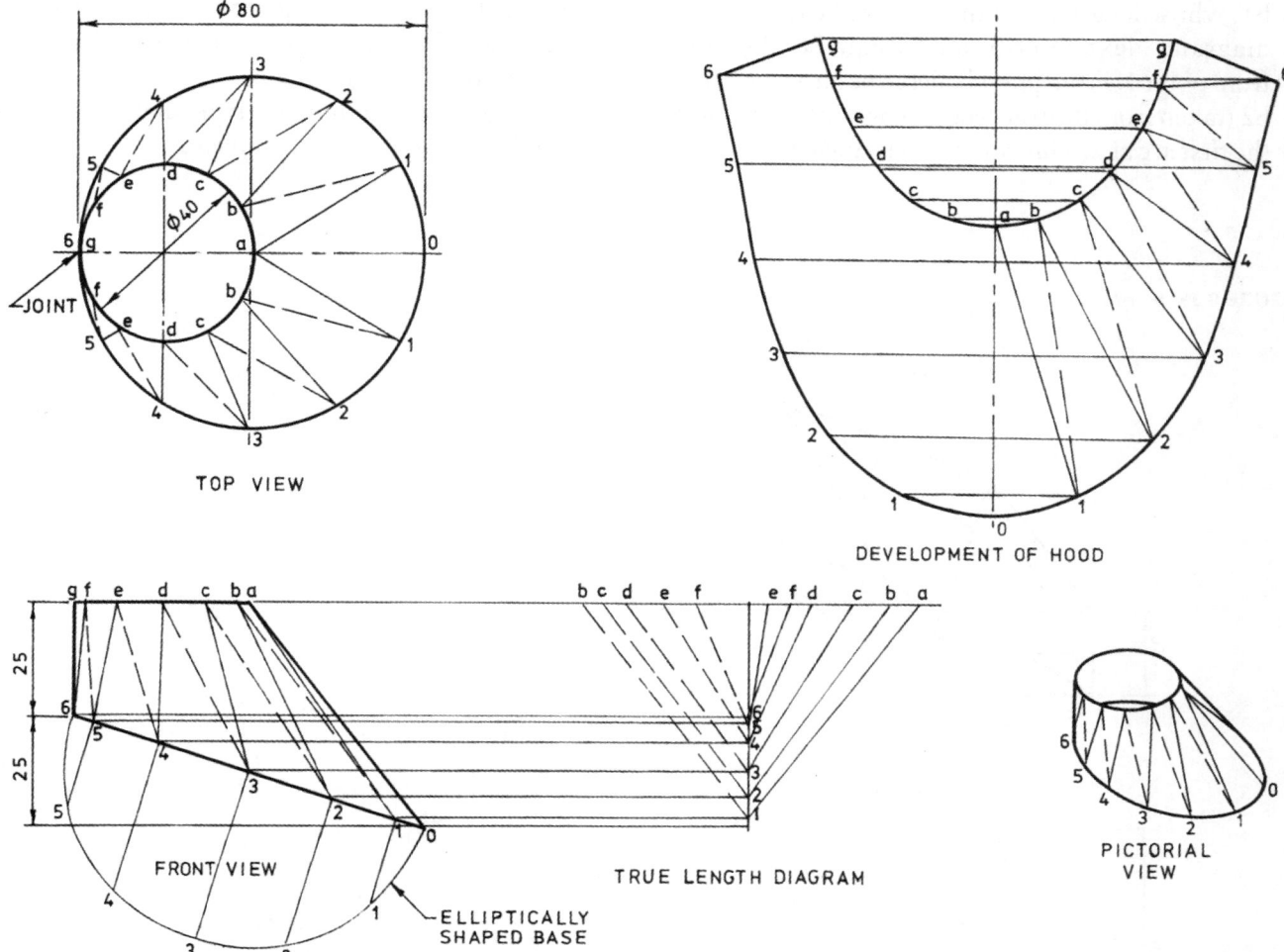

TOP VIEW

DEVELOPMENT OF HOOD

FRONT VIEW

ELLIPTICALLY SHAPED BASE

TRUE LENGTH DIAGRAM

PICTORIAL VIEW

3. Project points 1, 2, 3, 4, 5 and 6 across to the VD line, and join them to the appropriate base line point to give the true length of the surface element lines.

4. Set down vertically the centre line of the development, a0, equal to a0 on the front view. Now construct triangle a01 by describing an arc equal to a division of the elliptical base from 0 and intersecting it at 1 with another arc equal to a1 (full line) from the true length diagram described from a.

5. Continue constructing triangles until line g6 is laid down. Draw smooth curves through the apex points of the triangles to complete half the development.

6. The second half of the development is plotted quickly by projecting each point horizontally across from the first half and marking ordinates on the left of the centre line equivalent to those on the right. Draw a smooth curve through these points to complete the development.

## OFFSET RECTANGLE-TO-RECTANGLE TRANSITION PIECE

In order to make the transition piece using a series of flat triangular surfaces rather than twisted quadrangular surfaces, it is necessary to include four kinked edges (b1, c2, d3 and a4) as shown in Figure 8.35 on the front and top views.

1. Draw the front and top views in order to triangulate the surfaces by joining b1, c2, d3 and a4. Make the joint along the shortest kinked edge, b1.

2. Construct the true length diagram to the side of the front view by transferring plan lengths from the top view to the base of the true length diagram using both sides of the VD line and joining the ends to the top of the common vertical difference. As each true length is determined, mark it on the true length diagram to avoid confusion when taking off true lengths for the development. Note that a1 and c2 have the same true length because their plan lengths are identical.

3. The development is now set out commencing at line b1, whose length is obtained from the true length diagram. Next describe an arc equal to 340 mm from point 1. From point b, describe an arc equal to b2 (taken from the true length diagram) to intersect the first arc at 2. This completes triangle b12.

4. Next describe an arc from b equal to 150 mm. From point 2, describe an arc equal to c2 (taken from the true length diagram) to intersect the first arc at c. This completes triangle b2c.

5. Continue constructing the true shape triangles until the development is complete.

**FIGURE 8.35** ▶ Rectangle-to-rectangle, angular offset transition piece

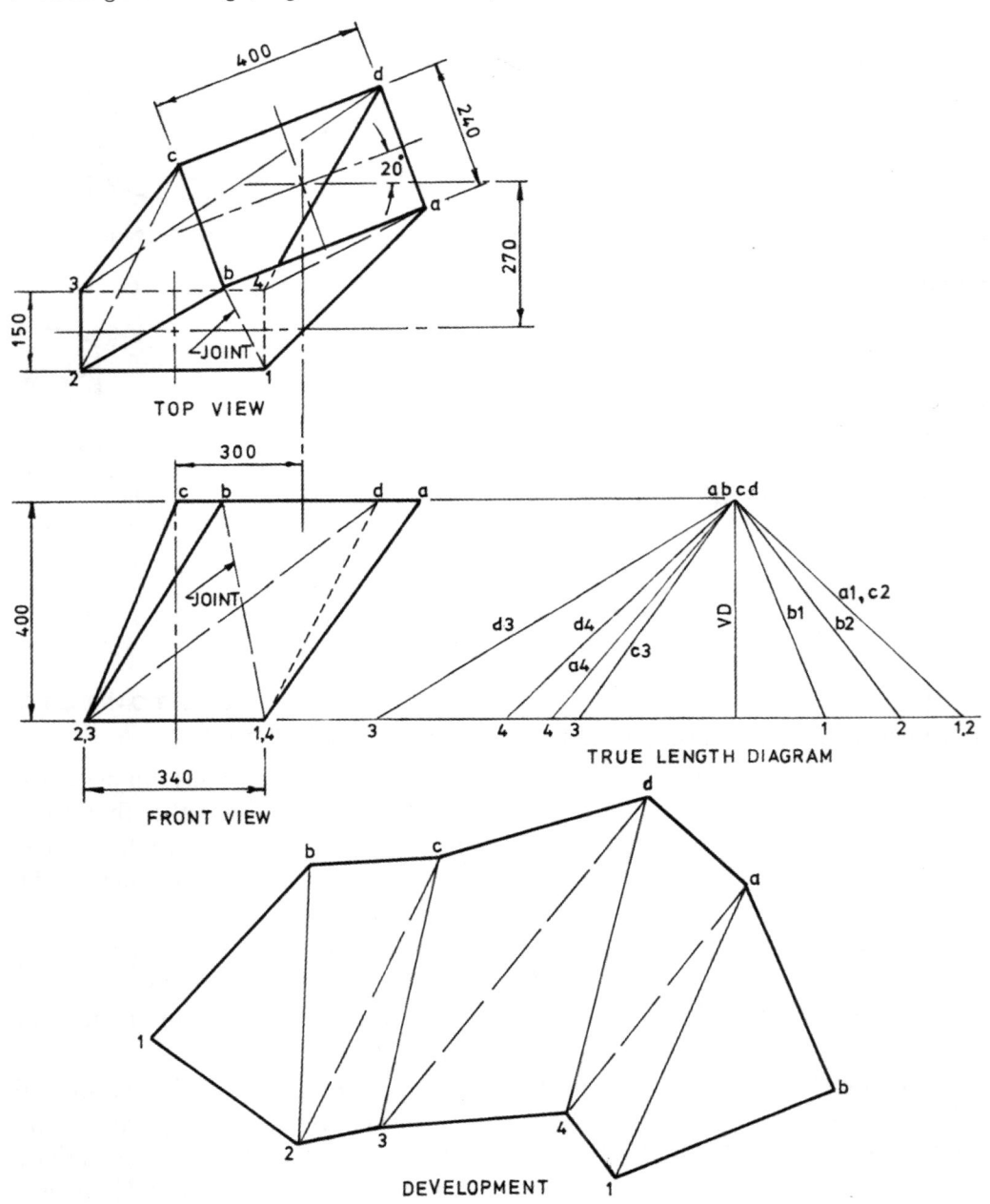

# 8.9 Problems (*development*)

## 8.1

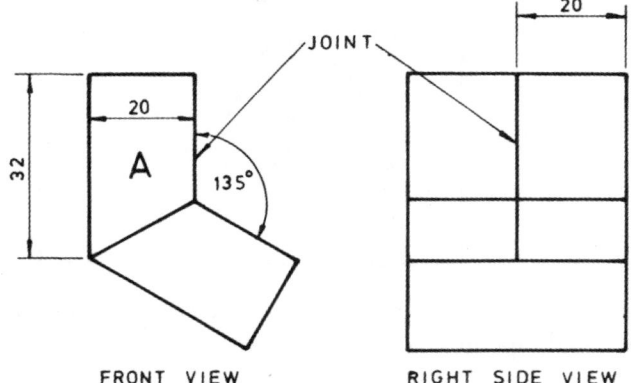

FRONT VIEW      RIGHT SIDE VIEW

Draw the development of part A of the rectangular pipe bend.
(scale 2:1)

## 8.2

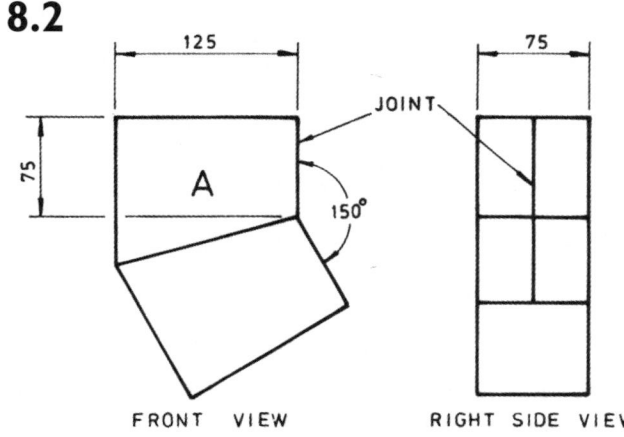

FRONT VIEW      RIGHT SIDE VIEW

Draw the development of part A of the rectangular pipe bend.
(scale 1:2)

## 8.3

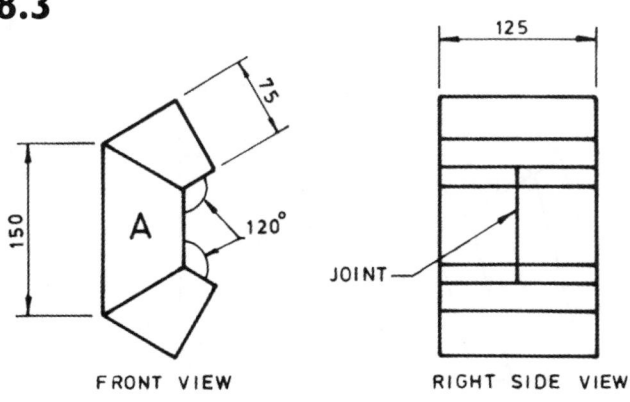

FRONT VIEW      RIGHT SIDE VIEW

Draw the development of part A of the rectangular pipe bend.
(scale 1:2)

## 8.4

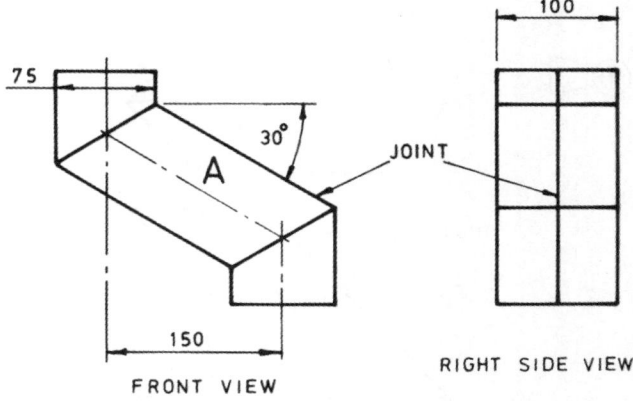

FRONT VIEW      RIGHT SIDE VIEW

Draw the development of part A of the rectangular pipe offset.
(scale 1:2)

## 8.5

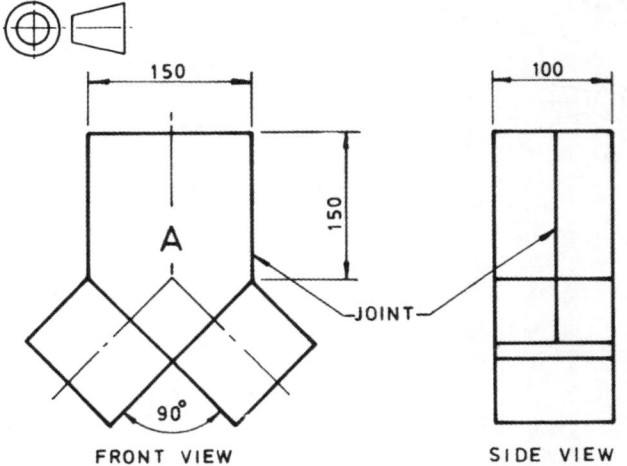

FRONT VIEW      SIDE VIEW

Draw the development of part A of the rectangular pipe junction.
(scale 1:2)

## 8.6

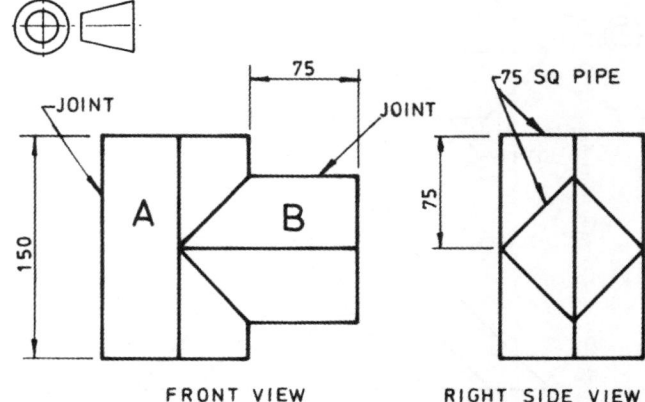

FRONT VIEW      RIGHT SIDE VIEW

Draw the development of parts A and B of the square pipe T piece.
(scale 1:2)

## 8.7

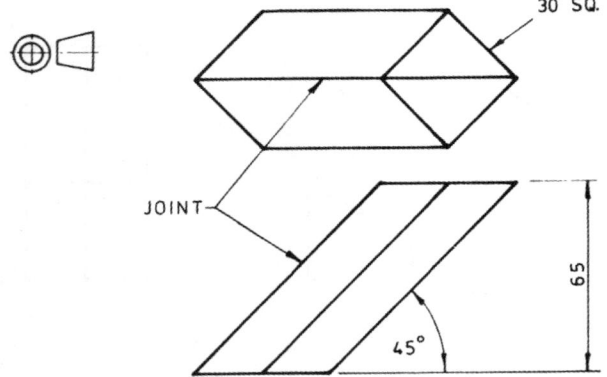

Draw the development of the open, oblique square-based prism shown above.
(scale 1:1)

## 8.8

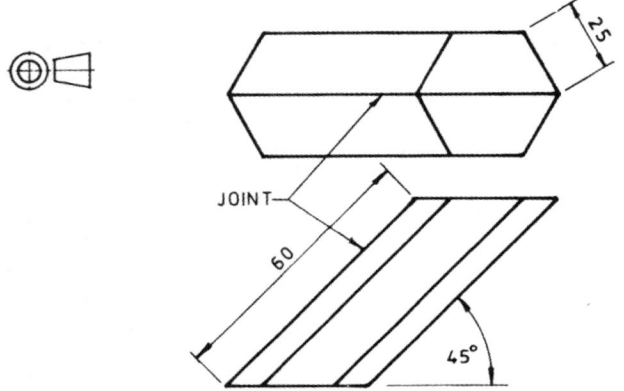

Draw the development of the open, oblique, hexagon-based prism shown above.
(scale 1:1)

## 8.9

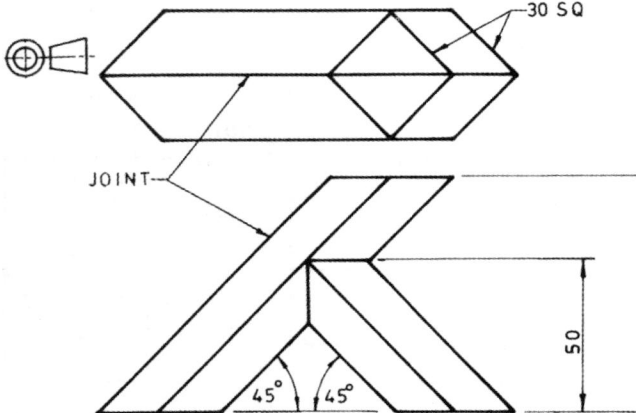

Draw the development of both branches of the oblique junction piece shown above.
(scale 1:1)

## 8.10

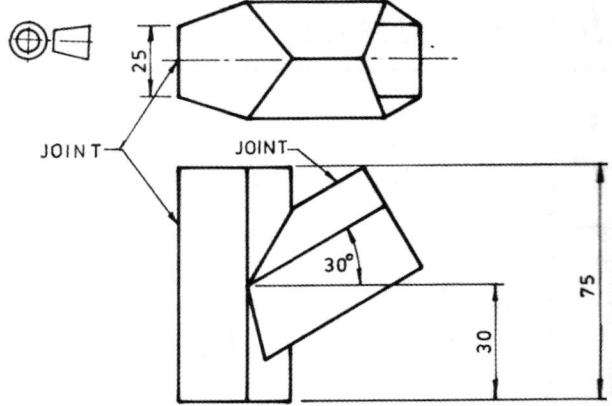

Draw the development of both parts of the pentagonal pipe junction shown above.
(scale 1:1)

## 8.11

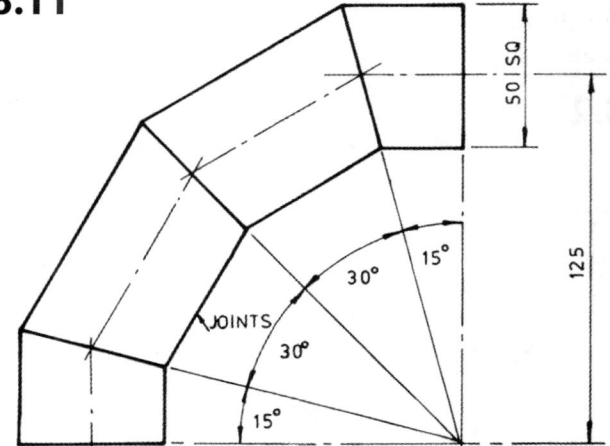

Draw the development of the four piece elbow required to make the 90° bend shown above.
(scale 1:1)

## 8.12

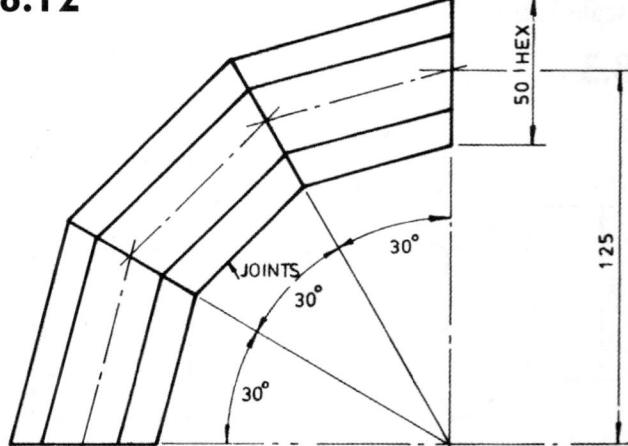

Draw the development of the segments required to make the 90° bend shown above.
(scale 1:1)

## 8.13

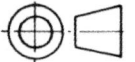

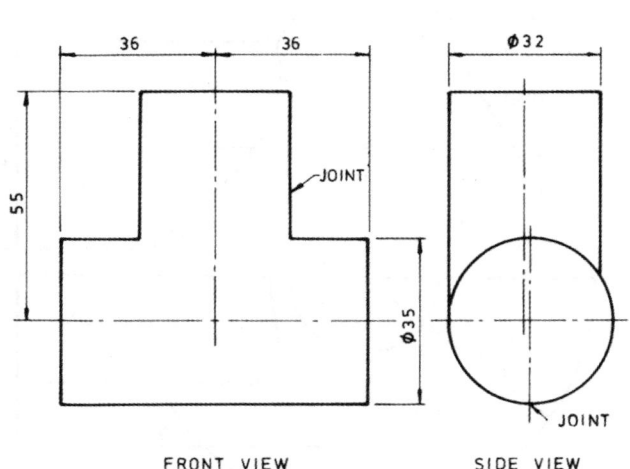

36    36    Ø32

55

Ø35

JOINT

JOINT

FRONT VIEW          SIDE VIEW

Complete the front view, showing the line of intersection of the branches of the T piece.
Draw the development of both branches.
(scale 1:1)

## 8.15

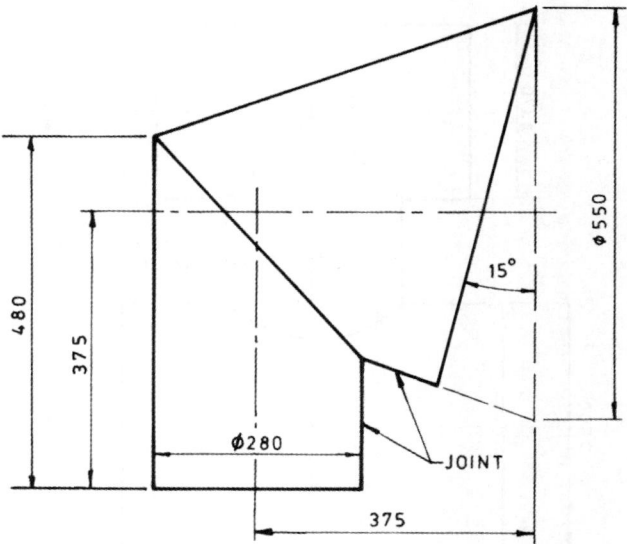

480    375    Ø550    15°

Ø280

JOINT

375

The front view of a conical sheet-metal hood is given.
Draw the development of each part of the hood.
(scale 1:5)

## 8.14

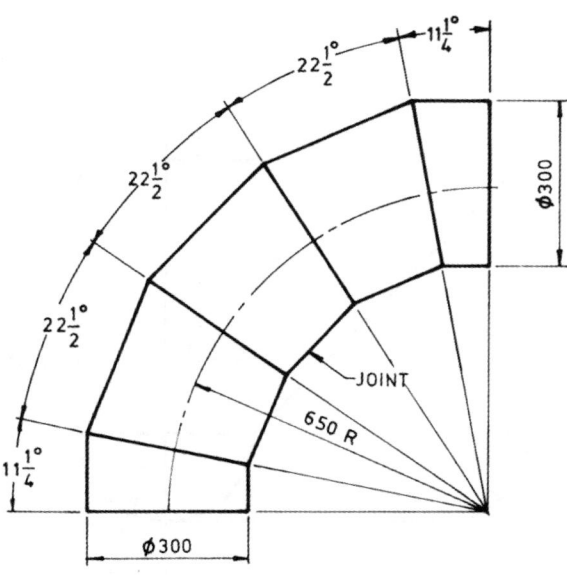

$22\frac{1}{2}°$    $11\frac{1}{4}°$

$22\frac{1}{2}°$

$22\frac{1}{2}°$

Ø300

JOINT

650 R

$11\frac{1}{4}°$

Ø300

Develop a pattern for the segments of the lobster-back quarter bend of round pipe shown above.
(scale 1:5)

## 8.16

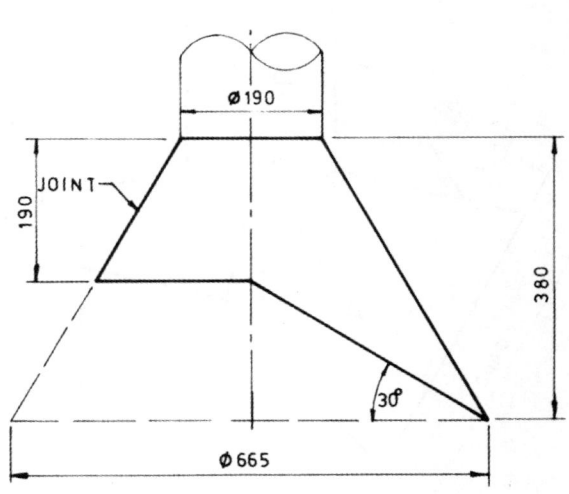

Ø190

JOINT

190    380

30°

Ø665

The front view of a conical fume extraction hood is shown above. Draw its development to a scale of 1:5.

## 8.17

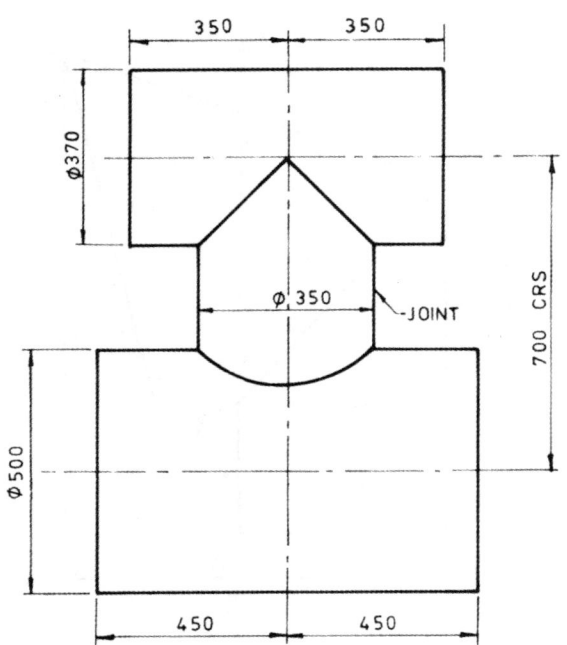

The view above is a front view of two parallel cylindrical pipes connected by another cylindrical pipe at right angles. Draw the front view, and complete the development of the connecting pipe.
(scale 1:10)

## 8.19

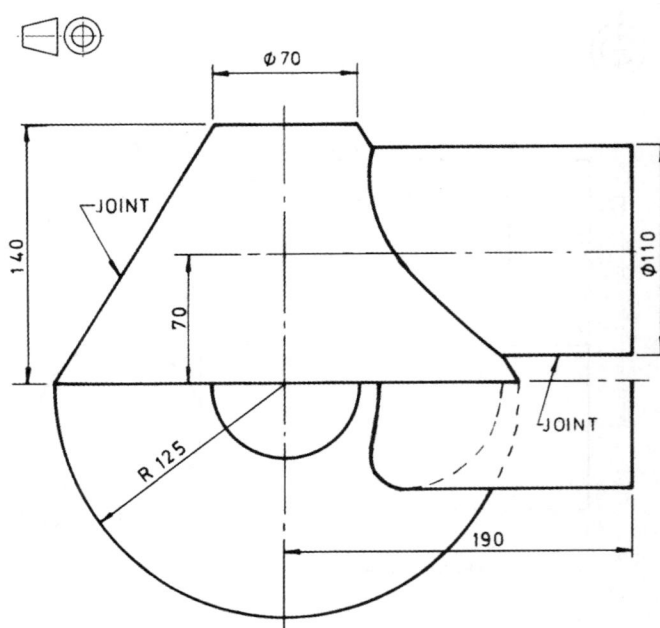

A combined front and half top view of a conical hood with a cylindrical branch is shown. Draw the development of both.
(scale 1:2)

## 8.18

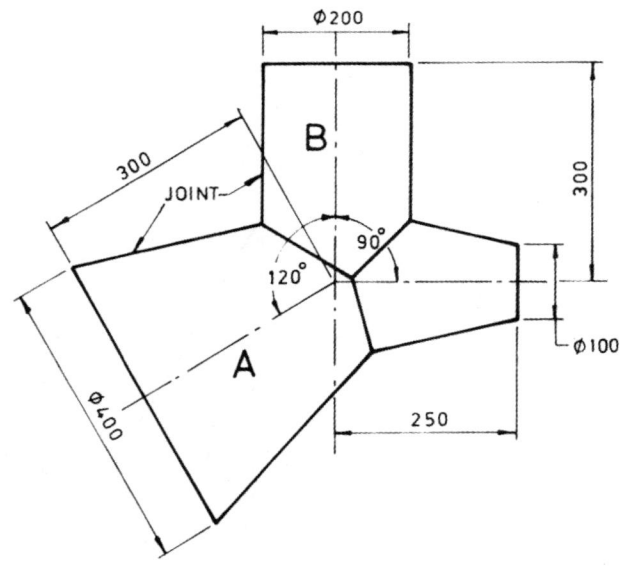

The front view of a breeches piece is given above. Draw this view, and complete the development of branches A and B.
(scale 1:5)

## 8.20

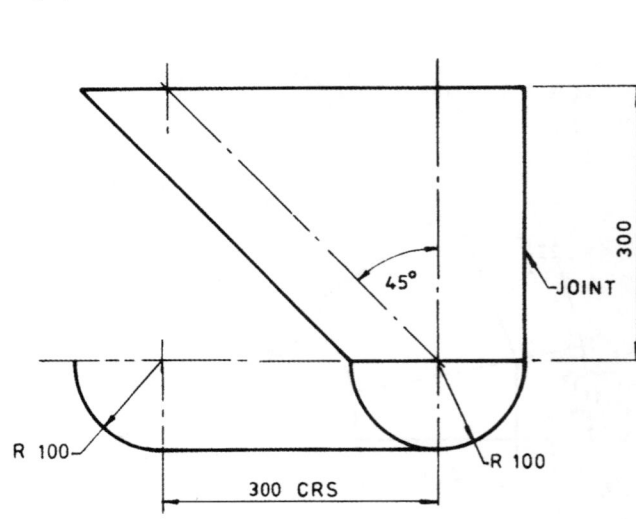

A combined front and half top view of a shoe-shaped funnel is shown above.
Develop the pattern for the funnel to a scale of 1:5.

## 8.21

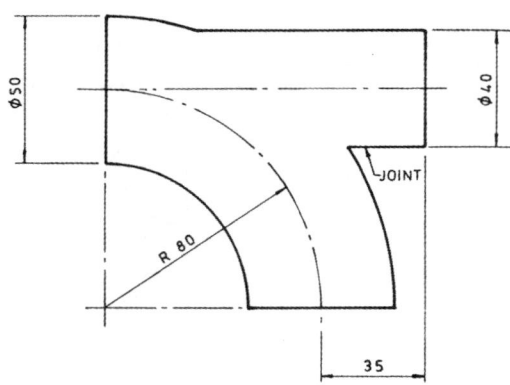

A view of a right-angled cylindrical bend with a cylindrical branch is given.

Complete this view, showing the line of intersection, and draw the development of the cylindrical branch. (scale 1:1)

## 8.22

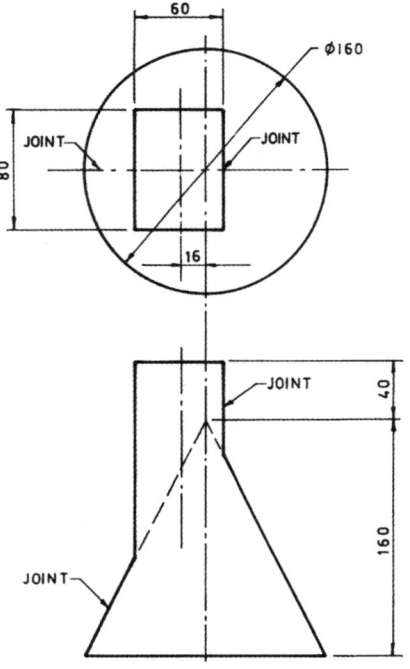

The front and top views of a conical hood fitted with a rectangular vent are given.

Draw the front view, showing the line of intersection, and complete the development of both the hood and the vent. (scale 1:2)

## 8.23

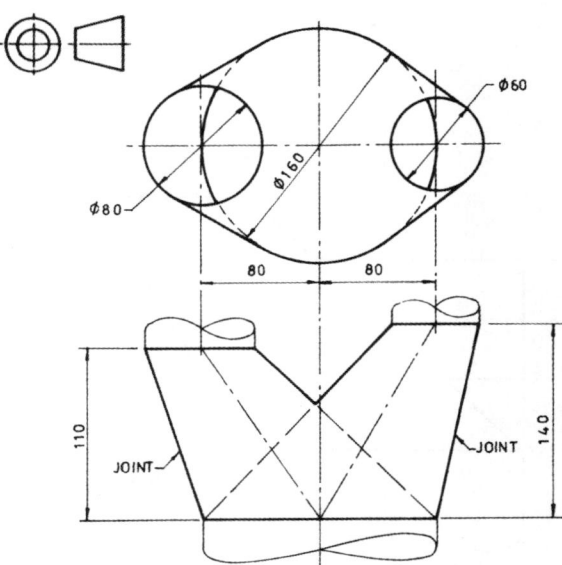

The front and top views of a junction piece for three unequal cylindrical pipes are shown.

(a) Draw the given front and top views, showing the line of intersection.

(b) Develop the patterns for the two arms of the junction piece.

(scale 1:2)

## 8.24

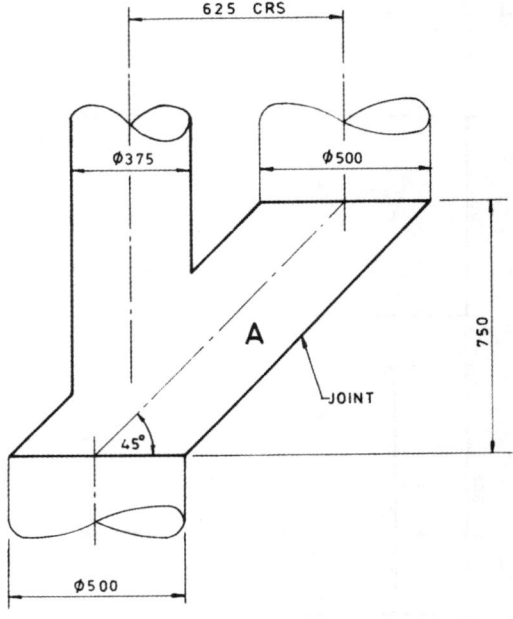

An oblique cylindrical connecting pipe fitted with a cylindrical branch is shown. Draw this view, showing the line of intersection of the two pipes, and draw the development of pipe A. (scale 1:10)

**8.25**

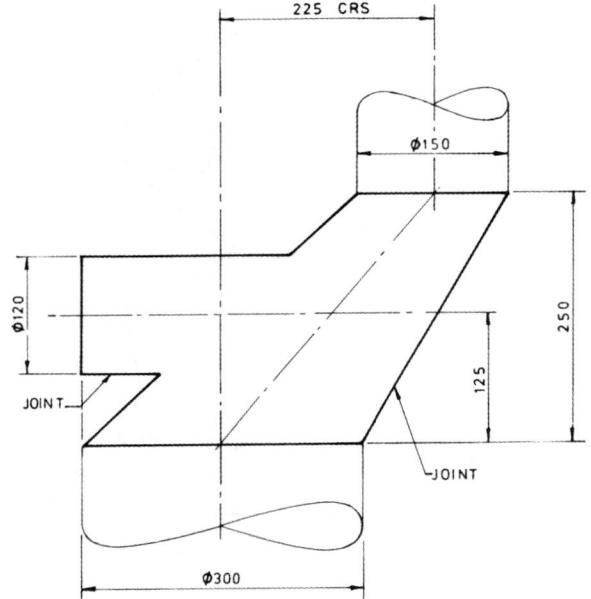

(a) Draw the complete front and top views of the cylinder and oblique cone intersection.
(b) Develop both the oblique cone and the cylindrical branch.
(scale 1:5)

**8.26**

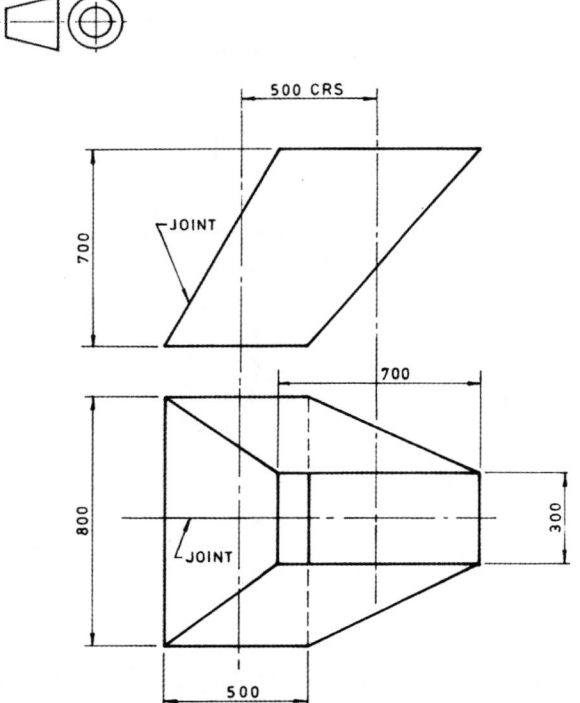

Draw the development of the transition piece, two views of which are given.
(scale 1:10)

**8.27**

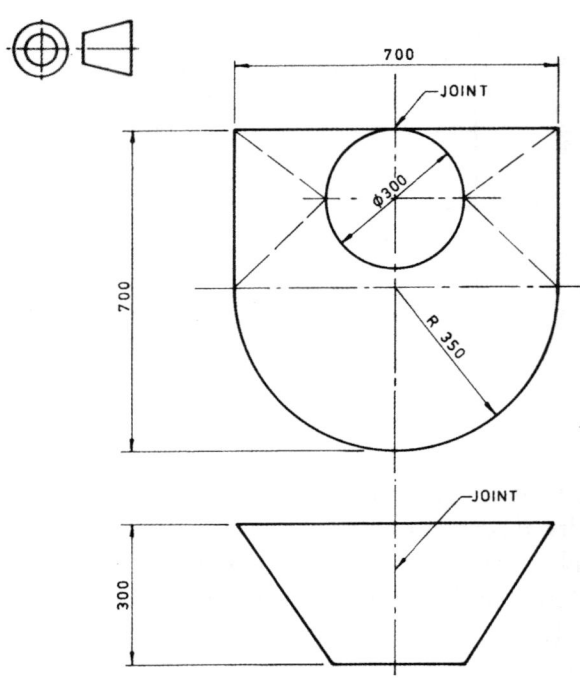

The front and top views of a grain hopper are given. Develop the pattern necessary to make the hopper.
(scale 1:10)

**8.28**

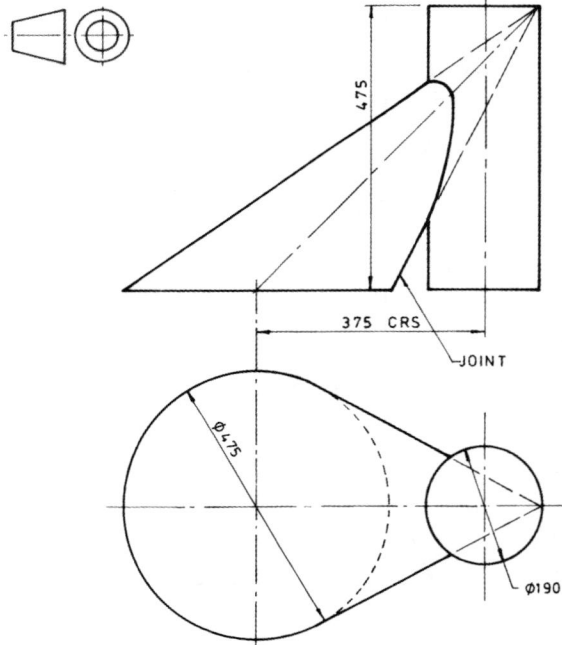

The front and top views of an oblique conical hood fitting into a cylindrical pipe are given.
Draw the front view, showing the line of intersection, and complete the development of the hood.
(scale 1:5)

## 8.29

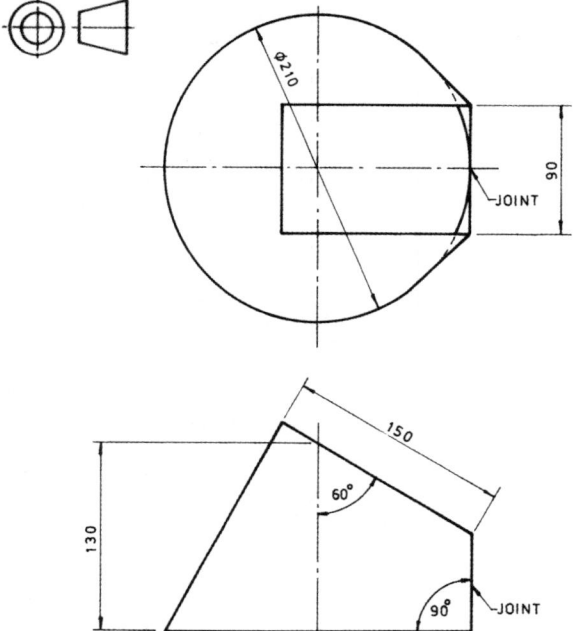

The front and top views of the sheet-metal transition piece are given.
Draw its development.

## 8.30

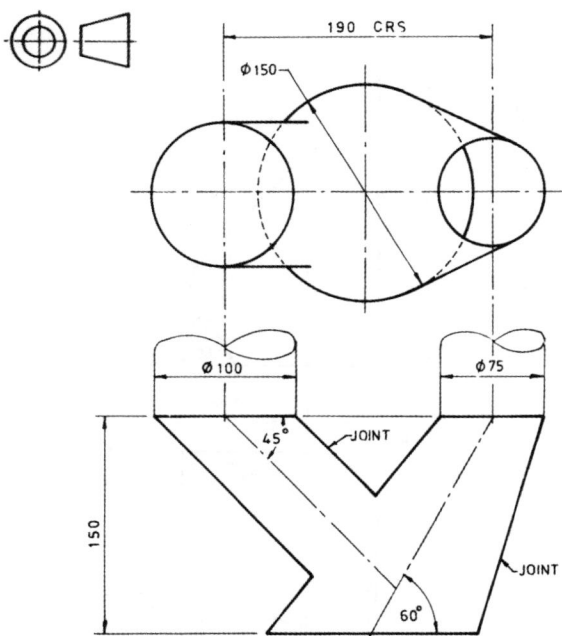

The front and top views of an oblique conical connecting pipe with an oblique cylindrical branch are given.

(a) Complete the front and top views, showing the line of intersection.

(b) Develop the connecting pipe and the branch. (scale 1:2)

## 8.31

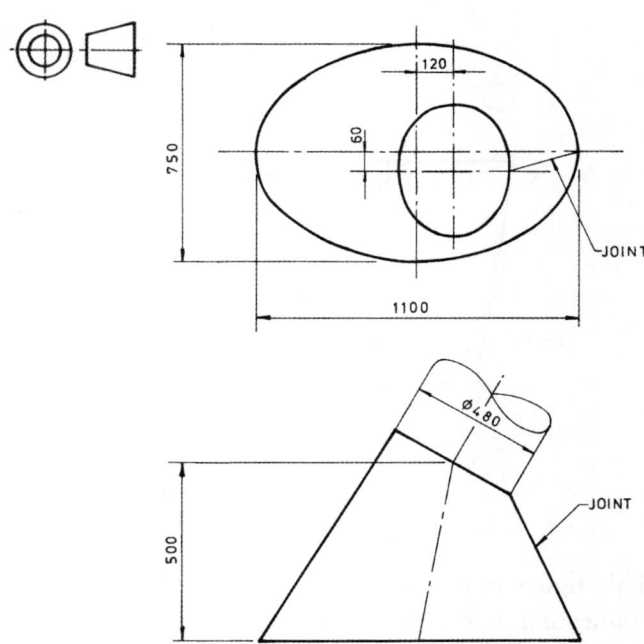

Two views of a hood with an elliptical base and an inclined offset circular top are given.

(a) Determine the angle which the axis of a cylindrical pipe must make with the base in order to fit it neatly.

(b) Draw the development of the hood to a scale of 1:10.

## 8.32

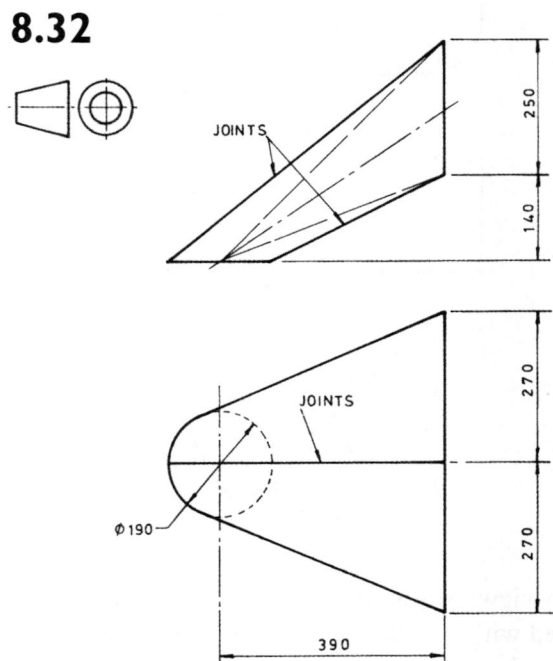

The transition piece shown is made from two equal sections and joins a cylindrical to a rectangular pipe. Using a scale of 1:5, draw the development of one of the sections.

## 8.33

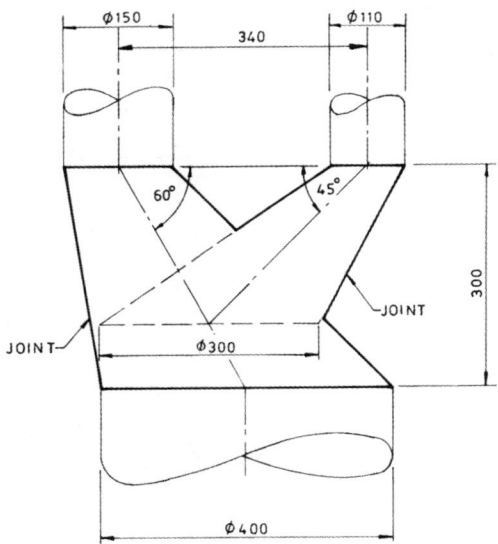

This figure is the front view of an oblique conical connecting pipe with an oblique conical branch pipe. Draw the front view, showing the line of intersection of the two pipes, and develop the patterns for both. (scale 1:5)

## 8.34

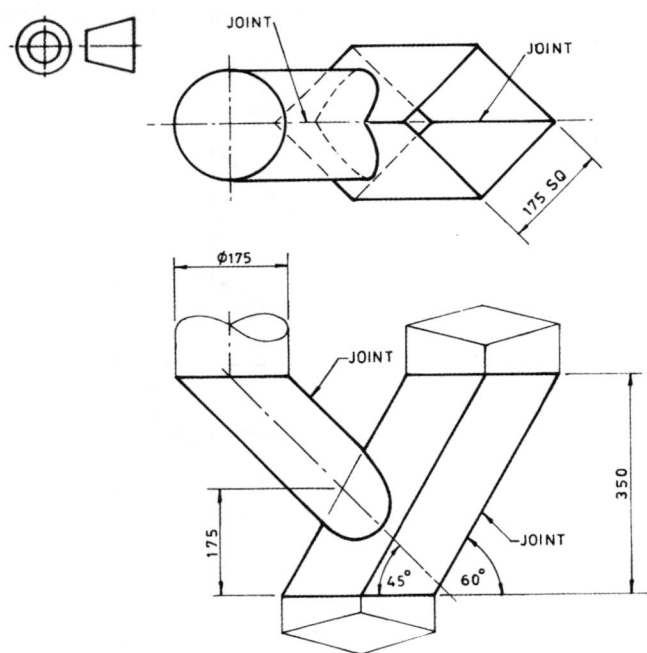

The two views show an oblique square connecting pipe fitted with an oblique cylindrical branch.
(a) Draw the two given views, plotting the line of intersection.
(b) Develop the pattern for both branches of the unit.
(scale 1:5)

## 8.35

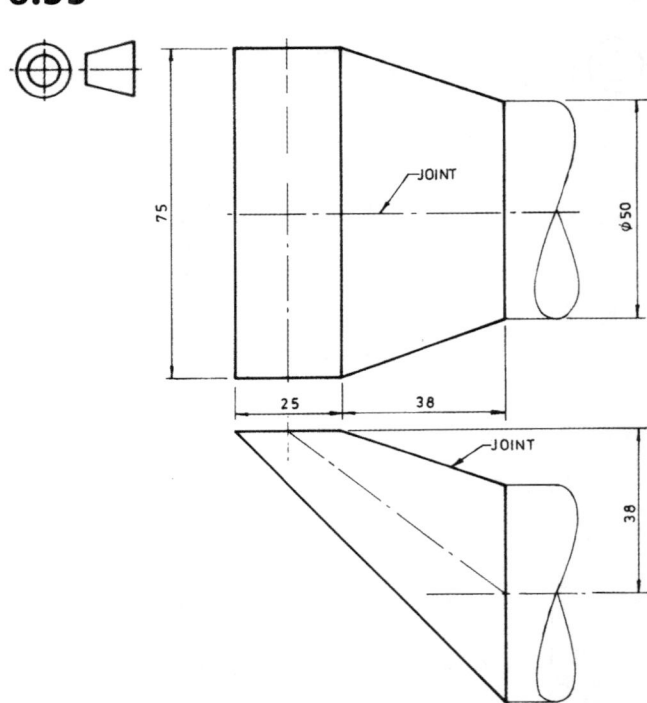

The figure shows two views of a transition pipe for joining a round to a rectangular pipe. Draw its development.
(scale 1:1)

## 8.36

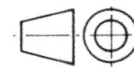

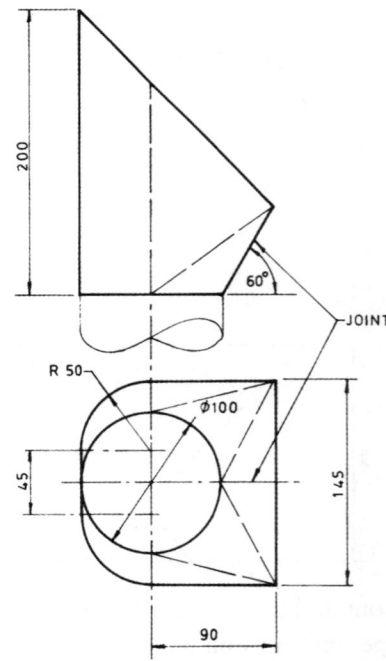

The front and top views of a sheet-metal transition are shown.
Develop the surface of the transition piece to a scale of 1:2.

# 8.10 CAD corner

CAD has a number of attributes, which can be used to advantage for solving problems in this chapter. In general CAD programs can:

- Construct geometry to a high degree of accuracy.
- 'Lock on' to existing geometry such as intersections, end points, centres, tangents etc., thus allowing accurate projection of construction lines.
- Divide drawing entities such as lines, arcs and circles into any number of equal segments.
- Orientate the coordinate system to align with angular geometry.
- Draw smooth curves through intersecting construction lines.

If a CAD program is capable of modelling in 3D complex joints, transitions etc. can be modelled and lines of intersection translated directly from the model. Some CAD programs are capable of flat pattern surface development directly from the solid/surface model.

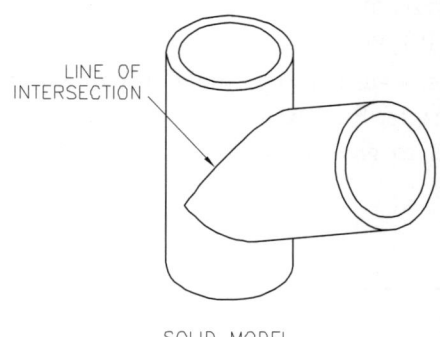

SOLID MODEL

Most CAD geometry is capable of being directly translated into CNC code for controlling profile cutting machines which eliminates the need for workshop mark out and allows flat surface production directly from data supplied from the CAD program.

**8.37** (a) Draw a complete front view of the horizontal (A) and oblique (B) cylinder connection shown below.

(b) Develop both cylinders required for the connection

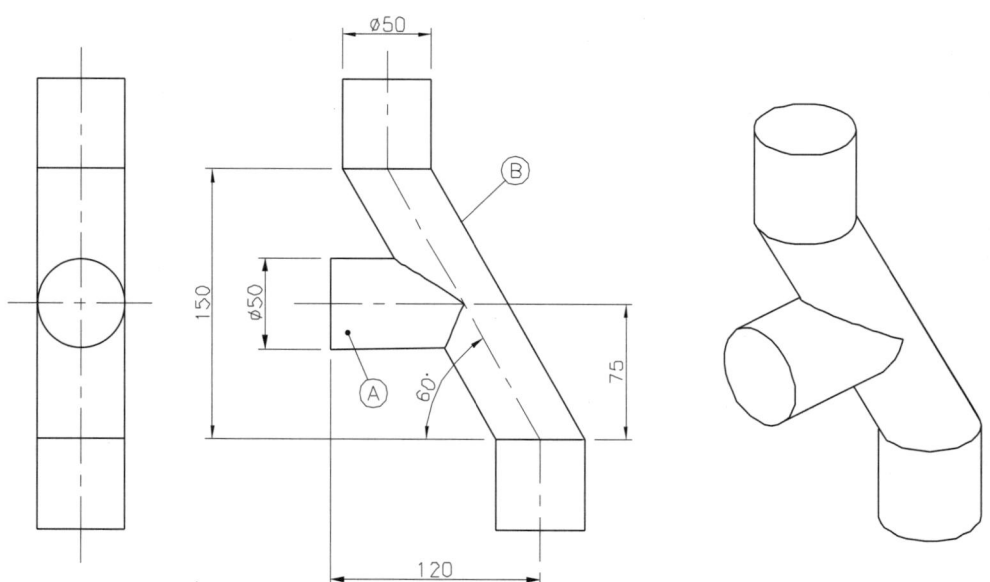

Horizontal/oblique cylinder connection

**8.38** If your CAD program has solid modelling capabilities complete (a) and (b) of this question. If your CAD program is not capable of modelling in 3D, complete (b) only.

(a) In 4 separate drawings, model each connection from the chapter 8 problems shown below. Orientate each model to give an orthogonal view of the line of intersection as shown on the right of each illustration below.

(b) Develop patterns for the square and cylindrical components of Problem 8.34.

Note: If you have not modelled Problem 8.34 you will need to answer the question in full. If you have modelled the connection you already have the line of intersection.

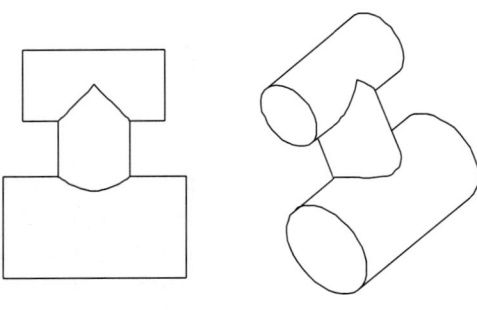 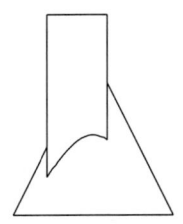

Problem 8.17

Problem 8.22

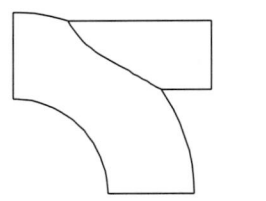 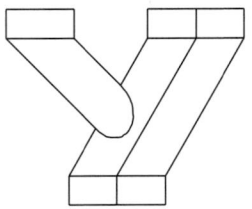

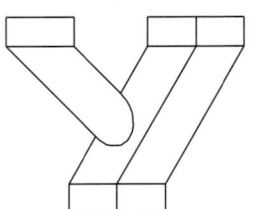 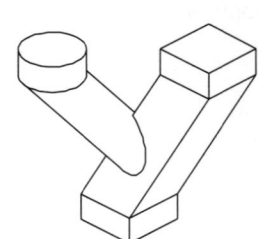

Problem 8.21

Problem 8.34

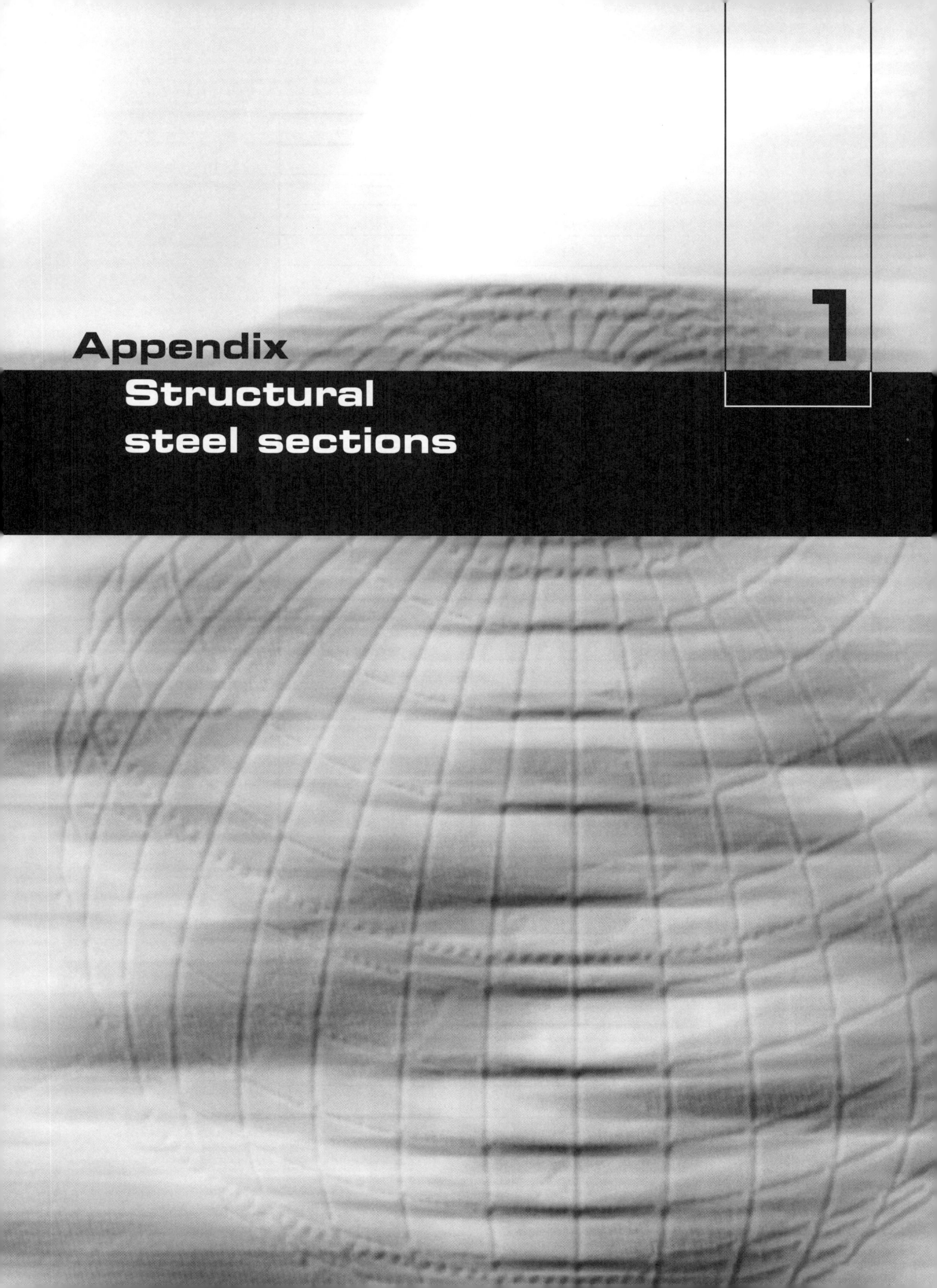

# Appendix
## Structural steel sections

1

**TABLE I** ► Rounds—size availability and mass

| DIAMETER (mm) | MASS (kg/m) |
|:---:|:---:|
| 10 | 0.616 |
| 12 | 0.887 |
| 13 | 1.04 |
| 14 | 1.21 |
| 15 | 1.39 |
| 16 | 1.58 |
| 17 | 1.78 |
| 18 | 1.99 |
| 19 | 2.23 |
| 20 | 2.46 |
| 21 | 2.72 |
| 22 | 2.98 |
| 23 | 3.26 |
| 24 | 3.55 |
| 26 | 4.17 |
| 27 | 4.49 |
| 30 | 5.55 |
| 32 | 6.31 |
| 33 | 6.71 |
| 36 | 7.99 |
| 39 | 9.38 |
| 42 | 10.9 |
| 45 | 12.5 |
| 48 | 14.2 |
| 50 | 15.4 |
| 56 | 19.3 |
| 60 | 22.2 |
| 65 | 26.0 |
| 75 | 34.7 |
| 80 | 39.5 |
| 90 | 49.9 |
| 100 | 61.7 |

Source: Hot Rolled and Structural Steel Products, 1998 Edition, BHP, Whyalla

**TABLE 2** ► Squares—size availability and mass

| THICKNESS(mm) | MASS (kg/m) |
|---|---|
| 10 | 0.790 |
| 12 | 1.13 |
| 16 | 2.01 |
| 20 | 3.14 |
| 25 | 4.91 |
| 40 | 12.5 |

Source: Hot Rolled and Structural Steel Products, 1998 Edition, BHP, Whyalla

**TABLE 3** ► Flats—size availability and mass (kg/m)

| WIDTH (mm) | THICKNESS (mm) | | | | | | | | | | |
|---|---|---|---|---|---|---|---|---|---|---|---|
| | 3 | 5 | 6 | 8 | 10 | 12 | 16 | 20 | 25 | 40 | 50 |
| 16 | | | | 1.00 | | | | | | | |
| 20 | 0.471 | 0.785 | 0.940 | | 1.57 | | | | | | |
| 25 | 0.589 | 0.981 | 1.18 | 1.57 | 1.96 | 2.36 | | | | | |
| 32 | 0.754 | 1.26 | 1.51 | 2.01 | 2.51 | | | | | | |
| 40 | 0.942 | 1.57 | 1.88 | 2.51 | 3.14 | 3.77 | 5.02 | | | | |
| 50 | 1.18 | 1.96 | 2.36 | 3.14 | 3.93 | 4.71 | 6.28 | 7.85 | 9.81 | | |
| 65 | 1.53 | 2.55 | 3.06 | 4.08 | 5.10 | 6.12 | 8.16 | 10.2 | | | |
| 75 | | 2.94 | 3.53 | 4.71 | 5.89 | 7.06 | 9.42 | 11.8 | 14.7 | 23.6 | |
| 90 | | 3.53 | 4.24 | 5.65 | 7.06 | 8.48 | | | | | |
| 100 | | 3.93 | 4.71 | 6.28 | 7.85 | 9.42 | 12.6 | 15.7 | 19.6 | | 39.3 |
| 110 | | 4.32 | 5.18 | 6.91 | 8.64 | 10.4 | | | | | |
| 130 | | 5.10 | 6.12 | 8.16 | 10.2 | 12.2 | 16.3 | 20.4 | 25.5 | | |
| 150 | | 5.89 | 7.06 | 9.42 | 11.8 | 14.1 | 18.3 | 23.6 | 29.4 | | 58.9 |
| 180 | | 7.07 | 8.48 | | 14.1 | 17.0 | | 28.3 | | | |
| 200 | | | 9.42 | 12.6 | 15.7 | 18.8 | 25.1 | 31.4 | 39.3 | | |
| 250 | | | 11.8 | 15.7 | 19.6 | 23.6 | | | | | |
| 300 | | | 14.1 | 18.8 | 23.6 | 28.3 | | | | | |

Source: Hot Rolled and Structural Steel Products, 1998 Edition, BHP, Whyalla

**TABLE 4** ▶ Universal beams—dimensions and properties

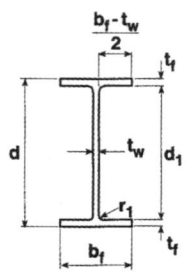 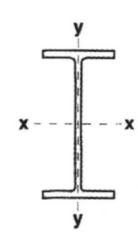

| DESIGNATION | DEPTH OF SECTION | FLANGE | | WEB THICK-NESS | ROOT RADIUS | DEPTH BETWEEN FLANGES | | | GROSS AREA OF CROSS-SECTION | ABOUT x-AXIS | ABOUT y-AXIS |
| | | WIDTH | THICK-NESS | | | | | | | | |
| | d | $b_f$ | $t_f$ | $t_w$ | $r_1$ | $d_1$ | $\frac{d_1}{t_w}$ | $\frac{b_f-t_w}{2t_f}$ | $A_g$ | $I_x$ | $I_y$ |
| kg/m | mm | mm | mm | mm | mm | mm | | | mm$^2$ | 10$^6$mm$^4$ | 10$^6$mm$^4$ |
| 610 UB 125 | 612 | 229 | 19.6 | 11.9 | 14.0 | 572 | 48.1 | 5.54 | 16000 | 986 | 39.3 |
| 113 | 607 | 228 | 17.3 | 11.2 | 14.0 | 572 | 51.1 | 6.27 | 14500 | 875 | 34.3 |
| 101 | 602 | 228 | 14.8 | 10.6 | 14.0 | 572 | 54.0 | 7.34 | 13000 | 761 | 29.3 |
| 530 UB 92.4 | 533 | 209 | 15.6 | 10.2 | 14.0 | 502 | 49.2 | 6.37 | 11800 | 554 | 23.8 |
| 82.0 | 528 | 209 | 13.2 | 9.6 | 14.0 | 502 | 52.3 | 7.55 | 10500 | 477 | 20.1 |
| 460 UB 82.1 | 460 | 191 | 16.0 | 9.9 | 11.4 | 428 | 43.3 | 5.66 | 10500 | 372 | 18.6 |
| 74.6 | 457 | 190 | 14.5 | 9.1 | 11.4 | 428 | 47.1 | 6.24 | 9520 | 335 | 16.6 |
| 67.1 | 454 | 190 | 12.7 | 8.5 | 11.4 | 428 | 50.4 | 7.15 | 8580 | 296 | 14.5 |
| 410 UB 59.7 | 406 | 178 | 12.8 | 7.8 | 11.4 | 381 | 48.8 | 6.65 | 7640 | 216 | 12.1 |
| 53.7 | 403 | 178 | 10.9 | 7.6 | 11.4 | 381 | 50.1 | 7.82 | 6890 | 188 | 10.3 |
| 360 UB 56.7 | 359 | 172 | 13.0 | 8.0 | 11.4 | 333 | 41.6 | 6.31 | 7240 | 161 | 11.0 |
| 50.7 | 356 | 171 | 11.5 | 7.3 | 11.4 | 333 | 45.6 | 7.12 | 6470 | 142 | 9.60 |
| 44.7 | 352 | 171 | 9.7 | 6.9 | 11.4 | 333 | 48.2 | 8.46 | 5720 | 121 | 8.10 |
| 310 UB 46.2 | 307 | 166 | 11.8 | 6.7 | 11.4 | 284 | 42.3 | 6.75 | 5930 | 100 | 9.01 |
| 40.4 | 304 | 165 | 10.2 | 6.1 | 11.4 | 284 | 46.5 | 7.79 | 5210 | 86.4 | 7.65 |
| 32.0 | 298 | 149 | 8.0 | 5.5 | 13.0 | 282 | 51.3 | 8.97 | 4080 | 63.2 | 4.42 |
| 250 UB 37.3 | 256 | 146 | 10.9 | 6.4 | 8.9 | 234 | 36.6 | 6.40 | 4850 | 55.7 | 5.66 |
| 31.4 | 252 | 146 | 8.6 | 6.1 | 8.9 | 234 | 38.4 | 8.13 | 4010 | 44.5 | 4.47 |
| 25.7 | 248 | 124 | 8.0 | 5.0 | 12.0 | 232 | 46.4 | 7.44 | 3270 | 35.4 | 2.55 |
| 200 UB 29.8 | 207 | 134 | 9.6 | 6.3 | 8.9 | 188 | 29.8 | 6.65 | 3820 | 29.1 | 3.86 |
| 25.4 | 203 | 133 | 7.8 | 5.8 | 8.9 | 188 | 32.3 | 8.15 | 3230 | 23.6 | 3.06 |
| 22.3 | 202 | 133 | 7.0 | 5.0 | 8.9 | 188 | 37.5 | 9.14 | 2870 | 21.0 | 2.75 |
| 18.2 | 198 | 99 | 7.0 | 4.5 | 11.0 | 184 | 40.9 | 6.75 | 2320 | 15.8 | 1.14 |
| 180 UB 22.2 | 179 | 90 | 10.0 | 6.0 | 8.9 | 159 | 26.5 | 4.20 | 2820 | 15.3 | 1.22 |
| 18.1 | 175 | 90 | 8.0 | 5.0 | 8.9 | 159 | 31.8 | 5.31 | 2300 | 12.1 | 0.975 |
| 16.1 | 173 | 90 | 7.0 | 4.5 | 8.9 | 159 | 35.3 | 6.11 | 2040 | 10.6 | 0.853 |
| 150 UB 18.0 | 155 | 75 | 9.5 | 6.0 | 8.0 | 136 | 22.7 | 3.63 | 2300 | 9.05 | 0.672 |
| 14.0 | 150 | 75 | 7.0 | 5.0 | 8.0 | 136 | 27.2 | 5.00 | 1780 | 6.66 | 0.495 |

Source: Hot Rolled and Structural Steel Products, 1998 Edition, BHP, Whyalla

**TABLE 5** ▶ Universal columns—dimensions

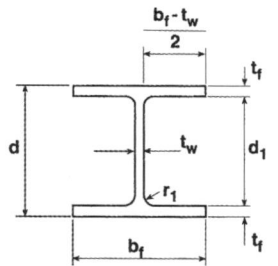 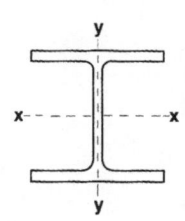

| DESIGNATION | DEPTH OF SECTION | FLANGE | | WEB THICK-NESS | ROOT RADIUS | DEPTH BETWEEN FLANGES | | | GROSS AREA OF CROSS-SECTION | ABOUT x-AXIS | ABOUT y-AXIS |
| | | WIDTH | THICK-NESS | | | | | | | | |
| | $d$ | $b_f$ | $t_f$ | $t_w$ | $r_1$ | $d_1$ | $\frac{d_1}{t_w}$ | $\frac{b_f-t_w}{2t_f}$ | $A_g$ | $I_x$ | $I_y$ |
| kg/m | mm | mm | mm | mm | mm | mm | | | mm$^2$ | 10$^6$mm$^4$ | 10$^6$mm$^4$ |
| 310 UC 158 | 327 | 311 | 25.0 | 15.7 | 16.5 | 277 | 17.7 | 5.91 | 20100 | 388 | 125 |
| 137 | 321 | 309 | 21.7 | 13.8 | 16.5 | 277 | 20.1 | 6.80 | 17500 | 329 | 107 |
| 118 | 315 | 307 | 18.7 | 11.9 | 16.5 | 277 | 23.3 | 7.89 | 15000 | 277 | 90.2 |
| 96.8 | 308 | 305 | 15.4 | 9.9 | 16.5 | 277 | 28.0 | 9.58 | 12400 | 233 | 72.9 |
| 250 UC 89.5 | 260 | 256 | 17.3 | 10.5 | 14.0 | 225 | 21.5 | 7.10 | 11400 | 143 | 48.4 |
| 72.9 | 254 | 254 | 14.2 | 8.6 | 14.0 | 225 | 26.2 | 8.64 | 9320 | 114 | 38.8 |
| 200 UC 59.5 | 210 | 205 | 14.2 | 9.3 | 11.4 | 181 | 19.5 | 6.89 | 7620 | 61.3 | 20.4 |
| 52.2 | 206 | 204 | 12.5 | 8.0 | 11.4 | 181 | 22.7 | 7.84 | 6660 | 52.8 | 17.7 |
| 46.2 | 203 | 203 | 11.0 | 7.3 | 11.4 | 181 | 24.8 | 8.90 | 5900 | 45.9 | 15.3 |
| 150 UC 37.2 | 162 | 154 | 11.5 | 8.1 | 8.9 | 139 | 17.1 | 6.34 | 4730 | 22.2 | 7.01 |
| 30.0 | 158 | 153 | 9.4 | 6.6 | 8.9 | 139 | 21.0 | 7.79 | 3860 | 17.6 | 5.62 |
| 23.4 | 152 | 152 | 6.8 | 6.1 | 8.9 | 139 | 22.8 | 10.7 | 2980 | 12.6 | 3.98 |
| 100 UC 14.8 | 97 | 99 | 7.0 | 5.0 | 10.0 | 83.0 | 16.6 | 6.71 | 1890 | 3.18 | 1.14 |

Source: Hot Rolled and Structural Steel Products, 1998 Edition, BHP, Whyalla

**TABLE 6** ▶ Taper flange beams—Dimensions

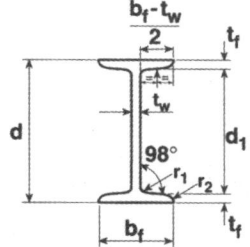 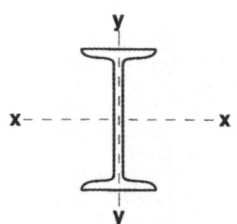

| DESIG-NATION | MASS PER METRE | DEPTH OF SECTION | FLANGE | | WEB THICK-NESS | RADII | | DEPTH BETWEEN FLANGES | | | GROSS AREA OF CROSS-SECTION | ABOUT x-AXIS | ABOUT y-AXIS |
| | | | WIDTH | THICK-NESS | | ROOT | TOE | | | | | | |
| | | $d$ | $b_f$ | $t_f$ | $t_w$ | $r_1$ | $r_2$ | $d_1$ | $\frac{d_1}{t_w}$ | $\frac{b_f-t_w}{2t_f}$ | $A_g$ | $I_x$ | $I_y$ |
| | kg/m | mm | mm | mm | mm | mm | mm | mm | | | mm$^2$ | 10$^6$mm$^4$ | 10$^6$mm$^4$ |
| 125 TFB | 13.1 | 125 | 65.0 | 8.5 | 5.0 | 8.0 | 4.0 | 108 | 21.6 | 3.53 | 1670 | 4.34 | 0.337 |
| 100 TFB | 7.20 | 100 | 45.0 | 6.0 | 4.0 | 7.0 | 3.0 | 88 | 22.0 | 3.42 | 917 | 1.46 | 0.0795 |

Source: Hot Rolled and Structural Steel Products, 1998 Edition, BHP, Whyalla

**TABLE 7** ▶ Parallel flange channels

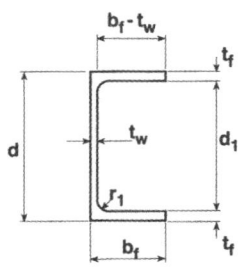 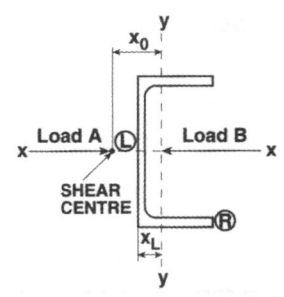

| DESIG-NATION | MASS PER METRE | DEPTH OF SECTION | FLANGE | | WEB THICK-NESS | ROOT RADIUS | DEPTH BETWEEN FLANGES | | | GROSS AREA OF CROSS-SECTION | CO-ORDINATE OF CENTROID | CO-ORDINATE OF SHEAR CENTRE | ABOUT x-AXIS | ABOUT y-AXIS |
| | | | WIDTH | THICK-NESS | | | | | | | | | | |
| | | $d$ | $b_f$ | $t_f$ | $t_w$ | $r_1$ | $d_1$ | $\dfrac{d_1}{t_w}$ | $\dfrac{b_f - t_w}{2t_f}$ | $A_g$ | $X_L$ | $X_o$ | $I_x$ | $I_y$ |
| | kg/m | mm | mm | mm | mm | mm | mm | | | mm$^2$ | mm | mm | $10^6$mm$^4$ | $10^6$mm$^4$ |
| 380 PFC | 55.2 | 380 | 100 | 17.5 | 10.0 | 14.0 | 345 | 34.5 | 5.14 | 7030 | 27.5 | 56.7 | 152 | 6.48 |
| 300 PFC | 40.1 | 300 | 90 | 16.0 | 8.0 | 14.0 | 268 | 33.5 | 5.13 | 5110 | 27.2 | 56.1 | 72.4 | 4.04 |
| 250 PFC | 35.5 | 250 | 90 | 15.0 | 8.0 | 12.0 | 220 | 27.5 | 5.47 | 4520 | 28.6 | 58.5 | 45.1 | 3.64 |
| 230 PFC | 25.1 | 230 | 75 | 12.0 | 6.5 | 12.0 | 206 | 31.7 | 5.71 | 3200 | 22.6 | 46.7 | 26.8 | 1.76 |
| 200 PFC | 22.9 | 200 | 75 | 12.0 | 6.0 | 12.0 | 176 | 29.3 | 5.75 | 2920 | 24.4 | 50.5 | 19.1 | 1.65 |
| 180 PFC | 20.9 | 180 | 75 | 11.0 | 6.0 | 12.0 | 158 | 26.3 | 6.27 | 2660 | 24.5 | 50.3 | 14.1 | 1.51 |
| 150 PFC | 17.7 | 150 | 75 | 9.5 | 6.0 | 10.0 | 131 | 21.8 | 7.26 | 2250 | 24.9 | 51.0 | 8.34 | 1.29 |
| 125 PFC | 11.9 | 125 | 65 | 7.5 | 4.7 | 8.0 | 110 | 23.4 | 8.04 | 1520 | 21.8 | 45.0 | 3.97 | 0.658 |
| 100 PFC | 8.33 | 100 | 50 | 6.7 | 4.2 | 8.0 | 86.6 | 20.5 | 6.84 | 1060 | 16.7 | 33.9 | 1.74 | 0.267 |
| 75 PFC | 5.92 | 75 | 40 | 6.1 | 3.8 | 8.0 | 62.8 | 16.5 | 5.94 | 754 | 13.7 | 27.2 | 0.683 | 0.120 |

Source: Hot Rolled and Structural Steel Products, 1998 Edition, BHP, Whyalla

**TABLE 8** ► Equal angles

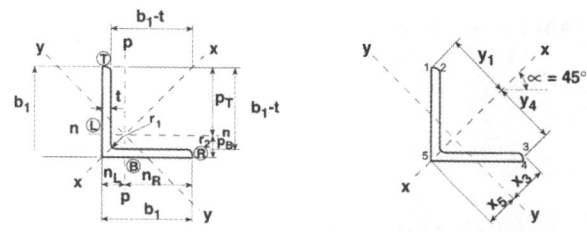

| DESIGNATION | | MASS PER METRE | ACTUAL THICK-NESS | RADII | | | GROSS AREA OF CROSS-SECTION | CO-ORDINATE OF CENTROID | | ABOUT x-AXIS | ABOUT y-AXIS |
|---|---|---|---|---|---|---|---|---|---|---|---|
| LEG-SIZE | NOMINAL THICKNESS | | | ROOT | TOE | | | $n_L =$ | $n_R =$ | | |
| $b_1 \times b_2$ | | | $t$ | $r_1$ | $r_2$ | $\dfrac{b_1-t}{t}$ | $A_g$ | $p_B$ | $p_T$ | $I_x$ | $I_y$ |
| mm mm | mm | kg/m | mm | mm | mm | | mm² | mm | mm | 10⁶mm⁴ | 10⁶mm⁴ |
| 200 × 200 × | 26 EA | 76.8 | 26.0 | 18.0 | 5.0 | 6.69 | 9870 | 59.3 | 141 | 56.8 | 14.9 |
| | 20 EA | 60.1 | 20.0 | 18.0 | 5.0 | 9.00 | 7660 | 57.0 | 143 | 45.7 | 11.8 |
| | 18 EA | 54.4 | 18.0 | 18.0 | 5.0 | 10.1 | 6930 | 56.2 | 144 | 41.7 | 10.8 |
| | 16 EA | 48.7 | 16.0 | 18.0 | 5.0 | 11.5 | 6200 | 55.4 | 145 | 37.6 | 9.72 |
| | 13 EA | 40.0 | 13.0 | 18.0 | 5.0 | 14.4 | 5090 | 54.2 | 146 | 31.2 | 8.08 |
| 150 × 150 × | 19 EA | 42.1 | 19.0 | 13.0 | 5.0 | 6.89 | 5360 | 44.2 | 106 | 17.6 | 4.60 |
| | 16 EA | 35.4 | 15.8 | 13.0 | 5.0 | 8.49 | 4520 | 43.0 | 107 | 15.1 | 3.91 |
| | 12 EA | 27.3 | 12.0 | 13.0 | 5.0 | 11.5 | 3480 | 41.5 | 108 | 11.9 | 3.06 |
| | 10 EA | 21.9 | 9.5 | 13.0 | 5.0 | 14.8 | 2790 | 40.5 | 109 | 9.61 | 2.48 |
| 125 × 125 × | 16 EA | 29.1 | 15.8 | 10.0 | 5.0 | 6.91 | 3710 | 36.8 | 88.2 | 8.43 | 2.20 |
| | 12 EA | 22.5 | 12.0 | 10.0 | 5.0 | 9.42 | 2870 | 35.4 | 89.6 | 6.69 | 1.73 |
| | 10 EA | 18.0 | 9.5 | 10.0 | 5.0 | 12.2 | 2800 | 34.4 | 90.6 | 5.44 | 1.40 |
| | 8 EA | 14.9 | 7.8 | 10.0 | 5.0 | 15.0 | 1900 | 33.7 | 91.3 | 4.55 | 1.17 |
| 100 × 100 × | 12 EA | 17.7 | 12.0 | 8.0 | 5.0 | 7.33 | 2260 | 29.2 | 70.8 | 3.29 | 0.857 |
| | 10 EA | 14.2 | 9.5 | 8.0 | 5.0 | 9.53 | 1810 | 28.2 | 71.8 | 2.70 | 0.695 |
| | 8 EA | 11.8 | 7.8 | 8.0 | 5.0 | 11.8 | 1500 | 27.5 | 72.5 | 2.27 | 0.582 |
| | 6 EA | 9.16 | 6.0 | 8.0 | 5.0 | 15.7 | 1170 | 26.8 | 73.2 | 1.78 | 0.458 |
| 90 × 90 × | 10 EA | 12.7 | 9.5 | 8.0 | 5.0 | 8.47 | 1620 | 25.7 | 64.3 | 1.93 | 0.500 |
| | 8 EA | 10.6 | 7.8 | 8.0 | 5.0 | 10.5 | 1350 | 25.0 | 65.0 | 1.63 | 0.419 |
| | 6 EA | 8.22 | 6.0 | 8.0 | 5.0 | 14.0 | 1050 | 24.3 | 65.7 | 1.28 | 0.330 |
| 75 × 75 × | 10 EA | 10.5 | 9.5 | 8.0 | 5.0 | 6.89 | 1340 | 22.0 | 53.0 | 1.08 | 0.282 |
| | 8 EA | 8.73 | 7.8 | 8.0 | 5.0 | 8.62 | 1110 | 21.3 | 53.7 | 0.913 | 0.237 |
| | 6 EA | 6.81 | 6.0 | 8.0 | 5.0 | 11.5 | 867 | 20.5 | 54.5 | 0.722 | 0.187 |
| | 5 EA | 5.27 | 4.6 | 8.0 | 5.0 | 15.3 | 672 | 19.9 | 55.1 | 0.563 | 0.147 |
| 65 × 65 × | 10 EA | 9.02 | 9.5 | 6.0 | 3.0 | 5.84 | 1150 | 19.6 | 45.4 | 0.691 | 0.183 |
| | 8 EA | 7.51 | 7.8 | 6.0 | 3.0 | 7.33 | 957 | 19.0 | 46.0 | 0.589 | 0.154 |
| | 6 EA | 5.87 | 6.0 | 6.0 | 3.0 | 9.83 | 748 | 18.3 | 46.7 | 0.471 | 0.122 |
| | 5 EA | 4.56 | 4.6 | 6.0 | 3.0 | 13.1 | 581 | 17.7 | 47.3 | 0.371 | 0.0959 |
| 55 × 55 × | 6 EA | 4.93 | 6.0 | 6.0 | 3.0 | 8.17 | 628 | 15.8 | 39.2 | 0.278 | 0.0723 |
| | 5 EA | 3.84 | 4.6 | 6.0 | 3.0 | 11.0 | 489 | 15.2 | 39.8 | 0.220 | 0.0571 |
| 50 × 50 × | 8 EA | 5.68 | 7.8 | 6.0 | 3.0 | 5.41 | 723 | 15.2 | 34.8 | 0.253 | 0.0675 |
| | 6 EA | 4.46 | 6.0 | 6.0 | 3.0 | 7.33 | 568 | 14.5 | 35.5 | 0.205 | 0.0536 |
| | 5 EA | 3.48 | 4.6 | 6.0 | 3.0 | 9.87 | 443 | 13.9 | 36.1 | 0.163 | 0.0424 |
| | 3 EA | 2.31 | 3.0 | 6.0 | 3.0 | 15.7 | 295 | 13.2 | 36.8 | 0.110 | 0.0289 |
| 45 × 45 × | 6 EA | 3.97 | 6.0 | 5.0 | 3.0 | 6.50 | 506 | 13.3 | 31.7 | 0.146 | 0.0383 |
| | 5 EA | 3.10 | 4.6 | 5.0 | 3.0 | 8.78 | 394 | 12.7 | 32.3 | 0.117 | 0.0303 |
| | 3 EA | 2.06 | 3.0 | 5.0 | 3.0 | 14.0 | 263 | 12.0 | 33.0 | 0.0790 | 0.0206 |
| 40 × 40 × | 6 EA | 3.50 | 6.0 | 5.0 | 3.0 | 5.67 | 446 | 12.0 | 28.0 | 0.0997 | 0.0265 |
| | 5 EA | 2.73 | 4.6 | 5.0 | 3.0 | 7.70 | 348 | 11.5 | 28.5 | 0.0801 | 0.0209 |
| | 3 EA | 1.83 | 3.0 | 5.0 | 3.0 | 12.3 | 233 | 10.8 | 29.2 | 0.0545 | 0.0142 |
| 30 × 30 × | 6 EA | 2.56 | 6.0 | 5.0 | 3.0 | 4.00 | 326 | 9.53 | 20.5 | 0.0387 | 0.0107 |
| | 5 EA | 2.01 | 4.6 | 5.0 | 3.0 | 5.52 | 256 | 8.99 | 21.0 | 0.0316 | 0.00839 |
| | 3 EA | 1.35 | 3.0 | 5.0 | 3.0 | 9.00 | 173 | 8.30 | 21.7 | 0.0218 | 0.00573 |
| 25 × 25 × | 6 EA | 2.08 | 6.0 | 5.0 | 3.0 | 3.17 | 266 | 8.28 | 16.7 | 0.0210 | 0.00600 |
| | 5 EA | 1.65 | 4.6 | 5.0 | 3.0 | 4.43 | 210 | 7.75 | 17.3 | 0.0173 | 0.00469 |
| | 3 EA | 1.12 | 3.0 | 5.0 | 3.0 | 7.33 | 143 | 7.07 | 17.9 | 0.0121 | 0.00319 |

Source: Hot Rolled and Structural Steel Products, 1998 Edition, BHP, Whyalla

**TABLE 9** ▶ Unequal angles

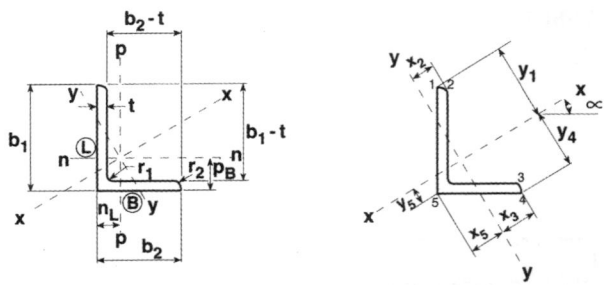

| DESIGNATION | | MASS PER METRE | ACTUAL THICK-NESS | RADII | | $\dfrac{b_1-t}{t}$ | $\dfrac{b_2-t}{t}$ | GROSS AREA OF CROSS-SECTION | CO-ORDINATE OF CENTROID | | ABOUT x-AXIS | ABOUT y-AXIS |
| LEG-SIZE | NOMINAL THICKNESS | | | ROOT | TOE | | | | | | | |
| $b_1 \times b_2$ | | | $t$ | $r_1$ | $r_2$ | | | $A_g$ | $p_B$ | $n_L$ | $I_x$ | $I_y$ |
| mm  mm | mm | kg/m | mm | mm | mm | | | mm² | mm | mm | $10^6 mm^4$ | $10^6 mm^4$ |
| 150 × 100 × | 12 UA | 22.5 | 12.0 | 10.0 | 5.0 | 11.5 | 7.33 | 2870 | 49.1 | 24.3 | 7.51 | 1.35 |
|  | 10 UA | 18.0 | 9.5 | 10.0 | 5.0 | 14.8 | 9.53 | 2300 | 48.1 | 23.3 | 6.11 | 1.09 |
| 150 × 90 × | 16 UA | 27.9 | 15.8 | 10.0 | 5.0 | 8.49 | 4.70 | 3550 | 52.5 | 22.7 | 8.80 | 1.32 |
|  | 12 UA | 21.6 | 12.0 | 10.0 | 5.0 | 11.5 | 6.50 | 2750 | 51.0 | 21.2 | 6.97 | 1.04 |
|  | 10 UA | 17.3 | 9.5 | 10.0 | 5.0 | 14.8 | 8.47 | 2200 | 50.0 | 20.2 | 5.66 | 0.847 |
|  | 8 UA | 14.3 | 7.8 | 10.0 | 5.0 | 18.2 | 10.5 | 1820 | 49.2 | 19.6 | 4.73 | 0.710 |
| 125 × 75 × | 12 UA | 17.7 | 12.0 | 8.0 | 5.0 | 9.42 | 5.25 | 2260 | 43.3 | 18.4 | 3.91 | 0.585 |
|  | 10 UA | 14.2 | 9.5 | 8.0 | 5.0 | 12.2 | 6.89 | 1810 | 42.3 | 17.5 | 3.20 | 0.476 |
|  | 8 UA | 11.8 | 7.8 | 8.0 | 5.0 | 15.0 | 8.62 | 1500 | 41.5 | 16.8 | 2.68 | 0.399 |
|  | 6 UA | 9.16 | 6.0 | 8.0 | 5.0 | 19.8 | 11.5 | 1170 | 40.7 | 16.0 | 2.10 | 0.315 |
| 100 × 75 × | 10 UA | 12.4 | 9.5 | 8.0 | 5.0 | 9.53 | 6.89 | 1580 | 31.8 | 19.4 | 1.89 | 0.401 |
|  | 8 UA | 10.3 | 7.8 | 8.0 | 5.0 | 11.8 | 8.62 | 1310 | 31.1 | 18.7 | 1.59 | 0.337 |
|  | 6 UA | 7.98 | 6.0 | 8.0 | 5.0 | 15.7 | 11.5 | 1020 | 30.3 | 17.9 | 1.25 | 0.265 |
| 75 × 50 × | 8 UA | 7.23 | 7.8 | 7.0 | 3.0 | 8.62 | 5.41 | 920 | 25.2 | 12.8 | 0.586 | 0.106 |
|  | 6 UA | 5.66 | 6.0 | 7.0 | 3.0 | 11.5 | 7.33 | 720 | 24.4 | 12.1 | 0.468 | 0.0842 |
|  | 5 UA | 4.40 | 4.6 | 7.0 | 3.0 | 15.3 | 9.87 | 560 | 23.8 | 11.5 | 0.370 | 0.0666 |
| 65 × 50 × | 8 UA | 6.59 | 7.8 | 6.0 | 3.0 | 7.33 | 5.41 | 840 | 21.1 | 13.6 | 0.421 | 0.0936 |
|  | 6 UA | 5.16 | 6.0 | 6.0 | 3.0 | 9.83 | 7.33 | 658 | 20.4 | 12.9 | 0.338 | 0.0743 |
|  | 5 UA | 4.02 | 4.6 | 6.0 | 3.0 | 13.1 | 9.87 | 512 | 19.8 | 12.4 | 0.267 | 0.0587 |

Source: Hot Rolled and Structural Steel Products, 1998 Edition, BHP Whyalla, 1998

**TABLE 10 ▶** Common types of steel sections

| 1<br>SECTION TYPE | 2<br>PICTORIAL VIEW | 3<br>ABBREVIATION<br>OR SYMBOL | 4<br>TYPICAL EXAMPLE OF DESIGNATION<br>(SEE NOTES 1 AND 3) |
|---|---|---|---|
| Universal beam<br>(depth greater than<br>flange width)<br>(See Note 2) | | UB | 200 UB 25<br>where<br>200 = nominal depth, $D$ (actual = 203)<br>25 = nominal mass, kg/m (actual = 25.4 kg/m) |
| Universal column<br>(depth and flange<br>width approx. equal)<br>(See Note 2) | | UC | 200 UC 46<br>where<br>200 = nominal depth, $D$ (actual = 203)<br>46 = nominal mass, kg/m |
| Rolled channel | | [ | 380 × 100 [<br>where<br>380 = $D$<br>100 = $B$ |
| Cold formed channel | | [ | 102 × 51 × 2.8 [<br>where<br>102 = $D$<br>51 = $B$<br>2.8 = $t$ |
| Rolled angle (leg<br>width, A, varies;<br>may be equal or<br>unequal) | | L | 64 × 51 × 5 L<br>where<br>64 = $A$<br>51 = $B$<br>5 = $t$ |
| Flats—<br>(a) flat bar<br><br>(b) plate | | (a) FL<br><br>(b) PL | (a) 200 × 10 FL<br>where<br>200 = $W$<br>10 = $t$<br>(b) 1200 × 20 PL<br>where<br>1200 = $W$<br>20 = $t$ |
| Round bar or rod | | RD | 16 RD<br>where<br>16 = $D$ |
| Square bar | | SQ | 25 SQ<br>where<br>25 = $D$ or $B$ |
| Circular hollow section | | CHS | 76 OD × 4.5 CHS<br>where<br>76 = nominal outside diameter $D$<br>4.5 = actual thickness $t$ |
| Hollow sections—<br>(a) rectangular<br><br>(b) square (D = B) | | (a) RHS<br><br>(b) RHS | (a) 152 × 76 × 4.9 RHS<br>where<br>152 = nominal depth $D$<br>76 = nominal width $B$<br>4.9 = actual thickness $t$<br>(b) 102 × 102 × 6.3 RHS |
| Structural decking | | — | Refer to manufacturers' literature |

**TABLE 10** ► Common types of steel sections (continued)

| 1<br>SECTION TYPE | 2<br>PICTORIAL VIEW | 3<br>ABBREVIATION<br>OR SYMBOL | 4<br>TYPICAL EXAMPLE OF DESIGNATION<br>(SEE NOTES 1 AND 3) |
|---|---|---|---|
| Taper flange beam | | TFB | Refer to AS 1131<br>e.g. 178 × 89 × 22 TFB |
| Cold formed purlins<br>(See Note 2) | | C<br>or<br>Z | Refer to manufacturers' literature<br>e.g. C20019, or Z20019<br>where<br>200 = nominal depth<br>19 = 1.9 mm thickness |
| Structural tees—<br>(a) from columns<br><br>(b) from beam | | (a) CT<br><br>(b) BT | (a) 155 CT 142<br>where<br>155 = nominal depth<br>142 = nominal mass, kg/m<br>(b) 380 BT 98<br>where<br>380 = nominal depth<br>98 = nominal mass, kg/m |

Notes:

1. Unless otherwise stated, dimensions are in millimetres.

2. Dimensions for the universal beam, universal column and cold formed purlins are nominal dimensions.

3. Nominal depth may vary appreciably from actual depth, $D$, which can be ascertained from AS 1131. Actual mass can also be determined from AS 1131. For example, for 200 UB 25, actual depth = 203 mm; actual mass = 25.4 kg/m.

# Gauge lines tables for structural steel (AISC)

**TABLE 11** ▶ Gauge lines for universal sections

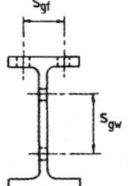 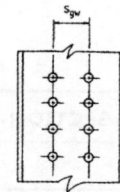

| SECTION | FLANGE $s_{gf}$ | | | | WEB $s_{gw}$ | | | | | |
|---|---|---|---|---|---|---|---|---|---|---|
| | **M20** | | **M24** | | **M20** | | | **M24** | | |
| **UNIVERSAL BEAMS** | | | | | | | | | | |
| 610UB | 140 | 90 | 140 | 90 | 140 | 90 | 70 | 140 | 90 | 70 |
| 530UB | 140 | 90 | 140 | 90 | 140 | 90 | 70 | 140 | 90 | 70 |
| 460UB | 90 | 140 | 90 | | 90 | 70 | 140 | 90 | 70 | 140 |
| 410UB | 90 | 70 | 90 | | 90 | 70 | 140 | 90 | 70 | 140 |
| 360UB, 310UB | 90 | 70 | 90 | | 90 | 70 | 140 | 90 | 70 | 140 |
| 310UB32.0 | 70 | | b | | 90 | 70 | 140 | 90 | 70 | 140 |
| 250UB | 70 | 90 | b | | 70 | 90 | 140 | 70 | 90 | |
| 250UB25.7 | 70* | | b | | 70 | 90 | | 70 | 90 | |
| 200UB | 70 | | b | | 70 | 90 | | 70 | 90 | |
| 200UB18.2 | 60* | | b | | 70 | 90 | | 70 | | |
| 180UB | 50** | | b | | 70 | | | 70 | | |
| 150UB | c | | b | | 60* | | | b | | |
| **UNIVERSAL COLUMNS** | | | | | | | | | | |
| 310UC | 140 | 90 | 140 | 90 | 90 | 70 | 140 | 90 | 70 | 140 |
| 250UC | 140 | 90 | 140 | 90 | 90 | 70 | 140 | 90 | 70 | |
| 200UC | 140 | 90 | 140 | 90 | 90 | 70 | | 90 | 70 | |
| 150UC | 90 | 70 | 90 | | 60* | | | b | | |
| 100UC | 60 | | b | | c | | | b | | |
| *Preference* | 1 | 2 | 1 | 2 | 1 | 2 | 3 | 1 | 2 | 3 |

\* Gauge listed are for M16 bolts.

\*\* Gauge listed are for M12 bolts.

b Indicates that the flange or web will not accommodate this size of bolt.

c Indicates that the flange or web will not accommodate two lines of bolts with a gauge of 50 mm or more for M12 and larger bolts.

Source: Australian Institute of Steel Construction

**TABLE 12** ▶ Gauge lines for taper flange beams and channels

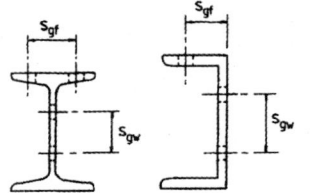
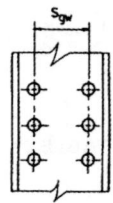

| SECTION | FLANGE $9_f$ | | | WEB $9_w$ | | | | | | | | |
|---|---|---|---|---|---|---|---|---|---|---|---|---|
| | **M16** | **M20** | **M24** | **M16** | | | **M20** | | | **M24** | | |
| Taper flange beams | | | | | | | | | | | | |
| 125TFB | b, bl | b | b | 50 | | | 50 | | | c | | |
| 100TFB | b, bl | b | b | c | | | c | | | c | | |
| Parallel flange channels | | | | | | | | | | | | |
| 380PFC | 55 | 55 | 55 | 140 | 90 | 70 | 140 | 90 | 70 | 140 | 90 | 70 |
| 300PFC | 55 | 55 | 55 | 140 | 90 | 70 | 140 | 90 | 70 | 140 | 90 | 70 |
| 250PFC | 55 | 55 | 55 | 140 | 90 | 70 | 140 | 90 | 70 | 140 | 90 | 70 |
| 230PFC | 45 | 45 | 45 | 140 | 90 | 70 | 90 | 70 | | 90 | 70 | |
| 200PFC | 45 | 45 | 45 | 90 | 70 | | 90 | 70 | | 90 | 70 | |
| 180PFC | 45 | 45 | 45 | 70 | | | 70 | | | c | | |
| 150PFC | 45 | 45 | 45 | 50 | | | 50 | | | c | | |
| 125PFC | 35 | 35 | b | 50 | | | c | | | c | | |
| 100PFC | 30 | b | b | c | | | c | | | c | | |
| 75PFC | b, bl | b | b | c | | | c | | | c | | |
| Preference | I | I | I | I | 2 | 3 | I | 2 | 3 | I | 2 | 3 |

b indicates that the flange will not accommodate this size of bolt.

bl indicates that the flange will not accommodate M12 bolts.

c indicates that web will not accommodate two lines of bolts with a gauge of 50 mm or more.

Source: Australian Institute of Steel Construction

**TABLE 13** ▶ Gauge lines for angles

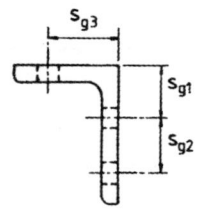

| NOMINAL LEG LENGTH | $g_1$ | $g_2$ | $g_3$ | BOLT |
|---|---|---|---|---|
| 200 | 65(75) | 70(75) | 100 | M24 |
| 150 | 50(55) | 60(55) | 75 | M24 |
| 125 | 45 | 50 | 65 (62) | M24 |

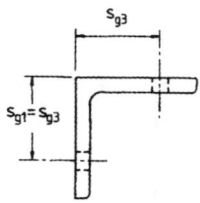

| NOMINAL LEG LENGTH | $g_3$ | BOLT |
|---|---|---|
| 100 | 50 | M20 |
| 90 | 45 | M20 |
| 75 | 40(38) | M20 |
| 65 | 35(32) | M16 |
| 55 | 30(28) | M16 |
| 50 | 30(25) | M16 |
| 45 | 25(22) | M12 |
| 40 | 25(20) | M12 |

Notes to all tables of gauge lines:

1. The gauges given are suitable for general use in member detailing. When angles are used as *components in connections*, gauge lines may be varied from the values given above in order to suit a particular connection.

2. The bolt diameters listed are the maximum that can be accommodated on the thickest angles of each leg length, using either—

   (a) high strength structural bolts with washers to AS/NZS 1252; or

   (b) commercial bolts to AS/NZS 1111 with 'large series' washers to AS 1237 (now superseded).

   For thinner legs and commercial bolts with 'normal series' washers, it may be possible to accommodate a larger bolt diameter.

Source: Australian Institute of Steel Construction

**FIGURE I** ▶ Detailing a universal beam—example

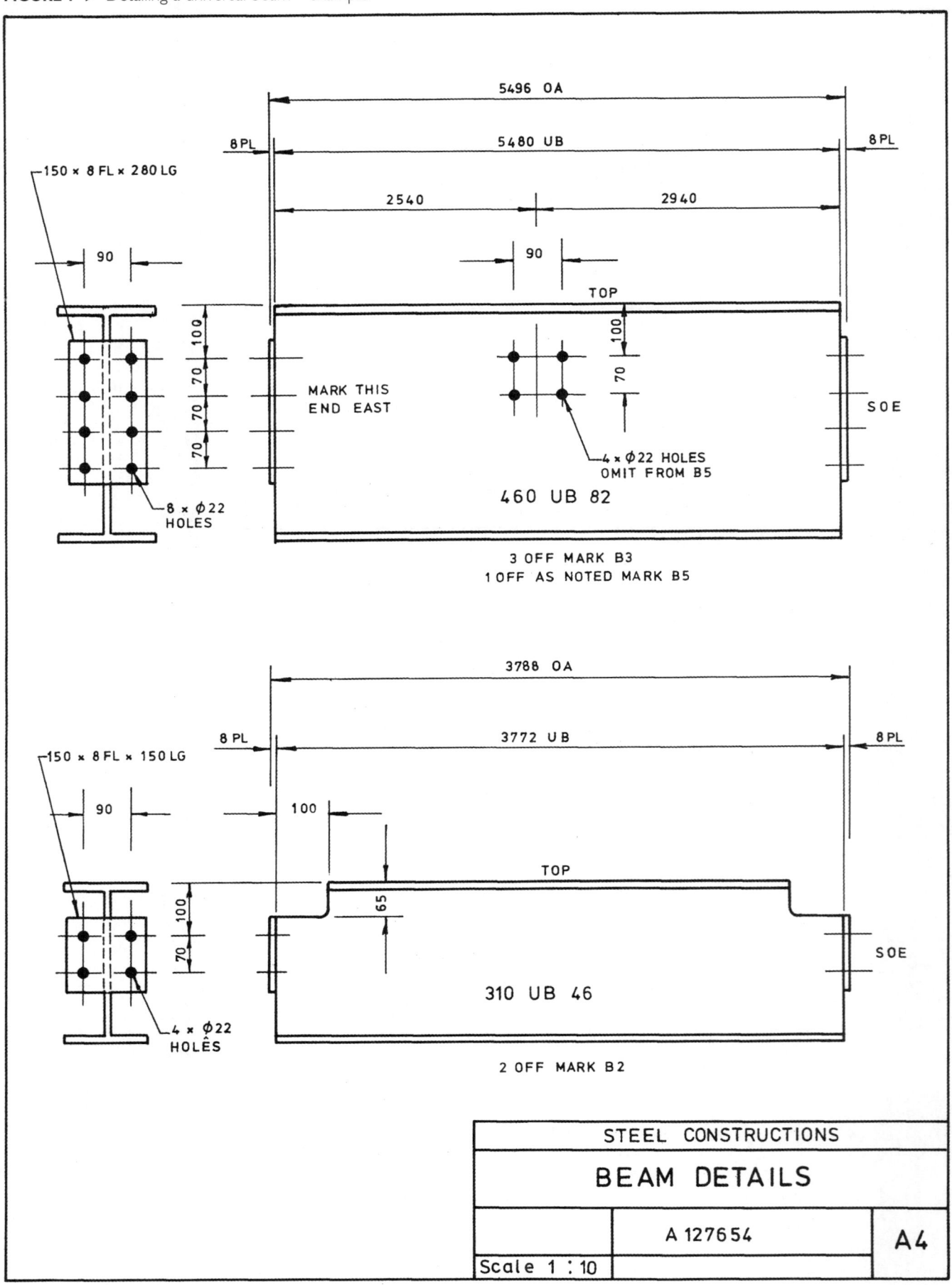

150 × 8 FL × 280 LG

5496 OA

8 PL                    5480 UB                    8 PL

2540                              2940

90

90

TOP

100

100

70

MARK THIS
END EAST

SOE

4 × ⌀22 HOLES
OMIT FROM B5

8 × ⌀22
HOLES

460 UB 82

3 OFF MARK B3
1 OFF AS NOTED MARK B5

3788 OA

150 × 8 FL × 150 LG

8 PL                    3772 UB                    8 PL

90                    100

100

TOP

65

70

SOE

4 × ⌀22
HOLES

310 UB 46

2 OFF MARK B2

| STEEL CONSTRUCTIONS | | |
|---|---|---|
| **BEAM DETAILS** | | |
| Scale 1 : 10 | A 127654 | A4 |

# Appendix
## Reinforced concrete

**2**

**TABLE 1** ► First preference bar bending shapes

| NAME | CODE | DIAGRAM | ESSENTIAL DIMENSIONS | | | | COMMENT |
|---|---|---|---|---|---|---|---|
| **Straight** | S | A | A | | | | A is also the length L |
| **L-shape** One 90° bend | L | | A A | B | | L L | A + B = L A would be critical B would be critical |
| | | | | B | | | |
| **Double L-shape** Two 90° bends | LL | | A | B | C | | A + B + C = L A and C should both be specified even if equal |
| **Hooked bar** One 180° bend | H | | A | B B B | | L | A + B = L When A is standard, it may be omitted |
| **Double hooked** Two 180° bends | HH | | A | B B B | C | L | A + B + C = L A = C = 0.5 [L-B] A and C standard |
| **Fabric** Only if flat sheet | F | | A | B | | | A is length of main wires (6000 mm max) B is extent of cross-wires including side laps |
| **V-shape** Bend less than 90° | V | | A A A A | B B | D C D C | L L | Give C when angle exceeds 70° A is critical always B may be omitted only if not critical |
| **U-shape** 180° bend only | U | D = bend dia | A A | B C | C D | | Specify U-shape not LL-shape when B is less than 20 bar diameters |
| **Tie** for beams or columns | T | | A A | B B | C | | L = 2[A + B + C] C is a cog length L may be omitted generally; C also |
| **Stirrup** for beams [hooks in] | SH | | A A | B B | C | | L = 2A + B + 2C C is a hook length L and C may be omitted generally |
| **Stirrup** for beams [cogs in] | SC | | A A | B B | C | | L = 2A + B + 2C C is a cog length L and C may be omitted generally |
| **Cranked column bar** for lap splice | CC | D = overall dimension | A A | B C | C D | D L | L = A + B where A is a lap-splice length, C ≥ 6 $d_b$ D ≥ 2 $d_b$ |

Notes:

1. All dimensions are to intersection of straight portions at the outside of all types of bends.

2. 'L' is the sum of the individual out-to-out dimensions 'A', 'B', etc.

Source: AS 1100, part 501, Standards Australia, 1985. Reproduced with permission.

**TABLE 2** ▶ Second preference bar bending shapes

| NAME | CODE | DIAGRAM | NAME | CODE | DIAGRAM |
|---|---|---|---|---|---|
| Right-angled crank shape | RC | | Right-angled truss shape | RT | |
| | | | 135° Hooked tie | HT | |
| V-plus L-shape | VL | | Diamond shaped tie for columns | DT | |
| Double V-shape | VV | | Cross-over tie | XT | |
| Joist bar or bent-up bar | J | | Circular tie for columns | CT | |
| L-bend added to J-shape | LJ | | Link for columns | LH | |
| Double J-shape or truss bar | JJ | | Spiral or Helix | SP | $B$ = pitch– number of turns $= E = A/B$ length $= \pi D (E + 3)$ |
| Acute angle bend more than 90° | A | | | | |
| Radiused bar | R | $A$ = cut length $R$ = ext radius | Non-standard shapes | NS | All shapes are to be drawn and dimensioned in full |

Notes:

1. All dimensions are to intersection of straight portions except where shown.

2. First preference shapes with hooks and cogs are to be included here.

Source: AS 1100, part 501, Standards Australia, 1985. Reproduced with permission.

**TABLE 3** ▶ Reinforcement schedule

# REINFORCEMENT SCHEDULE

For use with CIA preferred bending shapes

Construction by _____

Project _____

Part of project _____

**Material Type**
R 230R to AS1302
Y 410Y to AS1302
F Fabric to AS1304
W Wire to AS1303

Scheduled by _____ Date _____
Checked by _____ Date _____

Drawing No. _____ Schedule No. _____ Revision _____

| Item E | Location | Mark | Type-size | Total | Length | P | Shape h/c | A | B | C | D | E | h/c | Total lengths (m) | | | | |
|---|---|---|---|---|---|---|---|---|---|---|---|---|---|---|---|---|---|---|
| 1 | | | | | | | | | | | | | | | | | | |
| 2 | | | | | | | | | | | | | | | | | | |
| 3 | | | | | | | | | | | | | | | | | | |
| 4 | | | | | | | | | | | | | | | | | | |
| 5 | | | | | | | | | | | | | | | | | | |
| 6 | | | | | | | | | | | | | | | | | | |
| 7 | | | | | | | | | | | | | | | | | | |
| 8 | | | | | | | | | | | | | | | | | | |
| 9 | | | | | | | | | | | | | | | | | | |
| 10 | | | | | | | | | | | | | | | | | | |
| 11 | | | | | | | | | | | | | | | | | | |
| 12 | | | | | | | | | | | | | | | | | | |
| 13 | | | | | | | | | | | | | | | | | | |
| 14 | | | | | | | | | | | | | | | | | | |
| 15 | | | | | | | | | | | | | | | | | | |

**FIGURE I ▶** Example of reinforced concrete drawing

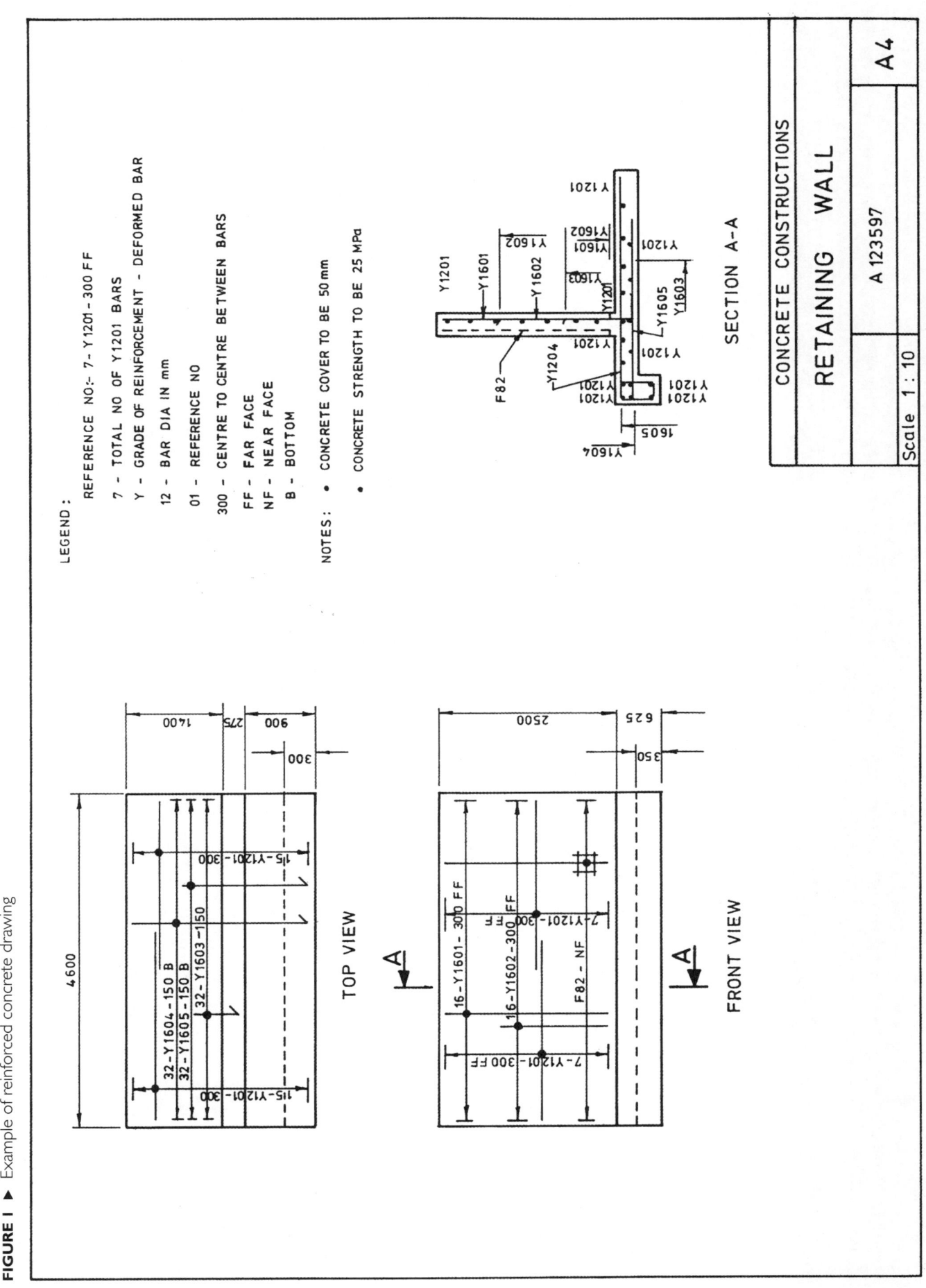

# Index